Cisco pyATS—Network Test and Automation Solution

Data-driven and reusable testing for modern networks

John Capobianco

Dan Wade

Cisco Press

221 River Street

Hoboken, NJ 07030 USA

Cisco pyATS—Network Test and Automation Solution

John Capobianco, Dan Wade

Copyright© 2025 Cisco Systems, Inc.

Cisco Press logo is a trademark of Cisco Systems, Inc.

Published by:
Cisco Press

$PrintCode

Library of Congress Control Number: 2024907000

ISBN-13: 978-0-13-803167-1

ISBN-10: 0-13-803167-3

Warning and Disclaimer

Trademark Acknowledgments

All terms mentioned in this book that are known to be trademarks or service marks have been appropriately capitalized. Cisco Press or Cisco Systems, Inc., cannot attest to the accuracy of this information. Use of a term in this book should not be regarded as affecting the validity of any trademark or service mark.

Special Sales

For information about buying this title in bulk quantities, or for special sales opportunities (which may include electronic versions; custom cover designs; and content particular to your business, training goals, marketing focus, or branding interests), please contact our corporate sales department at corpsales@pearsoned.com or (800) 382-3419.

For government sales inquiries, please contact governmentsales@pearsoned.com.

For questions about sales outside the U.S., please contact intlcs@pearson.com.

Feedback Information

At Cisco Press, our goal is to create in-depth technical books of the highest quality and value. Each book is crafted with care and precision, undergoing rigorous development that involves the unique expertise of members from the professional technical community.

Readers' feedback is a natural continuation of this process. If you have any comments regarding how we could improve the quality of this book, or otherwise alter it to better suit your needs, you can contact us through email at feedback@ciscopress.com. Please make sure to include the book title and ISBN in your message.

We greatly appreciate your assistance.

Please contact us with concerns about any potential bias at https://www.pearson.com/report-bias.html.

GM K12, Early Career and Professional Learning: Soo Kang

Alliances Manager, Cisco Press: Caroline Antonio

Director, ITP Product Management: Brett Bartow

Executive Editor: Nancy Davis

Managing Editor: Sandra Schroeder

Development Editor: Christopher Cleveland

Senior Project Editor: Mandie Frank

Copy Editor: Bart Reed

Technical Editors: Stuart Clark, Charles Greenaway

Editorial Assistant: Cindy Teeters

Designer: Chuti Prasertsith

Composition: codeMantra

Indexer: Timothy Wright

Proofreader: Barbara Mack

·¹|¹·¹|¹· CISCO.

Americas Headquarters	Asia Pacific Headquarters	Europe Headquarters
Cisco Systems, Inc.	Cisco Systems (USA) Pte. Ltd.	Cisco Systems International BV Amsterdam,
San Jose, CA	Singapore	The Netherlands

Cisco has more than 200 offices worldwide. Addresses, phone numbers, and fax numbers are listed on the Cisco Website at **www.cisco.com/go/offices.**

Cisco and the Cisco logo are trademarks or registered trademarks of Cisco and/or its affiliates in the U.S. and other countries. To view a list of Cisco trademarks, go to this URL: www.cisco.com/go/trademarks. Third party trademarks mentioned are the property of their respective owners. The use of the word partner does not imply a partnership relationship between Cisco and any other company. (1110R)

About the Authors

John Capobianco has a dynamic and multifaceted career in IT and networking, marked by significant contributions to both the public and private sectors. Beginning his journey in the field as an aluminum factory worker, Capobianco's resilience and dedication propelled him through college, earning a diploma as a Computer Programmer Analyst from St. Lawrence College. This initial phase set the foundation for a career underpinned by continuous learning and achievement, evident from his array of certifications, including multiple Cisco certifications as well as Microsoft certification.

Transitioning from his early educational accomplishments, Capobianco's professional life has spanned over two decades, featuring roles that showcased his technical prowess and strategic vision. His work has significantly impacted both the public and private sectors, including notable positions at the Parliament of Canada, where he served as a Senior IT Planner and Integrator, and at Cisco, where he began as a Developer Advocate. These roles have been instrumental in shaping his perspective on network management and security, leading to his recent advancement into a Technical Leader role in Artificial Intelligence for Cisco Secure, reflecting his commitment to integrating AI technologies for enhancing network security solutions.

In addition to his professional and technical achievements, Capobianco is also an accomplished author. His book *Automate Your Network: Introducing the Modern Approach to Enterprise Network Management*, published in March 2019, encapsulates his philosophy toward leveraging automation for efficient and effective network management. He is dedicated to lifelong learning and professional development, supported by a solid foundation in education and a broad spectrum of certifications, and now aims to share his knowledge with others through this book, YouTube videos, and blogs. John can be found on X using @john_capobianco.

Dan Wade is a Network and Infrastructure Automation Practice Lead at BlueAlly. As part of the Solutions Strategy team at BlueAlly, he is responsible for developing network and infrastructure automation solutions and enabling the sales and consulting teams on delivery of the developed solutions. Solutions may include infrastructure provisioning, configuration management, network source of truth, network observability, and, of course, automated testing and validation. Previous to this role, Dan worked as a consulting engineer with a focus on network automation.

Dan has more than seven years of experience in network automation, having worked with automation tooling and frameworks such as Ansible and Terraform, and Python libraries, including Nornir, Netmiko, NAPALM, Scrapli, and Python SDKs. Dan has been working with pyATS and the pyATS library (Genie) for the past four to five years, which has inspired him to embrace automated network testing. In 2021, Dan contributed to the genieparser library with a new IOS XE parser. He also enjoys creating his own open-source projects focused on network automation. Dan holds two professional-level certifications from Cisco: Cisco DevNet Professional and CCNP Enterprise. He is also a member of the Cisco DevNet 500 and Cisco Champions program.

Dan enjoys sharing knowledge and experience on social media with blog posts and YouTube videos as well as participating in podcast episodes. He's passionate about helping others explore network automation and advocating how network automation can empower, not replace, network engineers. You can find him on social media @devnetdan.

About the Technical Reviewers

Stuart Clark is a senior developer advocate, public speaker, author, and DevNet Expert No. 2022005. Stuart is a sought-after speaker, frequently gracing the stages of industry conferences worldwide, presenting on his expertise in programmability and DevOps methodologies. Passionate about fostering knowledge sharing, he actively creates community content and leads developer communities, empowering others to thrive in the ever-evolving tech landscape. In his previous role as a network engineer, he became obsessed with network automation and became a developer advocate for network automation. He contributed to the Cisco DevNet exams and was part of one of the SME teams that created, designed, and built the Cisco Certified DevNet Expert. He lives in Lincoln, England, with his wife, Natalie, and their son, Maddox. He plays guitar and rocks an impressive two-foot beard while drinking coffee. You can find him on social media @bigevilbeard.

Charles Greenaway, CCIE No. 11226 (R&S, Security, Datacenter), is a field CTO for BT (https://www.bt.com/about/bt). With more than 20 years of data networking experience across LAN/WAN/DC in multiple industry sectors across the globe, he ensures that his customers' use of technology is aligned with their business goals while developing and implementing the technology strategy.

His current focus is helping customers transition toward Global Fabric technologies that provide software-defined underlay and overlay networking to underpin secure multicloud connectivity.

As a member of the DevNet 500 and the Cisco Champions program, Charles promotes the use of programmability and automation to make it accessible to engineers at all levels of skill and experience. He has developed technical content through Greencodemedia Limited and in the public domain at events such as Cisco DevNet Create. Charles is a graduate of Loughborough University, holds a BSc in Computer Science, and lives in the United Kingdom.

Dedications

John: This book is dedicated to my wife and partner of more than 25 years, Michelle. Without her support and encouragement, I would likely still be driving a forklift. J'taime le.

Dan: I would like to dedicate this book to my wonderful wife Hailey and my two amazing kids. They are my foundation and have been patient during the entire writing process. I'd also like to dedicate this book to my parents, who have continued to push me to accomplish whatever I wanted in life. I love you all!

Acknowledgments

John: I would first like to acknowledge what a pivotal role St. Lawrence College has played in my life—first as a student, then as a professor. Thank you to Donna Graves, Janis Michael, and, of course, rest in peace, Carl Davis. To everyone I've ever worked with in my career from Empire Life to the Parliament of Canada, thank you. I am very proud of what we accomplished together and for the confidence you had in me to build, support, evolve, and, ultimately, automate your networks. Thanks also to Cisco for embracing me completely as one of your own—I've never had such a supportive culture.

Thanks as well to everyone involved with the publishing of this book, from Nancy Davis and Chris Cleveland and the Pearson team, to our editors Stuart Clark and Charles Greenaway for their dedication to the project, and last but not least, to Dan Wade for co-authoring the book. From joint live streams to collaborating on this book, I am really proud to call you a friend.

Finally, to JB and Siming, for inviting me to a private pyATS crash course. You have both been so giving and have provided me real guidance and direction and turned me onto Python. Thank you both.

Dan: First, I'd like to thank the Art of Network Engineering (AONE) community, specifically AJ Murray, for encouraging me to begin blogging and creating my own brand. I can confidently say there would be no DevNet Dan without AONE! I'd also like to thank NetCraftsmen for taking a chance on me in the beginning of my consulting career. It was my first time working in the consulting space, and they've consistently guided me to success. Thank you Terry, Shaffeel, Robert, Bill, Joel, and John for continuing to encourage me and push me to grow professionally.

I would like to thank Nancy Davis for giving me the opportunity to pursue this project. She continues to encourage and support me to pursue creative opportunities. A big thank you to Chris Cleveland, development editor, for providing the best support developing the book, and to Stuart and Charles for their unbiased and honest technical review of the book and its contents. I'd also like to thank my wonderful co-author John. It has been a pleasure working and writing this phenomenal book with him!

Finally, thanks to all the content creators, trainers, and authors who have influenced my narration style and contributed to my constant learning of networking and software development.

Contents at a Glance

Reader Services

Register your copy at www.ciscopress.com/title/ISBN for convenient access to downloads, updates, and corrections as they become available. To start the registration process, go to www.ciscopress.com/register and log in or create an account*. Enter the product ISBN 9780138031671 and click Submit. When the process is complete, you will find any available bonus content under Registered Products.

*Be sure to check the box that you would like to hear from us to receive exclusive discounts on future editions of this product.

Contents

Command Syntax Conventions

The conventions used to present command syntax in this book are the same conventions used in the IOS Command Reference. The Command Reference describes these conventions as follows:

- **Boldface** indicates commands and keywords that are entered literally as shown. In actual configuration examples and output (not general command syntax), boldface indicates commands that are manually input by the user (such as a **show** command).

- *Italic* indicates arguments for which you supply actual values.

- Vertical bars (|) separate alternative, mutually exclusive elements.

- Square brackets ([ ]) indicate an optional element.

- Braces ({ }) indicate a required choice.

- Braces within brackets ([{ }]) indicate a required choice within an optional element.

Foreword

In late 2013, I found myself seated at the end of a restaurant table in San Jose, celebrating the success of our latest Tcl language–based test automation feature release. Tibor Fabry-Asztalos, our visionary senior director, raised his glass in a toast: "We need to look at our next goal. It's time to transition to Python-based automation." With that, he gazed to his left, where coincidentally his chain of reports was seated in order—and each person looked further to their left, until there was just me, the final link in the chain, entrusted to shoulder that responsibility.

And so pyATS was born.

After two decades of Tcl/Expect-based automation and testing in Cisco, the call for a more modern, natively object-orienting infrastructure was undeniable—one that could scale forward, lower the barrier for adoption, and attract new talents as Tcl expertise waned.

2024 marks the 10-year anniversary for pyATS. Originally introduced as an internal testing solution, its 2017 public launch through Cisco DevNet marked a definitive, transformative moment. It enabled closer collaboration between Cisco engineering, customers, and their network engineers, unlocking a plethora of opportunities and use cases. Around that time, NetDevOps was in its infancy, and network engineers were seeking for their next career breakthrough. pyATS was ready just around the corner.

Rarely does one find themselves at the helm of opportunity to shape the next decade of network automation, a chance to redefine the landscape of network testing, and influence the careers of countless network engineers. It has been an exciting journey, filled with dedication, perseverance, and innovation. Most importantly, we took pride in what we have created and accomplished.

Looking back, could we have done better? Absolutely. Along the way, mistakes were made, and compromises became necessary. But as someone special to me once said, "every decision you make in life [sic] is always the best that you could, based on the limited knowledge you had at that time." We, the pyATS development team, gave it our best, and the community echoed positively.

It has been a privilege and an honor to be able to stand at the precipice of a new chapter in the history of test automation at Cisco. A heartfelt thank-you goes to our team, our community members, and everyone who supported us along the way. As pyATS continues to evolve, my sincerest wishes for its continued momentum and enduring legacy.

```
>>> from pyats import awesome
```

—Siming Yuan, pyATS Founder, Architect and Lead Developer

Reflecting on the inception of pyATS, it's astounding to see the journey from an ambitious project within Cisco engineering to a cornerstone of network automation. Born from the challenges we faced daily, it quickly grew beyond the initial scope, demonstrating the power of innovative solutions in a rapidly evolving field. I am filled with gratitude for the brilliant minds I worked with and the community that has grown around these tools. Your enthusiasm and support have been the driving force behind its success.

Now, looking back, I see the legacy of pyATS not just in the technical achievements, but in the community and collaboration it fostered. It has been a privilege to contribute to this chapter of network engineering, and I am proud of what we accomplished together.

A special thanks to all the pyATS team members who have worked on it. It wouldn't have been possible without you. Thank you to everyone who has joined us on this remarkable journey. Your contributions have made all the difference.

—*Jean-Benoit Aubin, Lead Developer and Architect, pyATS*

Introduction

This book was written to explore the powerful capabilities of automated network testing with the Cisco pyATS framework. Network testing and validation is a low-risk, yet powerful domain in the network automation space. This book is organized to address the multiple features of pyATS and the pyATS library (Genie). Readers will learn why network testing and validation are important, how pyATS can be leveraged to run tests against network devices, and how to integrate pyATS into larger workflows using CI/CD pipelines and artificial intelligence (AI).

Goals and Objectives

This book touches on many aspects of network automation, including device configuration, data parsing, APIs, parallel programming, artificial intelligence, and, of course, automated network testing. The intended audience for this book is network professionals and software developers wanting to learn more about the pyATS framework and the benefits of automated network testing. The audience should be comfortable with Python, as pyATS is built with the Python programming language.

Candidates who are looking to learn pyATS as it relates to the Cisco DevNet Expert Lab exam will find the use cases and examples throughout the book valuable for exam preparation.

How This Book Is Organized

Chapter 1, "Foundations of NetDevOps": This chapter introduces NetDevOps, outlining its benefits and how it merges with software development methodologies to enhance network automation. We compare key automation tools and detail the modern network engineer's toolkit, setting the stage for applying NetDevOps in practice.

Chapter 2, "Installing and Upgrading pyATS": The chapter shows how to install and upgrade pyATS and the pyATS library using Python package management tools and built-in pyATS commands.

Chapter 3, "Testbeds": This chapter covers YAML's basics, explores the concept of a testbed, and examines device connection abstractions. We discuss methods for testbed validation, the creation of dynamic testbeds, and how intent-based networking integrates with extended testbeds, providing a roadmap for their practical application.

Chapter 4, "AEtest Test Infrastructure": This chapter is one of the key chapters in this book. It goes in depth and reviews the different components that make up AEtest, the testing infrastructure that is the core of pyATS. Everything from defining testcases and individual test sections to running testscripts is covered in this chapter. After reading this chapter, you'll understand how to introduce test inputs and parameters, define test sections, control the flow of test execution, and review test results with the built-in reporting features.

Chapter 5, "pyATS Parsers": This chapter delves into pyATS parsers, emphasizing vendor-neutral automation strategies. It covers the essentials of pyATS learn and parse features, techniques for CLI parsing, and parsing with Python. Additionally, we explore how to perform dictionary queries and analyze differentials, equipping you with the necessary skills for effective network data handling.

Chapter 6, "Test-Driven Development": This chapter introduces test-driven development (TDD), its application in network automation, and an overview of pyATS. It further explores the pyATS framework, setting the foundation for incorporating TDD practices into network management.

Chapter 7, "Automated Network Documentation": This chapter explores automated network documentation, beginning with an introduction to pyATS jobs. It details executing pyATS jobs from the command-line interface (CLI), interpreting CLI logs, and utilizing the pyATS logs HTML viewer for enhanced analysis. We also delve into Jinja2 templating for document creation, culminating in the generation of business-ready documents.

Chapter 8, "Automated Network Testing": This pivotal chapter delves into automated network testing, the core focus of the book. It outlines a strategic approach to network testing, including software version testing, interface testing, neighbor testing, and reachability testing. Additionally, we explore intent-validation testing and feature testing, essential components for ensuring network reliability and performance.

Chapter 9, "pyATS Triggers and Verifications": This chapter reviews how to use triggers and verifications using the Genie Harness. Triggers and verifications allow you to build dynamic testcases, with a low-code approach, that can change with your network requirements.

Chapter 10, "Automated Configuration Management": In this chapter we will look at how to generate intent-based configuration using data models, Jinja2 templates, and Genie Conf objects. In addition to generating configurations, we will see how to push configuration to network devices using a file transfer server, Genie Conf objects, and pyATS device APIs.

Chapter 11, "Network Snapshots": This chapter looks at how to profile the network by creating and comparing snapshots of the network. Network snapshots can be helpful when you're troubleshooting a network issue or just learning about the network's operating state at a point in time.

Chapter 12, "Recordings, Playbacks, and Mock Devices": This chapter introduces pyATS recordings, covering the recording of pyATS jobs and the playback of these recordings. It explains how to create mock devices and simulate device interactions through the mock device CLI, offering practical insights into testing without the need for live network equipment.

Chapter 13, "Working with Application Programming Interfaces (API)": This chapter focuses on working with pyATS APIs, detailing the pyATS API framework, REST connector, YANG connector, and gNMI. It provides insights into how these tools and protocols can be utilized for efficient network automation and management through API interactions.

Chapter 14, "Parallel Call (pcall)": Testing in pyATS can be sped up using parallel processing (parallelism). In this chapter, we review the differences between parallelism and concurrency using asynchronous programming. Parallel call (pcall) in pyATS enables parallel execution and is built on the multiprocessing package in the Python standard library.

Chapter 15, "pyATS Clean": In this chapter, you will see how pyATS can reset devices during or after testing using the pyATS Clean feature.

Chapter 16, "pyATS Blitz": In this chapter, we will review pyATS Blitz, which creates a low-code approach to building pyATS testcases using YAML syntax.

Chapter 17, "Chatbots with Webex": This chapter explores integrating pyATS with Webex, including pyATS job and health check integrations. It delves into using Adaptive Cards within Webex for interactive content and outlines methods for setting up customized job notifications, enhancing communication and monitoring in network operations.

Chapter 18, "Running pyATS as a Container": This chapter introduces the concept of containers, focusing on the pyATS official Docker container. It guides you through the pyATS image builder and details the process of building a pyATS image from scratch, offering a comprehensive approach to deploying pyATS as a containerized application.

Chapter 19, "pyATS Health Check": This chapter dives into the different health checks that run to ensure devices under testing are operating correctly. Built-in health checks include checking CPU, memory, logging, and the presence of core dump files to ensure devices haven't malfunctioned or crashed during testing.

Chapter 20, "XPRESSO": This section covers pyATS XPRESSO, starting with installation instructions. It provides a beginner's guide to getting started with XPRESSO and details on running pyATS jobs within the XPRESSO environment, facilitating an easy entry into utilizing this powerful tool.

Chapter 21, "CI/CD with pyATS": The concept of CI/CD is a common practice in software development to build and test code before it's pushed to production. In this chapter, we see how to use multiple network automation tools, including GitLab, Ansible, and pyATS, to apply CI/CD practices when pushing configuration changes to the network.

Chapter 22, "Robot Framework": In this chapter, we review Robot Framework, an open-source test automation framework. Robot Framework allows you to use English-like keywords to define testcases. After we review Robot Framework, we see how the pyATS libraries—Unicon, pyATS, and the pyATS library (Genie)—are integrated into Robot Framework by providing test libraries that include keywords to interact with network devices and define testcases.

Chapter 23, "Leveraging Artificial Intelligence in pyATS": This chapter explores the integration of pyATS with artificial intelligence, focusing on leveraging the OpenAI API for enhanced network automation. It discusses the use of retrieval augmented generation (RAG) with Langchain for intelligent data handling and introduces rapid prototyping with Streamlit, showcasing the potential for AI to revolutionize network management processes.

Appendix A, "Writing Your Own Parser": This appendix covers how to contribute to the genieparser library (https://github.com/CiscoTestAutomation/genieparser) by creating a new parser for a Cisco IOS XE **show** command.

Appendix B, "Secret Strings": This appendix covers how to protect the sensitive data in your testbed.yaml files through secret strings.

Credits

Figure 1.1: chinnappa/123RF

Figures 5.2a, 5.7-5.9, 5.13, 5.14, 13.2a, 21.1, AppA 1, 2, 4-7: GitHub, Inc

Figures 5.15-5.17, 7.24-7.34: Microsoft Corporation

Figure 20.26a: Jenkins

Figures 21.3, 21.4: GitLab B.V.

Figures 23.3-23.6: Snowflake Inc

Chapter 1

Foundations of NetDevOps

The landscape of enterprise networking has changed dramatically over the past several years with an explosion of new tools, technologies, and methodologies for building and operating networks at any scale. Network automation and programmability have also matured to the point that the expectation is that modern solutions are designed and implemented with an automate-first, agile mindset. Networks today are undergoing an evolution similar to that of voice networks two decades ago with command-line interfaces and manual effort. Networks are rapidly being replaced with automation and programmability. Network engineers now must consider software development practices when planning, designing, building, and operating networks in their day-to-day activities. Gone are the days of manually drafting device-to-device configurations; the focus is now on a more holistic, software-driven, automated vision of network design, configuration, and testing. Methodologies and practices in NetDevOps are adopted and adapted from the broader DevOps movement, extending these practices to the network domain. Network automation and programmability are no longer the future—they are the present, and Python Automated Test Systems, or pyATS (pronounced *py A-T-S*), is crucial to accelerating your NetDevOps. pyATS was originally created by Jean-Benoit Aubin and Siming Yuan for internal Cisco testing, which has evolved into a free, public, mass-adoption network automation framework. Welcome to test-driven automation with Cisco pyATS; your transformative journey with pyATS begins here.

This chapter covers the following topics:

- Traditional network operations
- Software development methodologies
- Comparing network automation tools
- The modern network engineer toolkit

Traditional Network Operations

Historically, networks were predominantly driven by manual, human-centric efforts, although this did not undermine the significance of core principles, best practices, and foundational designs crucial for well-implemented and well-operated networks. With nearly 50 years of experience in the networking realm, the OSI and TCP/IP models remain relevant. Individuals boasting profound knowledge, expertise, and industry certifications often transition seamlessly into adept NetDevOps developers, as they meld their networking prowess with the contemporary tools and technologies essential for automation. However, the journey of automation wasn't straightforward. Initially, operations leveraged various tools for monitoring and managing configurations. These tools, primarily element managers, were adept at scrutinizing the active status of elements but fell short of monitoring the systems as a cohesive whole. Automation at a larger scale was a prerogative of those who could afford the coding outlay, as the high costs stemmed from the lack of standardized interfaces on vendor equipment and high-level language support. The business case for automation was hard to justify without a substantial scale that could offset the high initial costs. Fast-forward to recent times, the landscape has evolved favorably. Vendors have lowered the entry barriers by prioritizing their APIs, and the widespread support for languages like Python has catalyzed the adoption of automation technologies. Transitioning from a purely network-centric career to a hybrid developer role now entails a lesser extent of training and learning, thanks to the improved accessibility and support for automation tools. The ensuing sections delve into the typical duties of a network engineer and the application of these skills within the NetDevOps framework, reflecting the changes brought about by the more accessible automation solutions.

Architecture

The overall architecture of the network will still be required to establish the desired outcomes. Appropriate hardware selection for the physical layer of the network and the underlying connectivity model are vital to driving the design of the network. Architects also establish the best practices, validate designs, and work with vendors in product selection. These network architects are typically responsible for the overall network, handle high-level escalations, contribute to change management boards to review and approve operational changes, and interface with the enterprise leadership and management teams to establish roadmaps and long-term planning. Ancillary appliances that supply wireless access and identity, automation or software-defined network (SDN) controllers, and other required tools are necessary to meet the service level agreements (SLAs) and business requirements.

The modern high-level designs also often include the selection of closed- and open-source tools required to operate and monitor the network. A greenfield network requires a lot of planning, and network architects don't work in silos; in fact, there is often a great deal of collaboration with the vendor's experts and specialists, the Internet service provider (ISP), the chief technology officer (CTO), the chief information officer (CIO), and the chief information security officer (CISO) to translate business requirements, budgets, and SLAs. Ultimately, the network's purpose is to connect people with their systems and applications. Business needs are fulfilled by these applications and people, which in turn are underpinned by the network. Typically, the architect best understands the desired state, the expected baseline performance and services offered, and is an escalation point in the event of any unexpected issues or behaviors. In a modern NetDevOps environment, the architect often plays the role of the senior developer over the network. Code reviews, approving the merging of code via Git pull requests, developing or refactoring network automation code, ensuring the quality of continuous integration/continuous development (CI/CD) pipelines, holding daily standups, and playing a role in scrum teams are all additional expectations of the modern network architect, beyond being a highly certified and capable network engineer.

High-Level Design

After an architecture has been established, a high-level design is created, which is often a visual representation of the network. Various blocks of the modular architecture are connected and east-west traffic flows (that is, traffic remaining with a module horizontally) and north-south traffic flows (that is, traffic leaving a module northbound toward the public Internet and network egress to the ISP or southbound into the data center) are determined. Various protocols, network zoning (security perimeters and boundaries), layer-one interconnections, and levels of redundancy are visualized into a high-level design. The speeds of interconnections, optics and cabling types, wireless standards, and access point placement are also part of the high-level design. The architects or designated designers carry out these tasks, and the designs are often validated by the collaborative architecture teams. Vendor-validated designs, request for comments (RFCs), and industry best practices are used as reference models during the high-level design phase. Important artifacts such as the IP address scheme, network protocol design (such as OSPF areas, if OSPF is the selected routing protocol), and service requirements (Power over Ethernet [PoE] and 802.1x) accompany the high-level design. The high-level design should be a living document that evolves and changes as the network goes through its operational lifecycle in a continuous feedback loop.

Low-Level Design

Lower-level designs overlay the actual configuration derived from the high-level design. Internet addresses are assigned as subnets to areas of the network, and individual interfaces are assigned addresses. Routing and security protocol configurations are drafted as traffic flows are enforced across the network. The various spanning trees are mapped out and configurations established to ensure a loop-free topology with limited sized broadcast domains (VLANs). Redundant links are appropriately configured as interfaces are bundled and routers are set up in highly redundant pairs. Usually, low-level designs are developed outward from the center of the network. The vital high-speed backbone and core routing configurations are developed first with north-south traffic in mind. Distribution and aggregation layers, also typically routed, are drafted next, followed by access port configurations established for edge features like power, identity, security, and aggregation in the data center. Uplink port configuration standards are developed, and the Layer 2/Layer 3 boundaries are established. Firewalls, access control lists (ACLs), and other security boundaries are implemented. Load balancing, wireless controllers, access and identities, and other ancillary service configurations are also included in lower-level designs. Low-level designs are often impacted the most by NetDevOps because not only do the high-level designs have to be transformed into working functional models of the network configurations, but these configurations also need to be transformed into objects that can be used for programming and automation. Data models, such as YANG, configuration templates, tests, intended configurations, and code to push the initial images and day 0 configurations to the greenfield devices also need to be included as artifacts in the low-level design. There is also a growing popularity in virtual eXtensible LAN (VXLAN) technologies expanding on Layer 2 connectivity in the enterprise (software-defined access), data center (application-centric infrastructure/Nexus), and WAN networks (SD-WAN), adding even more to consider in the design phase of the modern network.

Day−1

Day−1 (day minus one) activities include the translation of business requirements into functional network building blocks, including creating initial architectures, high- and low-level designs, and topology diagrams. Procuring hardware and associated software licenses, establishing SLAs with the business, vendors, and supporting third parties, and preparing internal processes, procedures, and support models are all part of day−1 activities in preparation for standing up the network. All activities prior to actual device onboarding can be described as day−1 activity where enterprises prepare to deploy their new network. In this phase, the enterprise typically gathers the recommended software image releases from the Internet as well as transforms low-level designs into initial device configurations. These device configurations are typically the bare minimum to get a device up and running and often require design and operational knowledge of what will be connected to the device. Information is often entered via serial connection to the management console of each device using the serial number to identify the placement of the device in the new network. Day−1 configurations are prerequisite configurations in order to establish a minimum level of off-device connectivity, such as the hostname, VTY line configurations, a username and enable secret, possibly hostname resolution (DNS), and cryptography (SSH keys). A manual mapping from an IP address management (IPAM) and low-level design is often performed by a human operator who applies a minimal base connectivity, enabling configuration from a template of some kind or individual instructions per device. Often these configurations are performed in the central warehouse or location where the devices have been received from the vendor shipping them. Devices are potentially barcoded and their serial and part numbers recorded. Stacks of devices are often assembled and the basic stacking technology deployed in the event of stacked devices. Power supplies, inter-stack cabling, SFP (small form-factor pluggable) module insertion, and the assembly of all devices take place either within the warehouse or during the on-site truck roll. At the end of day−1, devices are typically flashed with the selected software image, basic testing has been performed, and minimal configuration has been applied. Devices are then repackaged with their unique identifiers (hostname, management IP address) and delivered to the data center, rack, or telecom closet for installation and day 0 configuration. The sections that follow look at some tasks that can be automated in day–1.

Offline Initial Configuration

Using a testbed file, covered in detail in Chapter 3, "Testbeds," new devices can be described in human-readable YAML format at scale. By extending the YAML file to include the intended state, the initial configuration can be derived from APIs or Jinja2 templates in an "offline" approach. The result is a simple, but validated, initial configuration file derived from the low-level design that operators can quickly and accurately apply via a serial connection in the warehouse. This is not only faster, which is a goal of the typically highly manual process of day–1 configuration, but it is more accurate, of higher quality, and it dramatically simplifies the process for human operators. Instead of having to manually adjust a set of default instructions per device or translate an IP address plan into management IP address configurations, pyATS can be leveraged to automate this process and provide human operators with ready-to-go initial configurations that are easily applied over the serial connection.

Software Images

All network devices and appliances should have either the latest or the vendor-recommended release applied to them during the day–1 process. Images can be distributed locally using USB drives or downloaded from a centralized image repository that is reachable after the initial minimal configuration is applied.

pyATS can play an important part of day–1 activity and automate both the initial base configurations as well as the software image management. The pyATS Clean framework, further explored in Chapter 15, "pyATS Clean," can be used to load new images and apply a base configuration.

Day 0

Day 0 is the onboarding process. Devices are racked, stacked, powered on, and interconnected in their location in the network depending on the role they will play. It is at this point where connectivity to the devices' neighbors, centralized management, orchestration, and monitoring systems are established. Access from distributed operator workstations or the management zone is also established, providing access to pyATS to perform additional configuration. The remaining configuration items that rely on the device being connected to the full network topology can now be pushed from pyATS. There are many advantages to using pyATS to complete the day 0 onboarding configurations, including the following:

- Intent-driven configuration from the testbed YAML file for a single device or an entire new topology at scale

- Templated configurations from APIs or Jinja2

- Pre-change state capture

- Automated configuration deployment

- Set of initial pyATS tests validating the onboarding

- Post-change state capture

- Differential output comparing day–1 and day 0 configuration states

- Automated state capture

- Automated business-ready documentation

With pyATS, network engineers can move towards "zero-touch provisioning" (ZTP) without using Plug and Play (PnP), DHCP, and TFTP/FTP/SCP server infrastructure to achieve automated onboarding of any size or scale. The sections that follow look at day 0 activities than can be automated with pyATS.

Layer 1

Layer 1 wiring and cabling of infrastructure is still required, unfortunately, and device interconnectivity (which ports connect to which ports) still needs to be mapped out from the low-level design. One major advantage of pyATS is that these interconnects can be quickly and automatically tested as part of the onboarding job. The presence of certain neighbor relationships, like Cisco Discovery Protocol (CDP) or its open standard counterpart Local Link Discovery Protocol (LLDP), or even OSPF or BGP neighbors, can be added as *tests* in the provisioning job. Engineers can review the job logs to quickly and easily determine if the provisioning was successful based on the results of these tests. Ping tests can be used to validate reachability from a given device to another destination in the network, and interface tests can be used to confirm the lack of errors or the presence of full-duplex connectivity, at scale, across entire testbeds. Network engineers can be reassured by pyATS that devices have been deployed correctly and that the required wiring is in place at the end of the onboarding process.

Initial Configuration

Initial configuration can be derived from the combination of the pyATS testbed and either pyATS application programming interfaces (APIs) or Jinja2 templates (expanded upon in Chapter 7, "Automated Network Documentation") that provide the Cisco IOS configuration code required to complete the devices' onboarding process. Much like the initial day–1 configuration, each device can use an intended configuration abstracted from Cisco OS configuration stanzas into human-readable structured data files. Every aspect of a device's configuration can be transformed in the simple testbed file and then the actual configuration called from either pyATS APIs or Jinja2 templated configurations. Using a NetDevOps approach and Git version and source control, the entire lifecycle of a device's configuration can be tracked from day–1 initial configuration and onboarding configuration all the way through to day N configuration. In fact, using this approach with pyATS, day–1 and day 0 initial configurations could be merged into a single intent-based, templated configuration, reducing the deployment lifecycle to a single stage of development. If you either apply initial minimal configurations and then complete the full configuration over two phases or merge the initial configurations to be totaled and completed in day–1, there is a major reduction of error-prone, human-driven, initial configuration management with pyATS.

Initial Testing and Validation

In the preceding "Layer 1" section, we discussed how pyATS opens up a myriad of testing possibilities after the initial onboarding phase. This encompasses reachability, neighbor relationships, interface configuration, and counter-information assessments—all crucial for ensuring a smooth onboarding process. However, as will be unraveled in the subsequent chapters of this book, pyATS's utility goes beyond just these preliminary tests. Any output from a **show** command can be rigorously scrutinized through pyATS tests. Once the initial onboarding configuration is deployed to the device by pyATS, you have the freedom to design tests tailored to your specific needs, aiding in validating the success of the onboarding processes.

Before the onsite technicians conclude their tasks at the deployment site, pyATS tests serve as a reliable tool to ascertain that no manual interventions or amendments are necessitated. Devices can be drop-shipped with preconfigurations and, upon arrival, be automatically tested to ensure they are in a healthy state, connected accurately, and configured in alignment with the intended design. This eradicates the traditional dependence on the command-line interface (CLI), sifting through the output of various **show** commands, and relying on human operators to undertake the challenging task of validating a device's configuration and state. Instead, pyATS automates these tasks, with its job logs and HTML log viewer significantly simplifying the most arduous aspect of day 0 configuration—ensuring accurate configuration and connectivity during the onboarding phase.

The narrative so far has revolved around the correctness of an individual device being onboarded and its interaction with neighboring devices. However, there's a broader perspective to be explored. pyATS is not confined to device-centric validation; it extends to scrutinizing the state of the entire network system. For instance, even if the configurations are accurate, interfaces are active, and routing adjacencies are established, there might be underlying routing faults or perhaps a newly defined prefix or ACL malfunctioning. These issues might manifest at an area border router or a peering router situated elsewhere in the network. pyATS empowers us to delve deeper and test for such network-wide anomalies too! It's not just about the correctness of individual device configurations but a holistic examination of the network system's state, ensuring everything operates cohesively and as intended across the network infrastructure.

Day 1

Day 1 can be demarcated by everything that occurs after network devices are onboarded, provided their initial configuration(s), tested, and validated as ready for service. Day 1 can be further broken down into day 2 and day N configurations for simplicity and provided further demarcation of tasks. pyATS plays a critical role in day 1, allowing network engineers and operators to update their intent files to reflect required changes to the network configurations. Intent should be the only place network staff members need to update, but in cases of provisioning new services, new APIs or Jinja2 templates will also need to be developed to support services that were not in scope of the original intended configurations. Using Git and a Git repository, these changes should also be part of a working branch, tested and validated, and ultimately merged back into the main branch and deployed to the network devices as part of a CI/CD pipeline, as covered in detail in Chapter 21, "CI/CD with pyATS." Human error is virtually eliminated in this CI/CD process, as only the easy-to-read, human-compatible YAML file is changed and pyATS takes care of the rest of the configuration management. The sections that follow cover some day 1 activities that can be automated with pyATS.

Incremental Configuration

As the network evolves over time and things change, the configuration of the network also evolves and changes. This could be something as simple as a new Network Time Protocol (NTP) source or Dynamic Host Control Protocol (DHCP) server that needs to be config-ured to make the network devices aware of it. Prior to network automation, operators would need to connect to each device in the network, possibly dozens or hundreds of devices, and manually apply the changes using the CLI. This was time-consuming, error-prone, and obviously not the best use of a person's time. Configuration drift between the intended configuration and state and the actual running configuration also occurs over time. Until every device is reconfigured, there is disparity between devices. Changes might take hours, days, weeks, or even months to complete, depending on the size of the network and available resources. These changes are also prone to human error, where the wrong device, incorrect information, or simple missed keystrokes could lead to disparity between the intent and the actual configurations. In a worst-case scenario, outages or interruptions to the flow of network traffic are introduced accidentally as a result of these human errors.

The reconfiguration is also only part of the story; testing and validating the changes are also required, compounding the time it takes to establish confidence the change was successful. Often, the validation can take longer than the actual change itself.

pyATS really shines in day 1 activities, as it not only can derive a configuration directly *from* the intended state but can also configure, test, and document these changes at scale, automatically, without a human ever having to log in to the CLI:

- Capture the pre-change configuration
- Test the pre-change state
  - Report errors and abort the change
  - Confirm an error-free state
- Push configuration to the device
- Test the post-change state
  - Report errors and roll back the change
  - Confirm an error-free state
  - Confirm the intent delivered

■ Capture the post-change state

■ Update the documentation

■ Perform a differential

   ■ Display changes to the configuration and state

Provisioning New Endpoints

Networks grow in size as capacity is required. Fortunately, with pyATS, all we need to do is add new devices' intended configuration to the testbed file. The initial configuration will be generated for offline installation, and the initial image and configuration will be automated. The new devices can be tested and existing device tests modified to accommodate the presence of a new device. So not only can new devices be quickly and accurately added to existing networks and tested as new devices, the existing network can be regression-tested to validate no unintended consequences or impacts occurred as the result of the presence of a new device.

Provisioning New Services

The only time anything is developed outside the testbed file is when new services are provisioned. The data model, the YAML intent in the pyATS testbed file, and the accompanying pyATS APIs or Jinja2 templates all will need to be created to provide these new services. If, for example, quality of service (QoS) was not deployed as part of the day 0 onboarding but needs to be configured to achieve the business SLAs, to support new devices such as voice over IP (VoIP) devices or IP cameras, or to reduce congestion in the new network topology, the QoS model needs to be mapped to the intent data model in YAML and then the appropriate APIs or Jinja2 templates developed. The pyATS tests to support the validation of the implementation also need to be developed. pyATS can reduce the time it takes to develop and deploy new services as networks become "agile," not "fragile."

Day N

Day N activities involve the day-to-day activities performed to maintain a healthy network state. In the past, network operations typically involved monitoring the network for system logging (syslog) events pushed from the network in response to activity on a device and Simple Network Monitoring Protocol (SNMP) polling (pulled from the device at intervals) or traps (events pushed from the device) and then responding according to these events. The sections that follow describe Day N network managing and monitoring activities in more detail.

Monitoring (and Now Testing)

Traditional network monitoring can be greatly augmented with pyATS testing. In addition to syslog and SNMP events, the network can now be *proactively* tested by pyATS to confirm that the intended configuration and state are maintained. pyATS also adds *context* to sometimes vague syslog or SNMP information. A typical alarm from syslog might indicate an interface has gone down, leaving operators to determine the impact of losing that interface. This analysis might take time, and low-level designs might need to be referenced to determine the impact of a particular interface indicating it is down. With pyATS, it can be quickly and easily determined that an interface has gone down as well as what other impacts that failed interface has had on a device or even the entire topology. Multiple pyATS instances might start to fail (such as neighbor relationships and establishments), routing protocol or routing table tests might start to fail, and ping and other connectivity tests might also start to fail. The scope of the impact can easily be determined using ongoing scheduled pyATS tests and the severity of the outage can be established rapidly without ever having to log in to a device or multiple devices to assess the impact. This is just one example of an interface going down, but every facet of a device or entire network's configuration, state, and health can be continuously tested by pyATS arming network operations with automated capabilities far beyond syslog and SNMP.

Responding to Events

In addition to traditional monitoring and the new continuous testing capabilities offered by pyATS, day N management of networks includes network operations responding to events encountered by network-generated syslog or SNMP traps or failed pyATS tests. One of the major benefits of pyATS is that it provides built-in *alerting* capabilities, including sending test reports by email or even via a messaging platform such as Webex, Slack, or Discord. Individual failed tests can also be sent as Webex alerts to the operators or engineers responsible for responding to and remediating failed tests. pyATS allows for *proactive* response to events, as failed tests can identify user-impacting failures immediately and alert those responsible for addressing the situation. This is a paradigm shift from either trying to make sense of multiple syslog or SNMP traps or, worse, responding to calls from the users impacted by the degraded network state.

Upgrading

Software image management is also a major part of day N operations and network management. Network vendors such as Cisco release frequent updates to address flaws in existing software releases or to release new features and components to the software image. Security patches are also a major consideration and should be addressed according to the severity of the security flaw. With the combination of pyATS testing and pyATS Clean framework, software image management becomes dramatically easier as operations are able to regression-test new images against a known working state. pyATS tests can be used to validate the software upgrade has not modified the configuration or state of the network compared to the previously working baselined tests. The pyATS Clean framework can be used to deploy software and related configurations, ensuring successful software upgrades. Should commands need to be modified as the result of a new software image, the pyATS testbed (human-readable intent), pyATS APIs, or Jinja2 templates can be modified and new configurations generated and deployed to respond to changes in the structure of device configurations, deprecated configurations, or net-new configuration requirements—all automatically, quickly (with agility), and with the highest quality. Once again, pyATS can be used to test individual devices in isolation as well as the entire topology before and after software images are upgraded.

Decommissioning

At the end of a device's lifecycle, it will need to be decommissioned and likely replaced with a newer model. The process starts over again from day–1 and the replacement device is onboarded. One major advantage of using pyATS is that the original device's intent and configuration APIs or templates can be used to provide a baseline and then reused to develop the replacement device's configuration. The data model and templates may need to be adjusted, but the operator is not starting from scratch. Both the configuration and test suite can be reused to provide rapid device replacement and decommissioning and reprovisioning of new devices.

Software Development Methodologies

The development practices, processes, and *culture* are as important, if not more important, than the code and network configurations and are often the most difficult changes to adapt. Traditional networking has mostly followed the waterfall methodology—legacy processes from almost 50 years ago! Software developers have adopted a more modern approach, known as Agile, which emerged in the early 2000s. This is a big reason the world around us is dominated by ever-evolving applications and innovation. Can we bring these Agile practices to the world of networking? Yes, it is known as *NetDevOps*. It is important to understand the evolution from waterfall practices to Agile practices and DevOps to see how well this new approach works with modern network engineering. The sections that follow contrast the various software methodologies.

Waterfall

Most networks have historically been managed using the waterfall methodology. In fact, if you review the traditional network section, the various stages of the waterfall process can be mapped to the architecture, high- and low-level designs, and day–1 to day N activities. Waterfall revolves around a linear progression from one phase to the next with an emphasis on gathering and defining requirements at the beginning of the project and building solutions in rigid phases of the project's lifecycle. Each phase builds upon the previous phase's deliverables as work flows down the waterfall:

1. Requirements are captured in a network requirements document.

2. High-level details lead into lower-level details. These details come from requirements that are analyzed to produce IP schemes, subnets and VLANs, routing, ACLs, and other network models, and schemas are produced.

3. A network architecture and high-level designs are created.

4. Network configurations are developed from the architecture and designs.

5. Devices are onboarded and configured.

6. Testing occurs and networks are debugged and validated.

7. Operations take over, and the resulting network is monitored.

According to the waterfall methodology, a new phase can only begin when the preceding phase has been reviewed and tested, leading to a potentially long and rigid overall project. The waterfall approach was widely used in the 1970s and 1980s, particularly in software development and project management.

Lean

In the 1980s, the Japanese automotive company Toyota developed a system known as "The Toyota Way," or the Toyota Production System (TPS), in an effort to improve efficiency by reducing waste. This system was coined "Lean" in 1988 in the John Krafcik article "Triumph of the Lean Production System" and defined in 1996 by American researchers James Womack and Daniel Jones. Lean production primarily focused on reducing production times and delivering "just-in-time" (JIT) manufacturing, matching production to demand. A major byproduct of Lean was the elimination of waste from the production processes and the use of automated quality controls. The five key principles of Lean, as outlined by Womack and Jones, are as follows:

- Precisely specify value by product.
- Identify value stream for each product.
- Make value flow without interruptions.
- Let customers pull value from the producer.
- Pursue perfection.

Ultimately, Lean was defined as a way to do more with less and maximize efficiency.

Agile

Stemming from the Lean changes in the manufacturing sector and management approaches in the early 1990s and seeking to define a new approach to software development and operations, 17 software developers met in 2001 to discuss lightweight development practices and released the "Manifesto for Agile Software Development," which defined what they valued as software developers. This new approach strived to bring in some of the Lean manufacturing principles, such as reducing times in the production process, JIT manufacturing, and the overall elimination of waste from the processes, and apply them to software development. This was in sharp contrast with previous heavyweight development approaches such as the predominate waterfall methodology. Agile is based on the following 12 principles:

1. Provide customer satisfaction by early and continuous delivery of valuable software.

2. Welcome changing requirements, even in late development.

3. Deliver working software frequently (weeks rather than months).

4. Close, daily cooperation between business people and developers.

5. Projects are built around motivated individuals, who should be trusted.

6. Face-to-face conversation is the best form of communication (co-location).

7. Working software is the primary measure of progress.

8. Sustainable development, able to maintain a constant pace. Agile processes promote sustainable development. The sponsors, developers, and users should be able to maintain a constant pace indefinitely.

9. Continuous attention to technical excellence and good design enhance agility.

10. Simplicity—the art of maximizing the amount of work not done—is essential.

11. Best architectures, requirements, and designs emerge from self-organizing teams.

12. Regularly, the team should reflect on how to become more effective and then adjust accordingly.

DevOps

DevOps is a set of practices, processes, and tools that bring the world of software development (Dev) and IT operations (Ops) together as one cohesive discipline. Much like Lean, the goals of DevOps are to shorten the systems development lifecycle (SDLC) while providing continuous improvement and high software quality. Many ideas in DevOps come directly from the Agile methodology. The key principle of DevOps is breaking down barriers and silos of developers and operations, bringing a shared sense of ownership over the product. Automation is a core component of DevOps, and the tools used to implement DevOps are critical to its success. Automated build and testing, continuous integration, continuous delivery, and continuous deployment, which originated in Agile, are pillars of DevOps.

Expanding into Networks

Seeing the obvious benefits of Agile and DevOps, the world of networking has adopted the practices, processes, and many of the tools into *NetDevOps*, especially after the availability of many new network automation tools, including pyATS. Hank Preston, principal engineer at Cisco, defines NetDevOps as follows: "NetDevOps brings the culture, technical methods, strategies, and best practices of DevOps to networking." The first point Preston makes is that the *culture* of DevOps, as well as the technical aspects, needs to be adopted by the networking team first to embrace NetDevOps. Networks need to become "agile," and not "fragile," and the impacts on the network due to the blast radius still need to be respected.

Infrastructure as Code

One of the first breakthroughs for NetDevOps was the widespread adoption of what has become known as infrastructure as code (IaC). Elements of infrastructure, like the network, are defined in structured data like YAML Ain't Markup Language (YAML) and JavaScript Object Notation (JSON) as well as configurations delivered with Python, RESTful APIs, or traditional CLI commands abstracted as code elements. Treating infrastructure as code enables NetDevOps and the application of software development practices like Agile to traditional waterfall network lifecycles. pyATS is an implementation of IaC allowing NetDevOps to use Python to program solutions for the network.

Test-Driven Development

By using a Python library and treating the infrastructure as code, not only can NetDevOps be applied and Agile principles enacted, but other software development practices can be utilized to maximize IaC quality, including an approach known as test-driven development (TDD). Testing is built into the process as requirements are broken down into small units known as test cases. Instead of building a full product and then testing it, testing occurs during the development cycle, where each unit is tested, and finally all tests are executed against the whole product. This approach fits very well with NetDevOps, and pyATS in particular, as network requirements can be broken down into consumable-sized test cases, developed and configured, and then tested for quality. TDD encourages simple designs and, according to Kent Beck, who is credited with developing the technique, "inspires confidence" in developers.

NetDevOps

NetDevOps, a philosophy blending the methodologies, cultural practices, tools, and Agile approach from DevOps with network operations, facilitates a seamless transition from day–1 to day N network activities within the DevOps model. Central to NetDevOps is the ethos of continuous improvement, characterized by delivering high-quality code through frequent releases and iterations, enhancing network performance and reliability. Figure 1-1 delineates the DevOps lifecycle, which is elucidated in greater detail in the subsequent sections.

Cisco pyATS, a pivotal tool in the NetDevOps arsenal, seamlessly integrates within this lifecycle, acting as a catalyst in automating and validating network states and configurations. Here's how Cisco pyATS interlaces with the NetDevOps process:

1. **Planning and coding:** In the initial phases of the DevOps cycle, network designs and configurations are conceived and coded. Cisco pyATS can be leveraged to script automated tests, ensuring the integrity and efficiency of network configurations right from the drawing board.

2. **Testing:** After the coding phase, Cisco pyATS shines in the testing domain. It automates the validation of network configurations, reachability, and the state of various network elements. This automation expedites the testing process, ensuring that any deviations from the intended configurations are promptly identified and rectified.

3. **Integration and deployment:** As new code and configurations are integrated and deployed, Cisco pyATS continues to play a crucial role. It aids in automating the deployment process while concurrently running validation tests to ensure seamless integration with existing network setups.

4. **Operation:** During the operation phase, Cisco pyATS facilitates continuous monitoring and validation of the network, ensuring it aligns with the defined operational standards and configurations. This continuous validation is instrumental in maintaining network reliability and performance.

5. **Monitoring and feedback:** Cisco pyATS provides a robust framework for monitoring network states and collecting valuable feedback. This feedback is crucial for identifying areas of improvement, which feeds into the planning phase for subsequent iterations, thus completing the cycle.

6. **Continuous improvement:** By providing insightful data and automation, Cisco pyATS fosters a culture of continuous improvement within the NetDevOps framework. It facilitates quicker iterations and releases, ensuring that the network is always optimized and aligned with evolving organizational needs.

Incorporating Cisco pyATS within the NetDevOps framework elevates the efficiency, reliability, and agility of network operations, embodying the continuous improvement spirit central to the NetDevOps philosophy. Through automation and validation, Cisco pyATS propels networks closer to the aspiration of self-operating and self-healing infrastructures, making it an invaluable asset in modern NetDevOps practices.

chinnappa / 123RF

Figure 1-1 *DevOps Lifecycle*

Plan

Fail to plan, plan to fail. For "greenfield" or "net-new" networks, we begin with planning. Business requirements, service level agreements, architectures, and high- and low-level designs drive the planning phase of NetDevOps.

Code

Instead of manually crafting configurations in NetDevOps, we start to create data models, templates, and *code* to generate and deliver configurations to the devices. pyATS *testbeds*, covered in depth in Chapter 3, describe devices and topologies that are modeled in YAML as intent. Jinja2 templates or pyATS APIs are also used to create the Cisco OS configuration code, substituting values from the YAML intent in the form of *variables*. Using pyATS, developers can approach the coding phase using test-driven development (TDD), creating tests that initially fail, and then are coded to pass, to validate intent and connectivity.

Build

Code is packaged up in builds as part of the continuous integration (CI) portion of CI/CD. Build often implies the creation of a delivery vehicle for the code—in the case of pyATS, this could be a pyATS *job*. pyATS jobs can also be further packaged up into Docker container images as part of the build process. Builds are automated using tools in the CI/CD process. In DevOps, builds are typically packaged software, but in NetDevOps builds represent intended configurations; templated configurations; connectivity, configuration, and integration tests; automated documentation; and other network-centric code in the form of pyATS jobs.

Test

As part of the CI/CD process, after we have a functional build and individual pieces of code tests using TDD, a larger testing process occurs in NetDevOps. Our build is tested, our full set of tests are executed, and larger, complete end-to-end tests are performed. The goal is to identify bugs or flaws in the code as early as possible. *Linting*, or programmatically checking code for syntax or stylistic errors, is also performed as part of the test phase to improve the quality of the code. Passed tests are also part of the version and source controls acting as a gate in the approval process used to merge code into the code base. Human approvals and quality assurance are also performed in the test phase prior to releasing code.

Release

Testing for infrastructure as code often includes the continuous *delivery* step of CI/CD, where the build is released to a virtual or physical non-production environment. This could be a mocked-up smaller-scale representation of the production network in a network simulation platform such as Cisco Modeling Labs (CML), or it could be a physical lab or pre-production environment. The build is delivered to this environment where more comprehensive integration testing can occur. Everything from connectivity to configuration and network state can be tested and bugs can be identified. In the event of failed tests at this phase, the process returns back to the planning, coding, building phases of NetDevOps, allowing for high-quality, nondisruptive, perfected builds before moving to the release, or continuous *deployment*, phase of the process. After passing all tests in the pre-production environments, the candidate release is moved into the deployment phase.

Deploy

DevOps deployments are software in nature, deploying the latest version of code to systems and users. In NetDevOps deployments, the continuous deployment portion of CI/CD pushes configuration changes to the production network. This might not be immediate, as in the case of software, because the impact on networks compared to the impact on software is much greater. Change management approvals and release scheduling are included in network deployments, and the actual release will need to be automatically triggered during the approved change windows. Automated deployments *run* the pyATS job within the Docker container, resulting in changes being pushed to the network devices. pyATS tests from the previous step are executed against the production environment to confirm and validate the change was successful and the intended configurations were integrated without impact or network degradation. Using IaC and the CI/CD pipeline, if the pyATS tests fail, a *rollback* could be triggered and the network returned to its previously known-good configuration state.

Operate

The Ops portion of NetDevOps takes over, and standard network operations continue to support the network, including the newly released configurations and features. The reduction of silos and the collaboration between developers and operators provide immediate and continuous feedback, validating that the network performance or recently released features and configurations are working as intended. There is no hand-off here as under the waterfall methodology; in fact, the operations team has been integrated with the development portions of this continuous cycle. Developers are also able to provide direct support and handle escalations from operations, thus ingraining themselves in the operations of the network. In a "brownfield" (existing) network, the NetDevOps cycle may actually start at the operations phase where developers and operators collaborate to tackle the automation of an existing network. Systems previously deployed manually under waterfall are analyzed by developers and operators and the configurations and state of the network are transformed into pyATS-driven, automated, tested, intent-based networks. The real-world experience from operations is used to provide developers the ability to start the planning phase of NetDevOps, and the cycle starts.

Monitor

Traditional network management systems (NMSs), syslog events, SNMP events, and, now, pyATS testing results are all used to monitor the health and performance of the network. The impact of the release is monitored as well as the overall state of the network, and metrics are used to provide feedback to the NetDevOps team to start the planning phase for the next release, either to remediate flaws from the previous release or to add additional enhancements, configurations, tests, or capacity. This process is followed in an endless cycle of continuous improvement following a version and source-controlled CI/CD pipeline driven by pyATS.

Additional Benefits of NetDevOps

The NetDevOps lifecycle itself will bring many benefits and advancements in your journey towards test-driven automation. Some of these benefits will be technologically based and others cultural. The primary benefit is the elimination of the silos between developers and operators, thus fostering an open and collaborative culture. Let's explore some ancillary benefits to NetDevOps.

Single Source of Truth

One of the major benefits, in addition to the Agile methodology, automated CI/CD pipeline, and test-driven development, of NetDevOps is the creation of a source of truth. Legacy networks are plagued by the lack of a central authoritative source of truth—what is the intended configuration and state of the network? Is it the collection of running configurations on each device in the network? Is it offline in a spreadsheet? Is it in an engineer or operator's head? NetDevOps creates version- and source-controlled intent files, templated configurations, tests and test results, and the mechanism (the CI/CD pipeline) to develop and release changes automatically. Configurations can also be stored and managed in a network source-of-truth platform, such as Netbox or Nautobot. Occasionally, under degraded circumstances, human operators *may* need to manually intervene and make changes directly at the CLI of some devices to restore critical network connectivity, which can lead to configuration drift and come into conflict with the source of truth. Immediately after manual remediation of the network, these changes need to be reflected in the code base, new intent and templates generated, and new or existing tests created and modified. Adoption of NetDevOps will reduce these "priority-one" events over time, and less and less human intervention will be required.

Intent-Based Configuration

A major component of the single source of truth is the *intent*: what the network engineers, developers, and operators agree *should* be the configuration and state of the network to fulfill the business requirements and establish a healthy, secure, redundant, resilient, high-performance network. Using pyATS, intent can be reflected in easy, human-readable YAML files, which are not concerned with a device's OS-specific configuration and abstract away the complexity of the working configuration. These intent files can be sourced from either templates or API calls to a source of truth that handles the generation of working device OS configuration code. The "compiled" composite output provided by pyATS is a working, valid, OS-specific configuration. pyATS testing can then be used to validate and confirm that the actual configuration matches the intended configuration. Configuration drift is eliminated, as is human intervention at the command line. What *should* be the configuration of a device is always known and can be referenced offline using a single source of truth. Version and source control are applied to the single source of truth, intended configuration files, and templates used to generate the code.

Version and Source Control

As an infrastructure as *code* approach is adopted, version and source control become vital to a successful NetDevOps culture. Working code, the known-good intent, templates, and tests are protected in a *Git* repository, and Git is used to provide the mechanism to safely develop iterations of the code base. Git also empowers NetDevOps collaboration, allowing individual members of the team to develop their own code inside *branches* as part of the NetDevOps code stage of the cycle. Code is tested and validated and then *merged* into the main repository. The entire lifecycle of a device's intent, test and test results, documentation, and configuration is preserved, and *versions* of the network are available from the initial onboarding to day N and end-of-life. Rolling back to a previous version of a known-good state can be done via the release process. Sometimes known as *GitOps*, this process drives the CI/CD pipeline. pyATS jobs, tests, intent, and templates are all protected by the version and source control process, and only validated code is ever used in the build, release, and deploy stages of NetDevOps.

GitOps

NetDevOps relies heavily on Git and Git repository systems. *GitOps* is the process of triggering the CI/CD pipelines when Git branches are merged into the main branch. NetDevOps developers make a working branch from the main branch in order to code a new feature or address deficiencies in code iterations, test their code, and submit a *pull request,* requesting their branch be merged into the main branch (with the single source of truth containing only validated, working intent). Once approved and completed, the pull request kicks off the CI/CD pipeline, and tests, builds, releases, and deployments occur *automatically*. Git is much more than just version and source control; it is a key component of the CI/CD process and NetDevOps.

Efficiency

NetDevOps eliminates wasted effort, and *efficiency* is achieved using automation. Humans reclaim the time previously spent drafting and configuring changes, device by device, manually. Everything from tests to documentation to configuration management is driven by automation, and people are empowered to spend their time on more *valuable* problem-solving. Quality is built into the process, and only validated changes are released to the production network reinforced by automated testing at every stage of the lifecycle.

In a well-defined NetDevOps process, only the intent, templates, and tests ever need updating by human beings, who rely on automation tools, version and source control, and the CI/CD pipeline to perform previously manual processes. Networks can be treated as cattle, and not pets, and "snowflakes" are eliminated as the single source of truth, ensuring uniformity across the network.

Speed

Automation, often synonymous with enhanced speed and performance, brings about a paradigm shift in network operations. Its prowess in exponentially outperforming manual processes, especially at larger network scales, is unequivocal. While a human operator is still launching their terminal client, logging in, authenticating, and navigating the CLI, most automation tools have already accomplished their tasks. The velocity of automation transcends just the configuration changes or network state captures; it significantly expedites the testing phases preceding and following these changes. Network operators seldom find the task of adding a few lines of configuration to a network device or multiple connected devices time-consuming; rather, the bulk of their time is invested in validating the impact of these changes. They need to ensure the desired outcomes are achieved without introducing any unintended repercussions that might degrade the network service.

Incorporating the NetDevOps approach alongside tools like pyATS can dramatically shrink the change windows from days or hours to mere minutes or seconds. This transformation propels networks from being fragile and static, where changes are a harbinger of potential problems, to becoming agile entities. In these agile networks, frequent alterations can be executed not only swiftly but with a high degree of quality.

Yet, the velocity afforded by automation carries a double-edged sword. It undeniably speeds up network operations, but concurrently, it has the potential to propagate errors or induce network issues at an equally accelerated pace. As the adage goes, "With great power comes great responsibility!" The discussion hitherto has skirted around a critical aspect—*risk*. The rapidity of automation can swiftly escalate into a network debacle if not wielded judiciously.

A robust risk mitigation strategy is indispensable to harness the full potential of automation while averting the pitfalls. Employing comprehensive unit and system tests, both pre- and post-implementation, is a prudent practice. These tests serve as a safeguard, ensuring that the automation scripts are functioning as intended and the network remains resilient after the changes. By meticulously managing the risks through rigorous testing, automation transitions from being a potential liability to a formidable asset in network operations. This balanced approach empowers network operators to exploit the efficiencies of automation via NetDevOps and pyATS, while keeping the associated risks at bay. Through a disciplined execution of testing protocols, automation in network operations transcends from being merely a tool for speed, to a well-oiled machinery driving speed, quality, and reliability in an ever-evolving network landscape.

Agility

The statement "the network has long been the bottleneck in responding to ever-increasing demands from the business" encapsulates the challenges faced by traditional network infrastructures in keeping pace with the rapid evolution of business needs. As businesses burgeon and diversify, the demand for more robust, scalable, and agile network systems escalates. However, legacy networks, often encumbered by manual configurations, scalability limitations, and slower adaptation to new technologies, struggle to meet these burgeoning demands swiftly. This lag in network adaptability hampers operational efficiency and the timely execution of business strategies, thus acting as a bottleneck in fulfilling the ever-increasing business requisites. Applications, servers (compute and storage), and security, having adopted DevOps, are rapidly made available while the organization waits for network services to enable these new capabilities. NetDevOps allows the network to respond rapidly to business demands by working with network infrastructure as code. Again, networks can become *agile* not *fragile*. NetDevOps simply needs to update their intent, develop new templates and validation tests, and perform the build/deploy/release stages iteratively. The business no longer waits possibly days or weeks for the network teams to respond manually in a waterfall process. Distributed, collaborative teams can rapidly develop and release the changes required to respond to business demands using the NetDevOps approach.

Quality

Above speed and agility, the number-one driver for NetDevOps is the *quality* of solutions. According to a recent Uptime Institute survey, "How to avoid outages: Try harder!," 70–75% of data center failures are caused by human error, producing a chain effect of downtime. Additionally, more than 30% of IT services and data center operators experience downtime or severe degradation of service. Also, 10% of the survey respondents reported that their most recent incident cost them more than $1 million. Elimination of human errors using NetDevOps methodology and automation tools like pyATS is the number-one driving factor for the adoption of IaC solutions. While moving faster and with more agility is definitely beneficial to the business, the reduction of human errors, and thus network outages, is paramount to a successful NetDevOps implementation.

Comparing Network Automation Tools

Along with culture and software development lifecycle methodology changes, NetDevOps brings new *tools* to the network engineer's toolkit. For 40 years, most engineers have had a few simple tools, including a terminal client and a text editor, to perform their day-to-day tasks to configure, validate, document, and test their networks. Over the past 5 years, the number of new tools has exploded, dramatically simplifying, and at the same time complicating, the role of a network engineer:

- **Agent-based tools:** Agent-based tools require software to be installed onto the network device, acting as an agent for remote commands. Tools like Puppet, Chef, and Salt Stack are examples of popular network automation tools that require an agent.

A centralized controller communicates using an agent protocol to the agents deployed into the network devices. Agent-based tools have lower bandwidth requirements, are perceived as more secure, and have a central point of management with the trade-off of performance, cost, and deployments times.

■ **Agentless tools:** In contrast with agent-based tools, agentless tools use protocols like SSH and HTTPS from distributed systems to perform their automation tasks. Nothing needs to be installed on the network devices to enable agentless automation tools. There is no centralized controller, and typically costs and complexity are drastically reduced. Ansible has become an extremely popular network automation tool, partly because it is agentless. Performance, costs, and deployment time are all benefits of agentless network automation tools. pyATS is an agentless tool requiring only SSH, Telnet, or HTTP/HTTPS access to a network device. pyATS testing can even be performed against offline JSON files without any connection to the device at all.

The Modern Network Engineer Toolkit

In addition to network automation tools, agent or agentless, the modern network engineer's toolkit includes many additional tools and technologies to perform NetDevOps tasks. Gone are the days of having a simple terminal client and text editor; network engineers must embrace the growing landscape of tools in order to approach problem-solving as a software developer that *simplifies* their roles and increases quality, time to delivery, and agility. The sections that follow look at the various tools and technologies that should be part of every modern network engineer's toolkit.

Integrated Development Environment

Infrastructure as *code* implicitly requires NetDevOps to write *code*. While a simple text editor *could* be used, an integrated development environment (IDE) will *help* network engineers write quality code with a superior experience. In this book, we will be using Visual Studio Code (VS Code) as the IDE of choice; however, many alternatives are available.

"Old School"

Historically, simple text editors like Notepad were used to draft configuration changes and to consume network state from the command line offline. The historical text editor is quite limited in its capabilities and functionality. With the advent of new structured data types such as JSON and YAML, as well as the need to write code in software languages like Python and JavaScript, using a plain text editor leads to poor quality code with syntax and indentation errors (particularly in YAML and Python) that will not compile or execute properly. Users are forced to *manually* detect these flaws, often using time-consuming trial and error approaches. Classic text editors are also not aware of version and source control, and cannot preview rendered output such as Markdown or Comma-Separated Values (CSV) files, which require another tool to view in their rendered state. Working directly with "raw" text is the only capability of a legacy text editor.

"New School"

IDEs such as VS Code take text editing and writing infrastructure as code to the next level. Built-in *linting* (syntax and code quality checks) provides immediate feedback. The IDE will proactively, automatically, help the author create quality code. Malformed code with indentation errors is highlighted and underlined in red, directly telling the code author the code has problems and will not compile. VS Code is highly *extensible* with thousands of extensions to enhance the experience of working with, for example, Python, adding linting for .py files. The editor can be split into vertical and horizontal panels, allowing for direct side-by-side comparison and the ability to work in multiple locations of the same file. VS Code also provides *previews* that render raw text, such as Markdown or CSV, directly in the editor, providing developers the ability to see what and how their code renders in its final state. One of the most important capabilities of VS Code is that it integrates with *Git* version and source control. Git commands are abstracted from the developer, who can use point-and-click capabilities to work with the version and source control system. VS Code also integrates with Windows Subsystem for Linux (WSL) and provides various terminals and shells, including Bourne Again Shell (bash), Windows terminal, and Ubuntu Linux shell. Remote SSH capabilities, allowing the IDE to connect to remote devices, can be added as an extension. Extensions for Docker and Kubernetes exist, making it easier to work with these platforms in the visually integrated IDE. Regardless of whether the developer uses VS Code or another platform, an IDE is *critical* and *foundational* to adopting NetDevOps.

Git

Git is a free, open-source, and distributed version control system created by Linus Torvalds in 2005 for development of the Linux kernel. Consider Git the glue that holds the NetDevOps lifecycle together, tracking all changes to files using commits. These tracked changes are included in the history of all Git repositories, and changes can be rolled back to any point in time. A distributed version control system, the full repository is *cloned* to a developer's local workstation. Git is very lightweight, portable, and integrates with VS Code. Version and source control is a key foundational element of automating a network with NetDevOps. Easy to learn yet extremely powerful, Git version-controls the lifecycle of pyATS jobs, tests, intent, and template code. Git commit history makes it easy to understand exactly what changed, under what branch, and by which developer. Artifacts can also be rolled back to a previous point in time using the commit history.

GitHub

GitHub is the largest collection of code, in the form of Git *repositories*, on the Internet. GitHub, the online central repository, should not be confused with Git, the version and source control software. GitHub offers Internet hosting for version and source control using Git. Both public and private repositories are available, where private repositories require access tokens to contribute to the code. Repositories can be *cloned* from GitHub locally. GitHub provides mechanisms to create *branches* from the *main* branch and the controls to *merge* code from branches into the main branch. The main branch in NetDevOps should be considered working, valid, and tested code representing the single source of truth. Issues with the code can be tracked and addressed in GitHub. GitHub has many free and paid options for CI/CD such as GitHub *Actions*.

GitLab

GitLab is a free, open-source alternative to GitHub that can be hosted privately. GitLab provides CI/CD capabilities and can be used as a central, web-based Git repository system. GitLab offers functionality to collaboratively plan, build, secure, and deploy software as a complete DevOps platform. GitLab can be hosted on-premises or in the cloud. GitLab comes with a built-in wiki, issue tracking, IDE, and CI/CD pipeline features. GitLab was originally written in Ruby but has since migrated to Go, and it offers extremely high performance as a Git repository platform.

Structured Data

Networks historically used command-line standard output—unstructured raw command-line output—which limited programmability and automation. Developers had to use *regular expressions (RegEx)* to tediously transform CLI output into more programmability-friendly structures. Structured data like JSON is also extremely easy to work with using programming languages like Python. One major feature of pyATS is the ability to *model* and *parse* commands into structured JSON, providing an easy path to programmability and automation.

JavaScript Object Notation (JSON)

JSON is a lightweight data-interchange format that is easy for humans to read and write as well as machines to parse and generate. Based on a subset of the JavaScript programming language, JSON is programming language independent. JSON is made up of two structures: key-value pairs and ordered lists, known as an array. An *object* is an unordered set of key-value pairs surrounded by curly braces. Keys are followed by colons, and objects are separated by commas. One of the major features and benefits of using pyATS is that the unstructured Cisco CLI output can be transformed into JSON using either pyATS *learn* modules or *parse* libraries, either from the CLI or as part of pyATS *jobs*. Parsers are covered in detail in Chapter 5, "pyATS Parsers." Interfaces could be represented by the JSON dictionary of objects demonstrated in Example 1-1.

Example 1-1 *show ip interface brief as JSON*

```
{
    "interface": {
        "GigabitEthernet1": {
            "interface_is_ok": "YES",
            "ip_address": "10.10.20.175",
            "method": "TFTP",
            "protocol": "up",
            "status": "up"
        },
        "GigabitEthernet2": {
            "interface_is_ok": "YES",
            "ip_address": "172.16.252.21",
            "method": "TFTP",
            "protocol": "up",
            "status": "up"
        }
    }
}
```

eXtensible Markup Language (XML)

XML uses *tags*, similar to Hypertext Markup Language (HTML), to create structured data. XML encodes data that is both human- and machine-readable and is used for both storing and transmitting structured data. XML was designed to be simple, general, and usable across the Internet. Tags in XML represent the data structure and can also contain metadata about the structure. Example 1-2 show the same interface example but in XML instead of JSON.

Example 1-2 *show ip interface brief as XML*

```
<?xml version="1.0" encoding="UTF-8" ?>
<interface>
  <GigabitEthernet1>
    <interface_is_ok>YES</interface_is_ok>
    <ip_address>10.10.20.175</ip_address>
    <method>TFTP</method>
    <protocol>up</protocol>
    <status>up</status>
  </GigabitEthernet1>
  <GigabitEthernet2>
    <interface_is_ok>YES</interface_is_ok>
    <ip_address>172.16.252.21</ip_address>
```

```
        <method>TFTP</method>
        <protocol>up</protocol>
        <status>up</status>
    </GigabitEthernet2>
</interface>
```

YAML Ain't Markup Language (YAML)

YAML is a recursive acronym that stands for YAML Ain't Markup Language. YAML is a superset of JSON, and all valid JSON files can be parsed with YAML. YAML is another human- and machine-readable data serialization language, but with minimal syntax. Like Python, YAML uses whitespace indentation to indicate nesting. Much like JSON, YAML uses key-value pair objects, using a colon to separate the key and the paired value, and it also supports lists (or arrays) using hyphens to indicate a list of objects. In the context of pyATS, testbeds (Chapter 4), Clean (Chapter 15), and Blitz (Chapter 19) all require YAML of some form. The interface JSON/XML can be expressed in YAML as shown in Example 1-3.

Example 1-3   *show ip interface brief as YAML*

```
    ---
interface:
  GigabitEthernet1:
    interface_is_ok: 'YES'
    ip_address: 10.10.20.175
    method: TFTP
    protocol: up
    status: up
  GigabitEthernet2:
    interface_is_ok: 'YES'
    ip_address: 172.16.252.21
    method: TFTP
    protocol: up
    status: up
```

YANG

YANG, initially published in October 2010 as Request for Comment (RFC) 6020 and superseded by RFC 7950 in August 2016, is a hierarchical data modeling language. It is used in conjunction with network protocols like NETCONF and RESTCONF to interact with network devices. YANG allows for the modeling of network state, notifications, remote procedure calls (RPCs), and configuration data in protocol-independent XML or JSON formats. To work with YANG models, network automation tools like pyATS can be employed, which facilitate the interaction with YANG-modeled data using Python.

Application Programing Interface (API)

Network engineers are familiar with interfaces, which could be physical interfaces, console interfaces, and virtual interfaces. An *application programing interface* is exactly that—an interface on a device that we can program. Some platforms have NETCONF or RESTCONF APIs, and other platforms have their own dedicated API such as Cisco Nexus NX-API. Like connecting to a virtual terminal interface (VTY interface) or console port, network engineers can connect to an API and interact with a device. APIs provide the ability to perform *CRUD* activities (that is, *Create, Read, Update,* and *Delete*) using various *methods* in the form of *verbs*. These verbs include GET (read), POST (create), PUT (update), PATCH (update), and DELETE (delete). APIs also provide back *status codes* indicating the reason for an API activity's success or failure:

- **1xx – Informational:** Transfer protocol level information.

- **2xx – Success:** Client request was accepted successfully.

- **3xx – Redirection:** Client must take additional action to complete the request.

- **4xx – Client error:** Error codes that indicate there is a problem on the client side with the sending request.

- **5xx – Server error:** Error codes that indicate there is a problem on the server side processing the request.

Representational State Transfer (REST)

Simple Object Access Protocol (SOAP) and what was known as a service-oriented architecture (SOA) was the original and predominate API in the late 1990s into the early 2000s. SOAP is a *protocol* and uses XML and RPCs to interact with web services. In 2000, Roy Fielding's dissertation "Architectural Styles and Design of a Network-based Software Architecture" introduced and outlined Representational State Transfer (REST), which is an architectural *style* and not a full-blown protocol. There are certain criteria that make an API a REST API:

- A client/server architecture managed through HTTP relying on HTTP methods like GET, POST, DELETE, and so on, to perform operations.
- *Stateless* communication:
  - No client information is stored between requests.
  - Each request is separate and not related.
- High performance in component interactions, resulting in efficiency.
- Scalability, allowing for larger numbers of components and interactions.

- Simplicity in a uniform interface.

  - Resource identification in the request.

  - Server can respond with HTML, XML, or JSON, which are not necessarily the server's internal representation of state.

  - Layered system that could involve multiple "hops," including security and load balancing, which are transparent to the client issuing the request.

Today, API is associated with RESTful APIs, which have become the standard in both networks and web applications typically responding with JSON payloads. pyATS includes a connection class implementation, REST, which allows pyATS jobs and scripts to connect to the device via REST using the topology/YAML format. Chapter 13, "Working with APIs," explores pyATS and APIs.

GraphQL

GraphQL is a user-friendly and efficient query language for APIs that simplifies the process of fetching and manipulating data. Imagine it as a bridge between your application and a server, allowing you to precisely request only the data you need and nothing more. Unlike traditional REST APIs, where each endpoint corresponds to a fixed set of data, GraphQL enables you to construct flexible queries, tailoring responses to your specific requirements. It empowers developers to avoid over-fetching and under-fetching issues, providing a more streamlined and performant data retrieval experience. With GraphQL, you can access multiple resources with a single request, making it incredibly efficient for modern web and mobile applications. Its versatility, simplicity, and ability to adapt to various data sources have made it a popular choice for building data-driven applications.

cURL

Client URL (cURL) is a CLI tool used to interact with APIs. cURL is used in command lines or scripts to transfer data specified in the URL syntax using various network protocols. cURL can be used to confirm connectivity to a URL or more advanced API integrations. cURL will default to an HTTP request but can be used against HTTPS URLs as well. Downloading single or multiple files, inspecting HTTP headers, following redirects, transferring files with FTP, sending cookies, using proxies, and saving the output to a file are all supported advanced options using cURL. Example 1-4 shows a verbose cURL request.

Example 1-4   *Verbose cURL Request Example*

```
$ curl -v cisco.com
*   Trying 72.163.4.185:80...
* TCP_NODELAY set
* Connected to cisco.com (72.163.4.185) port 80 (#0)
> GET / HTTP/1.1
```

```
> Host: cisco.com
> User-Agent: curl/7.68.0
> Accept: */*
>
* Mark bundle as not supporting multiuse
< HTTP/1.1 301 Moved permanently
< Location: https://cisco.com/
< Connection: close
< Cache-Control: no-cache
< Pragma: no-cache
<
* Closing connection 0
```

Postman

Postman is a GUI client platform for building, testing, and using APIs. API interactions can be saved as reusable requests or organized (grouped) into *collections*. Postman supports variables in *environments* that can be used to automate aspects of API interactions, such as saving authentication tokens, keys, and credentials. Pre- and post-request testing can be performed using JavaScript in Postman, which also includes a console. Working requests can quickly and easily be transformed into working *code* in a large variety of programming languages such as Python, C#, Java, and JavaScript. Developers working with APIs often start in Postman to develop working requests and then migrate into their programming language of choice. Postman is beginner friendly and provides thousands of open and free working API examples in public collections developers can download and install as working, preformed requests into their client.

Python

Python is a free, open-source, high-level, interpreted, general-purpose programming language. As the name implies, *py*ATS is written in Python. Python was released in February 1991 and has recently become popular for writing network automation and infrastructure as code because of its low barrier for entry, performance, and ability to work with REST APIs. pyATS allows for CLI commands and output to be transformed into REST-like APIs. JSON is easily written and parsed by Python, making it the best candidate programming language for interacting with structure data. Python is described as a "batteries-included" programming language stemming from the comprehensive standard library of functionality provided to developers. External libraries, known as *packages*, can be *imported* into Python code, extending the base capabilities of the language. Python is object-oriented, with an emphasis on code readability and simplicity, and therefore is an excellent language for beginners and advanced developers alike. Python can be run *interactively*, using the Python command line, or *non-interactively,* using .py files. Non-interactive pyATS has a *job* file, a control file, and the actual .py file, where the testing and Python operations are performed. Python code logic uses significant indentation, where the whitespace (indentation) indicates the logical flow of code. In 1999, software engineer Tim Peters wrote a set of 19 guiding principles for developers known as "The Zen of Python."

If you enter the interactive Python CLI and enter **import this**, Python will print "The Zen of Python" to the screen as demonstrated, in Example 1-5.

Example 1-5   *The Zen of Python*

```
C:\>python
Python 3.8.9 (default, Apr 13 2021, 15:54:59)  [GCC 10.2.0 64 bit (AMD64)] on win32
Type "help", "copyright", "credits" or "license" for more information.
>>> import this
The Zen of Python, by Tim Peters


Beautiful is better than ugly.
Explicit is better than implicit.
Simple is better than complex.
Complex is better than complicated.
Flat is better than nested.
Sparse is better than dense.
Readability counts.
Special cases aren't special enough to break the rules.
Although practicality beats purity.
Errors should never pass silently.
Unless explicitly silenced.
In the face of ambiguity, refuse the temptation to guess.
There should be one-- and preferably only one --obvious way to do it.
Although that way may not be obvious at first unless you're Dutch.
Now is better than never.
Although never is often better than *right* now.
If the implementation is hard to explain, it's a bad idea.
If the implementation is easy to explain, it may be a good idea.
Namespaces are one honking great idea -- let's do more of those!
>>>
```

pip

Package Installer for Python, or pip, is used to install packages from the Python Package index and other indexes. Many popular packages are stored on pypi.org, where the pip **install** command can install the package locally. Once a package is installed locally, its functionality can be *imported* into the interactive Python CLI or non-interactive .py files and then called by the code. Pip is preinstalled in Python as of versions 2.7.9 (for Python 2) and Python 3.4 (as pip3 for Python3) by default. pyATS is installed using pip.

Software Development Kits

Some platforms provide a Python (or other programming language) software development kit (SDK) users can import into their interactive or non-interactive Python code. Cisco Application Centric Infrastructure (ACI) Application Policy Infrastructure Controller (APIC) has a Python SDK called "Cobra" that provides an abstraction in the form of Python objects for every element in the ACI Management Information Tree (MIT). Meraki has a similar Python SDK. pyATS itself is a Python SDK providing testing, configuration management, REST APIs, differential capabilities, and much more, to users as an importable package for Python.

Virtual Environment

In order to avoid conflicts with package versions and to sometimes set up completely different versions of Python for development, the Python *virtual environment* module can be used to create fully isolated, self-contained directory trees that include a Python installation and additional packages. By default, virtual environments will install the latest version of Python available on a local host; however, the specific version of Python can be specified at virtual environment creation time. Virtual environments are first created using the **venv** package; then they are *activated* using a script provided by venv. Virtual environments allow developers to create isolated environments for creating new projects with specific Python or package dependencies or to test existing code with new versions of Python or dependency packages without damaging existing working environments. It is strongly recommended to run pyATS inside a virtual environment.

Virtual Machines

A virtual machine (VM) is a computer that instead of running on physical components is a logical construct that runs on a *hypervisor*. There are two types of hypervisors: Type-1, called *bare metal*, is a software layer installed directly on top of a physical server using the underlying hardware, and Type 2, called a *hosted* hypervisor, is installed in the operating system as software such as Windows, Linux, or macOS. Virtual machines can be hosted in private, public, or hybrid clouds and are a vital part of NetDevOps, where your infrastructure as code often resides in virtual machines to be executed.

Containers

Contrary to a virtual machine, containers do not require a hypervisor. Containers are standard units of packaged software that contain all of the required dependencies, making an application portable for multiple environments. A container *image* is the immutable static file that includes the executable code so it can run in an isolated process. Docker is the most popular platform for developing containers, and a Docker environment can be set up on top of a virtual or physical machine to run containers. pyATS provides a methodology to create pyATS containers. pyATS itself can be run as a container. XPRESSO, covered in depth in Chapter 20, "XPRESSO," is a pyATS scheduling platform that can also schedule and execute pyATS containers.

Kubernetes

With the proliferation of containerized applications, a need to orchestrate and manage these containers at scale emerged. Kubernetes, also known as K8s, is an open-source system for managing containerized applications. Kubernetes was originally developed by Google but is now maintained by the Cloud Native Computing Foundation (CNCF). Clusters of hosts can be used by Kubernetes to orchestrate containers, as *pods*, which can dynamically scale based on demand. Written in the Go programming language, Kubernetes enables infrastructure as a service (IaaS). Originally, Kubernetes only interfaced with Docker using a "Dockershim," which has since been deprecated and replaced with the containerd interface or Container Runtime Interface (CRI) as of May 2022.

CI/CD

Continuous integration and continuous deployment (CI/CD) are cornerstone practices in modern software development, and they find a significant place in DevOps culture. While traditionally associated with software development, these principles are increasingly being adopted in the realm of network infrastructure management to foster more reliable and efficient operational workflows.

- **Continuous integration (CI):** In a network-centric application, CI encompasses a practice where network configurations and scripts are frequently merged into a central repository. Following the merge, automated builds and tests are initiated to validate the new changes against the existing network configurations and operational states. The crux of CI in network management is to identify and rectify conflicts or bugs at an early stage, ensuring that the changes do not adversely impact the network's functionality. By integrating regularly, teams can swiftly detect and resolve errors, making it easier to maintain a stable network state.

- **Continuous deployment (CD):** Transitioning toward network operations, CD encapsulates an automated process of deploying validated configurations to the live network environment. Automated testing, leveraging tools like pyATS, validates the correctness and stability of configuration changes, ensuring they are ready for autonomous deployment to the production network. The objective is to maintain a network configuration that is always deployment-ready, facilitating a swift response to business or operational demands.

In the context of pyATS, CI/CD practices extend to testing network configuration files and evaluating outcomes in the infrastructure, such as verifying pre-change network states, interface statuses, or adjacency statuses. The CI/CD pipeline can automate the validation of network states before and after the deployment of configuration changes, ensuring the network operates as intended.

Incorporating CI/CD into network management establishes a consistent and automated pathway to build, package, and test network configurations. This uniformity in the integration and deployment process augments the ability to catch bugs and errors early on, thus reducing the debugging time and significantly enhancing the efficiency of network operations. The automation and continuous monitoring introduced by CI/CD span the lifecycle of network configurations, from integration and testing phases to delivery and deployment, mirroring the benefits seen in software development realms. This not only elevates the productivity of network operations teams but also accelerates the rate at which reliable, high-quality network services are delivered and maintained.

Jenkins

Jenkins is an open-source CI/CD tool enabling developers to build, test, and deploy software. Flexible and complex workflows can be created using Jenkins. A Git repository plugin is available for Jenkins that can be used not only as a Git repository but to automate workflows that start when code is merged into the repository. Jenkins testing can be used as part of continuous integration to test and validate builds. Jenkins is written in the Java programming language. A Jenkinsfile, a text file that contains the definition of a Jenkins pipeline, is checked into source control and is used to define CI/CD in Jenkins. Jenkinsfiles can be *declarative* (introduced in Jenkins Pipeline 2.5) or *scripted,* and they are broken up into *stages*, such as build, test, and deploy. Declarative Jenkinsfiles break down stages into individual parts that can contain multiple steps, while scripted pipelines reference the Jenkins pipeline domain-specific language within the stages without the need for steps.

GitLab CI/CD

GitLab CI/CD is a tool built into GitLab, a web-based DevOps lifecycle tool, which offers a continuous integration and deployment system to automate the pipeline for projects. Here's a brief explanation of how it works:

- **GitLab CI/CD pipeline:** The pipeline is the core component of GitLab CI/CD and represents the entire process, which is divided into multiple jobs. Each job has a specific role and responsibility in the pipeline. Jobs are organized into stages, and these stages execute in a particular order. Common stages include build, test, and deploy, but you can define as many stages as your project requires.

- **GitLab Runner:** GitLab Runner is an application that works with GitLab CI/CD to run jobs in your pipeline. It's responsible for receiving from GitLab the instructions for the jobs, executing them, and sending the results back to GitLab. Runners can be installed on various types of operating systems and can support multiple platforms. Runners can be installed closer to devices under management for remote code execution and can be deployed in a secure manner.

- **gitlab-ci.yml:** This is a YAML file that you create in your project's root. This file defines the structure and order of the pipelines and includes the definitions of the pipeline stages and the jobs to be executed. GitLab CI/CD looks for this file in your repository and uses its instructions to execute jobs.

When you commit and push the code to the repository, GitLab checks for the .gitlab-ci.yml file. If it finds the file, it triggers the CI/CD pipeline according to the instructions defined in the file. The jobs are then executed by the runners in the order defined by the stages. GitLab CI/CD is a powerful tool for automating the testing and deployment of your code. It's highly flexible and configurable, allowing you to tailor your pipeline to your project's specific needs. For example, if someone is working on a branch rather than the mainline code, they may want to execute a subset of tests instead of the full tests that would run when the code is merged into the main branch.

GitHub Actions

GitHub Actions is a CI/CD system built into GitHub, the popular web-based hosting service for Git repositories. It allows you to automate, customize, and execute your software development workflows right in your repository. Here's a brief explanation of how it works:

- **Workflows:** In GitHub Actions, a workflow is an automated procedure that you add to your repository. Workflows are made up of one or more jobs and can be scheduled or triggered by specific events. The workflow is defined in a YAML file (main. yml or any name you prefer) in the **.github/workflows** directory of your repository.

- **Jobs:** Jobs are sets of steps that execute on the same runner. By default, a workflow with multiple jobs will run those jobs in parallel. You can also configure jobs to depend on each other.

- **Steps:** Steps are individual tasks that can run commands in a job. A step can be either an action or a shell command. Each step in a job executes on the same runner, allowing the steps in a job to share data with each other.

- **Actions:** Actions are the smallest portable building blocks of a workflow. You can create your own actions, or you can use and customize actions shared by the GitHub community. Actions are reusable units of code that can be employed across different workflows.

- **Events:** Workflows are triggered by events. An event can be a push to the repository, a pull request, a fork, a release, a manual trigger by a user, and more.

- **Runners:** Runners are servers that have the GitHub Actions runner application installed. When you use a GitHub-hosted runner, machine maintenance and upgrades are taken care of for you. You can also host your own runners to run jobs on machines you own or manage.

GitHub Actions provides a powerful, flexible way to automate nearly any aspect of your development workflow. It's deeply integrated with the rest of GitHub, making it a convenient option for projects already hosted on GitHub.

Drone

Drone is an open-source CI/CD system built on container technology. It uses a YAML file for configuration and is known for its simplicity and ease of use. Drone integrates seamlessly with multiple source code management systems, including GitHub, GitLab, and Bitbucket. Here's a brief explanation of how it works:

- **Pipeline:** In Drone, a pipeline is a series of steps executed in a specific order to implement the CI/CD process. Each step in a pipeline is executed inside its own Docker container, providing an isolated and reproducible environment for each operation.

- **.drone.yml:** This is the configuration file where you define your pipeline. It's written in YAML and should be located in the root of your repository. The .drone.yml file specifies the steps to be executed, the order in which they should run, and the conditions under which they should be executed.

- **Steps:** Steps are individual tasks that make up a pipeline. Each step is executed in its own Docker container and can be used to build, test, or deploy your application. Steps are defined in the .drone.yml file and are executed in the order in which they appear in the file.

- **Plugins:** Drone has a rich ecosystem of plugins that can be used to extend its functionality. Plugins in Drone are simply Docker containers designed to perform specific tasks. For example, there are plugins to publish Docker images, deploy code to cloud providers, send notifications, and more.

- **Triggers:** Drone supports various types of triggers to start the execution of a pipeline. The most common trigger is a Git push event, but pipelines can also be triggered manually or on a schedule.

- **Runners:** Drone uses runners to execute pipeline tasks. Runners are lightweight, standalone processes that run on the host machine and execute tasks in Docker containers. Drone supports various types of runners, including Docker, SSH, and Kubernetes runners.

Drone is a simple, flexible, and powerful CI/CD system that leverages the power of container technology to provide isolated and reproducible environments for each step in your pipeline. Its plugin-based architecture makes it highly extensible and adaptable to a wide range of use cases.

Summary

Networks have evolved in complexity, size and scale, and criticality to the successful operations of any business or enterprise. The legacy waterfall methodologies, limited tools, and general approach network engineers and operators have relied on have also evolved in recent years. By adopting NetDevOps, networks can become "agile" not "fragile." The principles of "automate-first" have radically changed the way networks are planned, designed, built, tested, and deployed. Infrastructure as code, CI/CD, containers, clouds, and programming languages such as Python have revolutionized networking by making it more agile, scalable, and reliable. Collaborative culture is the key to success, as the silos and barriers between networking, developers, and operations are taken down in favor of one team solving problems using the team members' individual strengths and experiences. pyATS is the perfect tool for NetDevOps, as it is useful in all stages of the development lifecycle—from modeling, developing, testing, documenting, and deploying configuration to the network at scale with CI/CD integrations.

References

"Part 1: Embrace NetDevOps, Say Goodbye to a 'Culture of Fear,'" https://blogs.cisco.com/developer/embrace-netdevops-part-1

"Part 2: NetDevOps Goes Beyond Infrastructure as Code," https://blogs.cisco.com/developer/embrace-netdevops-part-2

"Part 2: NetDevOps Goes Beyond Infrastructure as Code," https://blogs.cisco.com/developer/embrace-netdevops-part-two

Uptime Institute Blog, "How to avoid outages: Try harder!" https://journal.uptimeinstitute.com/how-to-avoid-outages-try-harder/

"Manifesto for Agile Software Development," https://agilemanifesto.org/ https://daringfireball.net/projects/markdown/

Chapter 2

Installing and Upgrading pyATS

pyATS and the pyATS library (Genie), collectively referred to as pyATS, are both Python libraries that can be installed with the common Python package manager, pip. Currently, pyATS supports Python versions 3.7–3.10 on the following platforms: Linux (CentOS, RHEL, Ubuntu, Alpine) and macOS 10.13+. Windows is not officially supported. However, Windows provides Windows Subsystem for Linux (WSL), which allows you to run Linux on Windows. Using the WSL environment on Windows, you can install pyATS.

Installing pyATS is a two-part process. First, you'll need to install the pyATS core framework, followed by the pyATS library (Genie), which contains all the tools needed for automated network testing such as parsers, models, and a test harness. This can be accomplished in one or multiple commands. In the following sections, you'll see how to install and upgrade pyATS using pip and the pyATS command line.

This chapter covers the following topics:

- Installing pyATS

- Upgrading pyATS

- Troubleshooting pyATS

Installing pyATS

Before installing pyATS, or any Python package for that matter, it's highly recommended that you set up a Python virtual environment. A Python virtual environment allows you to install Python dependencies in an isolated environment. Using Python virtual environments is considered a best practice because they allow for installation of multiple Python packages on the same host, without you worrying about having dependency conflicts. Dependency conflicts arise when two packages have the same dependency but require different versions of that dependency. Another advantage to virtual environments is the ability to try different versions of the same Python package using different virtual environments. Once you validate the correct package version to use, you can save the current version of the locally installed packages to a requirements.txt file using the **pip freeze** command. Python virtual environments create isolation, which avoids dependency conflicts and allows you to easily manage the proper version of project dependencies.

Remember, PyATS does not support Windows. Therefore, if you are running Windows, make sure you're using a WSL environment, such as Ubuntu on WSL.

Setting Up a Python Virtual Environment

Python has a module called venv that's included in the standard library to create virtual environments. Example 2-1 shows the command to create a Python virtual environment using the venv module.

Example 2-1 *Creating a Python Virtual Environment*

```
dan@linux-pc# python3 | python -m venv { /path/to/venv }
```

Once the virtual environment is created, it needs to be activated. Example 2-2 shows how to activate the environment and confirm which Python interpreter is being used. In the example, the virtual environment was created in a local directory aptly named **.venv**.

Example 2-2 *Activating and Confirming the Python Virtual Environment*

```
dan@linux-pc# source .venv/bin/activate
(.venv)dan@linux-pc#
(.venv)dan@linux-pc# which python
~/.venv/bin/python
```

You can see the virtual environment's directory name in parentheses in the command prompt. Once you confirm your virtual environment has been activated, you're ready to install pyATS.

Installing pyATS Packages

Now for the fun part! Let's install pyATS and the pyATS library (Genie) using one command. Example 2-3 shows the *recommended* method to install both libraries. You may also install just the pyATS core framework using **pip install pyats**, but you will not have the pyATS library installed, which provides the necessary tools to learn and parse configuration and operational state data from network devices.

Example 2-3 *Installing pyATS and pyATS Library (Genie)*

```
(.venv)dan@linux-pc# pip install pyats[library]
```

This will install all the core pyATS framework and the pyATS library (Genie) packages. Example 2-4 shows how to confirm the packages are installed using the **pip list** command.

Example 2-4 *Confirming pyATS and the pyATS Library (Genie) Are Installed*

```
(.venv)dan@linux-pc# pip list
Package                     Version
--------------------------- --------
aiofiles                    23.1.0
aiohttp                     3.8.4
aiohttp-swagger             1.0.16
aiosignal                   1.3.1
async-lru                   2.0.3
async-timeout               4.0.2
attrs                       23.1.0
bcrypt                      4.0.1
certifi                     2023.5.7
cffi                        1.15.1
chardet                     4.0.0
charset-normalizer          3.2.0
cryptography                41.0.2
dill                        0.3.6
distro                      1.8.0
frozenlist                  1.4.0
genie                       23.8
genie.libs.clean            23.8.1
genie.libs.conf             23.8
genie.libs.filetransferutils 23.8
genie.libs.health           23.8
genie.libs.ops              23.8
genie.libs.parser           23.8
genie.libs.sdk              23.8.1
gitdb                       4.0.10
GitPython                   3.1.32
grpcio                      1.56.0
idna                        3.4
Jinja2                      3.1.2
```

jsonpickle	3.0.1
junit-xml	1.9
lxml	4.9.3
MarkupSafe	2.1.3
multidict	6.0.4
ncclient	0.6.13
netaddr	0.8.0
packaging	23.1
paramiko	3.2.0
pathspec	0.11.1
pip	22.0.4
prettytable	3.8.0
protobuf	4.23.4
psutil	5.9.5
pyats	23.8
pyats.aereport	23.8
pyats.aetest	23.8
pyats.async	23.8
pyats.connections	23.8
pyats.datastructures	23.8
pyats.easypy	23.8
pyats.kleenex	23.8
pyats.log	23.8
pyats.reporter	23.8
pyats.results	23.8
pyats.tcl	23.8
pyats.topology	23.8
pyats.utils	23.8
pycparser	2.21
pyftpdlib	1.5.7
PyNaCl	1.5.0
python-engineio	3.14.2
python-socketio	4.6.1
PyYAML	6.0.1
requests	2.31.0
ruamel.yaml	0.17.32
ruamel.yaml.clib	0.2.7
setuptools	60.10.0
six	1.16.0
smmap	5.0.0
tftpy	0.8.0
tqdm	4.65.0
typing_extensions	4.7.1

```
unicon                      23.6.1
unicon.plugins              23.6.1
urllib3                     2.0.3
wcwidth                     0.2.6
wheel                       0.40.0
xmltodict                   0.13.0
yamllint                    1.32.0
yang.connector              23.6
yarl                        1.9.2
```

Now that pyATS has been installed, you have access to the pyATS command line. Example 2-5 shows the available options in the pyATS command line using **–help** or **-h**.

Example 2-5 *pyATS Command-Line Options*

```
(.venv)dan@linux-pc# pyats { --help | -h }
Usage:
  pyats <command> [options]


Commands:
    clean               runs the provided clean file
    create              create scripts and libraries from template
    develop             Puts desired pyATS packages into development mode
    diff                Command to diff two snapshots saved to file or directory
    dnac                Command to learn DNAC features and save to file
                        (Prototype)
    learn               Command to learn device features and save to file
    logs                command enabling log archive viewing in local browser
    migrate             utilities for migrating to future versions of pyATS
    parse               Command to parse show commands
    run                 runs the provided script and output corresponding
                        results.
    secret              utilities for working with secret strings.
    shell               enter Python shell, loading a pyATS testbed file and/or
                        pickled data
    undevelop           Removes desired pyATS packages from development mode
    validate            utilities that help to validate input files
    version             commands related to version display and manipulation

General Options:
  -h, --help            Show help


Run 'pyats <command> --help' for more information on a command.
```

A quick command you can run to confirm the pyATS version installed is **pyats version check**. This command shows the current versions of your pyATS and pyATS library installations. Example 2-6 shows some sample output.

Example 2-6   *pyATS Command Line—Version Check*

```
(.venv)dan@linux-pc# pyats version check
You are currently running pyATS version: 23.8
Python: 3.9.13 [64bit]
  Package                     Version
  -------------------------   -------
  genie                       23.8
  genie.libs.clean            23.8.1
  genie.libs.conf             23.8
  genie.libs.filetransferutils 23.8
  genie.libs.health           23.8
  genie.libs.ops              23.8
  genie.libs.parser           23.8
  genie.libs.sdk              23.8.1
  pyats                       23.8
  pyats.aereport              23.8
  pyats.aetest                23.8
  pyats.async                 23.8
  pyats.connections           23.8
  pyats.datastructures        23.8
  pyats.easypy                23.8
  pyats.kleenex               23.8
  pyats.log                   23.8
  pyats.reporter              23.8
  pyats.results               23.8
  pyats.tcl                   23.8
  pyats.topology              23.8
  pyats.utils                 23.8
  rest.connector              23.8
  unicon                      23.8
  unicon.plugins              23.8
  yang.connector              23.8
```

Upgrading pyATS

Similar to how you installed pyATS, you can upgrade it, and all its dependencies, using pip or the pyATS command line. It is important to upgrade pyATS and the pyATS library (Genie) so that you always get the latest bug fixes, parsers, and any platform support that may be added to the pyATS libraries. To upgrade using pip, you simply need to add the **--upgrade** flag to the original **pip install** command (see Example 2-7). As a reminder, make sure your Python virtual environment is activated.

Example 2-7 *Upgrading pyATS Using pip*

```
(.venv)dan@linux-pc# pip install pyats[library] --upgrade
```

Alternatively, you may use the pyATS command line to upgrade pyATS. In Example 2-8, pyATS is updated using the **pyats version update** command. You'll notice that pyATS is already at the latest version, but the output provides a good example of what you should expect to see.

Example 2-8 *Upgrading pyATS Using the Command Line*

```
(.venv)dan@linux-pc# pyats version update
Checking your current environment...



The following packages will be removed:

  Package                         Version
  ----------------------------    -------
  genie                           23.8
  genie.libs.clean                23.8.1
  genie.libs.conf                 23.8
  genie.libs.filetransferutils    23.8
  genie.libs.health               23.8
  genie.libs.ops                  23.8
  genie.libs.parser               23.8
  genie.libs.sdk                  23.8.1
  pyats                           23.8
  pyats.aereport                  23.8
  pyats.aetest                    23.8
  pyats.async                     23.8
  pyats.connections               23.8
  pyats.datastructures            23.8
  pyats.easypy                    23.8
  pyats.kleenex                   23.8
  pyats.log                       23.8
  pyats.reporter                  23.8
  pyats.results                   23.8
  pyats.tcl                       23.8
  pyats.topology                  23.8
  pyats.utils                     23.8
  rest.connector                  23.8
  unicon                          23.8
```

```
   unicon.plugins              23.8
   yang.connector              23.8


Fetching package list... (it may take some time)

... and updated with:

  Package                    Version
  -------------------------- ------------
  genie                      latest (23.9)
  genie.libs.clean           latest (23.9)
  genie.libs.conf            latest (23.9)
  genie.libs.filetransferutils latest (23.9)
  genie.libs.health          latest (23.9)
  genie.libs.ops             latest (23.9)
  genie.libs.parser          latest (23.9)
  genie.libs.sdk             latest (23.9)
  genie.trafficgen           latest (23.9)
  pyats                      latest (23.9)
  pyats.aereport             latest (23.9)
  pyats.aetest               latest (23.9)
  pyats.async                latest (23.9)
  pyats.connections          latest (23.9)
  pyats.datastructures       latest (23.9)
  pyats.easypy               latest (23.9)
  pyats.kleenex              latest (23.9)
  pyats.log                  latest (23.9)
  pyats.reporter             latest (23.9)
  pyats.results              latest (23.9)
  pyats.tcl                  latest (23.9)
  pyats.topology             latest (23.9)
  pyats.utils                latest (23.9)
  rest.connector             latest (23.9)
  unicon                     latest (23.9)
  unicon.plugins             latest (23.9)
  yang.connector             latest (23.9)


Are you sure you want to continue [y/N]? y
Uninstalling existing packages...
Installing new packages...

Done! Enjoy!
```

Troubleshooting pyATS

Let's quickly touch on some of the common pitfalls and issues you may run into while installing pyATS and the pyATS library. One of the most common issues is that you must remember pyATS is only compatible with macOS, Linux, and WSL on Windows, meaning you can now install it on Windows. This might seem like a simple fact, but it's easy to forget. Another common issue involves version mismatches between pyATS and the pyATS library (Genie). Version mismatches can cause compatibility issues between the two libraries. To resolve this issue, simply run the **pyats version update** command. This command will ensure both libraries are up to date. Example 2-9 shows an instance where different versions of pyATS and the pyATS library (Genie) are installed and how to resolve the issue using the **pyats version update** command.

Example 2-9 *pyATS and pyATS Library Version Mismatch*

```
(.venv)dan@linux-pc# pyats version update
Checking your current environment...


The following packages will be removed:

  Package                       Version
  ----------------------------  -------
  genie                         23.8
  genie.libs.clean              23.8.1
  genie.libs.conf               23.8
  genie.libs.filetransferutils  23.8
  genie.libs.health             23.8
  genie.libs.ops                23.8
  genie.libs.parser             23.8
  genie.libs.sdk                23.8.1
  genie.trafficgen              23.9
  pyats                         23.9
  pyats.aereport                23.9
  pyats.aetest                  23.9
  pyats.async                   23.9
  pyats.connections             23.9
  pyats.datastructures          23.9
  pyats.easypy                  23.9
  pyats.kleenex                 23.9
  pyats.log                     23.9
  pyats.reporter                23.9
  pyats.results                 23.9
  pyats.tcl                     23.9
```

```
    pyats.topology            23.9
    pyats.utils               23.9
    rest.connector            23.8
    unicon                    23.9
    unicon.plugins            23.9
    yang.connector            23.8


Fetching package list... (it may take some time)

... and updated with:


    Package                   Version
    -------------------------- -------------
    genie                     latest (23.9)
    genie.libs.clean          latest (23.9)
    genie.libs.conf           latest (23.9)
    genie.libs.filetransferutils latest (23.9)
    genie.libs.health         latest (23.9)
    genie.libs.ops            latest (23.9)
    genie.libs.parser         latest (23.9)
    genie.libs.sdk            latest (23.9)
    genie.trafficgen          latest (23.9)
    pyats                     latest (23.9)
    pyats.aereport            latest (23.9)
    pyats.aetest              latest (23.9)
    pyats.async               latest (23.9)
    pyats.connections         latest (23.9)
    pyats.datastructures      latest (23.9)
    pyats.easypy              latest (23.9)
    pyats.kleenex             latest (23.9)
    pyats.log                 latest (23.9)
    pyats.reporter            latest (23.9)
    pyats.results             latest (23.9)
    pyats.tcl                 latest (23.9)
    pyats.topology            latest (23.9)
    pyats.utils               latest (23.9)
    rest.connector            latest (23.9)
    unicon                    latest (23.9)
    unicon.plugins            latest (23.9)
    yang.connector            latest (23.9)


Are you sure you want to continue [y/N]?
```

Summary

In this chapter, you learned how to install and upgrade pyATS. In future chapters, we will cover optional packages that can be installed with pyATS, such as the Robot framework. Now that we have pyATS installed, let's get started with adding devices to a pyATS testbed!

Testbeds

A fundamental and foundational part of pyATS is the testbed. Testbeds can be, and often are, structured text stored in a YAML file, but they can also be dynamically created at pyATS job runtime. Other structured text formats such as XML and JSON can be used, but the traditional format for most testbed automation frameworks such as pyATS and XPRESSO is YAML. Testbeds describe the topology, devices, and even intent, and they abstract the complexity of connecting to our devices using Python. With nothing more than a simple testbed.yaml file and pyATS installed in a virtual Python environment network, engineers can use the pyATS command-line interface (CLI) to interact with the devices and topology within the testbed. In this chapter, we will explore testbeds and begin the journey into test-driven development (TDD) with pyATS.

This chapter covers the following topics:

- What is YAML?

- What is a testbed?

- Device connection abstractions

- Testbed validation

- Dynamic testbeds

- Intent-based networking with extended testbeds

What Is YAML?

YAML, which stands for "YAML Ain't Markup Language" (or sometimes "Yet Another Markup Language"), is a human-readable data serialization format. It is often used for configuration files and data exchange between languages with different data structures.

YAML is a superset of JSON, which means that any valid JSON file is also a valid YAML file. Here are some key characteristics and features of YAML:

- **Human-readable:** YAML is designed to be easily readable by humans. Its indentation-based structure helps in representing hierarchical data in a clear manner.

- **Indentation:** Unlike JSON, which uses braces ({ }) and brackets ([ ]) to denote objects and arrays, respectively, YAML relies on indentation (usually spaces) to represent nesting.

- **Scalars:** YAML has support for string, integer, and floating-point types. Strings in YAML don't always require quotation marks.

- **Data structures:** YAML supports both lists (arrays) and associative arrays (hashes or dictionaries).

- **Multiline strings:** YAML provides multiple ways to represent strings that span multiple lines.

- **Comments:** YAML allows for comments using the # symbol.

- **No explicit end delimiter:** YAML, unlike some formats, does not require an explicit end delimiter.

- **Aliases and anchors:** YAML supports referencing, which allows for creating references to other items within a YAML document.

- **Three hyphens (---):** Indicate the beginning of a valid YAML file.

Example 3-1 demonstrates a simple YAML file.

Example 3-1 *A Simple YAML Example*

```
---
name: John Capobianco
age: 16
is_student: false
courses:
  - Math
  - Physics
  - Chemistry
address:
  street: 123 Main St
  city: Ottawa
  province: Ontario
```

This YAML snippet represents a person with some personal details, a list of courses, and an address. The same data in JSON would require more punctuation and might be less readable.

YAML is commonly used in various applications, including configuration for software tools, data exchange between languages, and as the data format for certain applications like Ansible and Kubernetes. pyATS expects testbed data, either in file format or dynamic creation, to be YAML. Often when people say, "infrastructure as code," they typically are referring to some form of YAML, depending on the context.

What Is a Testbed?

Testbeds are a core component of pyATS that describe and abstract network topologies, devices, and links. Testbeds can be expressed as YAML files but can also be objects in memory that are dynamically created. A testbed in the context of pyATS refers to the environment in which tests are executed. This environment includes devices, servers, connections, credentials, and other elements that are part of the network topology. The testbed provides pyATS with all the necessary information to connect to and interact with devices in the network. There are many benefits to using a testbed, including the following:

- **Consistency:** By defining the test environment in a structured testbed file, you ensure consistency across test runs.

- **Flexibility:** Testbed files can be easily shared, modified, or swapped, allowing for flexibility in testing different environments.

- **Scalability:** pyATS can scale from testing a few devices in a lab to testing a large, complex production network.

The topology module is designed to provide an intuitive and standardized method for users to define, handle, and query testbed/device/interface/link descriptions, metadata, and their interconnections. The topology module has two major functions:

- Define and describe testbed metadata using YAML, standardize the format of the YAML file, and load it into corresponding testbed objects.

- Query testbed topology, metadata, and interconnect information via testbed object attributes and properties.

As opposed to creating a module where the topology information is stored internally and then asking users to query that information via API calls, the pyATS topology module approaches the design from a completely different angle:

- Using objects to represent real-world testbed devices

- Using object attributes and properties to store testbed information and metadata

- Using object relationships (references/pointers to other objects) to represent topology interconnects

- Using object references and Python garbage collection to clean up testbed leftovers when objects are no longer referenced

Figure 3-1 should give you a good high-level view of how topology objects are referenced and interconnected.

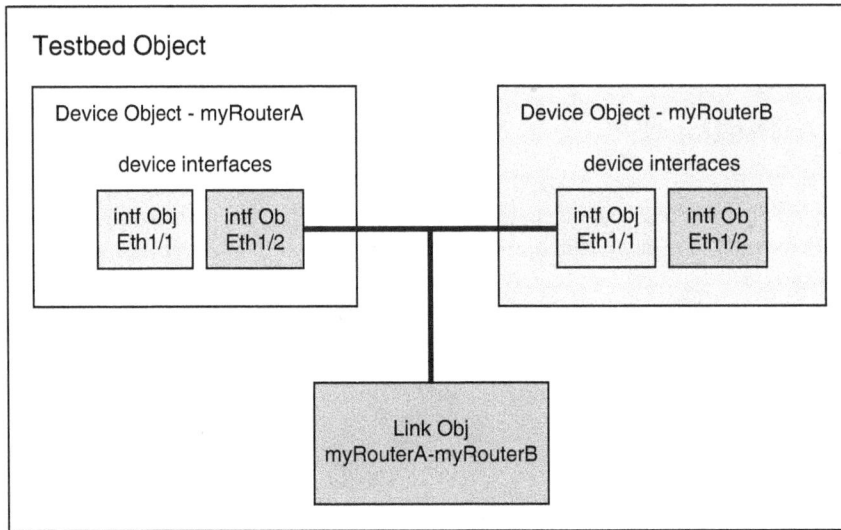

Figure 3-1   *Testbed Topology Objects*

The testbed object is the top container object, containing all testbed devices and all subsequent information that is generic to the testbed. Within a testbed, links and device names must be unique. Table 3-1 shows the complete list of testbed object attributes that are available.

Table 3-1   *Testbed Object Attribute List*

Attributes	Description
name	Testbed name. Should be unique.
alias	Testbed alias. Defaults to testbed name.
devices	Dictionary of testbed devices (name:Device).
tacacs	Dictionary of TACACS information common to the testbed.
passwords	Dictionary of password information common to the testbed.
credentials	Dictionary of credentials common to the testbed.
servers	Dictionary of testbed servers (name:dict). Testbed servers are those that service the entire testbed, such as FTP, TFTP, and NTP servers.
clean	Dictionary of clean parameters (name:value). Clean parameters are those used to clean up (reload) testbed devices.

Attributes	Description
custom	Dictionary of custom fields (name:value). Nonstandard testbed object metadata goes here.
testbed_file	Full path and name of the testbed file used to create this testbed object (only available through YAML load).

Properties	Description
links	Returns the set of unique Link objects connected to this testbed's device interfaces.

Methods	Description
add_device	Adds a device (Device object) to this testbed.
remove_device	Removes a device (Device object) from this testbed.
squeeze	Removes all unwanted devices, interfaces, and links from this testbed.
connect	Connects all or multiple devices in the testbed in parallel.
disconnect	Disconnects all or multiple devices in the testbed in parallel.
destroy	Destroys all or multiple device connections in the testbed in parallel.
execute	Executes commands against all or multiple devices in the testbed in parallel.
configure	Configures commands against all or multiple devices in the testbed in parallel.
parse	Parses commands against all or multiple devices in the testbed in parallel.

Building a Simple Testbed

Let's create a simple network testbed with a single device called **testbed.yaml**. First, we will define **devices:** as the parent key, followed by our single device description as YAML. We'll start with another key, nested and indented to make the YAML valid, called **csr1000v-1**. Inside this device parent key, we will add the necessary fields to allow pyATS to automatically connect and interact with the device. We will need the following keys and values in YAML:

- **alias:** This is an optional field you can use as an alias for the hostname of the device.

- **type:** Another optional field that can classify the device type, such as router or switch.

- **os:** The operating system field is required and is very important because it is used by pyATS to correctly identify the *parsing library* used for this device's operating system. Some valid choices include ios, iosxe, nxos, asa, junos, and others. For a full list of supported operating systems, visit https://pubhub.devnetcloud.com/media/genie-feature-browser/docs/#/parsers.

- **platform:** Another required key that also helps pyATS select the valid parsing library. This represents the hardware platform of the device, such as c9300 or isr in the case of Catalyst 9300 or Cisco ISRs, respectively.

- **credentials:** A device can have multiple credentials, including a default set used by pyATS to authenticate and log in to devices. It is strongly recommended that you use secret strings to encrypt your password (at a minimum). Secret strings are covered in Appendix B, "Secret Strings."

- **connections:** Connections indicate the various ways pyATS can connect to a device such as CLI or REST. Within the connection method there are sub keys that indicate the protocol, IP address, port, as well as arguments such as connection timeout.

Example 3-2 demonstrates a simple testbed with a single device.

Example 3-2 *A Simple Testbed Example with a Single Device*

```
---
devices:
    csr1000v-1:
        alias: 'DevNet_Sandbox_CSR1000v'
        type: 'router'
        os: 'iosxe'
        platform: isr
        credentials:
          default:
            username: developer
            password: C1sco12345
        connections:
          cli:
            protocol: ssh
            ip: sandbox-iosxe-latest-1.cisco.com
            port: 22
            arguments:
              connection_timeout: 360
```

Edge Cases

A few important optional additions that can be used in special situations should be noted. If you are connecting to legacy devices or get the following error, you can add **ssh_options** to the CLI settings to handle SSH options:

```
Unable to negotiate with <device ip> port 22: no matching key
exchange method found. Their offer: diffie-hellman-group-exchange-
sha1,diffie-hellman-group14-sha1
```

Adding **ssh_options** to the CLI settings to handle SSH options is demonstrated in Example 3-3.

Example 3-3   *Adding ssb_options to Your CLI Settings*

```
connections:
  cli:
    protocol: ssh
    ip: sandbox-iosxe-recomm-1.cisco.com
    port: 22
    ssh_options: -o KexAlgorithms=+diffie-hellman-group-exchange-sha1
 -o HostKeyAlgorithms=+ssh-rsa
    arguments:
      connection_timeout: 360
```

Another very important consideration is the initial **exec** commands and initial configuration commands pyATS executes upon successful connection to a device. By default, pyATS will adjust certain terminal settings when it first connects to a device. Depending on your device terminal settings, when you connect to a device using a CLI and execute a command, you will sometimes see "Press any key to continue." For humans, this breakpoint gives an opportunity to analyze output. However, from an automation point of view, it would break parsers, as they change output data. To avoid those, Unicon (the pyATS connection implementation) issues the following commands upon a connection being established:

- no logging console

- terminal width 511

- Possibly vty settings, depending on the implementation

All of these commands affect the terminal behavior, not your device's functionality. To disable default configuration in your testbed, override the **init exec** and **init config** commands, as demonstrated in Example 3-4.

Example 3-4   *Adding Additional Arguments to Specify Not to Run Any Initial exec or config Commands*

```
connections:
  cli:
    protocol: ssh
    ip: sandbox-iosxe-recomm-1.cisco.com
    port: 22
    arguments:
      connection_timeout: 360
      init_exec_commands: []
      init_config_commands: []
```

Using the CLI is only one way of connecting to a device with a testbed. Example 3-5 demonstrates another example, this time, using RESTCONF.

Example 3-5 *A Testbed Example Using RESTCONF*

```
---
devices:
    csr1000v-1:
        alias: 'sandbox'
        type: 'router'
        os: 'iosxe'
        platform: csr1000v
        connections:
            rest:
                # Rest connector class
                class: rest.connector.Rest
                ip: sandbox-iosxe-latest-1.cisco.com
                port: 443
                credentials:
                    rest:
                        username: developer
                        password: C1sco12345
```

External Sources of Truth

If you are thinking to yourself at this point, "But I already have all of my devices in a source of truth," and want to avoid the manual effort of creating a testbed for your topology, you have several approaches available. If you are using software-defined network (SDN) controllers or a Netbox or IPAM solution, you could use that as your source of truth to create the testbed.yaml file. First, you would convert your source of truth into a pyATS testbed-ready (comma-separated values CSV, XLS, and so on) file format and use the following steps to convert it to a valid YAML testbed file. The **pyats create testbed** command automatically converts the input and creates an equivalent YAML file. Follow these guidelines to create a valid YAML file:

- Separate the IP and port with either a space or a colon (:).

- The password column is the default password used to log in to the device.

- If you leave the password blank, the system prompts you for the password when you connect to the device.

- To enter privileged EXEC mode with the **enable** command, add a column with the header **enable_password**. The value can be the same as or different from the default password.

- Any additional columns you define, such as platform, alias, or type, are added to the YAML file as key-value pairs.

- The columns can be in any order, as long as you include the required columns.

- When creating the CSV file, separate the fields using a comma (,). If you need a text qualifier, use double quotes (").

When you're ready to create the YAML file, from your virtual environment, run the following command:

```
(pyats) $ pyats create testbed file --path my_devices.xls --output
yaml/my_testbed.yaml
```

Add the **--encode-password** option to hide the password in the YAML file as a secret string. Note that this only obfuscates the password—it does not make the password cryptographically secure.

Note One of the authors, John Capobianco, has published a Cisco DNAC-to-testbed conversion tool you can use if you have a DNAC (Digital Network Architecture Center) as your source of truth.

For Python code that generates a pyATS testbed from DNAC as a source of truth, visit https://github.com/automateyournetwork/dnac_pyats_testbed.

Other sources of truth like Netbox (http://netboxlabs.com/oss/netbox/) can also be used in a similar fashion using their API system to extract the necessary keys and values to create a testbed file.

In pyATS, everything is an object. Like testbeds, individual devices are also objects with their own accessible properties. Device objects represent any piece of physical and/or virtual hardware that constitutes an important part of a testbed topology:

- Each device may belong to a testbed (added to a Testbed object).

- Each device may host an arbitrary number of interfaces (Interface objects).

- Interface names must be unique within a device.

Table 3-2 outlines the complete list of Testbed object attributes available.

Table 3-2 *Device Object Attribute List*

Attributes	Description
name	Device name (aka hostname).
alias	Device alias. Defaults to device name.
os	Device OS, such as iosxe, iosxr, nxos, etc.
type	Device type (string).
testbed	The parent Testbed object. Internally this is a weak reference.

Attributes	Description
interfaces	Dictionary of device interfaces (name:Interface).
tacacs	Dictionary of TACACS information unique to this device.
passwords	Dictionary of password information unique to the device.
credentials	Dictionary of credentials for the device.
connections	Dictionary of connection descriptions (name:Dictionary). This is a description of connection methods to this device (e.g., telnet, ssh, netconf, etc.).
connectionmgr	Connection manager (ConnectionManager object). Manages all the connections to this device.
clean	Dictionary of clean parameters (name:value). Clean parameters are those used to clean up (reload) this device.
custom	Dictionary of custom fields (name:value). Nonstandard device object metadata goes here.

Properties	Description
links	Returns the set of unique Link objects connected to this device's interfaces.
remote_devices	Returns the set of unique devices connected to this device via its interface links.
remote_interfaces	Returns the set of unique interfaces connected to this device's interfaces via interface links.

Methods	Description
add_interface	Adds an interface (Interface object) to this device.
remove_interface	Removes an interface (Interface object) from this device.
find_links	Finds and returns a set of links connected to the provided destination object (Device/Interface).

Interfaces are also objects. Interface objects represent any physical/virtual interface/port that connects to a link of some sort (for example, Ethernet, SVI, or Loopback). It is important that you understand the following:

- Each interface connects to a single link (Link object).

- Each interface should belong to a parent device (Device object).

- Within a parent device, each interface name needs to be unique.

- Interfaces can be treated as objects, with the attributes listed in Table 3-3.

Table 3-3 *Interface Object Attribute List*

Attributes	Description
name	Interface name.
alias	Interface alias. Defaults to interface name.
type	Interface type (string).
device	Parent device object. Internally this is a weak reference.
link	The link this interface is connected to (Link object).
ipv4	IPv4 address information (ipaddress.IPv4Interface).
ipv6	IPv6 address information (ipaddress.IPv6Interface or a list of ipaddress. IPv6Interface).

Properties	Description
remote_devices	Returns the set of unique devices connected to this interface via its connected link.
remote_interfaces	Returns the set of unique interfaces connected to this interface via its connected link.

Finally, device objects can be linked together via interface objects using Link objects. Link objects represent the connection between two or more interfaces within a testbed topology. Note that in the case of a link connected to more than two interfaces, the link can also be interpreted as a Layer 2 switch. The following points are important to understand:

- Links may contain one or more interfaces (Interface object).
- Link names within a testbed must be unique.

Much like interfaces, the actual links themselves can be treated as objects with attributes, as displayed in Table 3-4.

Table 3-4 *Link Object Attribute List*

Attributes	Description
name	Link name.
alias	Link alias. Defaults to link name.
interfaces	List of interfaces connected to this link. Note that the interface objects are stored as weak references.

Properties	Description
connected_devices	Returns the set of unique devices connected to this link.

Methods	Description
connect_interface	Adds an interface (Interface object) to this link.
disconnect_interface	Removes an interface (Interface object) from this link.

Using these Python objects, you can describe an entire network topology using structured data in YAML as a single testbed.

We briefly mentioned Unicon as the pyATS connection implementation. We also mentioned some important keys in the testbed, such as os and platform. Let's take a look at Unicon briefly as well as device connection abstractions.

Device Connection Abstractions

pyATS abstracts the complexity of connecting to network devices, making it easy and accessible for those new to network automation. Let's take a look at the simple steps required to connect to a device from the Python interpreter. You can run the commands in the following examples on real devices, if you have them available. If you don't have a real device to practice with, pyATS offers a mock device that you can use with most of the pyATS Library examples.

Download the .zip file that contains the mock data and YAML file here:

> https://pubhub.devnetcloud.com/media/pyats-getting-started/docs/_downloads/04c1 c0ffd3a875e85db16c7408c0f784/mock.zip

Extract the files to a location of your choice and keep the .zip file structure intact. This example uses the directory **mock**.

Activate your virtual environment (refer to Chapter 2, "Installing and Upgrading pyATS," for reference), and change to the directory that contains the mock YAML file. The mock feature is location-sensitive. Make sure you change to the directory that contains the mock.yaml file and keep the .zip file structure intact.

First, let's assume you have the following folder structure:

```
my_project/
|
├── mock/
|    └── mock.yaml
|
└── venv/
```

Here, **my_project/** is your main project directory, **mock/** contains your mock testbed file mock.yaml, and **venv/** is your Python virtual environment directory.

If you haven't already created a virtual environment in your project, do so with the following command:

```
$ python -m venv venv
```

Activate the virtual environment on Windows like so:

```
$ venv\Scripts\activate
```

On macOS or Linux, use the following command:

```
$ source venv/bin/activate
```

Now, you can change directories into the mock folder:

```
(pyats) $ cd mock
```

Open the Python interpreter as follows:

```
(pyats) $ python
```

Load the pyATS Library testbed API so that you can create the testbed and device objects:

```
from genie.testbed import load
```

Create a testbed object (**tb**) based on your testbed YAML file. Specify the absolute or relative path (in this case, **mock/mock.yaml**):

```
tb = load('mock.yaml')
```

The result is that the system creates a variable named **tb** that points to the testbed object. This command also creates **tb.devices**, which contains the YAML device information in the form of key-value pairs.

Create an object (**dev**) for the device that you want to connect to:

```
dev = tb.devices['nx-osv-1']
```

The result is that the pyATS Library finds the device named nx-osv-1 in **tb.devices** and stores the information in the **dev** object.

Now, connect using the values stored in the device object:

```
dev.connect()
```

The result is that the system connects to the device and displays the connection details. Once you're connected, you can run **show** commands and parse the output. To exit the Python interpreter, use the following command:

```
exit()
```

You can put all of these commands into a single Python script! How does pyATS achieve this? Unicon is the pyATS connection implementation that is handling the SSH connection to the device in the preceding example. Unicon is a library developed by Cisco as part of the pyATS framework. Here's a brief overview of Unicon and its role within the pyATS framework:

■ **Purpose:** Unicon provides a unified connectivity interface to network devices. It abstracts the underlying connection mechanisms (like SSH, Telnet, and so on) and provides a consistent interface for interacting with devices, regardless of the connection method or device type.

- **Device independence:** One of the main features of Unicon is its ability to work with a wide range of network devices, regardless of the vendor or platform. This is achieved through plugins that cater to specific device types.

- **State machine:** Unicon uses a state machine model to understand and manage the different states a device can be in (exec mode, config mode, and so on). This allows for intelligent interactions with the device, ensuring that commands are executed in the appropriate context.

- **Ease of use:** With Unicon, users can easily establish connections, execute commands, and retrieve results without having to deal with the intricacies of different connection methods or device peculiarities.

- **Integration with pyATS:** While Unicon can be used as a standalone library, it is tightly integrated with the pyATS framework. This means that when you're using pyATS for network testing or automation, Unicon handles the device connectivity and interactions seamlessly in the background.

Let's delve deeper into the technical aspects of Unicon:

- **Architecture:**

  - **Core:** The core of Unicon provides the basic building blocks for device connectivity, including the state machine, connection mechanisms, and basic command execution.

  - **Plugins:** The extensibility of Unicon is achieved through plugins. Each plugin is tailored for a specific device or platform, encapsulating the nuances and peculiarities of that device. This allows Unicon to support a wide range of devices without bloating the core.

- **State machine:**

  - Unicon's state machine is a representation of the different modes or states a device can be in (for example, exec mode, config mode, or shell mode).

  - Transitions define how to move from one state to another, often involving sending specific commands or sequences.

  - The state machine ensures that commands are executed in the correct context and provides mechanisms to recover from errors or unexpected states.

- **Connection providers:**

  - Unicon supports multiple connection methods, such as SSH, Telnet, and console connections.

  - The connection provider abstracts the underlying connection mechanism, ensuring that the user interacts with the device in a consistent manner, regardless of the connection method.

- **Service framework:**

  - Services in Unicon are high-level operations that users might want to perform on a device, such as executing a command, transferring a file, or reloading the device.

  - Each service is implemented as a callable object, making it easy to extend and customize.

- **Logging and debugging:**

  - Unicon provides extensive logging capabilities, capturing all interactions with the device. This is invaluable for debugging and understanding device behavior.

  - The logs can be configured to capture different levels of detail, from high-level operations to the raw bytes sent and received.

- **Patterns and dialogs:**

  - Interacting with devices often involves recognizing specific prompts or messages and responding appropriately. Unicon uses regular expressions (patterns) to identify these.

  - Dialogs are sequences of expected patterns and responses. They allow Unicon to handle complex interactions, such as logging in, handling prompts, and navigating through device menus.

- **Exception handling:** Unicon is designed to handle exceptions gracefully. If an unexpected event occurs (for example, a timeout, unrecognized prompt, or connection drop), Unicon can attempt to recover the session or raise a meaningful exception to the user.

- **Performance:** Unicon is optimized for performance, ensuring that interactions with devices are fast and efficient. This is especially important in large-scale network testing scenarios. Performance is a big reason why pyATS is preferred on large-scale network topologies over other, much slower, network automation frameworks.

- **Extensibility:** One of the strengths of Unicon is its extensibility. Users can easily add support for new devices, customize existing behaviors, or add new services by extending the core classes and leveraging the plugin architecture.

In essence, Unicon provides a robust and flexible framework for device connectivity, abstracting the complexities and ensuring that users can focus on their automation tasks rather than the intricacies of device interactions.

Testbed Validation

Testbeds can easily be validated using either YAML lint or the built-in pyATS testbed validation command. Software linting, often simply referred to as "linting," is the process of running a program (called a "linter") to analyze source code for potential errors, bugs, stylistic issues, and suspicious constructs. The term "lint" originally referred to unwanted fluff or fuzz on clothing, and in the context of software, it refers to "unwanted" or "suspicious" parts of the code. Here are some key points about linting:

- **Static analysis:** Linting is a form of static code analysis, which means it examines the source code without executing it. This is in contrast to dynamic analysis, which analyzes software by executing it.

- **Code quality:** Linters not only identify potential errors but also enforce coding standards and styles. This helps maintain a consistent codebase, especially in projects with multiple contributors.

- **Customizability:** Most linters allow users to configure which rules to enforce, enabling teams to adopt their own coding standards.

- **Integration:** Linters can be integrated into the software development workflow in various ways:

  - **IDE/editor integration:** Many integrated development environments (IDEs) and code editors have built-in support or plugins for linting, providing real-time feedback as developers write code.

  - **Pre-commit hooks:** Linters can be set up as pre-commit hooks in version control systems, ensuring code is linted before it's committed.

  - **Continuous integration (CI):** Linting can be a step in the CI process, preventing code that doesn't meet the linting criteria from being merged.

- **Common linters:** Linters are available for almost every programming language. Some popular ones include:

  - ESLint for JavaScript

  - Pylint for Python

  - RuboCop for Ruby

  - golint for Go

  - TSLint (now deprecated in favor of ESLint) for TypeScript

- **Benefits:**

  - **Bug detection:** Linters can catch common programming errors, such as undeclared variables, unused variables, or mismatched parentheses.

  - **Code readability:** By enforcing a consistent style, linters help make code more readable for all team members.

  - **Learning:** Especially for beginners, linters can be educational, pointing out best practices and potential pitfalls.

- **Limitations:**

  - **False positives:** Linters can sometimes flag code that is technically correct but appears suspicious. It's up to the developer to determine whether the warning is relevant.

■ **Not a replacement for testing:** While linting can catch certain types of errors, it's not a substitute for thorough testing, including unit tests, integration tests, and end-to-end tests.

yamllint is a Python package used to validate YAML files. To install yamllint, use pip:

```
$ pip install yamllint
```

Then you can "yamllint" your testbed.yaml file to check it for errors prior to executing pyATS.

```
$ yamllint testbed.yaml
```

If yamllint returns "nothing," it means your testbed.yaml file is valid; otherwise, yamllint will display the warnings or errors that are problems with the testbed.yaml file. An alternative to yamllint is to use the built-in pyATS testbed validation command. This example shows a warning indicating that the device has no interface definitions:

```
$ pyats validate testbed testbed.yaml
```

Example 3-6 shows a sample pyATS testbed validation report.

Example 3-6 *Testbed Validation Report*

```
Loading testbed file: testbed.yaml
-------------------------------------------------

Testbed Name:
        testbed

Testbed Devices:
.
`- - csr1000v-1 [iosxe/csr1000v]
YAML Lint Messages
--------------------------

Warning Messages
-----------------------
Device 'csr100v-1' has no interface definitions
```

It is good practice to validate your testbeds after creation or modification in order to prevent issues with pyATS as well as to confirm you are committing valid YAML to your code base. There are certain situations where you will not need to manually create a testbed.yaml file. Testbeds are objects that follow structured YAML syntax; however, they can also be built *dynamically* at runtime.

Dynamic Testbeds

Testbeds are objects that follow a particular syntax and do not necessarily need to be saved as a YAML file. The official pyATS documentation on testbeds suggests that the YAML file is the easier and preferred method of using testbeds; however, there could be situations where you already have a source of truth and the information required to build a testbed already populating another source. Or you might be simply testing and want to build the testbed object dynamically in your code. There are cases where this might be necessary. Example 3-7 demonstrates pyATS manual testbed creation.

Example 3-7   *pyATS Manual Testbed Creation*

```
# Example
# -------
#
#    creating a simple testbed topology from scratch

# import testbed objects
from pyats.topology import Testbed, Device, Interface, Link

# create your testbed
testbed = Testbed('manuallyCreatedTestbed',
                  alias = 'iWishThisWasYaml',
                  passwords = {
                    'tacacs': 'lab',
                    'enable': 'lab',
                  },
                  servers = {
                    'tftp': {
                        'name': 'my-tftp-server',
                        'address': '10.1.1.1',
                    },
                  })

# create your devices
device = Device('tediousProcess',
                alias = 'gimmyYaml',
                connections = {
                    'a': {
                        'protocol': 'telnet',
                        'ip': '192.168.1.1',
                        'port': 80
                    }
                })
```

```
# create your interfaces
interface_a = Interface('Ethernet1/1',
                         type = 'ethernet',
                         ipv4 = '1.1.1.1')
interface_b = Interface('Ethernet1/2',
                         type = 'ethernet',
                         ipv4 = '1.1.1.2')

# create your links
link = Link('ethernet-1')

# now let's hook up everything together
# define the relationship.
device.testbed = testbed
device.add_interface(interface_a)
device.add_interface(interface_b)
interface_a.link = link
interface_b.link = link
```

Here is another real-world example from a Django project. Django, like pyATS, is a Python framework, but instead of focusing on network automation, Django focuses on web development. In this Django project, all of the required network device information is stored in a PostgreSQL database. The database table is loaded into Python, and then we assemble the testbed dynamically at runtime. This approach also scales, and the testbed could have 500 devices from the Django PostgreSQL database! Example 3-8 demonstrates how to build dynamically, at runtime, a pyATS testbed from a PostgreSQL database.

Example 3-8   *Dynamic Testbed from Django PostgreSQL*

```
from catalyst.models import Devices
def main(runtime):
    # Query the database for All Devices
    device_list = Devices.objects.all()
    # Create Testbed
    testbed = Testbed('dynamicallyCreatedTestbed')
    # Create Devices
    for device in device_list:
        testbed_device = Device(device.hostname,
                alias = device.alias,
                type = device.device_type,
                os = device.os,
                credentials = {
                    'default': {
```

```
                        'username': device.username,
                        'password': device.password,
                    }
                },
                connections = {
                    'cli': {
                        'protocol': device.protocol,
                        'host': device.ip_address,
                        'port': device.port,
                        'ssh_options': device.ssh_options,
                        'arguements': {
                            'connection_timeout': device.connection_timeout
                        }
                    }
                })
        # define the relationship.
        testbed_device.testbed = testbed
```

Testbeds can represent more than just your topology, devices, interfaces, and links; they can be *extended* to express *intent*. Let's take a look at how to extend and customize your testbeds.

Intent-based Networking with Extended Testbeds

Intent-based networking relies on a source of truth as an absolute gold standard that describes the intended configuration or state of an individual device or an entire topology. Intent can be enforced through testing and configuration management with pyATS by *extending* the base testbed object to include *customized* keys and values. Create another file, called intent.yaml, and include the following lines at the top of this file to extend your original testbed.yaml file:

```
---

extends: testbed.yaml
devices:
    csr1000v-1:
```

Continue with a **custom** key that then describes your intent. Example 3-9 demonstrates setting some global parameters as well as some per-interface intentions.

Example 3-9 *Extending a Testbed to Include Intent*

```
extends: testbed.yaml
defaults:
    domain_name: "lab.devnetsandbox.local"

    ntp_server: 192.168.100.100

devices:
    csr1000v-1:
        custom:
            interfaces:
                GigabitEthernet1:
                    type: ethernet
                    description: "Link to ISP"
            enabled: True
                GigabitEthernet2:
                    type: ethernet
                    description: "Unused"
                    enabled: False
                GigabitEthernet3:
                    type: ethernet
                    description: "Unused"
                    enabled: False
                Loopback100:
                    type: ethernet
                    description: "Primary Loopback"
                    enabled: True
```

When you run pyATS and specify the **--testbed-file** parameter or include a testbed file in a job, you will now point to and specify the intent.yaml file, which extends testbed.yaml, to ensure your intent is loaded.

Now that you are describing your *intent*, you can test the state or configuration of your network device or topology and validate that the running state of the device matches your intended configuration. Example 3-10 provides an example of a pyATS test, using **@aetest** decorator (which will be covered in Chapter 4, "AEtest Test Infrastructure") to verify that the interface description matches the intended description.

Example 3-10   *AEtest Testing the Actual Interface Description Matches the Intended Description*

```
@aetest.test
def test_interface_description_matches_intent(self):
    for actual_interface, actual_value in self.parsed_interfaces.info.items():
        actual_desc = value.get('description', None)
        for intent_interface, intent_value in self.device.custom.interfaces.items():
            if actual_interface == intent_interface:
                intended_desc = intent_value['description']
                if actual_desc != self.intended_desc:
self.failed("The interface description does not match the intended description)
```

We can include code that enforces our intent as well using the **.configure()** device method, as demonstrated in Example 3-11.

Example 3-11   *Enforcing Intent*

```
if actual_desc != self.intended_desc:
    self.update_interface_description()
    self.failed("The interface description does not match the intended description)
def update_interface_description(self):
    self.device.configure(f'''interface { self.intf }
                            description { self.intended_desc }
                            ''')
```

Intent-based networking using pyATS and extended testbeds is the foundation for a modern network automation CI/CD pipeline. Once all intent tests are passed and enforced, the infrastructure as code artifacts are stored in a Git repository. All future changes to the network are done using Git branches, code reviews, testing, and pull requests that merge the code into the code base. This process kicks off a series of automated tasks such as running the pyATS job to deploy and test the updated intended network configuration. CI/CD pipelines will be covered in Chapter 21, "CI/CD with pyATS."

Summary

Testbeds are foundational to network automation with pyATS. This chapter delved into the intricacies of network testing and configuration, starting with an introduction to YAML. YAML, which stands for "YAML Ain't Markup Language," is a human-readable data serialization format. It is often used for configuration files and data exchange between languages with different data structures.

Next, the chapter introduced the concept of a testbed. A testbed is a controlled environment where network devices and systems can be tested and validated before being deployed to production. This environment ensures that new configurations or software won't adversely affect the existing network setup.

The discussion then shifted to device connection abstractions. This section emphasized the importance of abstracting device connections, allowing for a more streamlined and consistent approach to connecting various devices, regardless of their underlying differences.

Testbed validation is another crucial topic covered. It underscores the importance of ensuring that the testbed environment accurately represents the intended production environment. This validation ensures that tests conducted in the testbed will yield results relevant to the real-world scenario.

The chapter then explored the concept of dynamic testbeds. Unlike static testbeds, dynamic testbeds can adapt and change based on the requirements of the tests being conducted. This flexibility ensures that the test environment is always optimized for the specific test scenario.

Lastly, the chapter delved into intent-based networking (IBN) with extended testbeds. IBN is a form of network administration that uses artificial intelligence and machine learning to automate administrative tasks. When combined with extended testbeds, IBN can lead to more efficient and accurate testing scenarios, ensuring that the network's intent aligns with its configuration and performance.

Chapter 4

AEtest Test Infrastructure

You may have heard about testing code. Code testing allows you to verify your code produces the results you're expecting. This is important, as it helps *minimize*, not remove, bugs in your code. Code testing also encapsulates the idea of regression testing. In the simplest terms, regression testing ensures new code updates do not introduce new bugs. Regression testing becomes more important as the codebase grows. The last concept you may hear about when it comes to testing code is code coverage. Code coverage is the amount of code in your codebase that is "covered" by a test. Many times, it's assumed that more code coverage equals fewer bugs; however, that's simply not true. Tests are only as good as they are written. If your tests are poorly written, then no amount of code coverage can save you from bugs.

Two of the most popular Python testing frameworks are unittest and pytest. unittest is included in the Python standard library and does not require any additional installation. pytest, on the other hand, is a separate library and requires installation. unittest and pytest are different in their own ways, but both have the same goal of allowing developers to write tests and verify their code is running as expected. Now substitute the word "code" with "network"—*write tests and verify your network is running as expected*. Doesn't that sound amazing? This defines Automation Easy Testing (AEtest)—the testing framework for the network.

In this chapter, we will cover the following topics:

- Getting started with AEtest
- Testscript structure
- AEtest object model
- Runtime behavior
- Test parameters
- Test results

- Running testscripts

- Processors

- Testscript flow control

- Reporting

- Debugging

Getting Started with AEtest

The goal of AEtest is to standardize the definition and execution of testcases against the network. In this section, we are going to cover the basics: ensuring the AEtest module is installed and reviewing the design features and core concepts of the framework.

Installation

The AEtest module is included as part of the default pyATS installation. To ensure the module is installed, run the **pip list** command within your Python environment. In the list of installed packages, you should see **pyats.aetest** listed, along with many other pyATS modules. If you don't see the pyats.aetest module listed, I recommend reinstalling pyATS as described in Chapter 2, "Installing and Upgrading pyATS."

Design Features

AEtest drew its design from two popular Python testing tools: unittest and pytest. If you're familiar with either library, the structure and design of AEtest may be familiar. With that said, let's review the design features of AEtest.

AEtest is built with a Pythonic object-oriented approach. This comes from the infamous object-oriented programming (OOP) programming paradigm, where the design is centered around classes and objects rather than functions. From a network perspective, think about all the individual components that make up a network—interfaces, links, devices, and so on. These are all considered "objects" and can be implemented as such in Python using an OOP approach. Moving on, another design feature is using a block-based approach to test sections. We are going to review each test section, but here's a quick breakdown of the approach:

- Common Setup with subsections

- Testcases with setup/tests/cleanup

- Common Cleanup with subsections

Each block listed has a purpose that will be explained further in the chapter. The next design feature is two-fold. AEtest was built to be highly modular and extensible, which, in turn, allows testcase inheritance, dynamic testcase generation, customer runner for testable objects, and a customizable reporter. The last design feature is what allows the tool to scale and cover multiple network use cases. AEtest provides enhanced looping and testcase parametrization. Enhanced looping allows the same test(s) to be reused with different parameters. This is huge, as looping cuts down on the need to write multiple tests that only require slight variation. Looping and testcase parametrization will be covered in further detail later in the chapter. I hope going through these design features helps set expectations and creates a mold for AEtest. Next, let's look at some of the core concepts of AEtest.

Core Concepts

The core concepts of AEtest are brief but promote boundaries of the framework. Here are the three core concepts:

- Main sections must be subdivided

- Sections must be explicitly declared

- Import, inspect, and run

Subdividing the main sections enhances the readability of the code. You can also quickly identify the section that failed in results. Imported sections (that is, testcases) must be inherited in the script for them to be included. Inheriting an imported section explicitly tells AEtest to include the imported testcase. Simply importing the testcase into the script does nothing. The last core concept (import, inspect, and run) is interesting and requires a little more explanation. When a testcase or testscript is imported, Python discovers the test sections (classes), which are instantiated and then run by AEtest. This might seem confusing, as you probably expect any test section that is imported, or included in a testscript, to run in the order provided, but that is not the case. The discovery process is similar to how pytest and unittest discovers testcases and will be covered in more detail further in the chapter. With the design features and core concepts covered, let's move on to the structure of testscripts.

Testscript Structure

Testscripts are the foundation to the AEtest test infrastructure. Testscripts are made up of three main "containers": Common Setup, Testcases, and Common Cleanup. Each container is a Python class with smaller sections (subsections, setup, test, and cleanup) that are defined as methods within the container class. Each method is decorated with a Python decorator that identifies the section type. Python decorators modify the behavior of functions. They can be complex to understand, but they work by passing a function into another function as an argument. The returned output of the function that's passed in as an argument is modified, which ultimately modifies the function's behavior. In AEtest, the different decorators help identify the execution order and resulting rollup of each section based on the decorator. We will dive into these smaller sections later in this chapter, but let's begin by reviewing the main containers.

Common Setup

The Common Setup container is where pyATS connects to testbed devices, applies base configuration for testing, and other initialization actions. This container essentially sets the stage for testing. Common Setup is not a required container in your testscript, but it's highly recommended. If Common Setup is defined, it will always run first. This is due to the discovery process performed by AEtest, which we will take a look at later on. One key feature of Common Setup is that it's also used to validate any script inputs (arguments) provided to the testscript. This allows your testscript to fail fast before going too far into testing before realizing one of the script inputs is incorrect. To better organize your code, you can break down Common Setup into multiple subsections. Each subsection should perform a specific task—for example, one subsection for connecting to testbed devices, another subsection for applying base configurations, and so on. The goal of the Common Setup container is to house the code that prepares your testbed devices for testing, whether that be connectivity, configuration, or operational state.

Subsection

Subsections are smaller actionable sections that make up Common Setup and Common Cleanup. Subsections can be seen as independent, as results from a previous subsection do not affect execution of the current subsection. The user can control whether to skip, abort, or continue to the next subsection after an unexpected result. The results of a subsection are rolled up to the parent section (Common Setup/Common Cleanup). Example 4-1 shows what a Common Setup section with two subsections might look like in a testscript.

Example 4-1 *Common Setup*

```
from pyats import aetest

class MyCommonSetup(aetest.CommonSetup):
    """Common Setup"""

    @aetest.subsection
    def connect_to_devices(self):
    """Code to connect to testbed devices"""
    pass

    @aetest.subsection
    def apply_base_config(self):
    """Code to configure devices with base/initial config"""
    pass
```

Testcases

The testcase container is made up of smaller tests and is the focal point of testscripts. Testcases are designed to be self-contained, modular, and extensible, which allows network engineers to build a library of testcases for their network testing needs. Testcases can have their own setup and cleanup sections, with an arbitrary number of test sections. Each testcase has a unique identifier (UID), which defaults to the testcase name, which is used for result reporting and other job artifacts. Testcases are run as they are defined in the testscript.

Setup Section

The setup section within a testcase is optional. If it's defined, there can only be one setup section within a testcase, and it is automatically run before any other sections. If the setup section fails, all test sections within the testcase would be "blocked" from running. The purpose of the setup section is to configure/enable specific features being tested in a particular testcase. The setup section result is rolled up to the parent testcase result.

Test Section

The test section is a basic building block of testcases. Test sections define the tests run against the network. Each test should test for one specific, identifiable objective—don't try to stuff too much logic or checks within one test section! They are run in the order in which they are defined in the testcase. All test results are rolled up to the parent testcase result.

Cleanup Section

The cleanup section is an optional section, like the setup section, within a testcase. It removes all configurations and/or features enabled during the setup and test sections of the testcase. Whether tests pass or fail, the goal of the cleanup section is to return the testbed devices back to the same state they were in before the testcase. This allows the testscript to continue executing without any lingering issues from previous testcase manipulation. The cleanup section result is rolled up to the parent testcase result.

Now that we have discussed testcases, and the individual sections, let's take a look at an example. Example 4-2 shows code scaffolding for a testcase with a setup section, two test sections, and a cleanup section.

Example 4-2   *Testcase*

```
from pyats import aetest

class MyTestcase(aetest.Testcase):
    """Testcase"""
```

```
@aetest.setup
def testcase_setup(self):
    """Code to setup testbed devices for testcase"""
    pass


@aetest.test
def test1(self):
    """Code for first test"""
    pass


@aetest.test
def test2(self):
    """Code for second test"""
    pass


@aetest.cleanup
def testcase_cleanup(self):
    """Code to cleanup config on testbed devices"""
    pass
```

Common Cleanup

Common Cleanup is much like the cleanup section defined for testcases, but it applies to the entire testscript. It is not required in a testscript, but it's highly recommended. The Common Cleanup is always the last section to run, after all testcases, and removes any configuration and environment changes that occurred during the testscript run. You can think of it as reversing the actions that were done in Common Setup. Like Common Setup, Common Cleanup can be broken down into subsections, which define specific actions. The goal of the Common Cleanup section is to reset the state of the testbed back to what it was before the testscript run. The Common Cleanup result is a combined rollup of results of all of its subsections.

To wrap up, take a look at Figure 4-1, which helps visualize the testscript structure with the different containers and their corresponding sections.

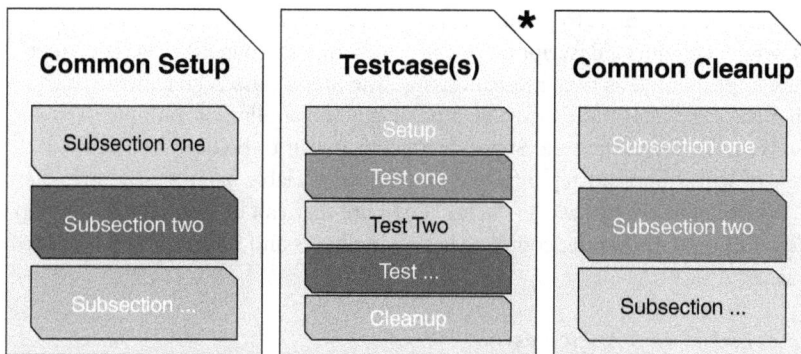

Figure 4-1   *AEtest Testscript Structure*

Section Steps

Previously, we discussed how container classes (Common Setup, Testcases, and Common Cleanup) are broken down into smaller sections to help better organize code and the overall testing workflow. However, we can go one step further. *Steps* allow you to break down your individual test sections into more granular actions. Steps are completely optional and should only be used if a test is larger and it makes sense to break it down further versus separating it out into smaller, individual tests.

Steps is a reserved parameter in the AEtest infrastructure and must be included as a test function argument in order to be used within the test. The **Steps** object is a Python context manager and is intended to be used by the **with** statement. Example 4-3 shows a simple example of a test section within a testcase broken down into multiple steps.

Example 4-3   *Steps in Test Section*

```
from pyats import aetest
class Testcase(aetest.Testcase):
        """Testcase with steps"""

        @aetest.test
        def test(self, steps):
        """Code for test section and steps"""
            # steps.start() begins the step
            with steps.start("The first step") as step:
              print("This is step one!")

            with steps.start("The second step") as step:
              print("This is step two!")
```

There are a couple of key points to point out from the example. The step begins with **steps.start()**, which contains a description of the step within the parentheses. The **Steps** object is really implemented as two internal classes. The **Steps** class is considered the base container class and allows the creation, reporting, and handling of multiple nested steps. The **Step** class inherits the base **Steps** class and is meant to be used as a context manager. We can access current step information with the variable name we set after the **as** keyword. For example, steps have an "index" attribute that can be accessed in the step via **step.index**. Table 4-1 shows the complete list of attributes and properties of the **Steps** and **Step** objects.

Table 4-1   *The steps and step Attributes*

Steps	
Attribute	**Description**
start	Starts a new step. Returns a **Step** instance.
result	Rollup result of all steps contained.
report	Reports current step details/results to a log file.
steps	List of **Step** objects representing each step taken.
Properties	**Description**
details	List of step details using **StepDetail namedtuple**.
Step (basecls: Steps)	
Attribute	**Description**
start	Starts a new child step. Returns a **Step** instance.
result	Rollup result of this step and all child steps contained.
report	Reports current step details/results to a log file.
steps	List of child **Step** objects.
description	Description of this step instance.
Properties	**Description**
details	List of step details using **StepDetail namedtuple**.
Result APIs	**Description**
passed	Provides a passed result to this step.
failed	Provides a failed result to this step.
aborted	Provides an aborted result to this step.
blocked	Provides a blocked result to this step.
skipped	Provides a skipped result to this step.
errored	Provides an errored result to this step.

Steps	
passx	Provides a passx result to this step.

Built-in	Description
__enter___	Method called with starting step through **with** statement.
__exit___	Method called with exiting step through **with** statement.

By inheriting the base **Steps** class, the **Step** object is able to nest steps, which provides more granularity during testing. Example 4-4 shows nested steps and the associated output, which expresses how the nested steps show up in the printed results. You'll notice the nested steps are separated from the parent step using "".

Example 4-4 *Nested Steps*

```
from pyats import aetest

class Testcase(aetest.Testcase):
    @aetest.test
    def test(self, steps):
        # demonstrating a step with multiple child steps
        with steps.start("test step 1") as step:
            with step.start("test step 1 substep a"):
                pass
            with step.start("test step 1 substep b") as substep:
                with substep.start("test step 1 sub-step b sub-substep i"):
                    pass
                with substep.start("test step 1 sub-step b sub-substep ii"):
                    pass
```

The results for each step roll-up to the parent test section. Let's use the previous example (Example 4-4) as an example. If Step 1.b.ii was the only step to fail, the entire test section, which includes the other four steps, would have a "Failed" result. That's why it's crucial to only include related test steps within a single test section. If one fails, the entire section is considered a failure. The roll-up nature of test results will be covered later in the chapter. For reporting, a testscript creates a steps report at the end of each test section that contains steps. During runtime, the steps report can be accessed with the **report()** attribute (**steps.report()**). For additional detail, the **details** attribute can be accessed (**steps.details**), which will return a list of StepDetail objects. Each StepDetail object is a named tuple containing the current step index, step name, and step result. The details attribute can be useful if you plan to parse and analyze the step results further. Steps are useful when needing to break down a lengthy test into smaller, granular chunks.

AEtest Object Model

An object model describes the classes and objects that make up a piece of software or system. In this section, we are going to go through the object model of AEtest. This will get into the implementation details of the classes that make up the testscript and different testscript sections previously described. If you are brand new to pyATS and Python, this section may be advanced, but try to stick with it! There's a lot of detail that explains how testscripts and their individual sections are constructed and why they behave the way they do.

TestScript Class

An AEtest testscript is a standard Python file, but without the .py extension. The file is considered a testscript because it imports the AEtest module from **pyats**. Testscripts are made up of three defined sections: Common Setup, Testcases, and Common Cleanup. During execution, the **aetest** infrastructure internally wraps the running testscript into a **TestScript** class instance. The important piece of the **TestScript** instance is that it stores script arguments as parameters in the instance to use throughout testing, and all the major sections (Common Setup, Testcases, Common Cleanup) point to the **TestScript** instance as their parent during testing (that is, Testcase.parent).

Container Classes

There are three container classes in AEtest: **CommonSetup**, **Testcase**, and **CommonCleanup**. Conveniently, each of these containers was covered in detail at the beginning of this chapter, but let's now focus on the implementation details. All the container classes are inherited from the **TestContainer** class, which is a base class. Base classes are internal to pyATS and will not be covered further in this book, as there is little reason to access or manipulate base classes. The purpose of container classes is simply to house other sections. Tests are not defined directly in a container class. They must be included in a test section, which is written inside of a container class. Table 4-2 shows the attributes and properties of a container class.

Table 4-2 *Container Class Attributes/Properties*

CommonSetup, CommonCleanup, Testcase (base cls: TestContainer)	
Attribute	**Description**
source	File/line information where the class was defined

CommonSetup, CommonCleanup, Testcase (base cls: TestContainer)

Properties	Description
uid	The common_setup/common_cleanup, or uid of Testcase
description	Class header (docstring)
parent	TestScript object
result	Rolled-up result of this section
parameters	Dictionary of parameters relative to this section

To dive deeper into the technical details, note that container class instances are callable iterables. Let's take a second to break that down. A callable allows you to run, or "call," the code. In Python, functions and classes can be called; hence, they can be referred to as callables. An iterable is an object that can be iterated or looped over. Example 4-5 shows how a Common Setup container instance can be looped over and directly called.

Example 4-5 *Container Class*

```
from pyats import aetest
# Define a container and two subsections
class MyCommonSetup(aetest.CommonSetup):
    @aetest.subsection
    def subsection_one(self):
        self.a = 1
        print("hello world")

    @aetest.subsection
    def subsection_two(self):
        assert self.a == 1

# Instantiate the class
common_setup = MyCommonSetup()

# Loop through to see what we get:
for i in common_setup:
    print(i)
```

Function Classes

Function classes are housed within container classes and are what carry out the actual tests. Function classes include the **Subsection**, **SetupSection**, **TestSection**, and **CleanupSection** classes. These class names may look familiar to the different section decorators we discussed previously in "Testscript Structure." Each function class is short-lived. They are instantiated during runtime and only live as long as the section runs. Table 4-3 shows the attributes and properties of function classes.

Table 4-3   *Function Class Attributes/Properties*

**Subsection, [Setup|Test|Cleanup]
Section (base cls: TestFunction)**

Attribute	Description
function	Function/method that was decorated to be this section
source	File/line information where the method was defined

Properties	Description
uid	Name of the function/method
description	Function header (docstring)
parent	Container (**CommonSetup/Testcase/CommonCleanup**)
result	Rolled-up result of this test function
parameters	Dictionary of parameters relative to this function

Any class method that has a section decorator is instantiated with their corresponding function class. For example, a class method with the decorator **@aetest.test** instantiates the **TestSection** class. This allows the AEtest infrastructure to manage each section's reporting context, enables result tracking, and other features specific to the test section methods. Example 4-6 shows the internals of each function class within a Testcase class instance.

Example 4-6   *Function Class*

```
from pyats import aetest

class MyCommonSetup(aetest.CommonSetup):

    # subsection corresponds to Subsection class
    @aetest.subsection
    def subsection_one(self):
        pass

class MyTestcase(aetest.Testcase):

    # setup corresponds to SetupSection class
    @aetest.setup
    def setup(self):
        pass
```

```
        # test corresponds to TestSection class
        @aetest.test
        def test_one(self):
            pass

        # cleanup corresponds to CleanupSection class
        @aetest.cleanup
        def cleanup(self):
            pass

# When container instances are iterated, the returned objects are function
# class instances
tc = MyTestcase()
for obj in tc:
    print(type(obj))
    print(obj.function)

# Printed results:
# <class 'pyats.aetest.sections.SetupSection'>
# <bound method MyTestcase.setup of <class 'MyTestcase' uid='MyTestcase'>>
# <class 'pyats.aetest.sections.TestSection'>
# <bound method MyTestcase.test_one of <class 'MyTestcase' uid='MyTestcase'>>
# <class 'pyats.aetest.sections.CleanupSection'>
# <bound method MyTestcase.cleanup of <class 'MyTestcase' uid='MyTestcase'>>
```

Runtime Behavior

The AEtest module provides access to objects and attributes that are only available
during runtime via the runtime object. The runtime object is available only while the
testscript is executing. Currently, **uids** and **groups** are the only two accessible variables;
however, this could change in future releases.

The runtime object can be useful for querying and possibly manipulating the execution
flow of your testscript. For example, you may want to ensure only certain testcases or
sections run during testing. This can be done by querying the testcase UID and/or group.
We will talk about groups later in the chapter, but in short, they allow you to arbitrarily
label testcases. This allows you to better organize which testcases to run in a testscript.
Example 4-7 shows how the testscript will only run testcases that belong to the "L3"
group but are not in the "L2" group.

Example 4-7   *AEtest Runtime*

```
from pyats import aetest

from pyats.datastructures.logic import And, Not

class CommonSetup(aetest.CommonSetup):
    # Allows testcases in "L3" and not in "L2" group to execute
    @aetest.subsection
    def validate_l3_testcases(self):
        aetest.runtime.groups = And("L3", Not("L2"))
        # Print runtime groups
        print(aetest.runtime.groups)
```

Although this example reads straightforward, I do want to touch on the logic expressions used to differentiate the group names. Notice at the beginning of the example that we imported the keywords **And** and **Not** from **pyats.datastructures.logic**. This is another hidden gem within pyATS. The logic module allows you to easily produce logic testing with English keywords. The logic module also allows us to use callables that accept arguments to perform additional logic before returning a value that is used for truth testing.

Self

In Python, when a class is instantiated, it has the ability to access its attributes and methods through the **self** keyword. The use of **self** is a convention, not a rule, in Python; however, there aren't many instances where you will see any other keyword used to represent an instance of a Python class. In AEtest, container classes (Common Setup, Testcases, and Common Cleanup) are all Python classes. This allows you to get and set class-level attributes during testing. For example, let's say you wanted to use a value from one test section in another, such as the MAC address table. You might have multiple tests that need the results of the **show mac-address table** command. Instead of running the command and collecting the results in each individual test, you can run it once and save it as a class attribute using **self**. Example 4-8 shows a testcase that executes the **show mac-address table** command, collects the output, and uses that output in a future test.

Example 4-8   *Self Testcase*

```
from pyats import aetest

class L2Testcase(aetest.Testcase):
    @aetest.setup
    def collect_l2_info(self, device):
```

```
    # Collect MAC address table entries
    self.mac_table = device.execute("show mac-address table")


@aetest.test
def confirm_mac_addresses(self):
    # Confirm the "important" MAC address is found in the MAC address
    important_mac = "0123.4567.0987"
    assert important_mac in self.mac_table
```

Obviously, this is not the best way to go about checking for a particular MAC address in the MAC address table, but you can clearly see how the MAC address table output is collected in the setup section and used in a separate test section within the testcase. Being able to get and set class attributes using **self** can be a powerful tool by allowing testcases to run more efficiently and remove redundancies.

Parent

At the beginning of the chapter, we touched on the testscript structure and the concept of parent-child object relationships. Let's dive a bit further into those topics. Besides the **TestScript** class, all other classes in the AEtest object model have a parent class. Figure 4-2 shows a graphical representation of the parent-child relationships among the different AEtest class objects.

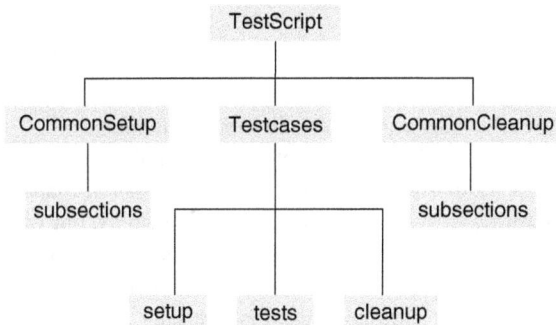

Figure 4-2 *Parent-Child Object Relationships*

The **parent** attribute is accessible during testing by using **self.parent** within the container/section. It's not used often but may be useful if you are trying to access a parent's parameters (for example, a **Testcase** class accessing the parameters assigned to the **TestScript** class).

Section Ordering

Testscripts have a logical, reproducible order in which container classes (Common Setup, Testcases, and Common Cleanup) and sections within the container classes (setup, test, cleanup) are discovered and executed. This allows for every testscript to run the same, regardless of the execution environment. Container classes execute in the following order:

- Common Setup always runs first.

- Testcases run in the order in which they appear in the script.

- Common Cleanup always runs last.

Within a container class, there can be multiple child methods: setup, subsection, test, and cleanup. The child methods execute in the following order:

- Setup always runs first (if defined).

- The subsection and test methods run as they appear in the script.

    If the parent class is inherited, the parent class's subsection and tests run first.

- Cleanup always runs last.

The execution order might seem apparent when looking at examples in this book or the library's documentation, but the guaranteed ordering provides uniformity to testscript execution and leaves nothing to chance.

Test Results

Now that we know about the structure of a testscript, all the different containers and sections, the order they are executed, and even the runtime behavior, let's talk about test results. Test results may seem straightforward—the test either passes or fails—but that's not all. Pass, fail, and error may work for traditional software testing frameworks, but we are dealing with a network infrastructure. As network engineers, we all know the engineering slogan: "It depends." To accommodate for potential unrelated network failures, poor design, and so on, AEtest has added some additional result types, such as skipped and errored, and exception handling to better describe and gain a better understanding of the test results. Along with test results, understanding how the results roll up and affect test reporting is crucial.

Result Objects

Before we dive deeper into understanding test results, let's list the different result objects with a short description of each:

- **Passed:** Test was successful.

- **Failed:** Test was not successful.

- **Aborted:** Something started but did not finish.

- **Blocked:** A test dependency was not met, and the test could not start.

- **Skipped:** A test was not executed and omitted.

- **Errored:** A mistake or unexpected exception occurred. The difference from Failed is that the test ran but did not meet expectations. Errored indicates something went wrong during testing and it could not be completed.

- **Passx:** Short for "pass with exception." Essentially remarks a Failed result with Passed based on an expected exception.

Result Behavior

By default, all test results are "Passed." A testcase could be empty, and as long as no exceptions are thrown, it will pass. This type of behavior is standard among other Python testing frameworks. However, when exceptions are thrown, AEtest has the capability to catch exceptions and assign a result to the corresponding section. AEtest can catch and handle the following exceptions:

- **AssertionError:** A built-in Python exception raised when an assert statement fails. AEtest will catch this exception and assign a Failed result to the test section.

- **Exception:** The base Exception class in Python. All built-in Python exceptions are derived from this class. AEtest will catch any exception and assign an Errored result, indicating an unhandled error to the developer. This should allow the developer to quickly locate any unhandled errors during testing and fix them or catch them properly.

To avoid these exceptions being caught by AEtest, you should use **try…except** blocks to catch any exceptions and handle them appropriately. For example, if you expect that a device or set of devices will not parse a particular **show** command due to no output, you should try to catch the SchemaParserEmpty Exception and handle it by either skipping or blocking the test from running for those devices. Otherwise, AEtest will catch the exception for you and assign an Errored result to the test section.

To be more granular and assign explicit results to a test section, you simply call the result object and provide a reason for the result. Example 4-9 shows an example of three tests marked with Passed, Skipped, and Errored.

Example 4-9 *Test Result Assignment*

```
from pyats import aetest

class ResultTestcase(aetest.Testcase):

    @aetest.test
```

```
def subsection_that_passes(self):
    self.passed("This test passed!")

@aetest.test
def subsection_that_skips(self):
    self.skipped("This test skipped!")

@aetest.test
def subsection_that_is_errored(self):
    self.errored("This test errored!")
```

Along with a reason that is printed with the result, results can also include a few more optional arguments:

- **Reason:** Describes to the user why a test result occurred (shown in Example 4-9).

- **Goto:** List of sections to "go to" after this section. This is essentially a one-way ticket to another test section in the testscript.

- **From_exception:** Accepts an exception object and will add the traceback to the reason.

- **Data:** A dictionary of data relevant to the result. This data is passed to and stored with the Reporter object for further processing.

Once a result is determined for the current section, AEtest moves on to the next test section. Any code that is included after the result is determined in a test section will not be executed.

Interaction Results

As a network engineer, you might still want or require some control of the testing that occurs. "What if I need to move patch cables during testing? I don't want to sit in a testing lab all day waiting around to pause testing, move one silly patch cable, and click a button to continue testing." Automation can't solve all our problems, but it can take them into consideration! AEtest offers the **WebInteraction** class, which pauses test execution and can notify a user (via email) that input is required via a web page. The web page is a form for the user to submit a test result for the test(s) that required intervention. You can even customize the email body and HTML web page using Jinja2 templates to really make it your own!

Result Rollup

Result rollup involves combining the results of many child sections into one summary result. The simplest example is a testcase with two test sections. If one of the two test sections fails, the testcase fails. The rollup concept is easy to understand, but it can be complex when dealing with many testcases and multiple test sections as well as adding a few steps in each test section. Table 4-4 shows a lookup table for a summary result when combining multiple results. This table can be used to check which result "wins" over the other. For example, let's say you have a testcase with three test section results: Passed, Skipped, and Passx, in that order. Starting from the top, you can compare the Passed and Skipped results and see in the table that the summary result for the testcase would be Passed. Next, compare the Passed and Passx results. This is interesting because now the summary result changes to Passx instead of Passed. This is due to the fact that the summary result is meant to inform the user of any negative results or exceptions caught during testing, as these issues would need fixed. If the summary result would have resulted in Passed, we wouldn't have seen the test section that passed with an exception (Passx result).

Table 4-4   *Result Roll-up Table*

Results	Failed	Passed	Aborted	Blocked	Skipped	Errored	Passx
Failed	Failed	Failed	Aborted	Failed	Failed	Errored	Failed
Passed	Failed	Passed	Aborted	Blocked	Passed	Errored	Passx
Aborted	Aborted	Aborted	Aborted	Aborted	Aborted	Aborted	Aborted
Blocked	Failed	Blocked	Aborted	Blocked	Blocked	Errored	Blocked
Skipped	Failed	Passed	Aborted	Blocked	Skipped	Errored	Passx
Errored	Errored	Errored	Aborted	Errored	Errored	Errored	Errored
Passx	Failed	Passx	Aborted	Blocked	Passx	Errored	Passx

Processors

Processors are functions that are executed before or after a given section. Processors are optional in AEtest but can be used to perform helpful checks before and after testing, such as collecting test environment information, taking snapshots, validating section results, or executing debug commands and collecting dump files. The possibilities are endless, as they are simply Python functions. In the following sections, we will cover the different processor types, including how to define and use them in a testscript.

Processor Types

There are three types of processors:

- Pre-processors
- Post-processors
- Exception processors

Pre-processors are executed before a given section and may be used to take snapshots of the current environment or determine whether a test should run. Post-processors are executed after a section and may be used to validate the test results or collect debug information or dump files. Exception processors are kicked off when an exception occurs. They may be used to collect debug information when an exception occurs or to suppress exceptions that are raised within a section and assign a proper result for the section. Because these processors are just Python functions, you can get creative with the logic and data collected for each processor type.

Processor Definition and Arguments

All the processor types (pre/post/exception) can be applied to test containers (Common Setup, Testcases, and Common Cleanup) and test sections (subsections, setup, test, and cleanup). Each section may have one or more processors, in the form of a list, that execute in the order they appear. A processor can be applied to a test container or section with the **@aetest.processors** decorator. Example 4-10 shows pre-, post-, and exception processors applied to a testcase with two tests.

Example 4-10 *Testcase Processors*

```
from pyats import aetest

# Print section uid
def print_uid(section):
    print("current section: ", section.uid)

# Print section result
def print_result(section):
    print("section result: ", section.result)

# Print the exception message and suppress the exception
def print_exception_message(section, exc_type, exc_value, exc_traceback):
    print("exception : ", exc_type, exc_value)
    return True

# Use the above functions as pre/post/exception processors to a Testcase
#    pre-processor   : print_uid
#    post-processor  : print_result
#    exception-processor : print_exception_message
@aetest.processors(pre = [print_uid],
                   post = [print_result],
                   exception = [print_exception_message])
class Testcase(aetest.Testcase):

    @aetest.test
    def test(self):
        print("First test section...)
```

```
@aetest.test
def testException(self):
    raise Exception("Exception raised during testing... ")
```

Processors may have parameters propagated to them via a datafile or parent containers/ sections. Processor arguments must have the same name as the parameters being passed in. Some default parameters included are section, processor, and steps. Exception processors have additional default parameters, which include the exception type (**exc_type**), exception value (**exc_value**), and exception traceback (**exc_traceback**). These default parameters can be very powerful when determining test section results or troubleshooting issues during testing. Example 4-10 shows the different attributes of the section parameter (**uid** and **result**).

Context Processors

Context processors are more advanced processors that act similar to Python context managers in the sense that they handle the pre-, post-, and exception-handling processors within a single class instead of a single Python function. When you're creating a context processor, the pre-processor actions are defined in the __enter__ method. Actions defined in the __exit__ method handle the post- and exception-processor logic. A context processor class has the results API available, so calling **self.failed()** within the context processor class would be like **processor.failed()** and fail the processor.

Global Processors

Global processors are processors that run automatically before and after each test container and section defined in a testscript. They do not require the **@aetest.processors** decorator to be applied to each container/section. To use global processors in your testscript, you must define a Python dictionary in your testscript called global_processors. In this dictionary, you must specify the following keys to represent the different processor types: pre, post, and exception. A list of processors can be specified as the value for each processor type key. Example 4-11 shows how global processors are defined in a testscript.

Example 4-11 *Global Processors*

```
from pyats import aetest

# Print section uid
def print_uid(section):
    print("current section: ", section.uid)

# Print section result
```

```
def print_result(section):
    print("section result: ", section.result)


# Print the exception message and suppress the exception
def print_exception_message(section, exc_type, exc_value, exc_traceback):
    print("exception : ", exc_type, exc_value)
    return True


# Use the above functions to define global pre/post processors
#    global pre-processor  : print_uid
#    global post-processor : print_result
#    global exception-processor : print_exception_message
global_processors = {
    "pre": [print_uid,],
    "post": [print_result,],
    "exception": [print_exception_message,],
}


class Testcase(aetest.Testcase):

    @aetest.test
    def test(self):
        print('running testcase test section')

<rest of testscript omitted for brevity>
```

Processor Results

Like test sections, processors have their own result and can be marked as passed, failed, skipped, and so on. The result will roll up to the parent object, the same as any other child result; however, the processor result can directly affect the parent test section's result. Because processors have access to the section object via the default parameters provided to processors, a processor can alter a section's result by calling the section results API. For example, to fail a test section, simply call **section.failed()** in a processor function. Once the processor is run, it will fail the parent test section. For pre-processors, this will block the execution of the test section and set the result as failed. For post-processors, this will override the existing results of the test section and mark it as failed.

Data-Driven Testing

AEtest testscripts and testcases are intended to be driven dynamically by data. Dynamic data that alters and affects the behavior of testscripts and testcases is called a parameter. Test parameters are meant to be dynamic in nature and can be provided to testscripts in the form of input arguments or generated during runtime. In this section, we will go over test parameter relationships, their properties, calling parameters, parametrization, and reserved parameters. Beyond test parameters, we will also take a look at datafile inputs and looping sections. If it isn't already apparent, the focus in these upcoming sections is how to *dynamically* affect the execution and runtime behavior of your testscript. By utilizing these concepts, your testscript organization and testing logic will greatly mature.

Test Parameters

Parameters are variables used to access input data (arguments) in Python functions and methods. In the context of AEtest, a parameter will take the value of a testbed argument, which is passed to a testscript, to instruct the testscript as to which testbed to connect to for testing. If the testbed argument is not available, a testscript would have to be hardcoded with the testbed name, which eliminates the dynamic nature of testing and makes it impossible to scale. This simple example, which outlines one way to pass data to AEtest through script arguments, allows you to understand the importance of test parameters. In the following sections, we are going to take a look at the relationships, properties, and ways to call different test parameters during testing.

Parameter Relationships

Test parameters are relative to the test section. The test parameters for a given test section are a combination of local parameters in the section and any parent parameters. Going back to the AEtest object model and how the container and sections relate to one another, we see that parameters are inherited the same way. Figure 4-3 shows a visual of the parameter relationship model and how parameters are inherited from their parent container/sections.

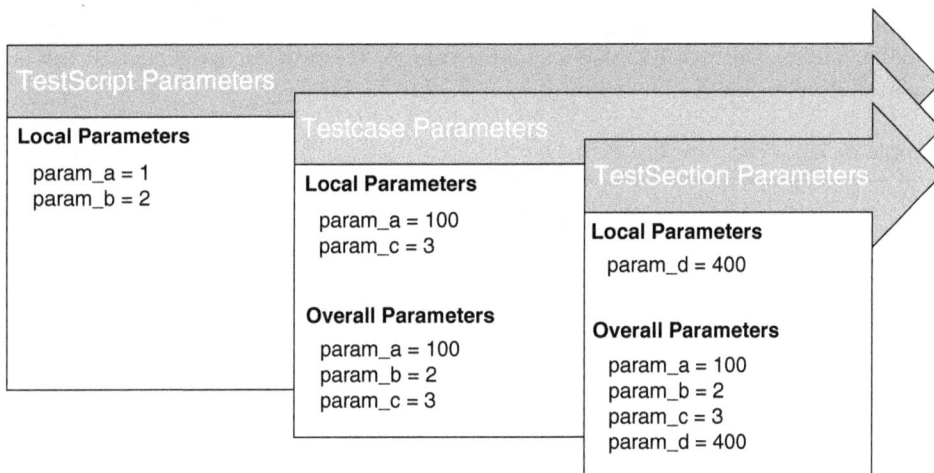

Figure 4-3 *Parameter Relationship Model*

You can see how the list of overall parameters continues to grow as you move from the TestScript parameters to the Testcase parameters, to the TestSection parameters, so that parameters defined at the TestScript level are made available at the TestSection level. Another key point is that test parameters can be overwritten. For example, **param_a = 1** at the TestScript level was changed to **param_a = 100** at the Testcase level and is presented as such to the TestSection. This can be key if you're planning to implement a parameter that you expect will change during testing. Initialize the parameter at the highest possible level to make it available throughout testing and then alter it as needed.

Parameter Properties

Each top-level object (TestScripts and Testcases) in AEtest has a special **parameters** property that represents the different test parameters for that particular object. The **parameters** property is a Python dictionary that stores the test parameters as key-value pairs. They can store default values for the test parameters, which can be changed by accessing the test parameter during runtime. Function classes such as Subsection, SetupSection, TestSection, and CleanupSection have a **parameters** property as well, but these class instances only exist briefly during runtime, so we cannot statically set a dictionary of default parameters for these sections. It's recommended that you consolidate test parameters for these sections and include them in the parent TestContainer **parameters** property.

Parameter Types

There isn't an official list of parameter "types," but it's important to understand the different ways in which a parameter can be included in testing. Test parameters can be added to a testscript as script arguments, function arguments, or callables. Let's take a look at each one.

Script arguments are any arguments passed directly to a testscript before startup. This includes arguments passed by a jobfile, command-line arguments, or even updating the **parameters** property of a TestScript object within the testscript code with a dictionary of parameters. Example 4-12 shows a simple example of how testscript parameters can be updated within code.

Example 4-12 *TestScript Test Parameters*

```
# The following parameters were already defined
parameters = {
    "arg_a": 1,
    "arg_b": 2,
}


# The following inputs were passed as arguments to the testscript
script_arguments = {
    "arg_a": 100,
    "arg_c": 3,
}
```

```
# The TestScript parameters would be built as follows
testscript.parameters = parameters
testscript.parameters.update(script_arguments)

# Result - you'll notice that arg_a was updated by the script arguments
testscript.parameters
# {"arg_a": 100,
#   "arg_b": 2,
#   "arg_c": 3}
```

Parameters can also be passed as function arguments. Input parameters passed to the testscript as script arguments can be explicitly passed to a function as an argument. This is due to the parent-child object model in AEtest. During runtime, all function arguments are filled with the corresponding parameter value, with the argument and parameter names matching. It is preferred to explicitly pass each parameter as function arguments, as it makes the code easier to understand and allows you to call each function with different arguments during testing or debugging. Example 4-13 shows how parameters can be passed down to child containers/sections and changed in a testscript.

Example 4-13 *Parameters—Function Arguments*

```
from pyats import aetest

# Script-level parameters
parameters = {
    "param_A": 1,
    "param_B": dict(),
}


class Testcase(aetest.Testcase):

    # "param_B" is passed to the setup section as a function argument
    @aetest.setup
    def setup(self, param_B):

        # param_B is a dictionary and can be changed (mutable)
        # Any changes are persist throughout the testscript
        param_B['new_key'] = "a key added during setup section"

    # "param_A" and "param_B" are passed to the test section
    @aetest.test
    def test_one(self, param_A, param_B):
        print(param_A)
        # 1
        print(param_B)
        # {'new_key': 'a key added during setup section'}
```

The last way a test parameter can be provided to a testscript is via a callable. As mentioned earlier in the chapter, functions and classes are considered callables. In the context of AEtest, many times we are dealing with functions when talking about callables. A callable parameter must evaluate to True to be valid, which means the callable can't return a boolean of False, None, empty strings, or a numeric value of 0. Callables are passed as function arguments and are "called" during runtime. The return value of the callable is used as the actual parameter. The one limitation to callables is that they cannot have arguments of their own, as AEtest will not pass any arguments to the callable.

Parameter Parametrization

Parametrized parameters are identical to callables, but they enable you to create "smarter" functions by allowing you to introduce more dynamic parameter values. Parametrized functions are declared using the **@aetest.parameters.parametrize** decorator. Unlike normal callables, arguments can be passed to the parametrized function with the **@aetest. parameters.parametrize** decorator. Along with allowing arguments, a special argument named **section** can be passed to a parametrized function that allows you to access the current section object. This includes access to the current section's properties, such as **uid** and **result**. By having access to the current section object, you can dynamically change the return value based on the parent section's result or in combination with the test parameters available to the current section.

Reserved Parameters

AEtest has reserved parameters that are generated during runtime. They are not available when you're accessing the parameters property but can be used to access internal objects to AEtest. They are only accessible if their name is provided as a keyword argument to a test section. Reserved parameters take precedence over normal parameters if there is a normal parameter with the same name. The purpose of reserved parameters is to provide a mechanism for engineers to access and dive deeper into the internals of AEtest within a testscript. It's highly recommended that you only access the reserved parameters if required and to never modify a reserved parameter, as it could lead to unexpected behavior.

Datafile Input

Up to this point, testscripts are defined as static files with multiple test containers/sections that can be altered by test parameters. However, what if we want more flexibility and the ability to dynamically update the testscript without having to manually change the code? *Datafiles* are YAML input files that can be passed to AEtest and allow you to dynamically update testscript parameter values. They are completely optional but allow you to easily change testscript values without having to modify the original testscript code. With datafiles being written in YAML, they are easily readable, even by nonprogrammers, which empowers users of pyATS to easily modify testscript values and feel confident doing so.

Datafile inputs directly update the testscript's module parameters before runtime. Only container classes can be updated via datafiles: **CommonSetup, Testcases,** and **CommonCleanup.** However, due to the parent-child relationship of these container classes to the individual test sections, the datafile values can be used in the individual test sections. The common and testcase container classes defined in the testscript must have matching names in the datafile. For example, a testcase defined in the testscript as **class BGPTestcase** that requires dynamic values from the datafile must include a **bgp-testcase:** section in the datafile. If a section is looped, which will be discussed in the next section, only the base class attributes are changed. If values must change on *n* iterations, you must pass those values as loop parameters. The last key point to datafiles is that only one can be provided to a testscript; however, you may *extend* another datafile. Much like Jinja2 template inheritance, a base datafile can be extended to create more modular data-files for testing. The extensibility reduces the amount of redundant datafiles that need to be created and can help promote others to build their own datafiles by simply extending a core datafile.

There is a defined datafile schema in the pyATS documentation, but for brevity, Example 4-14 shows two datafiles, base.yaml and datafile.yaml, that can be passed to a testscript (see Example 4-15). It's assumed that a pyATS testbed file (testbed.yaml) exists in the same directory.

Example 4-14   *The base.yaml and datafile.yaml Datafiles*

```
---
# base.yaml
# testscript parameters
parameters:
    cloudflare_dns: 1.1.1.1


...
---
# datafile.yaml
extends: base.yaml

# testscript parameters
parameters:
  google_dns: 8.8.8.8
  # Adds to existing cloudflare DNS from base.yaml

testcases:
  BGPTestcase:
    # testcase uid
    uid: routing_test_1
```

```
    # list of groups that testcase belongs to
    groups: [routing]

    # testcase parameters
    parameters:
      local_asn: 65000
      remote_asn: 65001

    # testcase class variable
    expected_routes: 5

  ExternalConnectivity:
    # testcase uid
    uid: ext_dns_test

    # list of groups that testcase belongs to
    groups: [routing]

    # testcase parameters
    parameters:
      alt_google_dns: 8.8.4.4
...
```

Example 4-15   *Testscript with Datafile Input*

```
⌐ Make into code block
import logging
from genie.utils import Dq
from pyats import aetest
from unicon.core.errors import ConnectionError


logger = logging.getLogger(__name__)
logger.setLevel("INFO")


class CommonSetup(aetest.CommonSetup):
    @aetest.subsection
    def connect_to_devices(self, testbed):
        """Connect to all testbed devices"""
        try:
            testbed.connect()
```

```
        except ConnectionError:
            self.failed(f"Could not connect to all devices in {testbed.name}")

        # Print log message confirming all devices are in a 'connected' state
        logger.info(f"Connected to all devices in {testbed.name}")

class BGPTestcase(aetest.Testcase):
"""Test BGP operational state"""

    @aetest.test
    def check_bgp_routes(self, testbed):
        """Check number of BGP neighbors equals expected number of routes in
        datafile."""

        # Print all class variables (as a Python dictionary)
        print(f"All class variables: {vars(BGPTestcase)}")
        # Example Output
# {'__module__': '__main__', 'check_bgp_routes': <function
# BGPTestcase.check_bgp_routes at 0x10bf30430>,
# '__parameters__': {'local_asn': 65000, 'remote_asn': 65001},
# '__doc__': None, 'source': <pyats.aetest.base.Source object at 0x10beeebb0>,
# '__uid__': 'routing_test_1', 'groups': ['routing'], 'expected_routes': 1,
# 'uid': <property object at 0x108195b30>}

# Print available test parameters (provided by datafile) - includes TestScript
# and Testcase-level parameters
        print(f"Available testcase parameters: {self.parameters}")
# Example Output
# ParameterMap({'local_asn': 65000, 'remote_asn': 65001},
# {'cloudflare_dns': '1.1.1.1', 'google_dns': '8.8.8.8',
# 'testbed': <Testbed object 'Cat8k Lab' at 0x108cfa490>})
        # Parse 'show up route bgp' command output using Genie parsers
        r1_bgp_routes = testbed.devices["cat8k-rt1"].parse("show ip route bgp")
        r2_bgp_routes = testbed.devices["cat8k-rt2"].parse("show ip route bgp")

# Capture the number of BGP routes in the routing table using the Genie Dq library
        self.r1_route_count = (
        (len(Dq(r1_bgp_routes).contains("routes").get_values("route")))
        )
```

```
            self.r2_route_count = (
            (len(Dq(r2_bgp_routes).contains("routes").get_values("route")))
            )

            # Confirm number of BGP routes equals the expected number of routes
            # provided as a class variable in the datafile
            if self.r1_route_count == self.expected_routes:
                self.passed(
                "There were the correct number of expected BGP routes on router 1."
                )
            else:
                self.failed(f"Router 1 does not have the expected number of BGP \
                routes ({self.expected_routes}). Instead, there are \
                {self.r1_route_count} BGP routes.")
            if self.r2_route_count == self.expected_routes:
                self.passed("There were the correct number of expected BGP routes \
                on router 2.")
            else:
                self.failed(f"Router 2 does not have the expected number of BGP \
                routes ({self.expected_routes}). Instead, there are \
                {self.r2_route_count} BGP routes.")

class ExternalConnectivity(aetest.Testcase):
"""Test external connectivity by pinging external DNS servers (Google and
Cloudflare) using pyATS device Ping API.
There are no pass/fail conditions in this testcase, as the goal is to
illustrate the use of datafile input parameters. All tests will pass.
"""

    @aetest.test
    def ping_cloudflare_dns(self, testbed):
        """Ping Cloudflare DNS servers"""

# Print all TestScript-level parameters - you'll notice the BGPTestcase
# parameters are not included, as they are Testcase-level parameters
# You'll also notice the addition of the 'alt_google_dns' parameter, as that
# is a Testcase-level parameter
        print(self.parameters)
# Example Output
# ParameterMap({'alt_google_dns': '8.8.4.4'}, {'cloudflare_dns': '1.1.1.1',
```

```
# 'google_dns': '8.8.8.8', 'testbed': <Testbed object 'Cat8k Lab' at
# 0x10a485310>})

# Use 'cloudflare_dns' TestScript-level parameter to ping Cloudflare DNS servers
# (found in base.yaml)
        testbed.devices["cat8k-rt1"].api.ping(self.parameters["cloudflare_dns"])


    @aetest.test
    def ping_google_dns(self, testbed):
        """Ping Google DNS servers"""

# Use 'google_dns' TestScript-level parameter and 'alt_google_dns' Testcase-level
# parameter to ping Google DNS servers (both found in datafile.yaml)
        testbed.devices["cat8k-rt1"].api.ping(self.parameters["google_dns"])
        testbed.devices["cat8k-rt1"].api.ping(self.parameters["alt_google_dns"])

class CommonCleanup(aetest.CommonCleanup):
    @aetest.subsection
    def disconnect_from_devices(self, testbed):
        """Disconnect from all devices"""
        testbed.disconnect()
        logger.info(f"Disconnected from all devices in {testbed.name}")

if __name__ == "__main__":

    from pyats.topology.loader import load

    # Load testbed object from testbed file
    tb = load("../testbed.yaml")
    # Run with standalone execution
    aetest.main(datafile="datafile.yaml", testbed=tb)
```

The datafile in Example 4-14 shows a few different values that can be changed, including the testcase UID, groups, test parameters, and class-level variables. The class variables are accessible via the **self** keyword within the respective testcase. For example, in the **BGPTestcase** class definitions, you can access **the expected_routes** class variable via **self. expected_routes**. It's important to verify the number of expected routes being received from BGP because if there are not enough or too many BGP routes being received, it can lead to abnormal routing in your network. If you're a service provider, it can be more detrimental to your business, as this can lead to outages across multiple customers.

Datafiles are extremely powerful and allow pyATS testscripts to be more modular and dynamic in nature without altering any testscript code. The only requirement is to be able to read and update a YAML-based file. Being able to hand over the keys to the testing framework to the engineer testing their changes increases adaptability and confidence of test-driven network automation.

Looping Sections

AEtest provides the ability to loop over test sections with different parameters for each loop iteration. This is another feature of AEtest, along with datafiles, that allows the testing infrastructure to be dynamic. You can reuse test section code without having to edit the code. Only certain test sections can be looped: subsections within **CommonSetup/CommonCleanup** and **Testcases** and test sections within **Testcases**.

Defining Loops

Sections that are decorated with the **@aetest.loop** decorator are marked for looping. The looping parameters are provided as decorator arguments. During runtime, if a test section is marked for looping, an instance of the test section is created for each loop iteration. As a convenience, you may also use the following decorators on the subsection and test sections, respectively: **@aetest.subsection.loop** and **@aetest.test.loop**. These decorators essentially combine the two decorators you normally would have to mark each section with—**@aetest.{subsection | test}** and **@aetest.loop**.

Loop Parameters

Looping over a test section is only useful if different test parameters are provided. These parameters are passed in as arguments to the **@aetest.loop** decorator. The test parameters are propagated to the test section as local parameters. There are two methods to providing loop parameters:

- Providing a list of parameters and another list of parameter values (uses args and argvs)

- Providing each parameter as a keyword argument and a list of the parameter values as the value to the argument

There isn't a suggested method, as both methods produce the same results, but it's up to the specific use case as to whether one method should be used over the other. Example 4-16 shows both methods being used on two different test sections.

Example 4-16   *Looping Parameters*

```
from pyats import aetest


class Testcase(aetest.Testcase):

    # Method 1 - args and argvs - the positions of each value match to its arg name
```

```
    @aetest.test.loop(args=('a', 'b', 'c'),
                      argvs=((1, 2, 3),
                             (4, 5, 6)))
    def test_one(self, a, b, c):
        print("a=%s, b=%s, c=%s" % (a, b, c))


    # Method 2 - keyword args - each argument in the lists are provided indepen-
    dently
    @aetest.test.loop(a=[1,4],
                      b=[2,5],
                      c=[3,6])
    def test_two(self, a, b, c):
        print("a=%s, b=%s, c=%s" % (a, b, c))

# OUTPUT GENERATED IF TESTCASE IS EXECUTED:
# testcase output:
#   a=1, b=2, c=3
#   a=4, b=5, c=6
#   a=1, b=2, c=3
#   a=4, b=5, c=6
#
#  SECTIONS/TESTCASES                                             RESULT
# ----------------------------------------------------------------------------
#  .
#  '-- Testcase                                                   PASSED
#      |-- test_one[a=1,b=2,c=3]                                  PASSED
#      |-- test_one[a=4,b=5,c=6]                                  PASSED
#      |-- test_two[a=1,b=2,c=3]                                  PASSED
#      '-- test_two[a=4,b=5,c=6]                                  PASSED
```

Along with test parameters, you may also pass in alternative UIDs to identify each looped section. When you're using loop parameters, the number of iterations depends on a couple different factors. If alternative UIDs are provided, the number of iterations is equal to the number of UIDs provided. If there are more loop parameter values than UIDs, the extra values are discarded. If there aren't any alternative UIDs provided, the number of iterations is equal to the number of loop parameter values.

Loop parameters can also be a callable, iterable, or generator. If the argument value is a callable, the return value from the callable is used as the loop argument value. If the argument value is an iterable or generator, only one element is used at a time for each loop iteration until the iterable or generator is exhausted. Example 4-17 shows a callable (function) and generator being used as loop parameter values.

Example 4-17 *Loop Parameters: Callable and Generator*

```python
from pyats import aetest

# callable function
def my_function():
    value = [1, 2, 3]
    print("returning %s" % value)
    return value

# generator
def my_generator():
    for i in [4, 5, 6]:
        print("generating %s" % i)
        yield i

class Testcase(aetest.Testcase):

    # creating test section with parameter "a" as a function
    # note that the function object is passed, not its values
    @aetest.test.loop(a=my_function)
    def test_one(self, a):
        print("a = %s" % a)

    # creating a test section with parameter "b" as a generator
    # note that the generator is a result of calling my_generator(), not
    # the function itself.
    @aetest.test.loop(b=my_generator())
    def test_two(self, b):
        print("b = %s" % b)

# OUTPUT GENERATED IF TESTCASE IS EXECUTED:
#    returning [1, 2, 3]
#    a = 1
#    a = 2
#    a = 3
#    generating 4
#    b = 4
#    generating 5
#    b = 5
#    generating 6
#    b = 6
```

You might notice that the callable is run and the return value is captured before the looped sections are created, while the generator is only queried before the next section needs created. This is important because the generator is only queried before each test iteration; it allows a generator to dynamically generate loop iterations based on the current test environment instead of providing one return value before test iteration.

Dynamic Looping

Up to this point, we've discussed how to statically mark different test sections for looping and provide loop parameters in a testscript, but what if we wanted to loop a section based on a runtime variable? For example, what if we only want to mark a test for looping based on a certain condition or calculated value that can only be determined during runtime. *Dynamic looping* offers the ability to mark a specific test section for looping using the **loop.mark()** function. Example 4-18 shows how you can mark a test section for looping in the setup section of a testcase.

Example 4-18   *Dynamic Looping*

```
from pyats import aetest

class Testcase(aetest.Testcase):

    @aetest.setup
    def setup(self):
        # mark the next test for looping
        # provide it with two unique test uids.
        # (self.simple_test is the next test method)
        aetest.loop.mark(self.simple_test, uids=["test_one", "test_two"])

    # note: the simple_test section is not directly marked for looping
    # instead, during runtime, its testcase's setup section marks it for
    # looping dynamically.

    @aetest.test
    def simple_test(self, section):
        # print the current section uid
        # by using the internal parameter "section"
        print("current section: %s" % section.uid)

# OUTPUT GENERATED IF TESTCASE IS EXECUTED:
#   current section: test_one
#   current section: test_two
#
#   SECTIONS/TESTCASES                                             RESULT
#   -------------------------------------------------------------------------
#   .
#   '-- Testcase                                                   PASSED
#       |-- setup                                                  PASSED
#       |-- test_one                                               PASSED
#       '-- test_two                                               PASSED
```

The **loop.mark()** function is identical to the **@aetest.loop** decorator, with the exception that the first argument must be the target test section/class. For example, to mark a BGP testcase that uses different ASNs for each loop iteration, you would use the following syntax in a preceding class or section:

```
loop.mark(BGPTestcase, asn=[65000, 65001, 65002])
```

Running Testscripts

Now time for what you've been waiting for... running a testscript! Testscripts can be run using one of two execution methods: Standalone or Easypy execution. Before reviewing each execution method, let's dive into the AEtest Standard Arguments and how arguments are parsed and propagated from the command line.

Testing Arguments

Test arguments provide a way to supplement and influence the execution of your testscript. AEtest has a set of standard arguments, called Standard Arguments, along with the ability to accept arguments from the command line when running a testscript. In the following sections, you'll see the Standard Arguments provided by AEtest and how we can use the Python argparse standard library module to parse command-line arguments and propagate them to individual test sections.

Standard Arguments

AEtest has a number of standard arguments, referred to as Standard Arguments, used to influence/change testscript execution. Standard Arguments can be provided as command-line arguments or keyword arguments to **aetest.main()** in Standalone execution or **easypy.run()** in Easypy execution. Table 4-5 shows the available AEtest Standard Arguments.

Table 4-5   *AEtest Standard Arguments*

Keyword	Command Line	Description
n/a	-help	Display help information
uids	-uids	Specify the list of section UIDs to run (logic expression)
groups	-groups	Specify the list of testcase groups to run (logic expression)
datafile	-datafile	Input datafile/value for this script
random	-random	Flag to enable testcase randomization
random_seed	-random_seed	Testcase randomization seed
max_failures	-max_failures	Max acceptable number of failures
Pdb	-pdb	Start interactive debugger on failure

Keyword	Command Line	Description
step_debug	-step_debug	Step debug input file
pause_on	-pause_on	Pause on phrase input string/file
loglevel	-loglevel	AEtest logging level
submitter	-submitter	Submitter of this script (defaults to current user)

You may remember the **datafile** argument from the previous "Datafile Input" section. Datafiles are provided to a testscript as a standard argument. If you recall, datafiles provide dynamic test parameters and updates to test sections, which influences the behavior and execution of testscripts. That is the overall goal of Standard Arguments—to influence the execution of testscripts.

Argument Propagation

AEtest parses and propagates all command-line arguments using the Python standard library argparse module. The argparse module makes it easy to write command-line interfaces in Python. Using the argparse module, AEtest parses the argument values stored in sys.argv, which is a list of command-line arguments passed to a Python script. You may be wondering, what if you pass in only Standard Arguments? How are they parsed? Here's the process of parsing command-line arguments:

- All standard arguments are parsed and removed from the sys.argv list.

- All unknown arguments, which are arguments that aren't part of the Standard Arguments, are parsed by sys.argv and the argparse module.

Argument propagation allows users to pass in additional arguments to the testscript via the command line, but a custom argument parser must be created to use those arguments in test sections. The custom argument parser can simply use **argparse.ArgumentParser** in a Python script to parse known arguments passed to the script. We will look at examples in each execution method section.

Execution Environments

AEtest can execute testscripts using one of two methods: Standalone execution or Easypy execution. Standalone execution is meant for testing scripts and rapid development, while Easypy execution is meant for production script where proper reporting and log archiving are required. In the following sections, we will go over the different execution methods and how they are run.

Standalone Execution

Standalone execution is meant to be used during script development and allows the user to have full control of the execution environment, including logging and reporting. All logging is redirected to standard output (stdout) and standard error (stderr). Reporting is handled by the Standalone Reporter, which tracks results and prints a summary at the end of testing to standard output (stdout). Many of the examples in this chapter have shown results from Standalone execution. No TaskLog, result report, or archives are generated during Standalone execution.

Testscripts are run as standalone when one of the two following methods is used to execute the script:

- Directly calling **aetest.main()** within a user script

- Indirectly calling **aetest.main()** by invoking Python's **__main__** mechanism

Example 4-19 shows a testscript executed as standalone, and Example 4-20 shows the accompanying results printed to standard output (stdout).

Example 4-19   *Standalone Execution*

```
import logging
from pyats import aetest

class CommonSetup(aetest.CommonSetup):
    # Subsection 1
    @aetest.subsection
    def subsection_one(self):
        pass


    # Subsection 2
    @aetest.subsection
    def subsection_two(self):
        pass

class Testcase(aetest.Testcase):
    # Test 1
    @aetest.test
    def test_one(self):
        pass


    # Test 2
    @aetest.test
    def test_two(self):
        pass
```

```
    # Test 3
    @aetest.test
    def test_three(self):
        pass

# add the following as the absolute last block in your testscript
if __name__ == '__main__':

    # control the environment
    # e.g., change some log levels for debugging
    logging.getLogger(__name__).setLevel(logging.DEBUG)
    logging.getLogger('pyats.aetest').setLevel(logging.DEBUG)

    # aetest.main() api starts the testscript execution.
    # defaults to aetest.main(testable = '__main__')
    aetest.main()
```

Example 4-20 *Standalone Execution Results*

```
    +------------------------------------------------------------------------+
    |                         Starting common setup
    |
    +------------------------------------------------------------------------+
    +------------------------------------------------------------------------+
    |                     Starting subsection subsection_one
    |
    +------------------------------------------------------------------------+
    The result of subsection subsection_one is => PASSED
    +------------------------------------------------------------------------+
    |                     Starting subsection subsection_two
    |
    +------------------------------------------------------------------------+
    The result of subsection subsection_two is => PASSED
    The result of common setup is => PASSED
    +------------------------------------------------------------------------+
    |                         Starting testcase Testcase
    |
    +------------------------------------------------------------------------+
    +------------------------------------------------------------------------+
    |                         Starting section test_one
    |
```

```
+------------------------------------------------------------------------+
The result of section test_one is => PASSED
+------------------------------------------------------------------------+
|                          Starting section test_two
|
+------------------------------------------------------------------------+
The result of section test_two is => PASSED
+------------------------------------------------------------------------+
|                          Starting section test_three
|
+------------------------------------------------------------------------+
The result of section test_three is => PASSED
The result of testcase Testcase is => PASSED
+------------------------------------------------------------------------+
|                          Detailed Results
|
+------------------------------------------------------------------------+
 SECTIONS/TESTCASES                                               RESULT
 -----------------------------------------------------------------------
    .
 |-- common_setup                                                 PASSED
 |   |-- subsection_one                                           PASSED
 |   '-- subsection_two                                           PASSED
 '-- Testcase                                                     PASSED
     |-- test_one                                                 PASSED
     |-- test_two                                                 PASSED
     '-- test_three                                               PASSED
+------------------------------------------------------------------------+
|                          Summary
|
+------------------------------------------------------------------------+
 Number of ABORTED                                                     0
 Number of BLOCKED                                                     0
 Number of ERRORED                                                     0
 Number of FAILED                                                      0
 Number of PASSED                                                      2
 Number of PASSX                                                       0
 Number of SKIPPED                                                     0
 -----------------------------------------------------------------------
```

The **aetest.main()** function provides the entry point and is what starts the script execution. Standard Arguments can be passed to **aetest.main()** as keyword arguments. Any other unknown keyword arguments are propagated as script arguments. If any unknown keyword arguments are passed as command-line arguments, you'll need to create a custom argument parser. This might sound like a lot, but you may use the argparse module to create an **ArgumentParser** object, add arguments, parse the arguments, and add them to **aetest.main()** as keyword arguments. Example 4-21 shows how to pass two command-line arguments, **testbed** and **vlan**, as keyword arguments to **aetest.main()**, which in turn makes them testscript parameters.

Example 4-21   *Standalone Execution—Input Arguments*

```
from pyats import aetest

class Testcase(aetest.Testcase):

    # defining a test that prints out the current parameters
    # in order to demonstrate argument passing to parameters
    @aetest.test
    def test(self):
        print('Parameters = ', self.parameters)

# do the parsing within the __main__ block,
# and pass the parsed arguments to aetest.main()
if __name__ == '__main__':

    # local imports under __main__ section
    # this is done here because we don't want to pollute the namespace
    # when the script isn't run under standalone
    import sys
    import argparse
    from pyats import topology

    # creating our own parser to parse script arguments
    parser = argparse.ArgumentParser(description = "standalone parser")
    parser.add_argument('--testbed', dest = 'testbed',
                        type = topology.loader.load)
    parser.add_argument('--vlan', dest = 'vlan', type = int)

    # do the parsing
    # always use parse_known_args, as aetest needs to parse any
    # remainder arguments that this parser does not understand
    args, sys.argv[1:] = parser.parse_known_args(sys.argv[1:])
```

```
    # and pass all arguments to aetest.main() as kwargs
    aetest.main(testbed = args.testbed, vlan = args.vlan)



# Let's run this script with the following command
#    example_script.py --testbed /path/to/my/testbed.yaml --vlan 50


# output of the script:
#
#    +----------------------------------------------------------------------+
#    |                        Starting testcase Testcase                     |
#.   |                                                                       |
#    +----------------------------------------------------------------------+
#    +----------------------------------------------------------------------+
#    |                          Starting section test                       |
#.   |                                                                       |
#    +----------------------------------------------------------------------+
#    Parameters = {'testbed': <Testbed object at 0xf717578c>, 'vlan': 50})
#    The result of section test is => PASSED
#    The result of testcase Testcase is => PASSED
```

Standalone execution provides the user ultimate control and is great when going through the trial-and-error process of writing code. However, what do we do if we want to run our testscripts in production and require proper logging and reporting? Easypy execution provides the answer.

Easypy Execution

Easypy execution is used when testscripts are executed with the Easypy runtime environment. With this execution method, the Easypy runtime environment controls the environment and provides the following features:

- Multiple testscripts can be run together in a job file.

- Logging configuration is done by Easypy.

- TaskLog, result reporting, and archives are generated.

- Reporter is used for reporting and result tracking as well as generating a YAML result file, results details file, and a summary XML file.

Example 4-22 shows an Easypy job file running two testscripts. Each testscript that's run within a job file is called a *task*.

Example 4-22   *Easypy Execution*

```
from pyats.easypy import run

# job file needs to have a main() definition
# which is the primary entry point for starting job files
def main():

    # run testscript 1
    run(testscript='/path/to/your/script1.py')

    # run testscript 2
    run(testscript='/path/to/your/script2.py')
```

To run the Easypy job, you must run **pyats run job** *jobfile-name***.py --testbed-file** */path/to/***testbed.yaml** from the terminal. The **--testbed-file** loads the testbed as a testbed object and is propagated to the testscript as a script argument named **testbed**.

If no **--testbed-file** is passed to **pyats run job**, the **testbed** argument is set to **None**.

In addition to the **--testbed-file** option, all AEtest Standard Arguments are accepted as keyword arguments and propagated to the testscript as script arguments. Any unknown keyword arguments provided to **easypy.run()** are also propagated to the testscript as script arguments. Example 4-23 shows how unknown keyword arguments are propagated to the testscript as script arguments.

Example 4-23   *Easypy Execution—Script Arguments*

```
from pyats.easpy import run

def main():
    run(
        testscript="standalone_exec_input_args.py",
        pyats_is_awesome=True,
        aetest_is_legendary=True
    )

# Run the easypy job
# pyats run job easypy_script_args.py --testbed ../testbed.yaml


#    +------------------------------------------------------------------------+
#    |                         Starting testcase Testcase
#    |
#    +------------------------------------------------------------------------+
```

```
#     +----------------------------------------------------------------+
#     |                      Starting section test                     |
#.    |                                                                |
#     +----------------------------------------------------------------+
#     Parameters = {'testbed': <Testbed object at 0xf742f74c>,
#                   'pyats_is_awesome': True,
#                   'aetest_is_legendary': True}
#     The result of section test is => PASSED
#     The result of testcase Testcase is => PASSED
```

Along with having the ability to run multiple testscripts in a single job, another major benefit of Easypy is the standardization of logging, reporting, and archiving. After a job file is run, a zipped archive folder is created in the user home directory under .pyats/archive/YY-MM (~/.pyats/archive/YY-MM). You can specify archives to not be created by specifying the **–no-archive** option. Table 4-6 shows a list of the files generated by Easypy job files.

Table 4-6 *Easypy Job Files*

Filename	Purpose
<job-name>.py	Copy of the jobfile that ran.
<job-name>.report	Copy of the email notification sent to the submitter. This report looks very much like the Standalone Reporter results.
TaskLog.*<task-id>*	One TaskLog is generated per jobfile task and is where all messages generated in the task are stored.
JobLog.*<job-name>*	Overall pyats.easypy module log.
testbed.static.yaml	Contents of the testbed file provided by the user.
testbed.clean.yaml	Contents of the clean file provided by the user.
env.txt	A dump of environment variables and CLI arguments of the Easypy run.
reporter.log	Reporter server log file. Contains a trace of XML-RPC calls.
results.json	JSON result summary file generated by Reporter.
xunit.xml	xUnit-style results reports and information required by Jenkins. Only generated if **–xunit** argument is provided.
ResultsSummary.xml	XML result summary file generated by Reporter.
ResultsDetails.xml	XML result details file generated by Reporter.
CleanResultsDetails.yaml	YAML clean result details file generated by Kleenex.

Filename	Purpose
Kleenex.*<device-name>*.log	Job-scope clean details for this device.
Kleenex_*<task-id>*.*<device-name>*.log	Task-scope clean details for this device.

As you can see from Table 4-6, many different files are generated and archived from an Easypy job. The archived files can be used for additional regression and sanity testing.

Testable

A testable in AEtest is any object that can be loaded into a **TestScript** class instance by the aetest.loader module and executed without any errors. The following are acceptable as testables:

- Any path to a Python file ending with .py

- Any module name that is part of the current PYTHONPATH

- Any non-built-in module objects (instances of **types.ModuleType**)

Testables are not the same as testscripts. Testscripts run tests and generate results. Testables can be meaningless modules to AEtest, such as the urllib module. It is a valid testable but produces zero test results.

Testscript Flow Control

AEtest provides many mechanisms to control the execution flow of testscripts. Different mechanisms include skipping testcases, jumping ahead in the testscript, grouping testcases, and only executing testcases by UID.

Skip Conditions

AEtest comes with built-in preprocessors that can be used to skip test sections, sometimes based on a condition. The following decorators and functions can be used to skip a test section:

- **@aetest.skip(reason = 'message')**: Unconditionally skip the decorated section. **reason** should describe why that section is being skipped.

- **aetest.skip.affix(testcase, reason)**: Same as the skip decorator but can be used on the fly to skip other testcases depending on one testcase result.

- **@aetest.skipIf(condition, reason = 'message')**: Skip the decorated test section if **condition** is True.

- **aetest.skipIf.affix(testcase, condition, reason):** This can be used to assign the **skipIf** decorator to the testcases; **condition** can be a callable or a boolean.

- **@aetest.skipUnless(condition, reason = 'message'):** Skip the decorated test section unless **condition** is True.

- **aetest.skipUnless.affix(testcase, condition, reason):** Can be used on the fly to assign decorators to the testcases.

Example 4-24 shows some of these decorators and functions used to skip testcases.

Example 4-24   *AEtest Skip Conditions*

```
from pyats import aetest

# Custom library used for testing
class mylibrary:
    __version__ = 0.1

# skip testcase intentionally
@aetest.skip('because we had to')
class Testcase(aetest.Testcase):
    pass

class TestcaseTwo(aetest.Testcase):

    # skip test section using if library version < some number
    @aetest.skipIf(mylibrary.__version__ < 1,
                   'not supported in this library version')
    @aetest.test
    def test_one(self):
        pass

    # skip unless library version > some number
    @aetest.skipUnless(mylibrary.__version__ > 3,
                       'not supported in this library version')
    @aetest.test
    def test_two(self):
        pass

    @aetest.test
    def test_three(self):
        aetest.skip.affix(section = TestcaseTwo.test_four,
                          reason = "message")
        aetest.skipIf.affix(section = TestcaseTwo.test_five,
```

```
                                 condition = True,
                              reason = "message")
        aetest.skipUnless.affix(section = TestcaseThree,
                                condition = False,
                                reason = "message")


    @aetest.test
    def test_four(self):
        # will be skipped because of test_three
        pass


    @aetest.test
    def test_five(self):
        # will be skipped because of test_three
        pass


    @aetest.test
    def test_six(self):
        # will be skipped because of test_three
        pass

class TestcaseThree(aetest.Testcase):
    # will be skipped because of TestcaseTwo.test_three
    pass
```

Running Specific Testcases

You might want to run only specific testcases. To do that, you can specify a testcase UID (or UIDs) as a Standard Argument when running the script or by setting a runtime variable (**runtime.uids**) dynamically during execution. The **uids** argument accepts a callable (function) that returns a truthy value. The list of test section UIDs present in the testscript are passed as arguments to the callable. If the callable returns True, the respective test section is run. Logic testing can also be used to evaluate test section UIDs. The running section UIDs are also accessible via the **runtime.uids** variable during runtime. Runtime variables are only accessible during runtime, so the UIDs will have to be dynamically set in the testscript versus passed in as a Standard Argument. Example 4-25 shows how a callable can be used to determine whether a UID should be run.

Example 4-25 *Running Specific Testcases*

```
from pyats.easypy import run


# function determining whether we should run testcase_A
# currently executing uids is always a list of:
# [ <container uid>, <section uid>]
# e.g., ['common_setup', 'subsection_one']
# thus varargs (using *) is required for the function input.
def run_only_testcase_one(*uids):
    # check that we are running TestcaseOne
    return "TestcaseOne" in uids


# run only TestcaseOne and its contents (using callable)
# executing uids has TestcaseOne:
def main():
    run("example_script.py", uids=run_only_testcase_one)
```

Testcase Grouping

Testcase grouping allows you to tag testcases that are similar in nature and may be run together by adding them to a *group*. By default, testcases do not belong to any groups. You may add testcases to groups by adding them in the testscript itself by assigning a list of groups to a group variable within the testcase. Testcases can also be grouped by specifying groups in datafile input that is provided as a Standard Argument. Example 4-26 shows an example of assigning a group named "traffic" to Testcase One and "sanity" to a Testcase Two.

Example 4-26 *Testcase Grouping*

```
from pyats import aetest

class TestcaseOne(aetest.Testcase):
    """Testcase One"""

    groups = ["traffic"]

    <TestcaseOne tests...>


class TestcaseTwo(aetest.Testcase):
    """Testcase Two"""

    groups = ["sanity"]

    <TestcaseTwo tests...>
```

Once the testcases are grouped, you can specify which testcase groups run using Standard Arguments (**--group**) or using the **runtime.groups** variable dynamically in your testscript. Just like how you specify which testcases to run using their UID, you pass a callable to the **groups** argument to determine whether the group(s) should run. The callable accepts a list of each of the testcase's group values and will return True if the testcase group(s) should run. Logic testing can also be used to evaluate whether a group value should run. Groups can also be evaluated at runtime using the **runtime.groups** variable within the testscript. The **runtime.groups** variable is dynamically set by performing logic testing. Example 4-27 shows how to filter certain testcase groups using a callable in Standard Arguments and also dynamically using the runtime variable.

Example 4-27   *Testcase Group Filtering*

```
from pyats.easypy import run
# import the logic objects
from pyats.datastructures.logic import And, Not


# create a function that tests for testcase groups
# this api tests that a testcase belongs to sanity but not traffic.
# note that varargs (using *) is required, as the list of groups to each
# testcase is unknown.
def non_traffic_sanities(*groups):
    # Runs testcases in "sanity" group and not in "traffic" group
    return "sanity" in groups and "traffic" not in groups


# Runs the testscript as two tasks using different logic testing
def main(runtime):
    ### Using function testing to evaluate testcase groups ###
    # Only runs Testcase Two
    run(testscript="example_script.py", runtime=runtime, groups=non_traffic_sani-
ties)

    ### Using logic testing to evaluate testcase groups ###
    # Only runs Testcase One
    run("example_script.py", groups=And("sanity", Not("traffic")))
```

Must-Pass Testcases

If there are testcases that must pass during testing, AEtest allows you to set a class attribute called **must_pass** to True. If a must-pass testcase fails during testing, the testscript will immediately jump to the Common Cleanup section, using the **goto** statement, and block any remaining testcases. The **goto** statement was touched on earlier in the chapter, but to recap, it allows you to jump to another section within a testscript. The **goto** target must be further in the testscript—you can't go back to a previously executed section. The available targets include the testcase's cleanup section (**cleanup**), the next testcase (**next_tc**), the Common Cleanup section (**common_cleanup**), or exiting the testscript completely (**exit**). Example 4-28 shows how to use the **goto** statement directly, and Example 4-29 shows how to set a testcase as "must pass" and what happens if that testcase fails. You'll notice the following testcase is blocked and the testscript jumps to the Common Cleanup section using the **goto** statement under the hood.

Example 4-28 *The goto Statement*

```
from pyats import aetest

class CommonSetup(aetest.CommonSetup):
    @aetest.subsection
    def subsection(self):
        # goto with a message
        self.errored('setup error, abandoning script', goto = ['exit'])


# ---------------------------------------------------------------------
class TestcaseOne(aetest.Testcase):
    @aetest.setup
    def setup(self):
        # setup failed, go to cleanup of testcase
        self.failed('test failed', goto = ['cleanup'])


# ---------------------------------------------------------------------
class TestcaseTwo(aetest.Testcase):
    # test failed, move onto next testcase
    @aetest.test
    def test(self):
        self.failed(goto = ['next_tc'])


# ---------------------------------------------------------------------
class TestcaseThree(aetest.Testcase):
    @aetest.setup
    def setup(self):
        # setup failed, move onto cleanup of this testcase, then
        # jump to common_cleanup directly.
        self.failed(goto=['cleanup','common_cleanup'])
```

Example 4-29 *The Must-Pass Testcase*

```
from pyats import aetest

class TestcaseOne(aetest.Testcase):

    must_pass = True

    @aetest.test
    def test(self):
        self.failed('boom!')

class TestcaseTwo(aetest.Testcase):
    pass

class CommonCleanup(aetest.CommonCleanup):

    @aetest.subsection
    def subsection(self):
        pass


# output result
#
#   SECTIONS/TESTCASES                                          RESULT
#   -----------------------------------------------------------------------
#   .
#   |-- TestcaseOne                                             FAILED
#   |    '-- test                                               FAILED
#   |-- TestcaseTwo                                             BLOCKED
#   '-- common_cleanup                                          PASSED
#        '-- subsection                                         PASSED
```

Testcase Randomization

By default, AEtest runs the Common Setup section first, then each testcase in the order they are defined, and finally wraps up with the Common Cleanup section. Testcase execution can be randomized by setting the random standard argument to True (**random=True**). Common Setup and Common Cleanup are not randomized and will always be executed first and last, respectively. Example 4-30 shows a basic example of randomizing testcases.

Example 4-30   *Testcase Randomization*

```
from pyats import aetest

# define a couple testcases
class TestcaseOne(aetest.Testcase):
    pass


class TestcaseTwo(aetest.Testcase):
    pass


class TestcaseThree(aetest.Testcase):
    pass


if __name__ == "__main__":
    aetest.main(random = True)


# output result
#
#  SECTIONS/TESTCASES                                                    RESULT
#  -------------------------------------------------------------------------
#  .
#  |-- TestcaseTwo                                                       PASSED
#  |-- TestcaseOne                                                       PASSED
#  '-- TestcaseThree                                                     PASSED
```

Maximum Failures

Let's say you are testing many network features and have a long-running testscript. By default, AEtest will run each testcase sequentially and record the respective result. However, if testcases begin to fail, wouldn't you want the ability to stop testing and figure out what's going on without waiting for the testscript to finish executing? AEtest provides a method to set a maximum threshold for testcase failures during a testscript run. The **max_failures** standard argument can provide the number of testcase failures before aborting the rest of the testcases and jumping to the Common Cleanup section of the testscript. The **goto** statement is used once again to jump to the Common Cleanup section. Example 4-31 shows how when one testcase fails, the testscript blocks execution of the other testcases and jumps to the Common Cleanup section before exiting.

Example 4-31 *Maximum Failures*

```
from pyats import aetest

class TestcaseOne(aetest.Testcase):

    @aetest.test
    def test(self):
        self.failed()

class TestcaseTwo(aetest.Testcase):

    @aetest.test
    def test(self):
        self.failed()

class TestcaseThree(aetest.Testcase):
    pass

class CommonCleanup(aetest.CommonCleanup):
    pass

# set max failure to 1 and run the testscript
if __name__ == "__main__":
    aetest.main(max_failures = 1)

# output result
#
# Max failure reached: aborting script execution
#
#  SECTIONS/TESTCASES                                               RESULT
# -------------------------------------------------------------------------
#  .
#  |-- TestcaseOne                                                  FAILED
#  |-- TestcaseTwo                                                  BLOCKED
#  |-- TestcaseThree                                                BLOCKED
#  '-- common_cleanup                                               PASSED
```

Custom Testcase Discovery

Customizing testcase discovery is an advanced topic but should be covered at a high level. Testcases are discovered using the **ScriptDiscovery** class in the discover module of the AEtest package. You can customize the testcase discovery process at the following levels: script discovery, testcase discovery, and common discovery. The **ScriptDiscovery** class finds the testcases within a testscript, the **TestDiscovery** class finds the test sections (setup, test, cleanup) within a testcase, and the **CommonDiscovery** class finds subsections within the common sections (**CommonSetup** and **CommonCleanup**). To override the discovery process at each level, you can create a new class that inherits the respective default discovery class. The new discovery class must have specific methods to enable the custom discovery logic. The **runtime.discoverer** properties can be configured in the testscript to use the new discovery classes instead of the default classes. Along with custom discovery, you can also customize the ordering of sections. If you would like to customize the discovery or ordering of sections, it's recommended that you reference the pyATS documentation for more details.

Reporting

AEtest provides reporting of testscript results, including which tests ran during testing and their associated results. The format and level of details in the report depend on the execution mode used for testing (Standalone or Easypy execution). In the following sections, you'll see the different report options available and dive into the reporting details of each execution mode.

Standalone Reporter

The Standalone Reporter is used when testscripts are run directly from the command-line using Standalone execution (via **aetest.main()**). Testcase, section, and step results are printed to standard output (stdout) in a tree-like format. All examples in this chapter, and most examples in this book, use the Standalone Reporter to showcase testscript results. Example 4-32 shows testscript results presented by the Standalone Reporter.

Example 4-32   *Standalone Reporter*

```
+------------------------------------------------------------------------+
|                          Detailed Results                              |
+------------------------------------------------------------------------+
  SECTIONS/TESTCASES                                              RESULT
  ----------------------------------------------------------------------
  .
|-- common_setup                                                  PASSED
|    |-- sample_subsection_1                                      PASSED
|    '-- sample_subsection_2                                      PASSED
|-- tc_one                                                        PASSED
|    |-- prepare_testcase                                         PASSED
|    |-- simple_test_1                                            PASSED
|    |-- simple_test_2                                            PASSED
|    '-- clean_testcase                                           PASSED
```

```
|-- TestcaseWithSteps                                          ERRORED
|   |-- setup                                                  PASSED
|   |   |-- Step 1: this is a description of the step          PASSED
|   |   '-- Step 2: another step                               PASSED
|   |-- step_continue_on_failure_and_assertions                FAILED
|   |   |-- Step 1: assertion errors -> Failed                 FAILED
|   |   '-- Step 2: allowed to continue executing              FAILED
|   |-- steps_errors_exits_immediately                         ERRORED
|   |   '-- Step 1: exceptions causes all steps to skip over   ERRORED
|   '-- steps_with_child_steps                                 PASSED
|       |-- Step 1: test step one                              PASSED
|       |-- Step 1.1: substep one                              PASSED
|       |-- Step 1.1.1: subsubstep one                         PASSED
|       |-- Step 1.1.1.1: subsubsubstep one                    PASSED
|       |-- Step 1.1.1.1.1: running out of indentation         PASSED
|       |-- Step 1.1.1.1.1.1: definitely gone too far...       PASSED
|       |-- Step 1.2: substep two                              PASSED
|       |-- Step 2: test step two                              PASSED
|       |-- Step 2.1: function step one                        PASSED
|       |-- Step 2.2: function step two                        PASSED
|       '-- Step 2.3: function step three                      PASSED
'-- common_cleanup                                             PASSED
    '-- clean_everything                                       PASSED
```

AEtest Reporter

The AEtest Reporter, known as the Reporter, is used when testscripts are executed via Easpy execution mode. The Reporter creates a package of test result artifacts. It contains information such as the section hierarchy, section results, and even the amount of time each section took during testing. The main files in the package are results.json and results.yaml, which contains hierarchical information about the job. The top-level is TestSuite, which contains high-level information about the entire job. Under TestSuite is the Task level. The Task level represents each testscript that is executed in the job. If you remember, in an Easypy job, multiple testscripts can be executed. Each testscript executed is called a task. As you can imagine, below each Task are the different container classes—Common Setup, Testcase, and Common Cleanup. Following the AEtest section hierarchy, each container class has child sections, including SetupSection, TestSection, CleanupSection, and Subsection. Optionally, these child sections can contain steps represented as Step. Each level to the report has information relevant to that section. Example 4-33 shows the different fields for each level represented in the report.

Example 4-33   *AEtest Report Structure*

```
+---------------+-----------------+-------------------------------------------+
| Section       | Field           | Description                               |
+===============+=================+===========================================+
| TestSuite     | type            | Identifier that this section is the root  |
|               |                 |  TestSuite                                |
|               | id              | Unique ID for this job execution          |
|               | name            | Name from jobfile                         |
|               | starttime       | Timestamp when job execution began        |
|               | stoptime        | Timestamp when job execution ended        |
|               | runtime         | Duration of execution                     |
|               | cli             | Command that started Easypy               |
|               | jobfile         | Location of jobfile                       |
|               | jobfile_hash    | SHA256 hash of the jobfile contents       |
|               | pyatspath       | Python environment executing pyATS        |
|               | pyatsversion    | Version of pyATS installed                |
|               | host            | Name of host machine                      |
|               | submitter       | User that started execution               |
|               | archivefile     | Path to generated archive file            |
|               | summary         | Combined summary of all Tasks             |
|               | details         | Details about any exceptions or errors    |
|               | extra           | Map of extra info about the TestSuite     |
|               | tasks           | List of child Tasks                       |
+---------------+-----------------+-------------------------------------------+
| Task          | type            | Identifier that this section is a Task    |
|               | id              | Unique ID for this Task                   |
|               | name            | Name of TestScript                        |
|               | starttime       | Timestamp when execution began            |
|               | stoptime        | Timestamp when execution ended            |
|               | runtime         | Duration of execution                     |
|               | description     | Description of TestScript                 |
|               | logfile         | Path to logfile for this Task             |
|               | testscript      | Path to testscript                        |
|               | testscript_hash | SHA256 hash of the testscript contents    |
|               | datafile        | Path to the data file                     |
|               | datafile_hash   | SHA256 hash of the data file contents     |
|               | parameters      | Any parameters passed to this Task        |
|               | summary         | Summary of results                        |
|               | details         | Details about any exceptions or errors    |
|               | extra           | Map of extra info about the Task          |
|               | sections        | List of child Sections                    |
+---------------+-----------------+-------------------------------------------+
```

```
| Section        | type            | Specific type of section being           |
|                |                 |  represented                             |
|                | id              | Unique ID for this Section               |
|                | name            | Name of this Section                     |
|                | starttime       | Timestamp when this Section began        |
|                | stoptime        | Timestamp when this Section ended        |
|                | runtime         | Duration of execution                    |
|                | description     | Description of this Section              |
|                | xref            | XReference to the code defining this     |
|                |                 |  Section                                 |
|                |   source_hash   | SHA256 hash of the source code for       |
|                |                 |  this Section                            |
|                | data_hash       | SHA256 hash of the data file input       |
|                | logs            | Path to logfile showing execution of     |
|                |                 |  this Section, as well as the beginning  |
|                |                 |  byte and size in bytes                  |
|                | parameters      | Any parameters passed to this Section    |
|                | processors      | Lists of processors that ran for this    |
|                |                 |  section, both before and after          |
|                | result          | The test result of this Section          |
|                | details         | Details about any exceptions or errors   |
|                | extra           | Map of extra info about this Section     |
|                | sections        | Any child sections of this section       |
|                |                 |  (Testcases have TestSections, which can |
|                |                 |  have Steps, etc.)                       |
+----------------+-----------------+------------------------------------------+
```

All levels below the TestSuite level are considered *sections*, as the information gathered from each level is about the same. The **type** identifies the section type. If there are any unique differences between sections, they will be saved under the **extras** key. Example 4-34 shows an abbreviated results.yaml file that shows the different levels to the report—TestSuite, Task, and the Common Setup section of the first task. The section types are highlighted to help identify the different levels.

Example 4-34 *results.yaml*

```
version: '2'
report:
  type: TestSuite
  id: example_job.2019Sep19_19:56:06.569499
  name: example_job
  starttime: 2019-09-19 19:56:07.603283
```

```
stoptime: 2019-09-19 19:56:19.951458
runtime: 12.35
cli: pyats run job job/example_job.py --testbed-file etc/example_testbed.yaml
  --no-mail
jobfile: /Users/user/examples/comprehensive/job/example_job.py
jobfile_hash: 2a452a8683f4f5e5c146d62c78a9a5253198e19c3fb6c8c1771bdf0eea622086
pyatspath: /Users/user/env
pyatsversion: '19.11'
host: HOSTNAME
submitter: user
archivefile: /Users/user/env/users/user/archive/
19-09/example_job.2019Sep19_19:56:06.569499.zip
summary:
  passed: 13
  passx: 0
  failed: 1
  errored: 12
  aborted: 0
  blocked: 4
  skipped: 0
  total: 30
  success_rate: 43.33
extra:
  testbed: example_testbed
tasks:
  - type: Task
    id: Task-1
    name: base_example
    starttime: 2019-09-19 19:56:08.432390
    stoptime: 2019-09-19 19:56:08.617640
    runtime: 0.19
    description: |+
      base_example.py


      This is a comprehensive example base script that walks users through AEtest
      infrastructure features, what they are for, how they are used, how it
impacts
      their testing, etc.


    logfile: TaskLog.Task-1
    testscript: /Users/user/examples/comprehensive/base_example.py
    testscript_hash:
```

```
2938f2d2efbf9be144a9fe68667dd1c12753b84017a56e7d04caefe46edc0602
parameters:
  labels: {}
  links: []
  parameter_A: jobfile value A
  routers: []
  testbed: <pyats.topology.testbed.Testbed object at 0x106da92b0>
  tgns: []
summary:
  passed: 3
  passx: 0
  failed: 0
  errored: 3
  aborted: 0
  blocked: 0
  skipped: 0
  total: 6
  success_rate: 50.0
sections:
- type: CommonSetup
  id: common_setup
  name: common_setup
  starttime: 2019-09-19 19:56:08.434411
  stoptime: 2019-09-19 19:56:08.458939
  runtime: 0.02
  description: |+
    Common Setup Section

        This is the docstring for your common setup section.
        Users should document the number of common setup subsections
        so that by reading this block of comments, it gives a generic
        feeling as to how CommonSetup is built and run.

  xref:
    file: /Users/user/examples/comprehensive/base_example.py
    line: 191
    source_hash:
    c366a269e45838deb9bed54d28fef648b921c4f19a1753fc1e46e4c9ba3f9264
  logs:
    begin: 0
    file: TaskLog.Task-1
    size: 4317
  parameters:
    labels: {}
```

```
        links: []
        parameter_A: jobfile value A
        parameter_B: value B
        routers: []
        testbed: <pyats.topology.testbed.Testbed object at 0x105ea92b0>
        tgns: []
    result:
        value: passed
```

Along with the results.yaml and results.json files, the Reporter also generates XML files named ResultsDetails.xml and ResultsSummary.xml for the aggregated results. The AEtest Reporter also provides the ability to subscribe to live result updates. The Reporter uses a Unix socket client/server model to collect information about each section during the job run, allowing the Reporter Client to subscribe to the Reporter Server for live updates on runtime details of each section. The subscribe functionality only works as an async function, and you should be familiar with the Python asyncio library (https://docs.python.org/3/library/asyncio.html) before proceeding with testing this feature. The asyncio library is part of the Python standard library and is used to write concurrent code using the async/await syntax and is well-suited for I/O-bound tasks. The client subscribes to the server and runs a callback each time event data is received. Table 4-7 shows the different values that can be extracted from event data.

Table 4-7   *Reporter Event Data*

Key	Description
event	A string to specify what kind of event occurred, such as **start_task** or **stop_section**
type	The type of section that triggered this event
seq_num	A unique number specific to the section triggering this event
parent_seq_num	The **seq_num** of the parent section if there is one
id	The ID of the related section
name	The name of the related section
starttime	A timestamp for when the section started
stoptime	A timestamp for when the section ended
runtime	How many seconds the section ran for
result	The result of the section
logfile	The name of the log file

Key	Description
logs	A mapping of log file name and offset of the relevant section of logs
xref	The location of this section in the script

The last interesting piece of information the Reporter collects and adds to the report package is git information. Git information, including the repo, file, branch, and commit hash, are added to the report. This can be helpful for regression testing. Let's say the testsuite, or part of the testsuite, broke and you want to quickly figure out when it last worked. By recording the git information captured by the Reporter, including the commit hash, you can quickly identify the last commit when the test worked.

The reporting features in the AEtest test infrastructure provide options to quickly review test results with the Standalone Reporter or to a complete reporting package provided by the AEtest Reporter. The data points, metrics, and other rich data that can be extracted from the AEtest Reporter reporting package provide endless options for further data analysis and visualization of the Easypy job results. It's really up to you on how you want to utilize the captured results!

Debugging

As with all code, you will find yourself debugging your AEtest testscripts. Python has a debugger module as part of the standard library called Python Debugger (pdb). The pdb debugger (https://docs.python.org/3/library/pdb.html) is used to set breakpoints, step through source code, line by line, and provide other debugging functions in your Python code. The one caveat to using pdb is when multiprocessing is involved. When multiprocessing is used, child processes are forked and break the functionality of pdb. Since AEtest uses multiprocessing, most notably with Easypy execution, AEtest built pdb debugging functionality into the framework.

When running AEtest testscripts, you can pass **pdb=True** as a Standard Argument and whenever an error, failure, or exception occurs, the testing engine pauses and starts an interactive post_mortem debugging section. The post_mortem functionality is built natively into pdb. Also, pdb can be passed as a command-line flag (**--pdb**) when you're running a job via the pyATS **run job** command.

Another pdb debugging feature in AEtest is "pause on phrase," which is the ability to pause test execution based on any log messages generated by the script. The log messages include Python logs, CLI output from devices, and any other logs captured by the root logger (logging.root). The following actions are supported when a script is paused on phrase:

- **Email:** Creates a pause file and emails the user. The script may continue to run once the pause file is deleted or when the timeout limit has been reached.
- **Pdb:** Pauses and opens a pdb debugger.
- **Code:** Pauses and opens a Python interactive shell.

To enable this feature, you must pass the **pause_on** Standard Argument to the script run with a value that provides a path to a YAML pause file that follows a specific schema to define the actions to take when the script is paused. Example 4-35 shows a YAML pause file.

Example 4-35   *YAML Pause File*

```
timeout: 600         # pause a maximum of 10 minutes

patterns:
   - pattern: '.*pass.*'              # pause on all log messages including
                                      # .*pass.* in them globally

   - pattern: '.*state: down.*'   # pause whenever  'state: down' is found
     section: '^common_setup\..*$' # enable for all common_setup sections

   - pattern: '.*should pause.*'      # pause whenever 'should pause' is found
     section: '^TestcaseTwo\.setup$'  # pause on TestcaseTwo setup section
```

The default action is to email the user. When a log message matches a pattern defined in the YAML pause file, the user is notified via email with the log phrase captured and instructions on how to remove the pause and continue testing. To change the action to one of the other two supported actions (pdb or code), you simply need to specify an "action:" key with one of those values in the YAML pause file.

Summary

This chapter covered the different components that make up the AEtest test infrastructure. The AEtest test infrastructure is the core to pyATS and provides the foundation for all testing. We reviewed the structure of a testscript, which went through the different sections, including common setup, testcases, and common cleanup. The AEtest object model reviewed the Python classes that are the base classes of the different sections from the testscript structure. The base classes include the TestScript classes, container classes, function classes, TestItem classes, and TestContainer classes. After understanding the testscript structure and the base classes of the AEtest object model, we dove into the behavior of test results. Section test results can be determined automatically using assertions or manually using the different result APIs available (passed, failed, errored, skipped, blocked, aborted, and passx). AEtest allows functions or methods to run before or after testscripts using pre-processors, post-processors, or exception processors. Pre- and post-processors can be helpful for checking the environment before a test executes, validating section results, and even taking and comparing snapshots before and after a test section. Exception processors can take post-execution snapshots of the test environment if an exception is raised in a test section or execute debug commands and collect dump files when an exception occurs.

Once we reviewed the intricacies of the AEtest testscript structure, object model, and how the results are determined, we dove into the extensibility of the test infrastructure, including datafiles and test parameters. Datafiles are YAML files that can provide dynamic test parameters to a testscript, making them more robust and reusable. Datafiles can be a gamechanger and should be used when possible to avoid static test parameters. After seeing how we can make testscripts more extensible and dynamic, we reviewed how to run testscripts using Standalone execution or through Easypy. Standalone execution is recommended for development purposes with all logging outputs being sent to standard output (stdout). The Easypy execution environment is recommended for "official" test execution, running testscripts as tasks in a jobfile. The jobfile produces logs and archives and is best suited for sanity and regression testing where reporting and archiving are required.

AEtest provides many ways to control the flow of testscript execution, including running specific testcases (using a UID or a group of testcases), declaring testcases as "must pass," randomizing testcase execution, declaring a maximum number of failures per testscript execution, and even customizing testcase discovery. We then wrapped up the chapter by reviewing the reporting mechanisms and how to debug testscripts. AEtest reporting mechanisms include the Standalone Reporter, which is used with Standalone execution, and the AEtest Reporter, which is used with the Easypy execution environment and produces a results.json file that includes all the details of the test execution.

It's recommended that you continue referencing the information from this chapter as you go through the rest of the book, as many topics and features are built on the topics discussed in this chapter.

Chapter 5

pyATS Parsers

In software engineering and computer science, parsing is the mechanism of translating (and comprehending) unstructured data into a script-readable form. Parsers are the root of automation; without them, automation could not understand the device. There are multiple ways to parse the device output, with different packages, each with their own style. There also are multiple ways to communicate with the device (CLI, XML, REST, YANG, and so on), with each providing a different structure for the same information!

Imagine being able to translate unstructured CLI output into structured JSON with a simple command! This is where the true power of pyATS lies—in its parsers and models. The pyATS metaparser's role is to unify those packages into one location and one structure. It's a unified collection of parsers that works across multiple parser packages, and across multiple communication protocols, and still returns a common structure. Metaparser allows one script that works across multiple operating systems, multiple communication protocols, and multiple parsing packages.

This chapter covers the following topics:

- Vendor-agnostic automation
- pyATS **learn**
- pyATS **parse**
- Parsing at the CLI
- Parsing with Python
- Dictionary query
- Differentials

Vendor-Agnostic Automation

Cisco's pyATS framework stands out as a robust network automation and validation tool, not just for Cisco devices but for a wide array of network equipment from various vendors. This vendor-agnostic capability is largely attributed to its integration with the Genie parsing libraries. Genie, as a part of the pyATS ecosystem, provides a comprehensive set of parsers that can interpret and transform raw command outputs from different network devices into a structured data format such as JSON. The beauty of Genie lies in its extensive library that supports multiple vendors, ensuring that network professionals are not confined to a single brand or platform. By leveraging Genie's parsing libraries, pyATS offers a unified and consistent approach to network automation, irrespective of the underlying hardware or software vendor. This flexibility underscores the framework's commitment to providing scalable and adaptable solutions in an ever-evolving networking landscape. Imagine a parsing infrastructure that can do the following:

- Promote more easily maintainable platform/type/version-agnostic testing scripts by deferring operational data parsing to backend libraries.

- Harmonize parsing output among various interface categories, such as the CLI, XML, and YANG.

- Enforce only enough structure to give the script writer a consistent look and feel across interface categories. The parser helps the script writer create scripts that are consistent in terms of both style and formatting.

- Be future proof, allowing a multitude of existing and yet-to-be-imagined parsing implementations to coexist in the backend.

- Enable an elastic parsing ecosystem that is simple enough for the novice but feature-rich enough for the power user.

- Leverage the strength of the modern Python 3 language while still allowing bridging/reuse of Cisco's vast store of legacy TCL-based parsers.

While this is a Cisco Press book, and pyATS is provided by Cisco, it is rare to find a homogeneous network made up of only Cisco devices. This does not mean you cannot use pyATS for your network automation needs, as it provides support for many non-Cisco devices and even agnostic support for REST APIs from any platform. By selecting an appropriate command output *parser* for the supported operating system of choice, you can easily extend network automation to many vendors outside of Cisco. A quick look at the available parser library will demonstrate this vendor-agnostic approach to pyATS (see Figure 5-1).

```
        ALL
[IOS]   IOS
[XE]    IOSXE
[XR]    IOSXR
[NX]    NXOS
[ASA]   ASA
[LNX]   LINUX
[JUN]   JUNOS
[SR]    SROS
[BIGIP] BIGIP
[VPTL]  VIPTELA
[APIC]  APIC
[DNAC]  DNAC
[IRON]  IRONWARE
[AIR]   AIREOS
[CHE]   CHEETAH
[GAIA]  GAIA
[GEN]   GENERIC
[COM]   COMWARE
```

Figure 5-1 *Operating Systems Supported by pyATS Parsers*

pyATS learn

Provided your operating system is supported, pyATS provides platform-agnostic *learn models* that perform one or more **show** commands that are combined and structured into JSON output. Regardless of operating system, standardized structured output is returned to the user. There are 32 available learn models:

acl	arp	bgp
dot1x	eigrp	fdb
hsrp	igmp	interface

isis	l2vpn	lag
lisp	lldp	mcast
mld	msdp	nd
ntp	ospf	pim
platform	prefix_list	rip
route_policy	routing	segement_routing
static_routing	stp	vlan
vrf	vxan	

These high-level models provide abstractions for the commands they run on various platforms to collect and structure the JSON output. By clicking any model (such as BGP, OSPF, or interfaces), you can drill down into the details of the model, configuration, and operation, as illustrated in Figure 5-2.

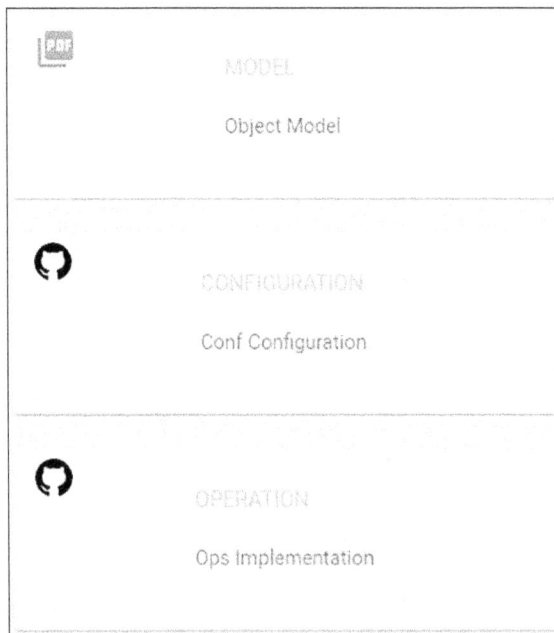

Figure 5-2   *Model Details*

The model link takes you to a PDF explaining how the model was built, including references to the related YANG models, the structure hierarchy, as well as the model's configuration and operations structure (including the **show** commands used), as illustrated in Figure 5-3 through Figure 5-6.

Interface

Created by Takashi Higashimura, last modified just a moment ago

- Referenced YANG models
  - IETF
  - OpenConfig
  - XE
  - XR
  - NX
- Structure Hierarchy
- Interface Conf Structure
- Interface Ops structure
  - show commands
  - Ops structure

Referenced YANG models

IETF

https://tools.ietf.org/html/rfc7223

OpenConfig

https://github.com/openconfig/public/blob/master/release/models/interfaces/openconfig-interfaces.yang

XE

https://github.com/YangModels/yang/blob/master/vendor/cisco/xe/1651/Cisco-IOS-XE-interfaces.yang

XR

https://github.com/YangModels/yang/blob/master/vendor/cisco/xr/621/cisco-xr-openconfig-interfaces-deviations.yang

https://github.com/YangModels/yang/blob/master/vendor/cisco/xr/621/cisco-xr-openconfig-interfaces-types.yang

Figure 5-3   *Interface Model Details—YANG References*

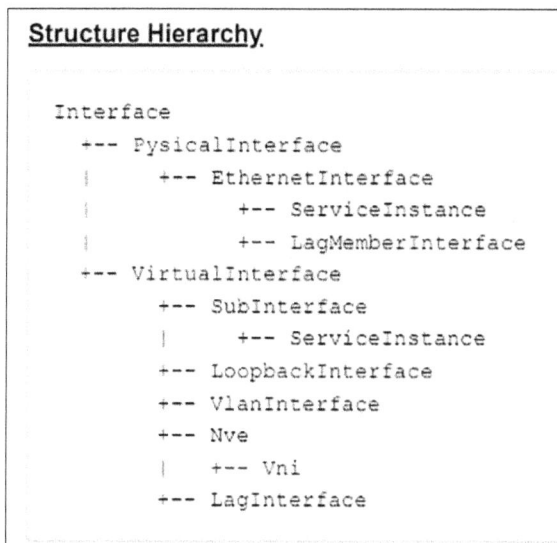

Structure Hierarchy

```
Interface
  +-- PysicalInterface
  |     +-- EthernetInterface
  |           +-- ServiceInstance
  |           +-- LagMemberInterface
  +-- VirtualInterface
        +-- SubInterface
        |     +-- ServiceInstance
        +-- LoopbackInterface
        +-- VlanInterface
        +-- Nve
        |   +-- Vni
        +-- LagInterface
```

Figure 5-4   *Interface Model Details—Structure Hierarchy*

Interface Conf Structure

	XE	XR	NX
Interface	<interface> (config)# interface <interface> (config-if)#	<interface>,<l2transport> : Bool (config)# interface <interface> [l2transport] (config-if)#	<interface> (config)# interface <interface> (config-if)#

Figure 5-5 *Interface Model Details—Config Structure*

Interface Ops structure

show commands

IOS-XE	IOS-XR	NX-OS
show interfaces	show interfaces detail	show interface
show vrf detail	show vlan interface	show vrf all interface
show ip interface	show vrf all detail	show ip interface vrf all
show ipv6 interface	show ipv4 vrf all interface	show ipv6 interface vrf all
show interface switchport	show ipv6 vrf all interface	show interface switchport
show etherchannel summary	show bundle	show routing ipv6 vrf all
show interfaces [intf] accounting	show interfaces [intf] accounting	show routing vrf all

Figure 5-6 *Interface Model Details—show Commands*

The configuration link from the model details takes you to the GitHub repository that contains the source code for the model (see Figure 5-7). If you need to look at the actual model code, you can review by operating system in this GitHub repository. The GitHub repository can be found at https://pubhub.devnetcloud.com/media/genie-feature-browser/docs/_models/interface.pdf.

Drilling down into the IOS XE interface.py file, you can see the actual code used, and you can even contribute to enhance the code, as it's open source, to transform the learn interface model into structured JSON, as illustrated in Figure 5-8.

Finally, the operation details take you to a different GitHub repository—this time for the Genie operations model. Figure 5-9 illustrates the corresponding IOS XE Genie operations model for interfaces. Here's the link to this GitHub repository: https://github.com/CiscoTestAutomation/genielibs/blob/master/pkgs/conf-pkg/src/genie/libs/conf/interface/iosxe/interface.py.

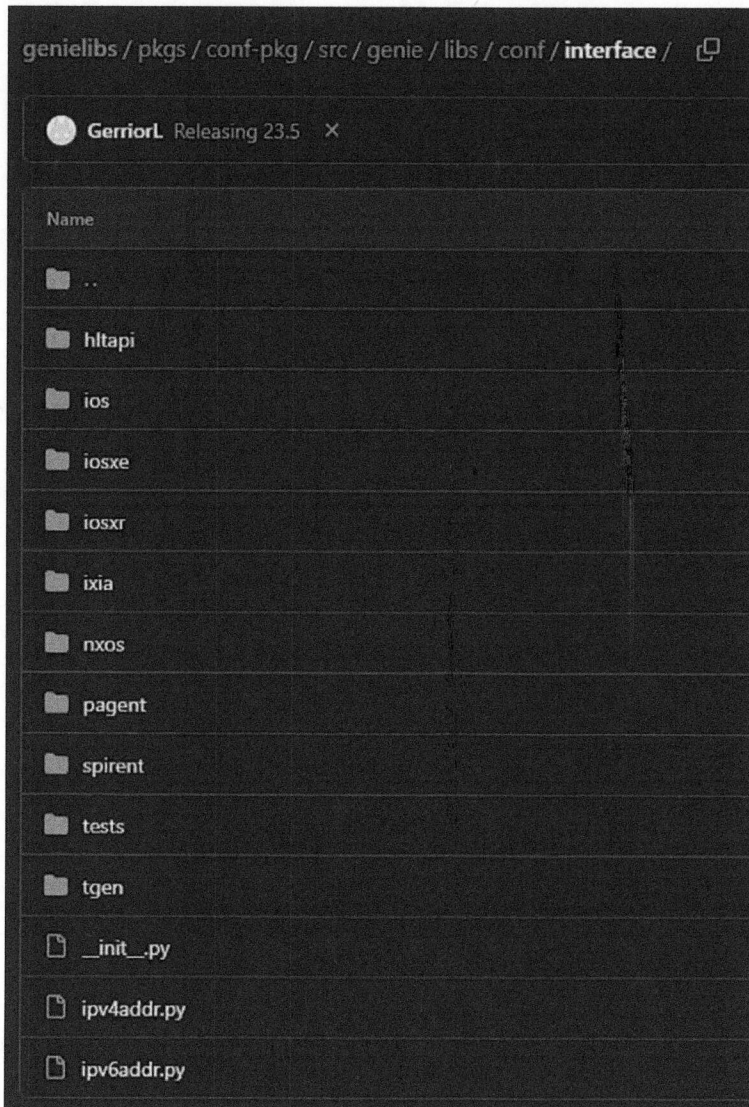

Figure 5-7 *Interface Model Configuration—Git Repository*

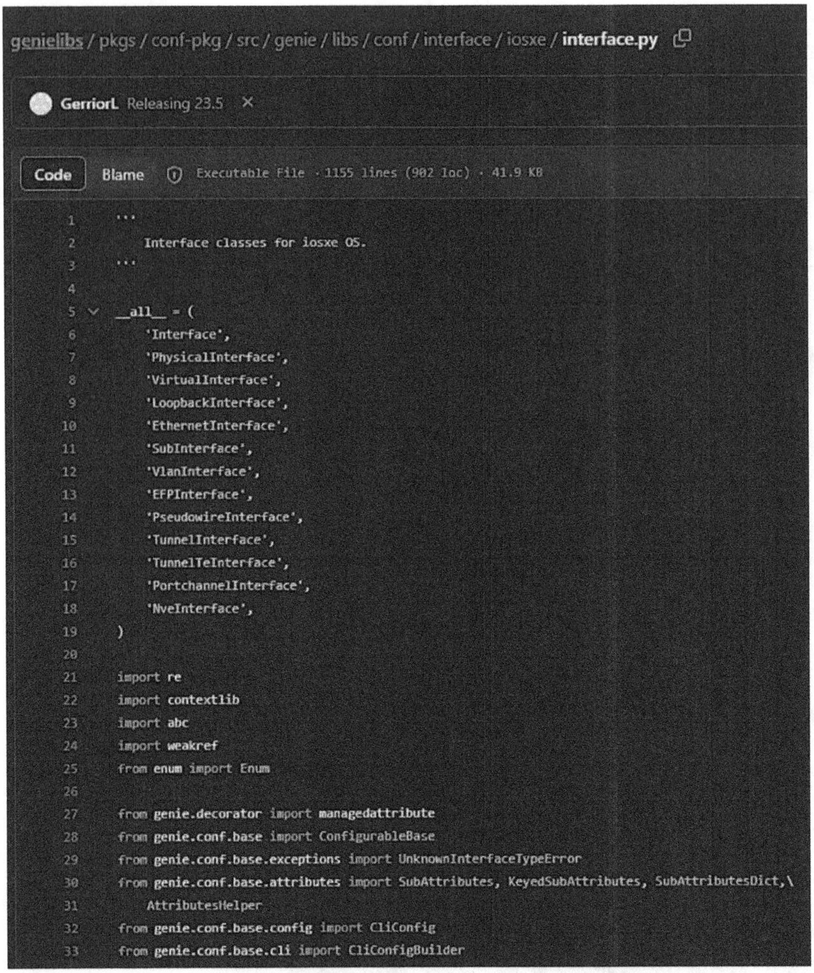

```
genielibs / pkgs / conf-pkg / src / genie / libs / conf / interface / iosxe / interface.py

    GerriorL  Releasing 23.5   ×

    Code    Blame    ⓘ  Executable File · 1155 lines (902 loc) · 41.9 KB

    1      ...
    2          Interface classes for iosxe OS.
    3      ...
    4
    5  ∨  __all__ = (
    6          'Interface',
    7          'PhysicalInterface',
    8          'VirtualInterface',
    9          'LoopbackInterface',
   10          'EthernetInterface',
   11          'SubInterface',
   12          'VlanInterface',
   13          'EFPInterface',
   14          'PseudowireInterface',
   15          'TunnelInterface',
   16          'TunnelTeInterface',
   17          'PortchannelInterface',
   18          'NveInterface',
   19      )
   20
   21      import re
   22      import contextlib
   23      import abc
   24      import weakref
   25      from enum import Enum
   26
   27      from genie.decorator import managedattribute
   28      from genie.conf.base import ConfigurableBase
   29      from genie.conf.base.exceptions import UnknownInterfaceTypeError
   30      from genie.conf.base.attributes import SubAttributes, KeyedSubAttributes, SubAttributesDict,\
   31          AttributesHelper
   32      from genie.conf.base.config import CliConfig
   33      from genie.conf.base.cli import CliConfigBuilder
```

Figure 5-8 *Interface Model Configuration—IOS XE interface.py*

In the realm of network automation and validation, pyATS has emerged as a beacon of adaptability and efficiency. Its "learn" feature, which is platform-agnostic, epitomizes the modern approach to network operations. Rather than being tethered to specific vendors or architectures, pyATS embraces a holistic model, ensuring that engineers and network professionals can seamlessly gather and analyze data across diverse network environments. This platform-neutral stance not only future-proofs network operations but also fosters an inclusive ecosystem where innovation isn't stifled by proprietary constraints. As networks continue to evolve and diversify, tools like pyATS, with their agnostic models, will be pivotal in ensuring that automation and validation remain consistent, efficient, and universally applicable. Figure 5-10 summarizes the pyATS **learn** model parsing.

```
genielibs / pkgs / ops-pkg / src / genie / libs / ops / interface / iosxe / interface.py  ⊡

⬤  GerriorL  Releasing 23.5   ✕

[ Code ]   Blame    ⓘ 349 lines (281 loc) · 14.9 KB

    1      ...
    2      Interface Genie Ops Object for IOSXE - CLI.
    3      ...
    4
    5
    6      # super class
    7      from genie.libs.ops.interface.interface import Interface as SuperInterface
    8
    9      # commands
   10      show_vrf = "show vrf"
   11      show_interfaces = "show interfaces"
   12      show_ip_interface = "show ip interface"
   13      show_ipv6_interface = "show ipv6 interface"
   14      show_interfaces_accounting = "show interfaces accounting"
   15
   16
   17  ∨  class Interface(SuperInterface):
   18          '''Interface Genie Ops Object'''
   19
   20  ∨      def learn(self, custom=None, interface=None, vrf=None, address_family=None):
   21              '''Learn Interface Ops'''
   22              ##############################################################################
   23              #                              info
   24              ##############################################################################
   25              # Global source
   26              src = '[(?P<interface>.*)]'
   27              dest = 'info[(?P<interface>.*)]'
   28              req_keys = ['[description]', '[type]', '[oper_status]',
   29                          '[phys_address]', '[port_speed]', '[mtu]',
   30                          '[enabled]', '[bandwidth]', '[flow_control]',
   31                          '[mac_address]', '[auto_negotiate]', '[port_channel]',
   32                          '[duplex_mode]', '[medium]', '[delay]']
   33
```

Figure 5-9 *Interface Model Configuration—Genie Ops for IOS XE interface.py*

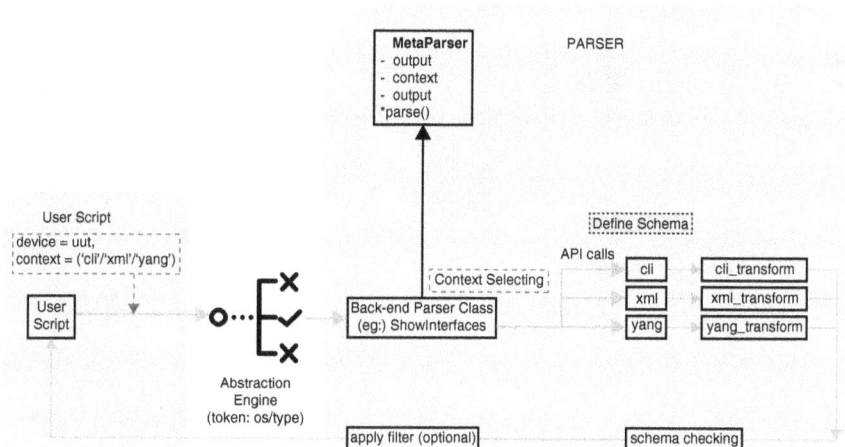

Figure 5-10 *pyATS learn Process*

pyATS Parsers

The learn models are abstractions that do not require any specific knowledge of underlying platform commands, but what if you know the command you want to transform into structured data? pyATS *parsers* provide this exact functionality. There are *thousands* of parsers available—over 4500 when this book was written!

Using the Genie documentation, you can filter by both operating system and command to find the appropriate parser. This will return exact matches and suggested matches, as illustrated in Figure 5-11.

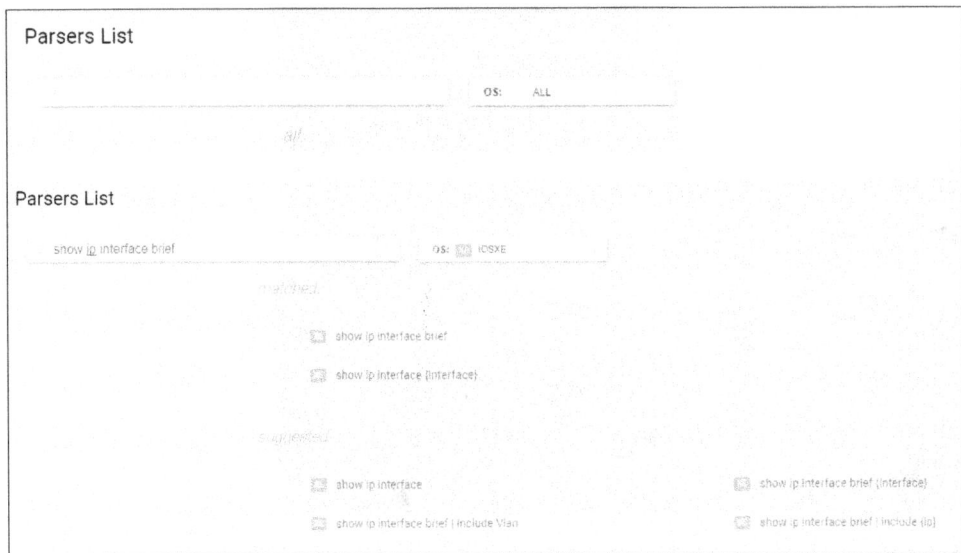

Figure 5-11 *The pyATS Parser Filter and the pyATS Parser Filter Applied*

You can click the results to see the structure of the JSON, including which fields are mandatory or optional, that the parser will return to the user (see Figure 5-12).

Should you want to view the source code on GitHub, you can click the View Source button in the right margin, as illustrated in Figure 5-13. Figure 5-14 shows an example of the results.

```
show ip interface brief

   <  Back



        Parser for: show ip interface brief


    Schema


        {
        'interface': {
         Any (str) *: {
           Optional (str) vlan_id: {
             Optional (Any) Any (str) *: {
               'ip_address': <class 'str'>,
               Optional (str) interface_is_ok: <class 'str'>,
               Optional (str) method: <class 'str'>,
               Optional (str) status: <class 'str'>,
               Optional (str) protocol: <class 'str'>,
               },
             },
           Optional (str) ip_address: <class 'str'>,
           Optional (str) interface_is_ok: <class 'str'>,
           Optional (str) method: <class 'str'>,
           Optional (str) status: <class 'str'>,
           Optional (str) protocol: <class 'str'>,
           },
         },
        }
```

Figure 5-12 *pyATS parser Details*

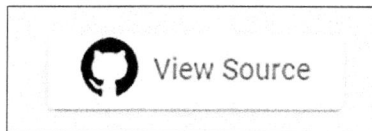

Figure 5-13 *pyATS parser View Source*

```
genieparser / src / genie / libs / parser / ios / show_interface.py

Code   Blame   ⓘ  Executable File · 155 lines (118 loc) · 4.74 KB

51      class ShowInterfaces(ShowInterfaces_iosxe):
57                          'input_queue_drops', 'out_interface_resets',
58                          'rxload', 'txload', 'last_clear', 'in_crc_errors',
59                          'in_errors', 'in_giants', 'unnumbered', 'mac_address',
60                          'phys_address', 'out_lost_carrier', '(Tunnel.*)',
61                          'input_queue_flushes', 'reliability', 'in_runts']
62
63          pass
64
65
66      class ShowIpInterfaceBrief(ShowIpInterfaceBrief_iosxe):
67          """Parser for: show ip interface brief"""
68          exclude = ['method', '(Tunnel.*)']
69          pass
70
```

Figure 5-14 *pyATS Parser Source*

The parsed output from either the learn model or command parsers is foundational to network automation with pyATS. The structured JSON means that, unlike with unstructured raw command-line output, Python can interact with it, transform it, and perform tests against it. The strength of pyATS as a network automation tool is significantly amplified by its robust parsing capabilities. These parsers, integral to the framework, transform the often verbose and unstructured output from network devices into a coherent, structured format, making data interpretation and subsequent automation tasks more efficient and error-free. Instead of manually sifting through lines of device output, engineers can leverage pyATS parsers to quickly extract the necessary information, streamlining their workflows. As the complexity of networks grows and the demand for swift, accurate automation escalates, the role of pyATS parsers becomes even more critical. They stand as a testament to the tool's commitment to simplifying and enhancing the network automation journey for professionals across the industry.

Parsing at the CLI

After installing pyATS and creating a valid testbed, you can immediately start using parsers from the command-line interface (CLI). Using an integrated development environment (IDE) that has an integrated Linux terminal, such as Visual Studio Code (VS Code), network engineers can rapidly adopt pyATS without writing a single line of code! Contrast this approach, using pyATS parsers and models from the CLI, against your current practices of gathering information from the network. Typically, network engineers do the following:

- Launch an SSH connection tool such as PuTTY.

- Set up session logging.

- Input the connection information or find a saved session.

- Connect and authenticate.

- Execute commands.

- Scrape the screen or open output file.

- Analyze raw device output.

Figure 5-15 illustrates how you can use the built in Terminal in VS Code to create a Python virtual environment and then install pyATS. Figure 5-16 demonstrates how you can then use the pyATS CLI to learn about a device's interfaces.

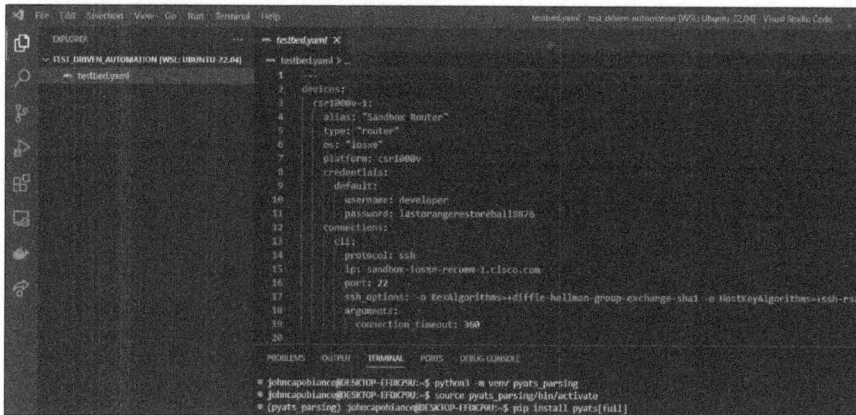

Figure 5-15 *VS Code Terminal Setup pyATS Environment*

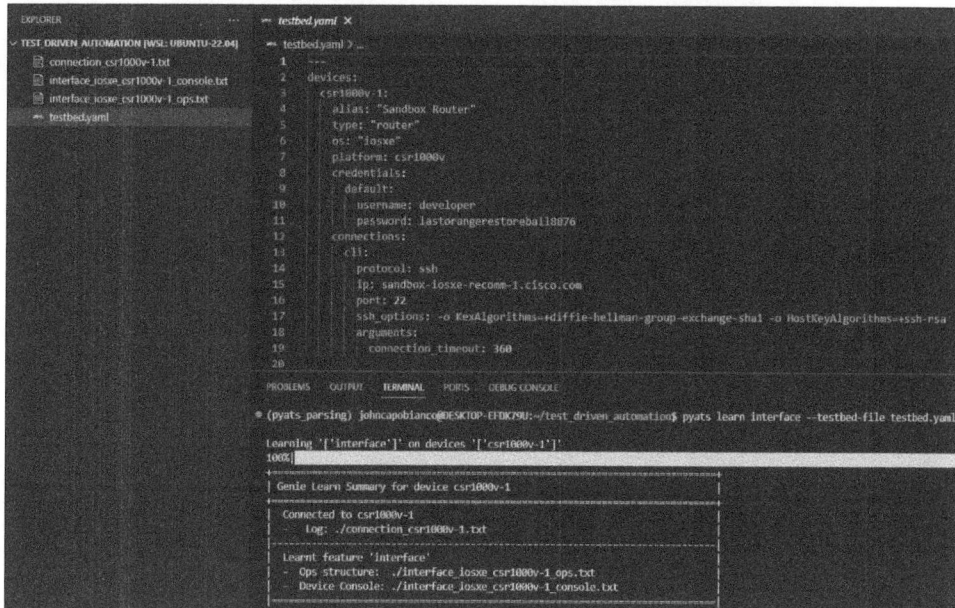

Figure 5-16 *VS Code Terminal with pyATS learn Command Example*

Contrast this with a simple one-time pyATS environment setup: a valid testbed object, followed by one-line commands in your terminal to *automatically* transform the state of the device into structured data. This works at scale, and you could learn *all devices* inside a testbed or pass the name of the single device you might want to filter on. As you might notice in Figure 5-16, the **learn** command creates three output files:

- connection_<hostname>.txt

- <model>_<operating system>_<hostname>_console.txt

- <model>_<operating system>_<hostname>_ops.txt

The connection file contains information about the initial connection. This is a good file to investigate if you have any connection issues if your command fails. There is some important information in this file, such as the commands pyATS has performed upon initial connection. This testbed is configured for the default initial connection commands, which you can see in the connection log. These commands can be suppressed in the testbed if you do not want to alter the device settings on initial connection. Example 5-1 demonstrates a successful connection file that pyATS saves automatically.

Example 5-1   *Example of a Successful Connection File*

```
2023-08-20 14:03:06,027: %UNICON-INFO: +++ csr1000v-1 logfile ./connection_
csr1000v-1.txt +++
2023-08-20 14:03:06,027: %UNICON-INFO: +++ Unicon plugin iosxe/csr1000v
(unicon.plugins.iosxe.csr1000v) +++
Welcome to the DevNet Always-On IOS XE Sandbox!
2023-08-20 14:03:06,681: %UNICON-INFO: +++ connection to spawn: ssh -1 developer
131.226.217.149 -p 22 -o KexAlgorithms=+diffie-hellman-group-exchange-sha1 -o
HostKeyAlgorithms=+ssh-rsa, id: 139886857701408 +++
2023-08-20 14:03:06,681: %UNICON-INFO: connection to csr1000v-1
(developer@131.226.217.149) Password:
csr1000v-1#
2023-08-20 14:03:07,074: %UNICON-INFO: +++ initializing handle +++
2023-08-20 14:03:07,140: %UNICON-INFO: +++ csr1000v-1 with via 'cli': executing
command 'term length 0' +++
term length 0
csr1000v-1#
2023-08-20 14:03:07,512: %UNICON-INFO: +++ csr1000v-1 with via 'cli': executing
command 'term width 0' +++
term width 0
csr1000v-1#
2023-08-20 14:03:07,787: %UNICON-INFO: +++ csr1000v-1 with via 'cli': executing
command 'show version' +++
show version
Cisco IOS XE Software, Version 16.09.03
```

```
Cisco IOS Software [Fuji], Virtual XE Software (X86_64_LINUX_IOSD-UNIVERSALK9-M),
Version 16.9.3, RELEASE SOFTWARE (fc2)
Technical Support: http://www.cisco.com/techsupport
Copyright  1986-2019 by Cisco Systems, Inc.
Compiled Wed 20-Mar-19 07:56 by mcpre
```

The console output is self-descriptive. Example 5-2 shows the raw console output of the session. This is similar to PuTTY session logging information.

Example 5-2 *Example of Console Output*

```
+++ csr1000v-1 with via 'cli': executing command 'show vrf' +++
show vrf
  Name                         Default RD            Protocols   Interfaces
  AAA                          <not set>             ipv4,ipv6
  CISCO                        <not set>                         Gi3
  CISCO2                       <not set>             ipv4,ipv6
  default1                     <not set>             ipv4,ipv6
csr1000v-1#
+++ csr1000v-1 with via 'cli': executing command 'show interfaces' +++
show interfaces
GigabitEthernet1 is up, line protocol is up
  Hardware is CSR vNIC, address is 0050.56bf.9379 (bia 0050.56bf.9379)
  Description: MANAGEMENT INTERFACE - DON'T TOUCH ME
  Internet address is 10.10.20.48/24
csr1000v-1#
+++ csr1000v-1 with via 'cli': executing command 'show interfaces accounting' +++
show interfaces accounting
GigabitEthernet1 MANAGEMENT INTERFACE - DON'T TOUCH ME
              Protocol    Pkts In    Chars In   Pkts Out   Chars Out
                 Other         80        4800         31        1860
                    IP     680109   102362922     587982   135048968
                   ARP         80        4800         31        1860
GigabitEthernet2 Wizkid wuz here
              Protocol    Pkts In    Chars In   Pkts Out   Chars Out
                 Other          0           0          5         300
                    IP         98        8362         48        4196
                   ARP          0           0          5         300
Interface GigabitEthernet3 is disabled
csr1000v-1#
+++ csr1000v-1 with via 'cli': executing command 'show ip interface' +++
show ip interface
```

```
GigabitEthernet1 is up, line protocol is up
  Internet address is 10.10.20.48/24
  Broadcast address is 255.255.255.255
csr1000v-1#
+++ csr1000v-1 with via 'cli': executing command 'show ipv6 interface' +++
show ipv6 interface
csr1000v-1#
Could not learn <class 'genie.libs.parser.iosxe.show_interface.ShowIpv6Interface'>
Show Command: show ipv6 interface
Parser Output is empty
+================================================================================
=================================================================+
| Commands for learning feature 'Interface'
|
+================================================================================
=================================================================+
| - Parsed commands
|
|--------------------------------------------------------------------------------
-----------------------------------------------------------------|
|   cmd: <class 'genie.libs.parser.iosxe.show_vrf.ShowVrf'>, arguments: {'vrf':''}
|
|   cmd: <class 'genie.libs.parser.iosxe.show_interface.ShowInterfaces'>,
arguments: {'interface':''}                                                      |
|   cmd: <class 'genie.libs.parser.iosxe.show_interface.ShowInterfacesAccounting'>,
arguments: {'interface':''}                                                      |
|   cmd: <class 'genie.libs.parser.iosxe.show_interface.ShowIpInterface'>,
arguments: {'interface':''}                                                      |
|================================================================================
=================================================================|
| - Commands with empty output
|
|--------------------------------------------------------------------------------
-----------------------------------------------------------------|
|   cmd: <class 'genie.libs.parser.iosxe.show_interface.ShowIpv6Interface'>,
arguments: {'interface':''}                                                      |
|================================================================================
=================================================================|
```

As you can see, pyATS models are capturing the output of several commands related, in this example, to interfaces. This alone is already more efficient and valuable than a single session running these commands and capturing them manually. Imagine gathering this information from an entire topology; it's rare to need data from a single device in the network. Next, Example 5-3 shows the structured data that pyATS is assembling from the raw output in Example 5-2.

Example 5-3 *Example of ops Output*

```
{
  "_exclude": [
    "in_discards",
    "in_octets",
    "in_pkts",
    "last_clear",
    "out_octets",
    "out_pkts",
    "in_rate",
    "out_rate",
    "in_errors",
    "in_crc_errors",
    "in_rate_pkts",
    "out_rate_pkts",
    "in_broadcast_pkts",
    "out_broadcast_pkts",
    "in_multicast_pkts",
    "out_multicast_pkts",
    "in_unicast_pkts",
    "out_unicast_pkts",
    "last_change",
    "mac_address",
    "phys_address",
    "((t|T)unnel.*)",
    "(Null.*)",
    "chars_out",
    "chars_in",
    "pkts_out",
    "pkts_in",
    "mgmt0"
  ],
  "attributes": null,
  "commands": null,
  "connections": null,
  "context_manager": {},
  "info": {
    "GigabitEthernet1": {
      "accounting": {
        "arp": {
          "chars_in": 4800,
          "chars_out": 1860,
          "pkts_in": 80,
          "pkts_out": 31
        },
```

```
    "ip": {
      "chars_in": 102362922,
      "chars_out": 135048968,
      "pkts_in": 680109,
      "pkts_out": 587982
    },
    "other": {
      "chars_in": 4800,
      "chars_out": 1860,
      "pkts_in": 80,
      "pkts_out": 31
    }
  },
  "auto_negotiate": true,
  "bandwidth": 1000000,
  "counters": {
    "in_broadcast_pkts": 0,
    "in_crc_errors": 0,
    "in_errors": 0,
    "in_mac_pause_frames": 0,
    "in_multicast_pkts": 0,
    "in_octets": 102366433,
    "in_pkts": 680174,
    "last_clear": "never",
    "out_errors": 0,
    "out_mac_pause_frames": 0,
    "out_octets": 135051811,
    "out_pkts": 588020,
    "rate": {
      "in_rate": 1000,
      "in_rate_pkts": 1,
      "load_interval": 300,
      "out_rate": 1000,
      "out_rate_pkts": 1
    }
  },
  "delay": 10,
  "description": "MANAGEMENT INTERFACE - DON'T TOUCH ME",
  "duplex_mode": "full",
  "enabled": true,
  "encapsulation": {
    "encapsulation": "arpa"
  },
```

```
    "flow_control": {
      "receive": false,
      "send": false
    },
    "ipv4": {
      "10.10.20.48/24": {
        "ip": "10.10.20.48",
        "prefix_length": "24",
        "secondary": false
      }
    },
    "mac_address": "0050.56bf.9379",
    "mtu": 1500,
    "oper_status": "up",
    "phys_address": "0050.56bf.9379",
    "port_channel": {
      "port_channel_member": false
    },
    "port_speed": "1000mbps",
    "switchport_enable": false,
    "type": "CSR vNIC"
  },
  "Loopback500": {...
  },
  "Nve1": {...
  },
  "VirtualPortGroup0": {...
  },
  "raw_data": false
}
```

Inside of VS Code, this .txt file's natural JSON structure allows for interactivity, such as collapsing or expanding individual interfaces. The information most engineers want from the ops.txt file is nested inside the **.info** parent key. This is important when you move on to start using these models Pythonically.

For help with pyATS **learn**, you can issue the following command at the CLI:

```
$ pyats learn --help
```

Example 5-4 shows the results of issuing this command.

Example 5-4   *pyATS learn Console Help*

```
(pyats_parsing) johncapobianco@Desktop:~/test_driven_automation$ pyats learn --help
Usage:
  pyats learn [commands] [options]


Example
-------
  pyats learn ospf --testbed-file /path/to/testbed.yaml
  pyats learn ospf --testbed-file /path/to/testbed.yaml \\
                   --output features_snapshots/ --devices "nxos-osv-1"
  pyats learn ospf config interface bgp platform \\
  --testbed-file /path/to/testbed.yaml --output features_snapshots/


Description:
  Learn device feature and parse into Python datastructures


      List of available features:
      https://pubhub.devnetcloud.com/media/genie-feature-browser/docs/#/models


Learn Options:
  ops                  List of Feature to learn,
                       comma separated, ospf,bgp
                   all can instead be
                       provided to learn all features
  --testbed-file TESTBED_FILE
                       specify testbed_file yaml
  --devices [DEVICES ...]
                       "DEVICE_1 DEVICE_2", spaceseparated,
                       if not provided it will learn on all
                       devices (Optional)
  --output OUTPUT      Which directory to store logs, by default it will be
current directory
                       (Optional)
  --single-process     Learn one device at the time
                       instead of in parallel (Optional)
 --via [VIA ...]       List of connection to use per device "nxos-1:ssh"
  --archive-dir        Directory to store a .zip of the output
  --learn-hostname     Learn the device hostname during connection (Optional)
  --learn-os           Learn the device OS during connection (Optional)


General Options:
  -h, --help           Show help
  -v, --verbose        Give more output, additive up to 3 times.
  -q, --quiet          Give less output, additive up to 3 times,
                       corresponding to WARNING, ERROR,
                       and CRITICAL logging levels
```

If you want to parse a specific command against your testbed instead of learning from the pyATS models at the CLI, simply run the **parse** command followed by the supported command and your testbed object, as demonstrated in Figure 5-17.

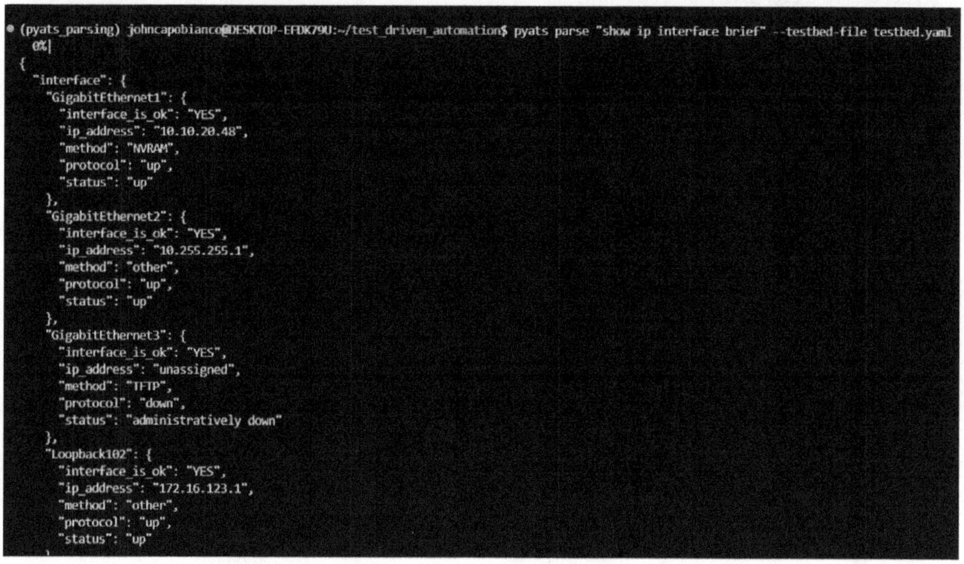

Figure 5-17 *VS Code Terminal with pyATS parse Command Example*

Unlike **learn**, pyATS **parse** immediately returns the JSON representation of the output from the **show** command parsed. There are no files created, and pyATS simply prints the JSON to the screen, as Example 5-5 demonstrates.

Example 5-5 *pyATS parse Console Help*

```
(pyats_parsing) johncapobianco@Desktop:~/test_driven_automation$ pyats parse --help
Usage:
  pyats parse [commands] [options]


Example
-------
  pyats parse "show interfaces" --testbed-file /path/to/testbed.yaml --devices uut
  pyats parse "show interfaces" --testbed-file \\
  /path/to/testbed.yaml --devices uut --output my_parsed_output/
  pyats parse "show interfaces" "show version" \\
  --testbed-file /path/to/testbed.yaml --devices helper


Description:
  Parse CLI commands into Pythonic datastructures
```

```
Parse Options:
  COMMANDS                  Show command(s) to parse,
                            all can instead be provided to parse all commands
  --testbed-file TESTBED_FILE
                            specify testbed_file yaml
  --devices [DEVICES ...]
                            Devices to issue commands to
  --output OUTPUT           Directory to store output files to.
                            When not provided, prints parsed JSON
                            output to screen. (Optional)
  --via [VIA ...]           List of connection to use per device "nxos-1:ssh"
  --fuzzy                   Enable fuzzy matching for commands
  --raw                     Store device output without parsing it
  --timeout TIMEOUT         Devices execution timeout
  --developer               Parser colored developer mode
  --archive-dir             Directory to store a .zip of the output
  --learn-hostname          Learn the device hostname during connection (Optional)
  --learn-os                Learn the device OS during connection (Optional)
  --rest                    run rest commands


General Options:
  -h, --help                Show help
  -v, --verbose             Give more output, additive up to 3 times.
  -q, --quiet               Give less output, additive up to 3 times,
                            corresponding to WARNING, ERROR,
                            and CRITICAL logging levels
```

As you can see from the help, you can add the **--output** flag and specify the JSON file to send the output of the parsed command. The traditional methods of network operations often involve manual configurations, tedious verifications, and error-prone troubleshooting. These methods not only consume a significant amount of time but also introduce potential risks to network stability.

With pyATS's parse feature, network engineers can effortlessly convert raw CLI outputs into structured data. This structured data can then be easily analyzed, compared, and integrated into other systems or tools. An evolution is taking place—a move away from manually sifting through pages of CLI output to find that one piece of information. With pyATS, it's all about efficiency and precision.

The learn feature, on the other hand, takes automation a step further. Instead of just parsing output, it understands the network's state and can provide insights into various operational aspects. Whether it's understanding interface states, routing tables, or device health, the learn functionality offers a comprehensive view without the manual labor.

In essence, pyATS's parse and learn capabilities represent a paradigm shift in how we approach network operations. Because these processes are automated, this not only saves us time and reduces errors but also frees up network engineers to focus on more strategic tasks, driving innovation and ensuring network resilience.

Parsing with Python

Both the learn and parse features are available not only at the Terminal CLI but also as Python functions we can use to easily transform raw output to JSON inside a script or pyATS job. It is essential to recognize that these capabilities are not confined to the CLI alone. In fact, pyATS offers a rich Python library that allows for the integration of **learn** and **parse** directly into Python scripts and applications. This means that network professionals can seamlessly incorporate these functionalities into more complex automation workflows and custom applications. They can even integrate them with other Python-based tools. By leveraging pyATS within Python, engineers can achieve a higher degree of flexibility, programmability, and scalability in their network operations. Example 5-6 shows how you can drop into an interactive pyATS Python shell directly from bash.

Example 5-6 *pyATS Shell*

```
bash$ pyats shell --testbed-file testbed.yaml
>>> testbed.devices['csr1000v-1'].connect()
>>> output = testbed.devices['csr1000v-1'].parse('show interfaces')
>>> print(output)
```

Parsers can also be included inside pyATS jobs (much like in Example 5-6) and are the basis for *test-driven automation*; keys and values make up the source of data tested against. Example 5-7 demonstrates how parsers and can be incorporated with the **.parse()** function.

Example 5-7 *pyATS Script Sample Showing Parsers*

```
    @aetest.test
    def capture_show_interfaces(self):
        self.parsed_interfaces = self.device.parse("show interfaces")
        with open(f'{self.device.alias}_Show_Interfaces.json', 'w') as f:
            f.write(json.dumps(self.interface_data_to_write, indent=4, sort_
keys=True))
```

The transformative potential of pyATS is not just limited to its standalone capabilities but is profoundly amplified when integrated into Python scripts. By harnessing the learn and parse functionalities within Python, network engineers can craft tailored solutions, automate intricate workflows, and ensure consistent network insights. Furthermore, with pyATS jobs, there's an added layer of automation and scheduling, allowing for periodic network checks, validations, and reporting. This synergy between Python and pyATS represents a new frontier in network operations, where precision meets automation, and proactive network management becomes the norm rather than the exception. Now that you have captured network state as JSON, there are even more powerful capabilities we can scaffold on top of.

A concrete example of how this normalization of data can benefit the network engineer might be an audit of all IP addresses in use across an enterprise's routers. This could take hours or days to manually document and analyze. Using pyATS learn modules and a simple testbed containing the target routes, the process is not only faster but easier; following up the structured JSON with a built-in Python tool such as the ipaddress module (https://docs.python.org/3/howto/ipaddress.html), the entire audit could be automated.

Dictionary Query

In the vast realm of programming, the dictionary stands out as one of the most versatile and powerful data structures, especially in languages like Python. At their core, dictionaries are collections of key-value pairs, offering a unique way to store and organize data. However, as these structures grow in complexity and size, efficiently retrieving specific information becomes a challenge. This is where dictionary querying comes into play. Dictionary querying allows developers to sift through these intricate structures, pinpointing the exact data they need with precision and speed. Filtering can be done based on specific criteria, searching for nested keys, or trying to extract a subset of the dictionary. Querying techniques can simplify these tasks and enhance the efficiency of data retrieval. Dictionary query (Dq) is a Python library for querying Python dictionaries using a very intuitive syntax, as demonstrated in Example 5-8.

Example 5-8 *Dq Examples with pyATS Parsed Data*

```
# Find all the bgp neighbors which are Established
>>> output = device.parse('show bgp neighbors')
>>> output.q.contains('Established').get_values('neighbor')


# effectively, device.parse() API returns a modified parsed dictionary
# enabling you to make quick accesses to the Dq object without having to be
# explicit. This is equivalent to:

>>> from genie.utils import Dq
>>> output = device.parse('show bgp neighbors')
>>> Dq(output).contains('Established').get_values('neighbor')
['10.2.2.2']

Find all the routes which are connected.
output = device.parse('show ip route')
# Find all the routes which are Connected
output.q.contains('connected').get_values('routes')
['10.0.1.0/24', '10.1.1.1/32', '10.11.11.11/32', '172.16.21.0/24']
```

```
Find all the ospf routes.
# Find all the routes for Ospf
output.q.contains('ospf').get_values('routes')
['10.0.2.0/24', '10.2.2.2/32']
```

Typically, to perform the same queries with Python, you would need **for** loops, **if** statement, and so on. Dq simplifies the whole process! Dq also supports RegEx (regular expression), as demonstrated in Example 5-9.

Example 5-9   *Dq RegEx Example*

```
# Check if the module in line card #4 contains a status
# and its value is ok or active
output.q.contains('lc').contains('4').contains_key_value('status', 'ok|active',
  value_regex=True)
{'lc': {'4': {'NX-OSv Ethernet Module': {'status': 'ok'}}}})
```

Dq supports the following chained actions:

- contains
- not_contains
- get_values
- contains_key_value
- not_contains_key_value
- value_operator
- sum_value_operator
- count
- raw
- reconstruct
- query_validator
- str_to_dq_query
- Timeout
- TempResult

In the evolving landscape of network automation, the ability to swiftly and accurately extract relevant data is paramount. This is where Dq (Dictionary Query) shines. Within the framework of pyATS, Dq is a powerful tool for querying the structured JSON collected by the parsers and models. By allowing engineers to delve deep into structured data returned from network devices, Dq streamlines the process of data extraction and manipulation. Instead of wading through vast amounts of information, engineers can pinpoint the exact data they need, whether it's a specific interface status, routing information, or device health metrics. This precision, combined with the automation capabilities of pyATS, ensures that network operations are not only more efficient but also more accurate. In essence, Dq bridges the gap between raw data and actionable insights, propelling network automation to new heights.

Differentials

Differential, or Diff, is an extremely powerful pyATS library that can be used to compare parsed JSON data and return Linux-like +/- differentials against our dictionaries. In the dynamic world of network automation, staying ahead means not just capturing data but understanding its evolution. Enter the Diff library from pyATS—a tool that's revolutionizing the way we perceive changes in our network environments. No longer are network engineers confined to manually comparing vast datasets or configurations. With Diff, they can effortlessly spot differences, track alterations, and monitor transitions in their network data. This isn't just about identifying changes; it's about understanding the story behind them. Whether it's tracking configuration drifts, validating network changes post-deployment, or ensuring compliance, the Diff library improves visibility into network changes.

Example 5-10 demonstrates some basic, abstract examples of Diff.

Example 5-10 *Differential Abstract Examples*

```
from genie.utils.diff import Diff
a = {'a':5, 'b':7, 'c':{'ca':8, 'cb':9}}
b = {'a':5, 'f':7, 'c':{'ca':8, 'cb':9}}

dd = Diff(a,b)
dd.findDiff()
print(dd)
+f: 7
-b: 7
# It also supports an exclude key, for the keys that shouldn't be compared.
from genie.utils.diff import Diff
a = {'a':1, 'b':2, 'c':{'ca':9}}
c = {'a':2, 'c':3, 'd':7, 'c':{'ca':{'d':9}}}

dd = Diff(a, c, exclude=['d'])
dd.findDiff()
print(dd)
-b: 2
+a: 2
-a: 1
```

```
c:
+ ca:
+   d: 9
-  ca: 9


# You can also only see which one were added
dd = Diff(a, c, mode='add')
dd.findDiff()
print(dd)
+d: 7
# Or Removed

dd = Diff(a, c, mode='remove')
dd.findDiff()
print(dd)
-b: 2


# Or modified, which mean it existed, but the value was modified
dd = Diff(a, c, mode='modified')
dd.findDiff()
print(dd)
+a: 2
-a: 1
c:
+ ca:
+   d: 9
-  ca: 9


# If you need a string representation of added items without diff labeling, you can do

a = {'a': 1, 'w': 5, 'p': {'q': {'a': 6}}}
c = {'b': 2, 'c': {'d': {'e': {'f': 2, 'g': 5}}}}
dd = Diff(a, c)
dd.findDiff()
print(dd.diff_string('+'))
b 2
c
d
  e
   f 2
   g 5
```

```
# Similarly, you can get a string for the removed items

dd = Diff(a, c)
dd.findDiff()
print(dd.diff_string('-'))
a 1
p
q
   a 6
w 5


# To print unchanged entries in a list or tuple, you can specify the verbose option
like so

a = { 'key': {'value': [1, 2, 3, 4]}}
b = { 'key': {'value': [1, 3, 3, 4]}}
dd = Diff(a, b, verbose=True)
dd.findDiff()
print(dd)
key:
value:
   index[0]: 1
-  index[1]: 2
+  index[1]: 3
   index[2]: 3
   index[3]: 4
```

Example 5-11 is an end-to-end differential example in Python where we first capture the network state, make an arbitrary change (in this case, add a loopback), capture the new state of the device, and finally perform a Diff.

Example 5-11 *Differential Network Automation Example*

```
import time
import difflib
import logging
from pyats import aetest
from genie.utils.diff import Diff


## Get Logger
log = logging.getLogger(__name__)
```

```
## AE TEST SETUP
class common_setup(aetest.CommonSetup):
    """Common Setup Section"""
    #Connect to testbed
    @aetest.subsection
    def connect_to_devices(self,testbed):
        testbed.connect()

    #Mark test case for loops in case there are more than 1 device in the testbed
    @aetest.subsection
    def loop_mark(self,testbed):
        aetest.loop.mark(Chat_With_Catalyst, device_name=testbed.devices)

class Chat_With_Catalyst(aetest.Testcase):
    """A sample differential"""

    # set individual device as current iteration if there is a loop
    @aetest.test
    def setup(self,testbed,device_name):
        self.device = testbed.devices[device_name]

    @aetest.test
    def capture_show_run(self):
        self.show_run = self.device.execute("show run")

    @aetest.test
    def capture_show_ip_interface(self):
        self.show_interfaces = self.device.parse("show interfaces")

    @aetest.test
    def capture_show_ip_interface_brief(self):
        self.show_ip_interface_brief = self.device.parse("show ip interface brief")

    @aetest.test
    def capture_show_ip_route(self):
        self.show_ip_route = self.device.parse("show ip route")

    @aetest.test
    def make_change(self):
        self.device.configure(f'''interface loopback100
                              description "This is a new loopback
                              ip address 192.168.100.100 255.255.255.0
                              no shut''')
```

```
    @aetest.test
    def recapture_show_run(self):
        self.new_show_run = self.device.execute("show run")


    @aetest.test
    def recapture_show_ip_interface(self):
        self.new_show_interfaces = self.device.parse("show interfaces")


    @aetest.test
    def recapture_show_ip_interface_brief(self):
        self.new_show_ip_interface_brief = self.device.parse
("show ip interface brief")


    @aetest.test
    def recapture_show_ip_route(self):
        time.sleep(10)
        self.new_show_ip_route = self.device.parse("show ip route")


    @aetest.test
    def perform_show_run_diff(self):
        pre_change = self.show_run
        post_change = self.new_show_run
        diff = difflib.ndiff(pre_change.splitlines(), post_change.splitlines())
        show_run_diff_output = '\n'.join(line for line in diff if line.
startswith('-') or line.startswith('+'))
        with open(f'Show_Run_Diff.txt', 'w') as f:
            f.write(show_run_diff_output)


    @aetest.test
    def perform_show_interface_diff(self):
        interface_pre_change = self.show_interfaces
        interface_post_change = self.new_show_interfaces
        interface_diff = Diff(interface_pre_change, interface_post_change)
        interface_diff.findDiff()
        with open(f'Show_Interfaces_Diff.txt', 'w') as f:
            f.write(str(interface_diff))


    @aetest.test
    def perform_show_ip_interface_brief_diff(self):
        ip_interface_brief_pre_change = self.show_ip_interface_brief
        ip_interface_brief_post_change = self.new_show_ip_interface_brief
        ip_interface_brief_diff = Diff(ip_interface_brief_pre_change,
ip_interface_brief_post_change)
        ip_interface_brief_diff.findDiff()
        with open(f'Show_IP_Interface_Brief_Diff.txt', 'w') as f:
            f.write(str(ip_interface_brief_diff))
```

```
    @aetest.test
    def perform_show_ip_route_diff(self):
        ip_route_pre_change = self.show_ip_route
        ip_route_post_change = self.new_show_ip_route
        ip_route_diff = Diff(ip_route_pre_change, ip_route_post_change)
        ip_route_diff.findDiff()
        with open(f'Show_IP_Route_Diff.txt', 'w') as f:
            f.write(str(ip_route_diff))

class common_cleanup(aetest.CommonCleanup):
    """Common Cleanup Section"""
    @aetest.subsection
    def disconnect_from_devices(self,testbed):
        testbed.disconnect()
```

As demonstrated in Example 5-11, Diff can be used for change validation. Point-in-time snapshots could be used with Diff. You have unlimited potential and possibilities with the scaffolding tools and libraries like Dq and Diff with pyATS learn models and parsed output.

Summary

In this chapter, we took an enlightening journey through the expansive area of network automation. We began by underscoring the key role of vendor-agnostic automation, highlighting its promise of flexibility and adaptability in a diverse networking landscape. Our exploration then led us to the dynamic capabilities of pyATS, with the "learn" feature offering an automated lens into the network's state and the "parse" functionality transforming raw CLI outputs into structured, actionable data. This versatility was further showcased as we discussed direct parsing from the command-line interface and its seamless integration within Python scripts. The chapter also introduced Dictionary Query (Dq), a tool that simplifies the extraction of specific information from intricate data structures. We concluded our exploration by unveiling the transformative potential of the Diff library in pyATS, emphasizing its ability to track and understand nuanced changes in network data. Together, these insights paint a vivid picture of a future where network automation is not just about management but also improved comprehension.

Chapter 6

Test-Driven Development

In the realm of network engineering and operations, the deficiency of a unified methodology for creating, examining, and preserving solutions is quite palpable. Each enterprise tends to navigate its unique path concerning change management and infrastructure solution development. This fragmented approach is most noticeable when organizations venture into the automation domain, which predominantly hinges on standardized configurations and templates. Traditionally, networks are conceived, erected, and then scrutinized through manual processes. This scrutiny typically comes at the tail end of the delivery cycle, set against the backdrop of large and intricate topologies. The toolkit available to network engineers for this purpose has been rather primitive, mostly confined to outdated tools like syslog and Simple Network Management Protocol (SNMP). Moreover, the prevailing ethos has been more reactive than proactive when it comes to incident response.

However, amid this scenario, there's a silver lining. The domain of software development presents a methodical approach known as *test-driven development (TDD)* that could be a game changer for network engineers. When synergized with Cisco's pyATS, TDD lays down a structured pathway, inclusive of guidelines, scientific methodologies, and the necessary instrumentation for architecting robust, modern, automated solutions.

This venture into a more standardized approach becomes even more pertinent in the context of larger networks where an entire industry thrives on monitoring and management tooling. Herein, the proof-of-concept (POC) or proof-of-value (POV) scenarios have been instrumental. Cisco has tailored its own Center of Proof of Concept (CPOC) for this endeavor, fostering extended engagements with clientele. Notably, entities with substantial stakes, like banks, often opt for pre-stage environments owing to the hefty costs associated with potential failures. These POCs are envisioned to have well-defined outcomes, and given the high stakes, scripts are meticulously crafted to run tests. However, the caveat has been the high-level skill requirement and the extensive timeframe, stretching over weeks or even months, necessary for planning and logistics. The tools of the trade have traditionally been RegEx, bash, and similar technologies.

Yet, the landscape is shifting with the advent of pyATS, which has emerged as a catalyst in simplifying this intricate process. The utilization of pyATS substantially diminishes the hurdles, rendering the process less complex, more cost-effective, and quicker. Moreover, it opens the doors to a wider audience, democratizing the once highly specialized work. This transition not only accelerates the pace at which network solutions are tested and deployed but also elevates the accessibility and efficiency of network management practices, paving the way for a more proactive and standardized approach to network engineering in contemporary enterprise environments.

This chapter covers the following topics:

- Introduction to TDD

- Applying TDD to network automation

- Introduction to pyATS

- The pyATS framework

Introduction to Test-Driven Development

Test-driven development (TDD), at its core, is a process that converts *requirements* into *test cases*. Kent Beck, creator of the "extreme programming" approach to software development, is credited with "rediscovering" the test-driven development technique. One of the 17 original signatories on the revolutionary "Manifesto for Agile Software Development" (agilemanifesto.org), Beck stressed the importance of simplicity over complexity. It is important to understand the 12 principles behind the Agile Manifesto, and you can see where the test-driven approach emerged:

> Our highest priority is to satisfy the customer through early and continuous delivery of valuable software.

> Welcome changing requirements, even late in development. Agile processes harness change for the customer's competitive advantage.

> Deliver working software frequently, from a couple of weeks to a couple of months, with a preference to the shorter timescale.

> Business people and developers must work together daily throughout the project.

> Build projects around motivated individuals. Give them the environment and support they need, and trust them to get the job done.

> The most efficient and effective method of conveying information to and within a development team is face-to-face conversation.

Working software is the primary measure of progress.

Agile processes promote sustainable development.
The sponsors, developers, and users should be able
to maintain a constant pace indefinitely.

Continuous attention to technical excellence
and good design enhances agility.

Simplicity—the art of maximizing the amount
of work not done—is essential.

The best architectures, requirements, and designs
emerge from self-organizing teams.

At regular intervals, the team reflects on how
to become more effective, then tunes and adjusts
its behavior accordingly.

Along with other fundamental principles that define Agile, "simplicity—the art of maximizing the amount of work *not done* —is essential" is foundational to the test-driven development approach to network automation. Another important figure in test-driven development is "Uncle Bob," that is, Robert C. Martin (cleancoder.com). Robert is a programmer, speaker, and teacher who contributed the concise set of rules that govern TDD. There are only three rules to follow,[1] and they are things you are *not* permitted to do as a TDD developer:

1. You are **not** allowed to write any *production* code unless it is to make a failing unit test **pass.**

2. You are **not** allowed to write any more of a unit test than is sufficient to **fail** (and compilation failures *are* failures).

3. You are **not** allowed to write any more production code than is sufficient to **pass** the **one** failing unit test.

These rules directly relate to the Agile Manifesto's principle of simplicity and will guide your journey into the world of test-driven automation with pyATS. Along with the rules there is an equally simple three-step approach to development:

1. Write a **failing** test.

2. Make the test **pass.**

3. Refactor.

First, you will create a new test. Run all tests, including the new test, which should **fail.** This might seem counterintuitive, but we want to be as scientific as possible, meaning we need to validate the failed state of the test first. Next, write the simplest code possible that will make the new test pass. All tests should now pass. Refactor as needed, improving the code and repeating the testing cycle after **each** refactor to ensure refactoring quality. Next, add a new test and repeat the cycle until all requirements and use cases are covered by tests.

Here are some best practices to keep in mind as you adopt this new approach:

- **Keep the unit tests small:** For example, do not write a test to check if an interface is healthy based on all the counters; write an interface CRC error test, a half-duplex test, and individual counters.

- **The general test structure:**

  - Setup

  - Execution

  - Validation

  - Cleanup

- **Limit, or eliminate, dependencies between tests.**

- **Complex is fine; complicated is not.**

- **Avoid "all-knowing" tests:** See "Keep the unit test small."

Applying Test-Driven Development to Network Automation

Test-driven development (TDD) seamlessly integrates with network automation, particularly through the use of Cisco pyATS, by elevating the conventional network engineering process that is driven by business requirements. By deriving unit tests from these requirements, TDD establishes a clear link between them and individual test cases, enhancing the precision and reliability of network designs and implementations. This methodology is particularly effective in practice, as pyATS enables the comprehensive testing of both network state—via the output of show commands—and configuration. Additionally, it supports the evaluation and enforcement of network intent as defined in a source of truth through configuration automation. This approach not only ensures alignment with business objectives but also facilitates the creation of Continuous Integration and Continuous Delivery (CI/CD) pipelines. These pipelines incorporate test cases into a fully automated solution, further augmented by tools such as Cisco XPRESSO—a web-based GUI dashboard featured in Chapter 20, "XPRESSO," for scheduling and orchestrating pyATS jobs. This synergy between TDD and network automation tools like pyATS and Cisco XPRESSO paves the way for more efficient, reliable, and hands-off network management solutions.

Delving into a specific example, let's unravel a TDD workflow utilizing pyATS to validate that the count of Cisco Discovery Protocol (CDP) neighbors is indeed four:

1. **Requirements gathering:** Ensure that the network device has exactly four CDP neighbors, which is a business requirement.

2. **Test case derivation:** Derive a test case to confirm that the count of CDP neighbors is four.

3. **Initial test execution:** Run the test using pyATS against the current network configuration. At this juncture, the test is likely to fail since the requisite configurations have not been implemented yet.

4. **Network configuration:** Configure the network device to meet the business requirement of having four CDP neighbors.

5. **Test execution:** Execute the test again using pyATS. If the configuration is accurate, the test should pass. If not, it will fail, indicating that further configuration adjustments are necessary.

6. **Refinement (if necessary):** If the test fails, refine the network configuration and rerun the test until it passes, thereby confirming that the network configuration aligns with the business requirement.

7. **Integration:** Integrate this test into a CI/CD pipeline using tools like Cisco XPRESSO for continuous monitoring and validation.

8. **Automation:** Automate the entire process from configuration to testing, ensuring that any future modifications adhere to the business requirement of having exactly four CDP neighbors.

This workflow exemplifies how TDD, fortified with pyATS, provides a structured, iterative, and automated methodology for validating network configurations against business requirements. Through continuous testing and automation, network engineers can ensure that the network's state perpetually aligns with organizational objectives, thereby fostering a more robust, reliable, and efficient network infrastructure.

There is a particularly nuanced application of TDD principles in network engineering, as listed next. This is especially the case in brownfield environments, where integration with existing architectures is a common scenario. Unlike greenfield settings, where network engineers have the liberty to start afresh, brownfield scenarios demand a meticulous integration approach to ensure that existing business and operational requisites are not compromised.

- **Adherence to three laws in network engineering context:** The three laws of TDD, originally crafted for software development, find a broader spectrum of application in network engineering, extending beyond unit tests to system tests. The essence of these laws remains intact—to ensure that every piece of code or configuration is validated through tests.

- **Brownfield environments:** In brownfield settings, a growing catalog of unit and system tests becomes imperative. These tests, mirroring the existing active business and operational requisites, form a baseline that must always be met. While you're integrating new configurations or systems, it's crucial that these baseline tests pass affirm that the integration has not disrupted the existing setup. This scenario does present edge cases/exceptions to the first law of TDD (writing a failing test first), as there will be tests that should and will pass due to pre-existing configurations.

■ **Pseudocode:** Here is an example using pseudocode. Consider a brownfield scenario where the task is to ensure a specific VLAN configuration on a switch while ensuring that existing configurations are not disrupted. Here's a simplified TDD workflow with pseudocode:

Existing Tests

Run existing unit and system tests to ensure the current setup is sound:

```python
Copy code
def test_existing_setup():
    # Existing tests
    pass
```

New Test Case

Write a new test case to check the desired VLAN configuration:

```
def test_vlan_configuration():
    vlan_config = pyats parse ('show vlan')
    assert 'VLAN10' in vlan_config
```

Run the new test; it may fail if the VLAN isn't configured yet:

```
def configure_VLAN():
    pyats configure ('vlan 10')
```

Rerun Tests

Re-run all tests (existing and new) to ensure both the new and existing configurations are correct:

```
pyats run job test_existing_setup
pyats run job test_vlan_configuration
```

This TDD workflow exemplifies how both existing and new requirements can be validated in a brownfield environment. Although the pseudocode is simple, it gives a gist of how tests for configuration and state can be structured and run to ensure the robustness of network configurations amid evolving requirements. This practical approach, coupled with tools like pyATS, fosters a disciplined testing culture, bridging the traditional network engineering practices with modern, automated, and test-driven methodologies.

Introduction to pyATS

Now that NetDevOps, test-driven development, and the Agile methodology have been introduced, let's look at Cisco's Python Automation Test solution, pyATS. pyATS is made up of both a *framework* and *libraries* that can be used for automation, testing, network assurance, and much more. pyATS considers everything an object, including network state and configurations, which are made available as structured data that can be acted upon by either consuming or delivering commands programmatically. In a way, pyATS acts as an API toolkit that abstracts the complexity of the underlying parsing and connectivity layers of network automation.

pyATS was originally developed for internal Cisco engineering and was made available to the public for general use in 2017. Cisco runs *millions* of internal tests monthly with pyATS, which is the de facto standard testing framework at Cisco. Holistically, pyATS can be considered an automation ecosystem made up of both the core framework and the software development kit (SDK), or libraries, that extend the core functionality. Multiplatform and multivendor, pyATS has become an integral part of CI/CD, connectivity, configuration, regression, scale, and overall solution testing for thousands of developers globally, both internally at Cisco and externally around the world. As illustrated in Figure 6-1, the main components of pyATS include the core framework, or toolbox; the SDK and libraries, formerly known as Genie; and upper-layer business logic integrations. pyATS is an all-purpose generic framework, while the libraries extend the capabilities and specialize in network device automation and validation.

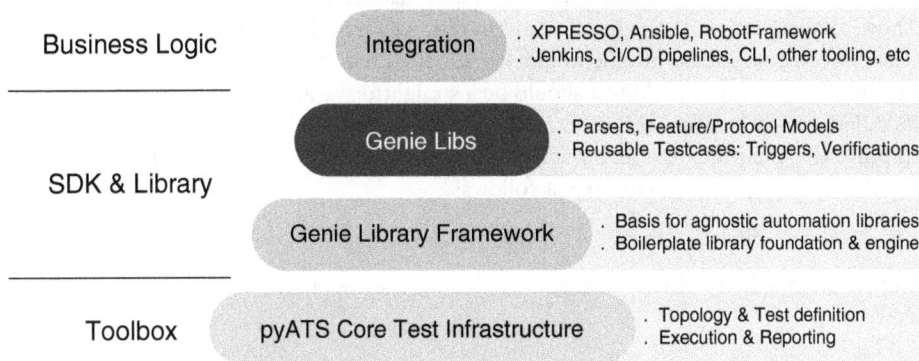

Business Logic	Integration	. XPRESSO, Ansible, RobotFramework . Jenkins, CI/CD pipelines, CLI, other tooling, etc
SDK & Library	Genie Libs	. Parsers, Feature/Protocol Models . Reusable Testcases: Triggers, Verifications
	Genie Library Framework	. Basis for agnostic automation libraries . Boilerplate library foundation & engine
Toolbox	pyATS Core Test Infrastructure	. Topology & Test definition . Execution & Reporting

Figure 6-1   *Layers of pyATS*

The pyATS Framework

The pyATS framework can be broken down into the main components and supporting components. The main components include the following:

- AEtest test infrastructure

- Easypy runtime environment

- Testbed and topology information

- Testbed and device cleaning

The supporting components of pyATS include the following:

- Asynchronous library

- Data structures

- TCL integration

- Logging

- Result objects

- Reporter

- Utilities

- Robot Framework support

- Manifest

AEtest

Automation Easy Testing (AEtest) is the pyATS standard test engineering automation harness. The test cases translate into this simple and straightforward foundation of pyATS testing jobs. AEtest is implemented as aetest and was designed to be fully object-oriented. Those familiar with Python's unittest and pytest, which both inspired the architectural design of AEtest, should be able to quickly adopt AEtest. The high-level design features dictate that working with AEtest should be a straightforward, Pythonic experience with its object-oriented design. AEtest is included with the pyATS full installation using pip (**pip install pyats[full]**) or is also available as a standalone package. To install AEtest as a standalone library, you can use pip as follows:

```
pip install pyats.aetest
```

Part of the simplicity of easy testing is the block-based approach to test section break-downs:

- Common Setup with subsections

- Testcases with setup/tests/cleanup

- Common Cleanup with subsections

You can import AEtest from pyATS and create a common setup, where you can establish connectivity with your testbed, as demonstrated in Example 6-1.

Example 6-1 *Using AEtest to Connect to a Device*

```
from pyats import aetest

class CommonSetup(aetest.CommonSetup):
    @aetest.subsection
    def connect_to_device(self, testbed):
        # connect to testbed devices
        for device in testbed:
            device.connect()
```

Once connectivity has been established, tests can be performed against the network state, as demonstrated in Example 6-2.

Example 6-2 *Parsing show interfaces and Printing the JSON Output*

```
class SimpleTestcase(aetest.Testcase):
    @aetest.test
    def print_interface(self, testbed):
        # print each device interface
        for device in testbed:
            interface = device.parse("show interface")
            print(interface)
```

Finally, you can tear down and gracefully disconnect from the devices in the testbed as demonstrated in Example 6-3.

Example 6-3 *Disconnecting Gracefully from the Device*

```
class CommonCleanup(aetest.CommonCleanup):
    @aetest.subsection
    def disconnect_from_devices(self, testbed):
        # disconnect_all
        for device in testbed:
            device.disconnect()
```

To allow this to run as its own Python executable, we can also include **aetest.main()**, as demonstrated in Example 6-4.

Example 6-4 *Using main() to Allow the Executable to Run as a Standalone Function*

```
# for running as its own executable
if __name__ == '__main__':
    aetest.main()
```

It is recommended to run pyATS testscripts using standard execution only, running aetest directly (**aetest.main()**) during script development. This allows for a quick turnaround when testing code.

You can modify the **print** statement to test network state as demonstrated in Example 6-5.

Example 6-5 *Testing Interfaces for CRC Errors*

```
class SimpleTestcase(aetest.Testcase):
    @aetest.test
    def test_for_input_crc_per_interface(self, testbed):
        # configure each device interface
        self.failed_interface = {}
```

```
        for interface,value in self.interfaces.items():
            print(f"Testing interface { interface } for Input CRC Errors")
            if value['counters']['in_crc_errors'] > 0:
                print(f"{ interface } failed crc test")
                self.failed_interface = interface
            else:
                print(f"{ interface } passed crc test")


    @aetest.test
    def pass_or_fail(self):
        if self.failed_interface:
            self.failed()
        else:
            self.passed()
```

Adhering to this modular structure is as easy as keeping all testscripts broken down into the Common Setup, Testcase(s), and Common Cleanup sections. Common Setup allows for checking the validity of input scripts, verifying the connectivity to the targeted testbed and devices, bringing up the topology, loading the device configurations, and setting up any dynamic looping. This section always runs first, before testcases. Example 6-6 demonstrates the start of a pyATS script's Common Setup section.

Example 6-6   *A Typical Common Setup Example*

```
from pyats import aetest
# define a common setup section by inheriting from aetest
class ScriptCommonSetup(aetest.CommonSetup):
    @aetest.subsection
    def check_script_arguments(self):
        pass


    @aetest.subsection
    def connect_to_devices(self):
        pass


    @aetest.subsection
    def configure_interfaces(self):
        pass
```

Testcases can now be defined and executed knowing the environment connected and inputs are validated. Each testcase is defined by inheriting the **aetest.Testcase** class and defining one or more test sections inside. These testcases run in the order in which they are defined in the testscript. Each testcase is associated with a unique ID, which defaults to the class name but can be changed by setting the **testcase.uid** attribute. The unique ID is used for reporting purposes. Testcases are independent, and the code of a testcase should be self-contained such that it can run in isolation, with any number of other testcases. Each testcase's result is a combined rollup result of all its child sections. These results are counted as 1 in the summary table of results.

Finally, Common Cleanup is the last section to run within each testscript. Environments and network connections should be returned to the same state as prior to the script running. Removal of all Common Setup changes, removal of lingering changes, and gracefully disconnecting from network devices are all part of the Common Cleanup phase and ensure for a nondisruptive approach to testing. Common Cleanup should also be used as a catchall, regardless of whether individual testcases clean up after themselves. This section should be used as a safety net to ensure the testbed returns to a healthy state after the scripts are completed. Refer back to the AEtest test infrastructure in Chapter 4, "AEtest Test Infrastructure."

Easypy

Easypy provides a standardized runtime environment for testscript execution in pyATS. It offers a simple, straightforward way for users to aggregate testscripts together into jobs. It also integrates various pyATS modules together into a collectively managed ecosystem as well as archives all resulting information for post-mortem debugging:

- **Jobs:** Easypy aggregates multiple testscripts into one job.
- **TaskLog:** Easypy stores all runtime log outputs in TaskLog.
- **Email notification:** Easypy emails the user result information upon finishing.
- **Multiprocessing integration:** Easypy executes each job file task in a child process and configures the environment to allow for hands-off forking.
- **Clean:** Easypy cleans/brings up the current testbed with new images and a fresh configuration.
- **Plugins:** Easypy has a plugin-based design, allowing custom user injections to alter and/or enhance the current runtime environment.

In pyATS, a *job* involves aggregating multiple testscripts together and executing them within the same runtime environment.

The concept of Easypy revolves heavily around the execution of such jobs. Each job corresponds to a job file—a standard Python file containing the instructions of which testscripts to run and how to run them.

During runtime, these testscripts are launched as tasks and executed through a test harness (for example, aetest). Each task is always encapsulated in its own child process. The combined tasks in the same job file always share the same system resources, such as testbeds, devices, and their connections.

All logs, files, and artifacts generated by tasks during the job file runtime are stored in a runinfo folder. After execution has completed, they are archived into a .zip folder. The .zip folder is stored in the directory

./users/<userid>/archive/*YY-MM*/

where *YY-MM* represents the current year and month (using two digits each), thus providing some level of division/classification between jobs. The following files are stored in the .zip folder:

- **<job-name>.py**: A copy of the job file that ran.

- **<job-name>.report**: A copy of the email notification sent to the submitter.

- **TaskLog.<task-id>**: TaskLog (one per job file task) is where all messages generated in a task are stored.

- **JobLog.<job-name>**: Overall **pyats.easypy** module log.

- **Testbed.static.yaml**: Contents of the **--testbed-file**, if specified by the user.

- **Testbed.clean.yaml**: Contents of the **--clean_file**, if specified by the user.

- **Env.txt**: A dump of the environment variables and CLI arguments of this Easypy run.

- **Reporter.log**: Reporter server log file, which contains a trace of XML-RPC call sequences.

- **Results.json**: JSON result summary file generated by Reporter.

- **xunit.xm**: Files containing xUnit-style result reports and information required by Jenkins. These files are only generated if the **--xunit** argument is provided to Easypy.

- **ResultsSummary.xml**: XML result summary file generated by Reporter.

- **ResultsDetails.xml**: XML result details file generated by Reporter.

- **CleanResultsDetails.yaml**: YAML clean result details file generated by Kleenex.

- **Kleenex.<device-name>.log**: Job-scope clean details for this device.

- **Kleenex_<task-id>.<device-name>.log**: Task-scope clean details for this device.

Testbed and Topology

The topology module is designed to provide an intuitive and standardized method for users to define, handle, and query testbed/device/interface/link descriptions, metadata, and their interconnections.

The two major functionalities of the topology module are as follows:

1. Defining and describing testbed metadata using YAML, standardizing the format of the YAML file, and loading it into corresponding testbed objects.

2. Querying testbed topology, metadata, and interconnect information via testbed object attributes and properties.

YAML (short for "YAML Ain't Markup Language" or "Yet Another Markup Language") is a human-readable data serialization format designed to be both human- and machine-readable.

YAML is indentation and whitespace sensitive. Its syntax maps directly to most of the common data structures in Python, such as dict, list, str, and more.

As opposed to creating a module where the topology information is stored internally and then asking users to query that information via API calls, the pyATS topology module approaches the design from a completely different angle:

- Using objects to represent real-world testbed devices

- Using object attributes and properties to store testbed information and metadata

- Using object relationships (references/pointers to other objects) to represent topology interconnects

- Using object references and Python garbage collection to clean up testbed leftovers when objects are no longer referenced

Testbed and Device Cleaning with Kleenex

Device cleaning is the process of preparing physical testbed devices by loading them with appropriate images (recovering from bad images), removing unnecessary configurations, and returning devices to their default initial state by applying basic configurations such as console/management IP addresses and so on.

Kleenex Clean offers the base infrastructure required by all clean implementations:

- Integration with Easypy (runtime environment and testbed objects)

- Structured input format and information grouping through a clean file

- Automatic asynchronous device cleaning

- Runtime, exception, and logging handling

- All of the necessary guidelines and information required for users to develop their platform-specific clean methods.

Kleenex Clean standardizes how users implement platform-specific clean methods, providing the necessary entry points and subprocess management.

Asynchronous Library (Parallel Call)

Asynchronous (async) execution defines the ability to run programs and functions in a non-blocking manner. In pyATS, it's recommended that you execute in parallel using multiprocessing. The proper use of multiprocessing can greatly improve the performance of a program and is only bounded by the physical number of CPUs and I/O limits. Multiprocessing is recommended with pyATS and test-driven automation for a variety of reasons:

- Separate memory space, meaning no race conditions (except with external systems)

- Very simple, straightforward code

- No global interpreter lock, taking full advantage of multiple CPU/cores

- Easily interruptible/killable child processes

pyATS also provides an API from the async module known as Parallel Call, or pcall, to further abstract and simplify parallel calls. pcall supports calling procedures and functions in parallel using multiprocessing fork, without you having to write boilerplate code to handle the overhead of process creation, waiting, and terminations. pcall supports calling all procedures, functions, and methods in parallel, if the return of the called target is a pickleable object. In Python, "pickleable" refers to the capability of an object to be serialized into a byte stream (and subsequently deserialized back into an object). Serialization with the pickle module allows objects to be saved to a file, transmitted over a network, or, as in the case of pyATS's pcall, passed between processes.

Consider pcall as a shortcut library to multiprocessing, intended to satisfy most users' need for parallel processing. However, for more custom and advanced use cases, you should stick with direct usage of multiprocessing.

The pcall API allows users to call any functions/methods (a.k.a. targets) in parallel. It comes with the following built-in features:

- Builds arguments for each child process/instance

- Creates, handles, and gracefully terminates all processes

- Returns target results in their called order

- Reraises child process errors in the parent process

Data Structures

New data structures have been introduced and maintained as part of the pyATS infrastructure. These new data structures are used as part of the pyATS source code and may prove to be useful in our users' day-to-day coding:

- **Attribute dictionaries:**

  - **AttrDict:** This is the exact same as the Python native dictionary, except that in most cases you can use the dictionary key as if it was an object attribute instead.

- **NestedAttrDict:** This is a special subclass of **AttrDict** that recognizes when its key values are other dictionaries and auto-converts them into further **NestedAttrDict** instances.

- **Weak list references:** A standard list object stores every internal object as a direct reference. That is, if the list exists, its internally stored objects exist.

- **Dictionary represented using lists:** Accessing nested dictionaries often calls for recursive functions in order to properly parse and/or walk through them. This isn't always easy to code around. **ListDict** provides an alternative view on nested dictionaries, breaking down the value nested within keys to a simple concept of path and value. This flattens the nesting into a linear list, greatly simplifying the coding around nested dictionaries.

- **Orderable dictionary:**

  - Python's built-in **collections.OrderedDict** only remembers the order in which keys were inserted into it and does not allow users to reorder the keys and/or insert new keys into arbitrary positions in the current key order.

  - **OrderableDict** (Orderable Dictionary) is almost exactly the same as Python's **collections.OrderedDict**, with the added ability to order and reorder the keys inserted into it.

- **Logic testing:** Boolean algebra is sometimes confusing when used in the context of the English language. The goal of this module is to standardize how to represent and evaluate logical expressions within the scope of pyATS as well as to offer standard APIs, classes, and behaviors for users to leverage.

- **Configuration container:**

  - The **Configuration** container is a special type of **NestedAttrDict** intended to store Python module and feature configurations.

  - Avoid confusing Python configurations with router configurations. Python configurations tend to be key-value pairs that drive a particular piece of infrastructure, telling it how its behavior should be.

TCL Integration

This module effectively enables you to make Tcl calls in the current Python process and is 100% embedded. There's no child Tcl process. The actual Tcl interpreter is embedded within the current Python process, and the process ID (PID) of both Python and Tcl is the same.

Part of the goal of pyATS is to enable the testing community to leverage existing Tcl-based scripts and libraries. In order to make the integration easier, the Tcl module was created to extend the native Python-Tcl interaction:

- **Interpreter class:** Extends the native Tcl interpreter by providing access to ATS-tree packages and libraries.

- **Two-way typecasting:** APIs and Python classes enable typecasting Tcl variables to the appropriate Python objects and back. This includes, but is not limited to, int, list, string, array, and keyed lists.

- **Call history:** An historical record of Tcl API calls is maintained for debugging purposes.

- **Callbacks:** Callbacks from Tcl to Python code enable closer coupling.

- **Dictionary access:** You can access Tcl variables as if accessing a Python dictionary.

- **Magic Q calls:** You can call Tcl APIs as if calling a Python object method, with support for Python *args and **kwargs mapping to Tcl positional and dashed arguments.

Logging

A log is a log, regardless of what kind of prefixes each log message contains and in what format it ended up as, as long as it is human-readable and provides useful information to the user.

The Python logging module's native ability to handle and process log messages is more than sufficient for any logging needs and has always been suggested as the de facto logging module to use.

Therefore, for all intents and purposes, users of the pyATS infrastructure should always use just the native Python logging module as-is in their scripts and testcases. Example 6-7 demonstrates some simple logging functions.

Example 6-7 *pyATS Logging Functions*

```
#    import the logging module at the top of your script
#    setup the logger


import logging


# always use your module name as the logger name.
# this enables logger hierarchy
logger = logging.getLogger(__name__)


# use logger:
logger.info('an info message')
logger.error('an error message')
logger.debug('a debug message')
```

Result Objects

In most test infrastructures, such as pytest and unittest, test results are only available as pass, fail, or error. This works quite well in unit and simplistic testing. The downside of having only three result types, however, is the inability to describe testcase result relationships, or distinguish a test's genuine failure, versus a failure of the testscript caused by poor design/coding (for example, the testcase encountered a coding exception).

To accommodate complex test environments, pyATS supports more complicated result types such as "test blocked," "test skipped," "test code errored," and so on, and it uses objects and object relationships to describe them. These objects simplify the whole result tracking and aggregation infrastructure and grant the ability to easily roll up results together. Here are the available result objects:

- **Passed:** Indicates that a test was successful, passed, the result accepted, and so on.

- **Failed:** Indicates that a test was unsuccessful, fell short, the result was unacceptable, and so on.

- **Aborted:** Indicates something was started but was not finished, or was incomplete and/or stopped at an early or premature stage. For example, a script was killed via a **Ctrl+C** keypress.

- **Blocked:** Used when a dependency was not met and the following event could not start. Note that a "dependency" doesn't strictly mean order dependency or setup dependency. It could also mean cases where the next event to be carried out is no longer relevant.

- **Skipped:** Used when a scheduled item was not executed and was omitted. The difference between skipped and blocked is that skipped is voluntary, whereas blocked is collateral.

- **Errored:** A mistake or inaccuracy (for example, an unexpected exception). The difference between failed and errored is that failed represents carrying out an event as planned with the result not meeting expectation, whereas errored means something went wrong in the course of carrying out that procedure.

- **Passx:** Short for "passed with exception." Use passx with caution because you are effectively re-marking a failed result as passed, even though there was an exception.

Reporter

Reporter is a package for collecting and generating test reports in YAML format. This results file contains all the details about execution (section hierarchy, time, results, and so on).

The results.json report contains hierarchical information about the pyATS job executed. The top level is the TestSuite, which contains information about the job as a whole. Under the TestSuite are all of the tasks executed as a part of the job. Each task then has the various sections of testing underneath CommonSetup, CommonCleanup, and Testcases. These then have child sections, such as TestSection, SetupSection, CleanupSection, and Subsection. The children of these would be steps, which can be nested with their own children steps.

Being able to parse the generated test reports (results.json) allows you to further dig into and programmatically analyze the test results. This allows you to take further action based on the testing results using an automated workflow.

Utilities

pyATS comes with a variety of additional utilities to enhance and support the framework:

- **Find:** Used to search and filter against objects

- **Secret Strings:** Used to protect and encrypt strings (such as passwords)

- **Multiprotocol file transfer utilities:** Used to transfer files to/from a remote server

- **Embedded pyATS file transfer server:** Supports FTP, TFTP, SCP, and HTTP

- **Import utilities:** Used to translate "x.y.z"-style strings into "from x.y import z" and then return z

- **YAML File Markups:** pyATS-specific YAML markup; similar to Django template language

Robot Framework Support

Robot Framework is generic Python/Java test automation open-source framework that focuses on acceptance test automation using an English-like keyword-driven test approach. https://robotframework.org/

You can now freely integrate pyATS into the Robot Framework, and vice versa:

- You can run Robot Framework scripts directly within Easypy, save runtime logs under the runinfo directory, and aggregate results into Easypy report.

- You can leverage the pyATS infrastructure and libraries within Robot Framework scripts.

However, because Robot Framework support is an optional component under pyATS, you must install the package explicitly before being able to leverage it. For more information, see Chapter 22, "Robot Framework."

Manifest

The pyATS Manifest is a file with YAML syntax describing how and where to execute a script. It is intended to formally describe the execution of a single script, including the runtime environment, script arguments, and the profile(s) that define environment specific settings and arguments. Profiles allow the same script to be run against multiple environments or run with different input parameters—for example, using multiple test-beds representing different environments (testing/production) or different scaling numbers to test scalability and resiliency. A script can be executed via the manifest using the **pyats run manifest** command. Manifest files use the file extension .tem, which stands for Test Execution Manifest. Manifest files can be tracked via source control, which can help standardize testing environments across multiple testing scenarios.

Summary

Adopting network automation has never been easier with the introduction of a modern tool like pyATS, and its associated framework, and the test-driven development methodology. TDD emerged from the Agile manifesto and is a common form of software development that can be extended to network automation. Network engineers will gather business requirements and transform them into testcases. These testcases are small unit tests that at first fail when they are written. The smallest, simplest amount of code possible is applied to make the test pass, while all other tests remain passing, and is refactored until the developer is satisfied with the passing test. This iterative approach is performed for each use case until test coverage is established for all business requirements.

The iterative essence of TDD fosters a disciplined, incremental, and feedback-driven approach to both software and network automation development. This iterative process, initiated from gathering business requirements, which are then transmuted into test cases, is at the heart of promoting a robust and reliable network automation culture. The rhythm of writing a failing test, making it pass with the simplest code, and then refining the code, embodies a cycle of continuous improvement and validation:

■ **Evolution of tests:** As development progresses, tests evolve in tandem. Initially, tests are rudimentary, focusing on basic functionalities. Over time, as more features are integrated and complexities arise, tests become more comprehensive and nuanced. This evolution of tests is a natural reflection of the growing understanding and unfolding of business requirements.

■ **Improved code quality:** One of the stellar benefits of this iterative approach is the uplift in code quality. Each cycle of TDD pushes the code through a crucible of validation, ensuring that it not only meets the immediate requirement but is also robust and resilient to potential issues. The refactoring stage, an integral part of the TDD cycle, further polishes the code, enhancing its readability, efficiency, and maintainability.

■ **Prompt identification and resolution of issues:** The TDD cycle facilitates early detection of discrepancies and bugs. As each piece of functionality is validated through tests before being integrated, issues are spotted and rectified promptly. This proactive error detection significantly reduces the time and effort that would otherwise be expended in debugging and fixing problems later in the development process.

■ **Increased confidence:** The rigorous validation imbued by TDD instills a higher degree of confidence in the reliability and accuracy of network automation solutions. The iterative testing and refactoring provide a safety net that ensures each increment of development solidifies the solution rather than introducing regressions.

■ **Enhanced understanding and documentation:** The iterative testing process also acts as a documentation of what the code is supposed to achieve. Each test case elucidates the business requirements and the expected behavior of the system, thereby enriching the understanding of the system among the development and operations teams.

■ **Facilitation of change:** The iterative nature of TDD, coupled with the comprehensive suite of tests, provides a sturdy foundation for accommodating changes. Whether adapting to evolving business requirements or integrating new technologies like pyATS, the TDD approach ensures that the system remains robust and reliable.

By marrying the TDD methodology with modern tools like pyATS, network engineers are poised to harness a powerful synergy that accelerates the adoption of network automation, while concurrently elevating the quality, reliability, and adaptability of the solutions crafted. Through this iterative and validation-centric approach, the journey of network automation becomes a structured, manageable, and rewarding endeavor.

Endnotes

1. Robert C. Martin (Uncle Bob), "The Cycles of TDD," http://blog.cleancoder.com/uncle-bob/2014/12/17/TheCyclesOfTDD.html

Chapter 7

Automated Network Documentation

Early in my career, I was tasked by the finance department to capture the information of a new pair of Cisco 6500 Catalyst switches with various supervisors and line cards. The naive junior engineer that I was at the time brought them pages of printouts of the **show inventory** command! "We can't use this. *We need something that is business-ready*." Those that are forever etched in my memory. Nearly 15 years later I was able to *automatically* transform that **show inventory** command from raw unusable output into a "business-ready" comma-separated values (CSV) spreadsheet with pyATS jobs and a templating language called *Jinja2*. Automated network documentation is the recommended starting place for anyone who is brand new to network automation or pyATS, primarily because it is *safe*. No changes are being made; no modifications to the network or to the configuration. All we are doing is running and parsing **show** commands. As trivial as this seems, it can become the foundation for a source of truth in a Git-tracked repository showing state and configuration change history. The answer to the question "What has changed?" becomes extremely obvious using your IDE and reviewing the Git change history over the business-ready reports. All of these report formats have VS Code extensions that allow for direct viewing and integration with tools like Excel Preview, Markdown Preview, and Open in Browser for HTML pages.

This chapter covers the following topics:

- Introduction to pyATS jobs
- Running pyATS jobs from the CLI
- pyATS job CLI logs
- pyATS logs HTML viewer
- Jinja2 templating
- Business-ready documents

Introduction to pyATS Jobs

In pyATS, the aggregation of multiple testscripts that are executed within the same runtime environment is called a *job*. The concept behind Easypy revolves heavily around the execution of such jobs. Each job corresponds to a job file—a standard Python file containing the instructions of which testscripts to run and how to run them. During runtime, these testscripts are launched as tasks and executed through a test harness (for example, aetest). Each task is always encapsulated in its own child process. The combined tasks in the same job file always share the same system resources, such as testbeds, devices, and their connections.

Table 7-1 outlines the type and purpose of pyATS testbed, script, and job files.

Table 7-1   *pyATS Files*

File	Type	Purpose
testbed.yaml	YAML	Defines the network device(s) and topology
pyats_job.py	Python	Jobfile loads the pyATS environment, testbed file, and Python script
pyats.py	Python	Log in to connect, test, and disconnect from network devices in the testbed.

Job files are the bread and butter of Easypy. They allow aggregation of multiple testscripts to run under the same environment as tasks, sharing testbeds and archiving their logs and results together. A job file is an excellent method to batch and/or consolidate similar testscripts together into relevant result summaries. Job files are required to satisfy the following criteria:

- Each job file must have a **main()** function defined. This is the main entry point of a job file run.

- The **main()** function accepts an argument called **runtime**. When it's defined, the engine automatically passes the current runtime object in.

- Inside the **main()** function, **easypy.run()** or **easypy.Task()** can be used to define and run individual testscripts as tasks.

- The name of the job file, minus the .py extension, becomes this job's reporting name. This can be modified by setting the **runtime.job.name** attribute.

Figure 7-1 shows an example of a pyATS job file.

```
# Example
# -------
#
#   a simple, sequential job file

import os

from pyats.easypy import run

# main() function must be defined in each job file
#    · it should have a runtime argument
#    · and contains one or more tasks
def main(runtime):

    # provide custom job name for reporting purposes (optional)
    runtime.job.name = 'my-job-overwrite-name'

    # using run() api to run a task
    #
    # syntax
    # ------
    #   run(testscript = <testscript path/file>,
    #       runtime = <runtime object>,
    #       max_runtime = None,
    #       taskid = None,
    #       **kwargs)
    #
    #   any additional arguments (**kwargs) to run() api are propagated
    #"  to AEtest as input arguments.
    run(testscript = 'script_one.py', runtime = runtime)

    # each job may contain one or more tasks.
    # tasks defined using run() api always run sequentially.
    run(testscript = 'script_two.py', runtime = runtime)

    # access runtime information, such as runtime directory
    # eg, save a new file into runtime directory
    with open(os.path.join(runtime.directory, 'my_file.txt')) as f:
        f.write('some content')
```

Figure 7-1 *A Sample pyATS Job File*

Job files are provided as the only mandatory argument to the Easypy launcher.
As outlined earlier in Table 7-1, pyATS jobs typically have three files:

- The **<test>_job.py** file: This is the pyATS job file executed using the **pyats run job <job> --testbed-file <testbed>** command.

- The **Python <test>.py** file: The main script where all of the testing logic is contained.

- **A valid testbed file:** The target topology for the job and script.

The pyATS script inside a job is typically broken into three major areas as Python classes:

- Common Setup

  - Connections to the topology are made.

  - Any tests that should be marked for looping are marked.

  - Any customized common setup to occur before testing starts is indicated.

- Tests

  - N+1 tests cases

  - Happen after connectivity to topology is established

  - Marked as passed, failed, or skipped

- Common Cleanup

  - Disconnect from the topology gracefully.

  - Leave the environment in same state as when originally connected.

  - Any custom cleanup activities are performed.

Refer to Chapter 4, "AEtest Test Infrastructure" for a deep dive into pyATS jobs.

Figure 7-2 illustrates pyATS job structure, while Figure 7-3 illustrates the pyATS job structure as classes.

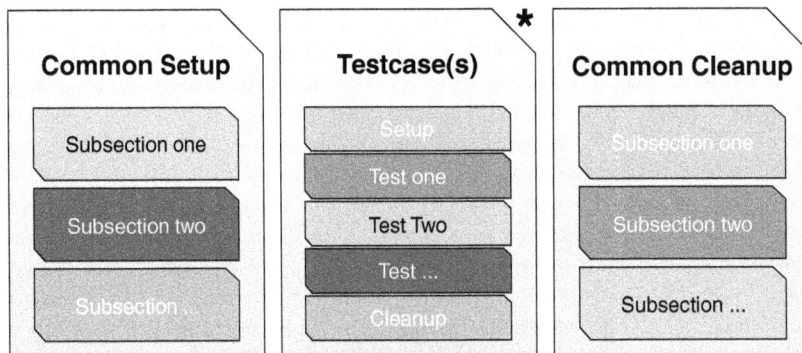

Figure 7-2   *pyATS Job Structure*

In the preceding example, we first, in the Common Setup section, connect to our topology and mark the testcase for looping in case there is more than one device in the testbed file we want to document. We will complete this first testcase next, and then we gracefully disconnect from the topology in the Common Cleanup section.

For our first test, let's capture the parsed JSON version of **show ip interface brief** and simply save it to a file locally (see Figure 7-4). From this JSON file, we will build all our other Jinja2 templates. pyATS includes many abstractions in the form of application programming interfaces (APIs). There is an API to save both raw text and dictionaries to files.

```
import logging
from pyats import aetest

## Get Logger
log = logging.getLogger(__name__)

## AE TEST SETUP
class common_setup(aetest.CommonSetup):
    """Common Setup Section"""
    #Connect to testbed
    @aetest.subsection
    def connect_to_devices(self,testbed):
        testbed.connect(log_stdout=False)

    @aetest.subsection
    def loop_mark(self,testbed):
        aetest.loop.mark(Automate_Show_IP_Interface_Brief, device_name=testbed.devices)

class Automate_Show_IP_Interface_Brief(aetest.Testcase):
    """Capture Show IP Interface Brief and transform to documentation"""
    @aetest.test
    def setup(self,testbed,device_name):
        self.device = testbed.devices[device_name]



    # Test case here

class common_cleanup(aetest.CommonCleanup):
    """Common Cleanup Section"""
    @aetest.subsection
    def disconnect_from_devices(self,testbed):
        testbed.disconnect()
```

Figure 7-3 *pyATS Structure as Classes*

Figure 7-4 *pyATS APIs – Save*

Using the pyATS **.parse()** and **save_dict_to_json_file()** commands, we can set up our first test, as shown in Example 7-1.

Example 7-1   *Example of Parsing JSON and Saving to a File*

```
class Parse_And_Save_Show_IP_Interface_Brief_to_JSON(aetest.Testcase):
    """Capture Show IP Interface Brief and transform to documentation"""
    @aetest.test
    def setup(self,testbed,device_name):
        self.device = testbed.devices[device_name]

        parsed_show_ip_interface_brief = self.device.parse\\
                                ("show ip interface brief")

        self.device.api.save_dict_to_json_file\\
        (data=list(parsed_show_ip_interface_brief.values()),\\
        filename="Show IP Interface Brief.json")
```

The complete automated_network_documentation.py script should now look like Example 7-2. Note that, for simplicity, we are simply using **testbed.connect()** and **.disconnect()** to establish and tear down connectivity to all devices in the testbed, respectively, without the need for a loop.

Example 7-2   *automated_network_documentation.py*

```
import json
import logging
from pyats import aetest


## Setup Logging
log = logging.getLogger(__name__)


## Common Setup
class common_setup(aetest.CommonSetup):
    """Common Setup Section"""
    #Connect to testbed
    @aetest.subsection
    def connect_to_devices(self,testbed):
        testbed.connect()

    #Mark tests for loops
    @aetest.subsection
    def loop_mark(self,testbed):
        aetest.loop.mark(Parse_And_Save_Show_IP_Interface_Brief_to_JSON,\\
                    device_name=testbed.devices)
```

```
# Test Cases
class Parse_And_Save_Show_IP_Interface_Brief_to_JSON(aetest.Testcase):
    """Capture Show IP Interface Brief and transform to documentation"""
    @aetest.test
    def setup(self,testbed,device_name):
        #Set current device in loop to self.device
        self.device = testbed.devices[device_name]

        #Parse show ip interface brief to JSON
        self.parsed_show_ip_interface_brief = \\
        self.device.parse("show ip interface brief")

        #Save JSON to file
        self.device.api.save_dict_to_json_file\\
        (data=list(self.parsed_show_ip_interface_brief.values()),\\
        filename="Show IP Interface Brief.json")

#Common Cleanup
class common_cleanup(aetest.CommonCleanup):
    """Common Cleanup Section"""
    @aetest.subsection
    #Disconnect from devices
    def disconnect_from_devices(self,testbed):
        testbed.disconnect()
```

Example 7-3 shows the testbed file for the Cisco DevNet Always-On IOS-XE Sandbox we can use to test.

Example 7-3 *testbed.yaml for the Cisco DevNet Always-On IOS-XE Sandbox*

```
---
devices:
  Cat8000V:
    alias: "Sandbox Router"
    type: "router"
    os: "iosxe"
    platform: Cat8000v
    credentials:
      default:
        username: admin
        password: C1sco12345
```

```
connections:
  cli:
    protocol: ssh
    ip: devnetsandboxiosxe.cisco.com
    port: 22
    arguments:
      connection_timeout: 360
```

And finally, Example 7-4 shows the pyATS job file—automated_network_documentation_
job.py.

Example 7-4   *pyATS Job File (automated_network_documentation_job.py)*

```
import os
from genie.testbed import load

def main(runtime):
    if not runtime.testbed:
        # If no testbed is provided
        testbedfile = os.path.join('testbed.yaml')
        testbed = load(testbedfile)
    else:
        testbed = runtime.testbed

    testscript = os.path.join(os.path.dirname(__file__),\\
    'automated_network_documentation.py')

    runtime.tasks.run(testscript=testscript, testbed=testbed)
```

Next, we will examine how to execute, or run, this pyATS job that captures the **show ip
interface brief** output as JSON and then, using a pyATS API, saves the JSON to a local
file, automatically, from the command-line interface (CLI). Note that pyATS jobs can
be scheduled and executed graphically inside of XPRESSO, the topic of Chapter 20,
"XPRESSO."

Running pyATS Jobs from the CLI

A pyATS job can be executed from the command line, which provides real-time connectivity details in the runtime logs, a summary of the job and its tasks outcomes, as well as the
command to start the HTML log viewer. By default, standard output logging is enabled
when you use the **testbed.connect()** function, which provides verbose logs for every step
of the job. This is an option you can toggle with the **log_stdout=False** option inside
testbed.connect(log_stdout=False) should you want less-verbose output from the logs.

From the command line, pyATS jobs are run using the following command:

```
$ pyats run job <job name> --testbed-file <testbed file>
```

However, it should be noted that a pyATS job has a *lot* of arguments and options that can be passed to it, as demonstrated in Figure 7-5 through Figure 7-9.

```
® (coding_with_capo_pyats) johncapobianco@DESKTOP-EFDK79U:~/chapter_seven$ pyats run job ?
Usage:
  pyats run job [file] [options]

Example
-------
  pyats run job /path/to/jobfile.py
  pyats run job /path/to/jobfile.py --testbed-file /path/to/testbed.yaml

Description:
  Runs a pyATS job file with the provided arguments, generating & report result.

Job Information:
  JOBFILE                 target jobfile to be launched
  --job-uid               Unique ID identifiying this job run
  --pyats-configuration
                          pyats configuration override file

Tasks:
  --task-uids LOGIC       Logic string to match task UIDs to run eg: "Or('Task-[12]')"

Mailing:
  --no-mail               disable report email notifications
  --mail-to               list of report email recipients
  --mail-subject          report email subject header
  --mail-html             enable html format report email

Reporting:
  --submitter             Specify the current submitter user id
  --image                 Specify the image under test
  --release               Specify the release being tested
  --branch                Specify the branch being tested
  --meta                  Specify some meta information as a dict. Supports a JSON string, a base64
                          encoded string, a URL, a file path, or an individual ("=" separated)
                          key=value pair. Can be used multiple times in one command.
  --source-id             XPRESSO source - eg. xpresso-dev-1@cisco.com
  --no-xml-report         Disable generation of the XML Report
```

Figure 7-5 *pyATS Job Arguments*

```
Runinfo:
  --no-archive            disable archive creation
  --no-archive-subdir     disable archive subdirectory creation
  --runinfo-dir           specify alternate runinfo directory
  --archive-dir           specify alternate archive directory
  --archive-name          specify alternate archive file name

Liveview:
  --liveview              Starts a liveview server in a separate process
  --liveview-host HOST    Specify host for liveview server. Default is localhost
  --liveview-port PORT    Specify port for liveview server.
  --liveview-hostname HOSTNAME
                          Displayed hostname for liveview.
  --liveview-displayed-url LIVEVIEW_DISPLAYED_URL
                          Displayed url for liveview, for example, http://<liveview_hostname>:<port>
  --liveview-keepalive    Keep log viewer server alive after the run finishes.
  --liveview-callback-url LIVEVIEW_CALLBACK_URL
                          Specify xpresso callback url for jenkins run.
  --liveview-callback-token LIVEVIEW_CALLBACK_TOKEN
                          Specify xpresso token for jenkins run.

Testbed:
  -t, --testbed-file      Specify testbed file location
```

Figure 7-6 *pyATS Job Arguments (Continued)*

```
Clean:
 --clean-file FILE [FILE ...]
                      Specify clean file location(s). Multiple clean files can be specified by
                      separating them with spaces.
 --clean-devices [ ...]
                      Specify list of devices to clean, separated by spaces. To clean groups of
                      devices sequentially, specify as "[[dev1, dev2], dev3]".
 --clean-scope {job,task}
                      Specify whether clean runs before job or per task
 --invoke-clean       Clean is only invoked if this parameter is specified.
 --clean-device-image  [ ...]
                      list of clean images per device in format device:/path/to/image.bin
 --clean-os-image  [ ...]
                      list of clean images per OS in format os:/path/to/image.bin
 --clean-group-image  [ ...]
                      list of clean images per group in format group:/path/to/image.bin
 --clean-platform-image  [ ...]
                      list of clean images per platform in format platform:/path/to/image.bin
 --clean-image-json  [ ...]
                      dictionary of clean images in JSON string

Bringup:
 --logical-testbed-file
                      Specify logical testbed file location

Rerun:
 --rerun-file FILE    rerun.results file that contains the information of tasks and testcases
 --rerun-task  [ ...] TASKID TESTSCRIPT [TESTCASES...] Details to identify a specific Task to
                      rerun. Can be used multiple times for multiple tasks.
 --rerun-condition  [ ...]
                      Results type list for the condition of rerun plugin.

xUnit:
 --xunit [DIR]        Generate xunit report in the provided location. If used as a flag, generates
                      xunit reports runtime directory

HTML Logging:
 --html-logs [DIR]    Directory to generate HTML logs in addition to any existing log files. Note
                      - will increase archive size due to log duplication.
```

Figure 7-7   *pyATS Job Arguments (Continued)*

```
mapleClean:
 --maple-testsuite TESTSUITE_FILE
                      Specify maple yaml file

TopologyUpPlugin:
 --check-all-devices-up
                      Enable/Disable checking for topology up pre job execution
 --connection-check-timeout CONNECTION_CHECK_TIMEOUT
                      Total time allowed for checking devices connectivity
 --connection-check-interval CONNECTION_CHECK_INTERVAL
                      Time to sleep between device connectivity checks

WebEx:
 --webex-token        Webex Bot AUTH Token
 --webex-space        Webex Space ID to send notification to
 --webex-email        Email of specific user to send WebEx notification to
```

Figure 7-8   *pyATS job Arguments (Continued)*

```
pyATS Health:
  --health-file HEALTH_FILE
                        Specify health yaml file
  --health-tc-sections HEALTH_TC_SECTIONS
                        Specify sections where to run pyATS Health Check. Regex is supported. You
                        can also filter based on class type. e.g. type:TestCase
  --health-tc-uids HEALTH_TC_UIDS
                        Specify triggers uids where to run pyATS Health Check. Regex is supported
  --health-tc-groups HEALTH_TC_GROUPS
                        Specify groups where to run pyATS Health Check. Regex is supported
  --health-config HEALTH_CONFIG
                        Specify pyATS Health Check configuration yaml file
  --health-remote-device [HEALTH_REMOTE_DEVICE ...]
                        Specify remote device information for copy files to remote
  --health-mgmt-vrf [HEALTH_MGMT_VRF ...]
                        Specify Mgmt Vrf which is reachable to remote device
  --health-threshold [HEALTH_THRESHOLD ...]
                        Specify threshold for cpu, memory
  --health-show-logging-keywords [HEALTH_SHOW_LOGGING_KEYWORDS ...]
                        Specify logging keywords to search
  --health-clear-logging
                        Specify logging keywords to search
  --health-core-default-dir [HEALTH_CORE_DEFAULT_DIR ...]
                        Specify directories where to search for core files
  --health-checks [HEALTH_CHECKS ...]
                        Specify checks to run
  --health-devices [HEALTH_DEVICES ...]
                        Specify devices which checks run against
  --health-notify-webex
                        Flag to send webex notification

General Options:
  -h, --help            Show help information
  -v, --verbose         Give more output, additive up to 3 times.
  -q, --quiet           Give less output, additive up to 3 times, corresponding to WARNING, ERROR,
                        and CRITICAL logging levels
```

Figure 7-9 *pyATS job Arguments (Continued)*

Make sure all three files (job file, script file, and testbed file) are saved, identify any optional arguments you want to try (such as email, Webex, or verbosity), and run the pyATS job, like so:

```
(virtual_environment)$pyats run job automated_network_documentation_
job.py --testbed-file testbed.yaml
```

The three files are automated_network_documentation_job.py (the job file), automated_network_documentation.py (the script), and testbed.yaml (the testbed file).

pyATS Job CLI Logs

pyATS will start the job run and start printing Easypy logs to the screen. The name of the job and the running directory are printed first. The Clean information (covered in Chapter 15, "pyATS Clean"), if any, is displayed first under its own logging banner. If the **–check-all-devices-up** argument was passed along, pyATS will check that all devices are up and ready first, or else log that this was disabled, before proceeding to the Common Setup section. The Common Setup banner will display and the subsections marked as PASSED, FAILED, SKIPPED, or ERRORED. Our testcases will then display their banners and perform the tests. Figure 7-10 shows the CLI logs from the start of a pyATS job.

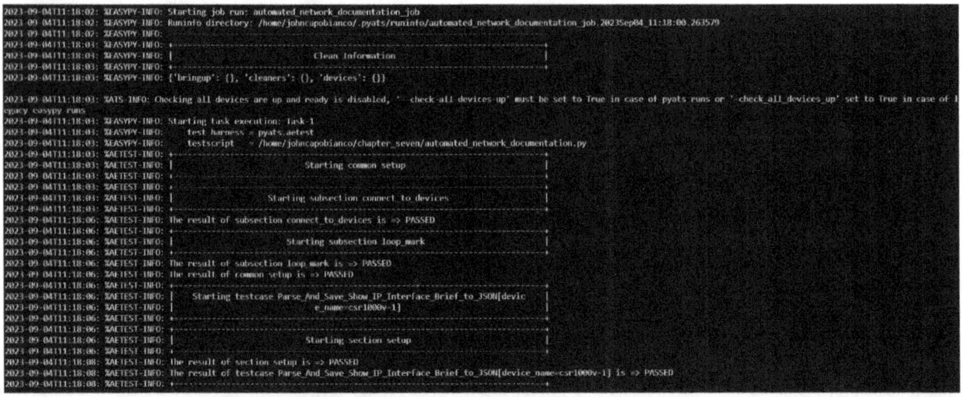

Figure 7-10 *pyATS job CLI Logs*

At this stage of the job, you should now have a local file called "Show IP Interface Brief.json" inside your local directory. If you open it and enable formatting in VS Code (https://code.visualstudio.com/docs/python/formatting), it should look like Figure 7-11.

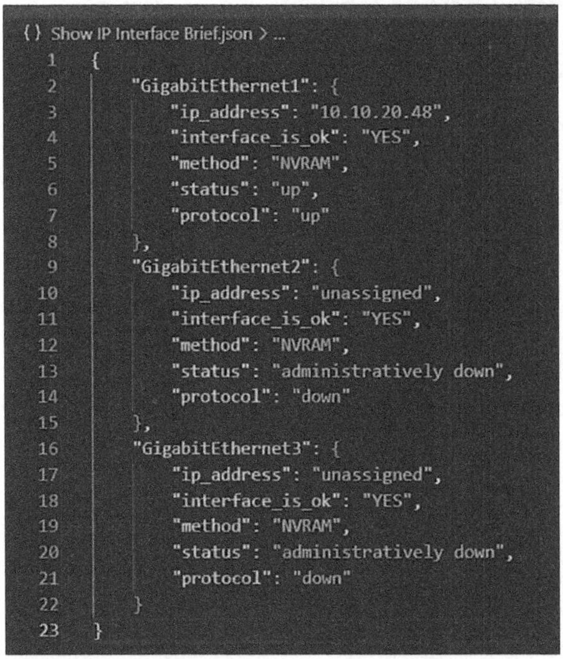

Figure 7-11 *Automated JSON Network State File*

Next, the Common Cleanup section banner appears in the logs and the devices are disconnected gracefully. If Webex arguments were included, the notification would also occur at this step. An archive of the job log is created and then an Easypy report is displayed (see Figure 7-12) containing the following information:

- The pyATS instance and version
- The CLI arguments passed
- The user and local environment (host and OS) information
- The name of the job
- Start and stop time
- Elapsed time
- Archive location
- Total tasks
- Overall stats
  - Passed
  - Passx
  - Failed
  - Aborted
  - Blocked
  - Skipped
  - Errored
  - Total tasks
  - Success rate %

Finally, the job summary information is displayed, as shown in Figure 7-13. A complete breakdown, step by step, task by task, class by class, is displayed with the results of each step in the job and task in the step. Notice the "pro tip" at the end of the output suggesting you can run the **pyats logs view** command to launch the HTML logs viewer.

```
2023-09-04T11:18:08: %AETEST-INFO: +---------------------------------------------------------------------+
2023-09-04T11:18:08: %AETEST-INFO: |                         Starting common cleanup                      |
2023-09-04T11:18:08: %AETEST-INFO: +---------------------------------------------------------------------+
2023-09-04T11:18:08: %AETEST-INFO: +---------------------------------------------------------------------+
2023-09-04T11:18:08: %AETEST-INFO: |                Starting subsection disconnect_from_devices           |
2023-09-04T11:18:08: %AETEST-INFO: +---------------------------------------------------------------------+
2023-09-04T11:18:19: %AETEST-INFO: The result of subsection disconnect_from_devices is => PASSED
2023-09-04T11:18:19: %AETEST-INFO: The result of common cleanup is => PASSED
2023-09-04T11:18:19: %CONTRIB-INFO: WebEx Token not given as argument or in config. No WebEx notification will be sent
2023-09-04T11:18:19: %EASYPY-INFO: -------------------------------------------------------------------
2023-09-04T11:18:19: %EASYPY-INFO: Job finished. Wrapping up...
2023-09-04T11:18:20: %EASYPY-INFO: Creating archive file: /home/johncapobianco/.pyats/archive/23-09/automated_network_documentation_job.2023Sep04_11:18:00.26
3579.zip
2023-09-04T11:18:20: %EASYPY-INFO: +---------------------------------------------------------------------+
2023-09-04T11:18:20: %EASYPY-INFO: |                            Easypy Report                             |
2023-09-04T11:18:20: %EASYPY-INFO: +---------------------------------------------------------------------+
2023-09-04T11:18:20: %EASYPY-INFO: pyATS Instance    : /home/johncapobianco/coding_with_capo_pyats
2023-09-04T11:18:20: %EASYPY-INFO: Python Version    : cpython-3.10.6 (64bit)
2023-09-04T11:18:20: %EASYPY-INFO: CLI Arguments     : /home/johncapobianco/coding_with_capo_pyats/bin/pyats run job automated_network_documentation_job.py --
testbed-file testbed.yaml
2023-09-04T11:18:20: %EASYPY-INFO: User              : johncapobianco
2023-09-04T11:18:20: %EASYPY-INFO: Host Server       : DESKTOP-EFDK79U
2023-09-04T11:18:20: %EASYPY-INFO: Host OS Version   : Ubuntu 22.04 jammy (x86_64)
2023-09-04T11:18:20: %EASYPY-INFO:
2023-09-04T11:18:20: %EASYPY-INFO: Job Information
2023-09-04T11:18:20: %EASYPY-INFO:       Name        : automated_network_documentation_job
2023-09-04T11:18:20: %EASYPY-INFO:       Start time  : 2023-09-04 11:18:03.600534-04:00
2023-09-04T11:18:20: %EASYPY-INFO:       Stop time   : 2023-09-04 11:18:19.398265-04:00
2023-09-04T11:18:20: %EASYPY-INFO:       Elapsed time : 15.797731
2023-09-04T11:18:20: %EASYPY-INFO:       Archive     : /home/johncapobianco/.pyats/archive/23-09/automated_network_documentation_job.2023Sep04_11:18:00.263579
.zip
2023-09-04T11:18:20: %EASYPY-INFO:
2023-09-04T11:18:20: %EASYPY-INFO: Total Tasks       : 1
2023-09-04T11:18:20: %EASYPY-INFO:
2023-09-04T11:18:20: %EASYPY-INFO: Overall Stats
2023-09-04T11:18:20: %EASYPY-INFO:       Passed      : 3
2023-09-04T11:18:20: %EASYPY-INFO:       Passx       : 0
2023-09-04T11:18:20: %EASYPY-INFO:       Failed      : 0
2023-09-04T11:18:20: %EASYPY-INFO:       Aborted     : 0
2023-09-04T11:18:20: %EASYPY-INFO:       Blocked     : 0
2023-09-04T11:18:20: %EASYPY-INFO:       Skipped     : 0
2023-09-04T11:18:20: %EASYPY-INFO:       Errored     : 0
```

Figure 7-12   *Easypy Report*

```
2023-09-04T14:34:01: %EASYPY-INFO: Overall Stats
2023-09-04T14:34:01: %EASYPY-INFO:       Passed      : 3
2023-09-04T14:34:01: %EASYPY-INFO:       Passx       : 0
2023-09-04T14:34:01: %EASYPY-INFO:       Failed      : 0
2023-09-04T14:34:01: %EASYPY-INFO:       Aborted     : 0
2023-09-04T14:34:01: %EASYPY-INFO:       Blocked     : 0
2023-09-04T14:34:01: %EASYPY-INFO:       Skipped     : 0
2023-09-04T14:34:01: %EASYPY-INFO:       Errored     : 0
2023-09-04T14:34:01: %EASYPY-INFO:
2023-09-04T14:34:01: %EASYPY-INFO:       TOTAL       : 3
2023-09-04T14:34:01: %EASYPY-INFO:
2023-09-04T14:34:01: %EASYPY-INFO: Success Rate      : 100.00 %
2023-09-04T14:34:01: %EASYPY-INFO:
2023-09-04T14:34:01: %EASYPY-INFO: +---------------------------------------------------------------------+
2023-09-04T14:34:01: %EASYPY-INFO: |                         Task Result Summary                          |
2023-09-04T14:34:01: %EASYPY-INFO: +---------------------------------------------------------------------+
2023-09-04T14:34:01: %EASYPY-INFO: Task-1: automated_network_documentation.common_setup              PASSED
2023-09-04T14:34:01: %EASYPY-INFO: Task-1: automated_network_documentation.Parse_And_Save_Show_IP_Inte... PASSED
2023-09-04T14:34:01: %EASYPY-INFO: Task-1: automated_network_documentation.common_cleanup            PASSED
2023-09-04T14:34:01: %EASYPY-INFO:
2023-09-04T14:34:01: %EASYPY-INFO: +---------------------------------------------------------------------+
2023-09-04T14:34:01: %EASYPY-INFO: |                         Task Result Details                          |
2023-09-04T14:34:01: %EASYPY-INFO: +---------------------------------------------------------------------+
2023-09-04T14:34:01: %EASYPY-INFO: Task-1: automated_network_documentation
2023-09-04T14:34:01: %EASYPY-INFO: |-- common_setup                                                  PASSED
2023-09-04T14:34:01: %EASYPY-INFO: |   |-- connect_to_devices                                       PASSED
2023-09-04T14:34:01: %EASYPY-INFO: |   `-- loop mark                                                PASSED
2023-09-04T14:34:01: %EASYPY-INFO: |-- Parse_And_Save_Show_IP_Interface_Brief_to_JSON[device_name=csr1... PASSED
2023-09-04T14:34:01: %EASYPY-INFO: |   `-- setup                                                    PASSED
2023-09-04T14:34:01: %EASYPY-INFO: `-- common_cleanup                                               PASSED
2023-09-04T14:34:01: %EASYPY-INFO:     `-- disconnect_from_devices                                  PASSED
2023-09-04T14:34:01: %EASYPY-INFO: Sending report email...
2023-09-04T14:34:01: %EASYPY-INFO: Missing SMTP server configuration, or failed to reach/authenticate/send mail. Result notification email failed to send.
2023-09-04T14:34:01: %EASYPY-INFO: Done!

Pro Tip
-------
  Try the following command to view your logs:
      pyats logs view
```

Figure 7-13   *Summary of pyATS job in the CLI Logs*

pyATS Logs HTML Viewer

As indicated by the "pro tip" in Figure 7-13, there is a built-in HTML-enriched log viewer from pyATS you can launch at the end of a job with the following command:

```
(virtual_environment)$ pyats logs view
```

Figure 7-14 shows launching the HTML viewer, and Figure 7-15 shows the results.

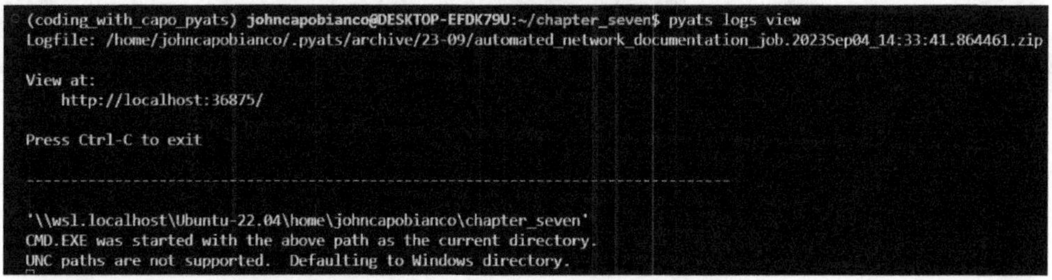
(coding_with_capo_pyats) johncapobianco@DESKTOP-EFDK79U:~/chapter_seven$ pyats logs view
Logfile: /home/johncapobianco/.pyats/archive/23-09/automated_network_documentation_job.2023Sep04_14:33:41.864461.zip
View at:
http://localhost:36875/
Press Ctrl-C to exit
'\\wsl.localhost\Ubuntu-22.04\home\johncapobianco\chapter_seven'
CMD.EXE was started with the above path as the current directory.
UNC paths are not supported. Defaulting to Windows directory.

Figure 7-14 *Launching pyATS Logs Viewer*

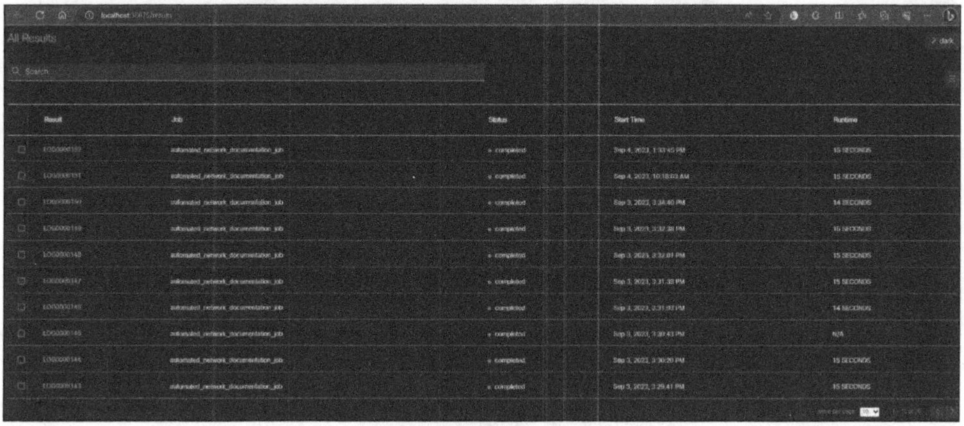
All Results
Search
Result
Job
Status
Start Time
Runtime

Figure 7-15 *pyATS Log Viewer Default Results Page*

A browser window will spawn inside the pyATSLiveView HTML logs viewer. The default page shows the results in chronological order. Light or dark mode can be set, and results can be searched. Clicking the results of our automated_network_documentation_job.py job brings us into the detailed logs from that job, as illustrated in Figure 7-16.

Figure 7-16 *pyATS Logs Viewer Details – Results*

Each result can be expanded and selected to see the CLI logs (see Figure 7-17). These results can be easily copied to the clipboard provided.

Figure 7-17 *pyATS Logs Viewer Details – Results (Highlighted)*

The Overview tab provides an overview of the pyATS job, as illustrated in Figure 7-18.

Figure 7-18 *pyATS Logs Viewer Details – Overview Tab*

The Files tab provides access to all the files related to the job, including console logs, as illustrated in Figure 7-19.

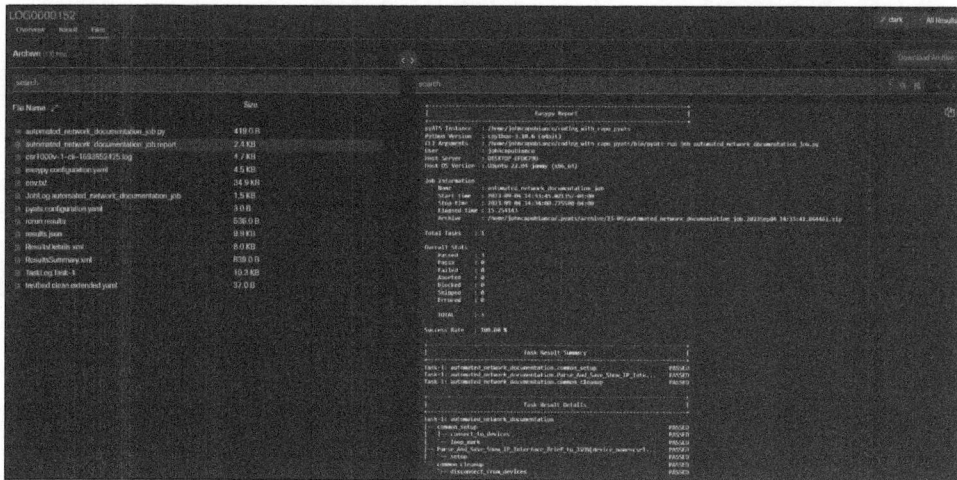

Figure 7-19 *pyATS Logs Viewer Details – Files Tab*

Both the verbose command line and HTML logs viewer capabilities of pyATS set it apart from other network automation frameworks. pyATS boasts exceptional logging capabilities that distinguish it from its counterparts. At the CLI, pyATS provides detailed and customizable logging outputs, ensuring that users receive precise feedback during test execution and troubleshooting. This granularity in logging is instrumental for engineers looking to track and pinpoint issues. Additionally, the framework offers an integrated HTML log viewer, which presents test results in an intuitive and visually appealing manner. This graphical representation not only simplifies the process of analyzing results but also enables users to quickly identify anomalies or areas of concern. Together, these logging features make pyATS a powerful tool for network professionals, setting it a notch above other frameworks in terms of debugging and results interpretation.

Jinja2 Templating

If every problem is a nail, Jinja2 templates can be your hammer! Jinja2 is a modern and designer-friendly templating engine for Python programming languages. It's utilized in various applications to generate content quickly and efficiently from data structures, such as rendering HTML pages in web applications. In the context of Jinja2 templates, the primary emphasis is on the provisioning of placeholders and control structures within templates, allowing dynamic content to be inserted or altered at runtime. This is achieved using a combination of template tags, encapsulated by {{ ... }} for expressions and {% ... %} for statements, which instruct the Jinja2 engine on how to process and render the final output.

In relation to pyATS, Jinja2 templates play a pivotal role in crafting test cases and configuration structures. Given that networks can be intricate with numerous device types, configurations, and protocols, having a static testing or configuration approach is impractical. Instead, pyATS leverages Jinja2 to create dynamic, data-driven templates. Engineers can create a base template for a network device configuration or test scenario and then use variables to adjust specific values based on the target device, protocol, or environment. Once the data is fed into the Jinja2 template, a fully rendered configuration or test case, tailored to the specific requirements, is produced. This integration allows for immense scalability and flexibility, ensuring that network testing and automation with pyATS can be as comprehensive and adaptable as necessary.

Jinja2 is one of Python's most popular templating engines and is extended to many network automation frameworks like Ansible and pyATS. pyATS, like parsing data or saving a file, has several Jinja2 API abstractions available, as Figure 7-20 illustrates.

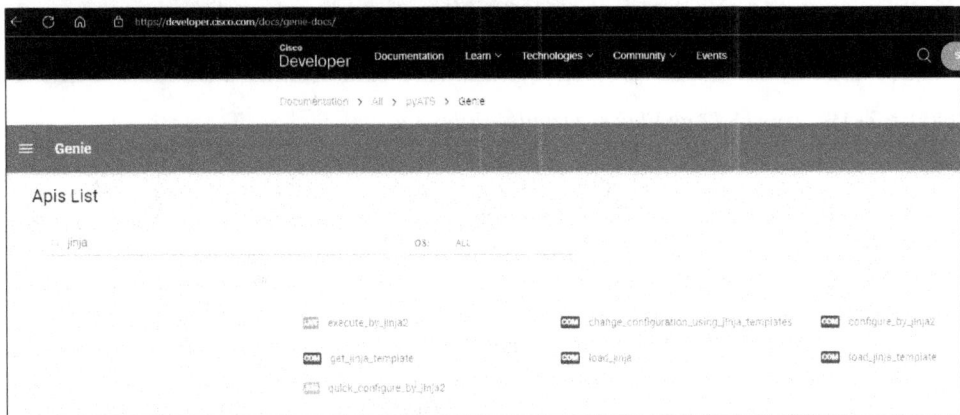

Figure 7-20   *pyATS Jinja2 APIs*

Jinja2 templates can be used easily to transform the structured JSON data captured by the pyATS parser. In our first pyATS job, we captured the JSON. Let's use various approaches, including templating with Jinja2, to create business-ready documents.

Business-Ready Documents

As I stated at the beginning of this chapter, early in my career, very early, I was asked by the Finance department to get an inventory from the new 6500 core routers, line cards, and various components for their records. With little thought, I printed the output of a **show inventory** command and provided the Finance department with the printouts. They were not very impressed and asked if I could provide something that is *business-ready* for them—a spreadsheet.

Almost 15 years later, I realized that using pyATS and Jinja2 templates, I could automate the process of creating these so-called business-ready documents—a self-documenting network, if you will. And one thing that always seems to be missing is good, current network documentation. In this section, we will explore various file types and ways to use pyATS to capture network state as structured JSON. Then we'll look at how we can work with that structured data to generate our automated business-ready documents.

JSON

Our job already captures the JSON representation of **show ip interface brief** for the Always-On DevNet IOS-XE Sandbox. It should be noted that this can be scaled vertically for more devices by simply scaling the number of devices in the testbed.yaml file. Horizontal scaling across more **show** commands is a simple matter of copying the existing testcase, updating the command and filename.

YAML

YAML Ain't Markup Language is another good network automation, infrastructure as code, data serialization format. Since we have the JSON, we can simply transform it into YAML and save the output to a YAML file. Add a new testcase that converts the JSON to a YAML file, as demonstrated in Example 7-5. Note that we are simply adding another loop marker and testcase inside the automated_network_documentation.py script.

Example 7-5   *Transform JSON to YAML*

```
import yaml
#Mark tests for loops
@aetest.subsection
def loop_mark(self,testbed):
    aetest.loop.mark(Parse_And_Save_Show_IP_Interface_Brief_to_JSON,\\
    device_name=testbed.devices)

    aetest.loop.mark(Parse_And_Save_Show_IP_Interface_Brief_to_YAML,\\
    device_name=testbed.devices)

# Test Cases - YAML file
class Parse_And_Save_Show_IP_Interface_Brief_to_YAML(aetest.Testcase):
    """Capture Show IP Interface Brief and transform to YAML documentation"""
    @aetest.test
    def save_yaml_file(self,testbed,device_name):
        #Set current device in loop to self.device
        self.device = testbed.devices[device_name]

        #Parse show ip interface brief to JSON
        self.parsed_show_ip_interface_brief = \\
        self.device.parse("show ip interface brief")
```

```
        #Convert to YAML
        yaml_show_ip_interface_brief = \\
        yaml.dump(self.parsed_show_ip_interface_brief,\\
        default_flow_style=False)

        #Save YAML to file
        with open("Show IP Interface Brief.yaml", "w") as yml_file:
            yml_file.write(yaml_show_ip_interface_brief)
```

Now, in addition to the JSON file, you should have a new YAML file and a new set of passing tests in your pyATS jobs! These YAML files are interactive in VS Code and can be expanded or collapsed, and they are generally easier to read than JSON files, as illustrated in Figure 7-21.

```
Show IP Interface Brief.yaml > {} dictitems > {} interface
  1    !!python/object/new:genie.conf.base.utils.QDict
  2    dictitems:
  3      interface:
  4        GigabitEthernet1:
  5          interface_is_ok: 'YES'
  6          ip_address: 10.10.20.48
  7          method: NVRAM
  8          protocol: up
  9          status: up
 10  >     GigabitEthernet2: ···
 16  >     GigabitEthernet3: ···
 22  >     Loopback203: ···
 28  >     Loopback3: ···
```

Figure 7-21   *show ip interface brief as YAML*

Comma-Separated Values

Arguably the most powerful business-ready document format is the CSV file. These files are supported in Excel or Excel Preview for VS Code, which allows you to sort, filter, reorder, and perform powerful visualizations, as well as being extremely simple to create from JSON structured data. Make a new testcase, as follows, to create the CSV file. This testcase will be added to the automated_network_documentation.py script. The **aetest. loop.mark** line should be within the common setup with the other loop markers. Note the use of the **load_jinja_template()** pyATS API to render the CSV file in Example 7-6.

Example 7-6   *Transform JSON to CSV Using a Jinja2 Template*

```
aetest.loop.mark(Parse_And_Save_Show_IP_Interface_Brief_to_CSV,\\
                 device_name=testbed.devices)

# Test Cases - CSV File
```

```
class Parse_And_Save_Show_IP_Interface_Brief_to_CSV(aetest.Testcase):
    """Capture Show IP Interface Brief and transform to CSV documentation"""
    @aetest.test
    def save_csv_file(self,testbed,device_name):
        #Set current device in loop to self.device
        self.device = testbed.devices[device_name]

        #Parse show ip interface brief to JSON
        self.parsed_show_ip_interface_brief = \\
        self.device.parse("show ip interface brief")

        # Load the Jinja2 template
        csv_show_ip_interface_brief = \\
        self.device.api.load_jinja_template \\
        (path="", file="csv.j2",\\
         to_parse_interfaces=self.parsed_show_ip_interface_brief['interface'])

        #Save CSV to file
        with open("Show IP Interface Brief.csv", "w") as csv_file:
            csv_file.write(csv_show_ip_interface_brief)
```

The Jinja2 template is only a few lines of code: the header row, separated by commas, and, inside a loop, a comma-separated row of the data fields aligned with their header row. Think of it as a grid of columns and rows with individual cells. The csv.j2 file looks like Example 7-7.

Example 7-7   *The csv.j2 Jinja2 Template*

```
Interface,IP Address,Status,Protocol,Method,Interface is OK
{% for interface in to_parse_interfaces %}
{{ interface }},{{ to_parse_interfaces[interface].ip_address }},
{{ to_parse_interfaces[interface].status }},
{{ to_parse_interfaces[interface].protocol }},
{{ to_parse_interfaces[interface].method }},
{{ to_parse_interfaces[interface].interface_is_ok }}
{% endfor %}
```

This results in an easy-to-read and universally appreciated spreadsheet, as illustrated in Figure 7-22.

Interface	IP Address	Status	Protocol	Method	Interface is OK
GigabitEthernet1	10.10.20.48	up	up	NVRAM	YES
GigabitEthernet2	unassigned	administratively down	down	NVRAM	YES
GigabitEthernet3	unassigned	administratively down	down	NVRAM	YES
Loopback3	128.0.0.8	up	up	other	YES
Loopback203	192.168.45.1	up	up	other	YES

Figure 7-22 *show ip interface brief as CSV*

With this base csv.j2 file, most other tabular formats can be created using **find/replace** in your IDE.

Markdown: Tables

Markdown is another lightweight data-encoding format that can be used to produce various visualizations of structured data. The csv.j2 template can be copied and modified as markdown_table.j2 by replacing the commas with pipes (|) and by adding a delimiter row. This markdown table renders inside VS Code as well as GitHub and other markdown-friendly environments. Example 7-8 shows how to add another loop marker, again to be added to the **def loop_mark** method inside the "class common setup." Most of the actual testcase can be reused from the CSV example.

Example 7-8 *Transform CSV to Markdown with Pipes and a Delimiter*

```
aetest.loop.mark(Parse_And_Save_Show_IP_Interface_Brief_to_MD_Table,\\
                device_name=testbed.devices)


# Testcases - Markdown Table File
class Parse_And_Save_Show_IP_Interface_Brief_to_MD_Table(aetest.Testcase):
    """Capture Show IP Interface Brief and transform
     to Markdown Table documentation"""
    @aetest.test
    def save_markdown_table_file(self,testbed,device_name):
        #Set current device in loop to self.device
        self.device = testbed.devices[device_name]

        #Parse show ip interface brief to JSON
        self.parsed_show_ip_interface_brief = \\
        self.device.parse("show ip interface brief")
```

```
# Load the Jinja2 template
md_table_show_ip_interface_brief = \\
self.device.api.load_jinja_template \\
(path="", file="markdown_table.j2", \\
 to_parse_interfaces= \\
 self.parsed_show_ip_interface_brief['interface'])


#Save Markdown Table to file
with open("Show IP Interface Brief.md", "w") as md_file:
    md_file.write(md_table_show_ip_interface_brief)
```

The markdown table has a title for the first row, followed by a leading and trailing pipe
(|), and all commas replaced with space-padded pipes (|). The second line in the file is a
header row, much like the CSV file, except padded with pipes. Next, we need a delimiter
row enclosed in pipes (| ----- |). These delimiters need to match the exact number of char-
acters as the header row. Finally, our data rows inside the loop are padded with pipes, as
Example 7-9 shows.

Example 7-9 *The markdown_table.j2 Jinja2 Template*

```
# Show IP Interface Brief
| Interface | IP Address | Status | Protocol | Method | Interface is OK |
| --------- | ---------- | ------ | -------- | ------ | --------------- |
{% for interface in to_parse_interfaces %}
| {{ interface }} | {{ to_parse_interfaces[interface].ip_address }} |
{{ to_parse_interfaces[interface].status }} |
{{ to_parse_interfaces[interface].protocol }} |
{{ to_parse_interfaces[interface].method }} |
{{ to_parse_interfaces[interface].interface_is_ok }} |
{% endfor %}
```

This renders as an easy-to-use table, as illustrated in Figure 7-23.

Markdown is also extensible with other tools such as Markmap, a markdown-to-mind-
map tool, and Mermaid, a special type of markdown used to create diagrams and more.
Let's make a Markmap mind map next with a few modifications to the base csv.j2 Jinja2
template. Markdown files can be rendered, or previewed, inside VS Code by clicking the
Open Preview button or right-clicking the show_IP_Interface_Brief.md file and selecting
Open with… and then **Markdown Preview**.

Show IP Interface Brief

Interface	IP Address	Status	Protocol	Method	Interface is OK
GigabitEthernet1	10.10.20.48	up	up	other	YES
GigabitEthernet2	unassigned	up	up	NVRAM	YES
GigabitEthernet2.2206	100.64.0.2	up	up	other	YES
GigabitEthernet2.2207	100.64.0.6	up	up	other	YES
GigabitEthernet2.2208	100.64.0.10	up	up	other	YES
GigabitEthernet3	unassigned	administratively down	down	NVRAM	YES
GigabitEthernet3.2216	100.64.1.2	administratively down	down	other	YES
GigabitEthernet3.2217	100.64.1.6	administratively down	down	other	YES
GigabitEthernet3.2218	100.64.1.10	administratively down	down	other	YES
Loopback44	44.44.44.1	up	up	other	YES
Loopback55	55.55.55.1	up	up	other	YES

Figure 7-23 *show ip interface brief as a Markdown Table*

Markdown: Markmap Mind Maps

Markmap mind maps are interactive, colorful, visual representation of your structured data with collapsing capabilities and zoom controls. Markmap can be used as a VS Code extension and allows you to render and export to HTML the markdown mind maps (see Example 7-10). Make a new testcase, copying and updating any of the previous examples. Make sure you change the name of this file by adding Mind Map.md. Otherwise, you will overwrite the previous markdown table example.

Example 7-10 *Transform JSON to Markmap Mind Maps*

```
aetest.loop.mark(Parse_And_Save_Show_IP_Interface_Brief_to_MD_Mindmap,\\
                 device_name=testbed.devices)
# Testcases - Markdown Mind Map File
class Parse_And_Save_Show_IP_Interface_Brief_to_MD_Mindmap(aetest.Testcase):
    """Capture Show IP Interface Brief and
       transform to Markdown Mind Map documentation"""
    @aetest.test
    def save_markdown_mindmap_file(self,testbed,device_name):
        #Set current device in loop to self.device
        self.device = testbed.devices[device_name]
```

```
#Parse show ip interface brief to JSON
self.parsed_show_ip_interface_brief = \\
self.device.parse("show ip interface brief")

# Load the Jinja2 template
md_mindmap_show_ip_interface_brief = \\
self.device.api.load_jinja_template \\
(path="", file="markdown_mindmap.j2", \\
 to_parse_interfaces= \\
 self.parsed_show_ip_interface_brief['interface'])

#Save Markdown Table to file
with open("Show IP Interface Brief Mind Map.md", "w") as md_file:
    md_file.write(md_mindmap_show_ip_interface_brief)
```

The markdown mind map has a title for the first row that will act as the "root" of the horizontally scaling mind map. Markmaps use levels of nested # symbols, up to six levels deep per nesting, which transform into "branches" off the root level above, creating a visual mind map of the information (see Example 7-11).

Example 7-11   *The markdown_mindmap.j2 Jinja2 Template*

```
# Show IP Interface Brief
{% for interface in to_parse_interfaces %}
## {{ interface }}
### IP Address: {{ to_parse_interfaces[interface].ip_address }}
### Status: {{ to_parse_interfaces[interface].status }}
### Protocol: {{ to_parse_interfaces[interface].protocol }}
### Method: {{ to_parse_interfaces[interface].method }}
### Is Interface OK: {{ to_parse_interfaces[interface].interface_is_ok }}
{% endfor %}
```

Click the **Extensions** tab in VS Code and search for **Markmap**, as illustrated in Figure 7-24.

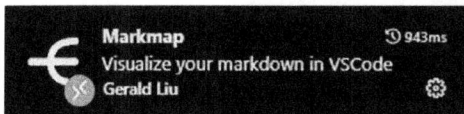

Figure 7-24   *Install Markmap for VS Code*

Then, on any valid markdown file (including our previous tabular markdown file), you can click the **Markmap** button in VS Code to preview the file as a mind map, as illustrated in Figure 7-25. Figure 7-26 shows the rendered Markmap mind map, whereas Figure 7-27 shows how even the tabular markdown can be rendered as a Markmap.

Figure 7-25   *Render as Markmap Mind Map Button*

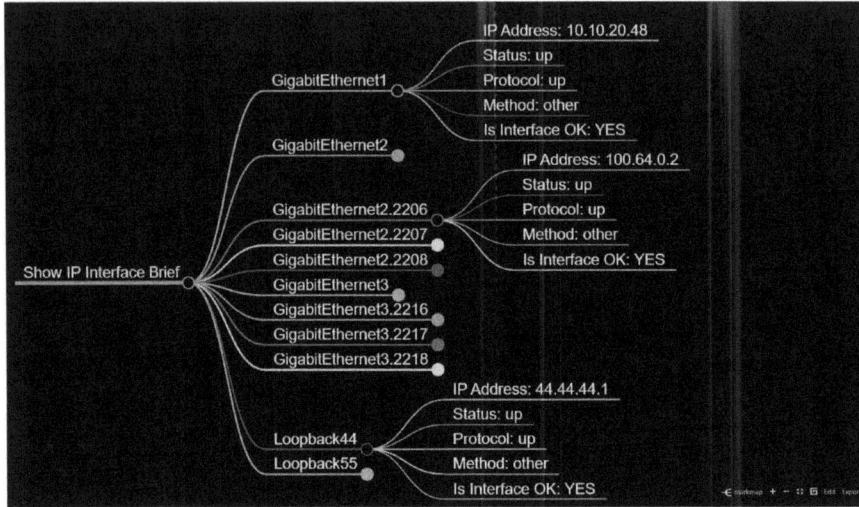

Figure 7-26   *show ip interface brief as a Markmap Mind Map*

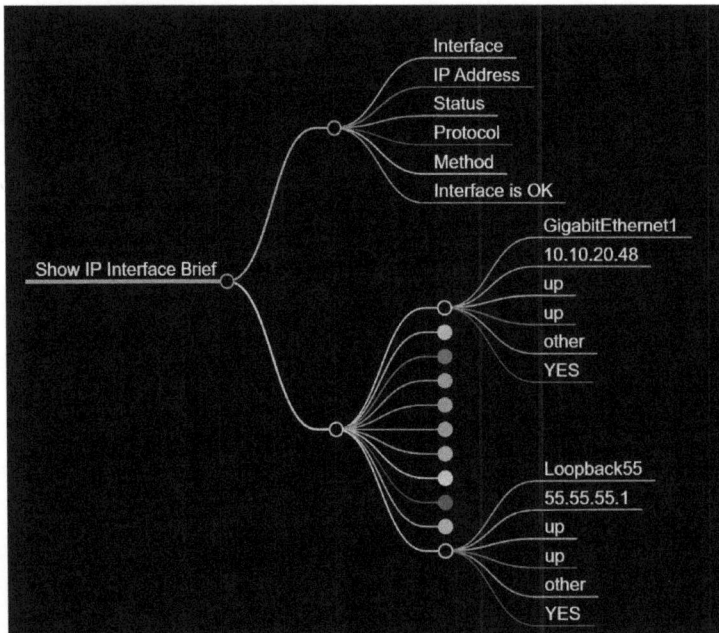

Figure 7-27   *show ip interface brief table as a Markmap Mind Map*

Mind maps are a whole new way to interactively visualize your pyATS network state with simple markdown formatting inside Jinja2 templates. Let's take a look at another type of markdown known as Mermaid.

Markdown: Mermaid Flowcharts

Mermaid (https://mermaid.js.org/) is a JavaScript-based diagramming and charting tool that enables developers and content creators to generate visualizations using simple text-based definitions. Integrated within Markdown, Mermaid allows for the creation of flow-charts, sequence diagrams, class diagrams, state diagrams, Gantt charts, and more, direct-ly within documentation, wikis, or other Markdown-supported platforms. Instead of embedding static images, users can embed live diagrams that are rendered on the fly. This integration offers a seamless way to incorporate visual aids into textual content, making complex ideas easier to convey and understand. The Mermaid syntax is both concise and readable, ensuring that even those unfamiliar with diagramming can quickly grasp its structure and start creating their own visual representations. Using the pyATS JSON, we can represent the **show ip interface brief** command in various ways using Mermaid, as demonstrated in Example 7-12.

Example 7-12 *Transform JSON Mermaid Flowchart*

```
aetest.loop.mark(MD_Mermaid_Flowchart,\\
                device_name=testbed.devices)
# Testcases - Markdown Mermaid Flowchart File
class MD_Mermaid_Flowchart(aetest.Testcase):
    """Capture Show IP Interface Brief and transform
       to Markdown Mermaid Flowchart documentation"""
    @aetest.test
    def save_markdown_mermaid_flowchart_file(self,testbed,device_name):
        #Set current device in loop to self.device
        self.device = testbed.devices[device_name]

        #Parse show ip interface brief to JSON
        self.parsed_show_ip_interface_brief = \\
        self.device.parse("show ip interface brief")

        # Load the Jinja2 template
        md_mermaid_flowchart_show_ip_interface_brief = \\
        self.device.api.load_jinja_template \\
        (path="", file="markdown_mermaid_flowchart.j2", \\
         to_parse_interfaces= \\
         self.parsed_show_ip_interface_brief['interface'])

        #Save Markdown Table to file
        with open("Show IP Interface Brief Mermaid Flowchart.md", "w") as md_file:
            md_file.write(md_mermaid_flowchart_show_ip_interface_brief)
```

Each Mermaid type has its own header and structure. When you're embedding Mermaid diagrams in a Markdown file, you typically wrap the Mermaid syntax in a code block with the language identifier **mermaid**. Example 7-13 shows the Jinja2 used to make a flowchart for **show ip interface brief**.

Example 7-13 *The markdown_mermaid_flowchart.j2 Jinja2 Template*

```
{% for interface in to_parse_interfaces %}
```mermaid
flowchart LR
    {{ interface }}[{{ interface }}]
    {{ interface }} -->
    {{ to_parse_interfaces[interface].ip_address }}[IP Address:
    {{ to_parse_interfaces[interface].ip_address }}]
    {{ interface }} -->
    {{ to_parse_interfaces[interface].status }}
    [Status: {{ to_parse_interfaces[interface].status }}]
    {{ interface }} -->
    {{ to_parse_interfaces[interface].protocol }}
    [Protocol: {{ to_parse_interfaces[interface].protocol }}]
    {{ interface }} -->
    {{ to_parse_interfaces[interface].method }}
    [Method: {{ to_parse_interfaces[interface].method }}]
    {{ interface }} -->
    {{ to_parse_interfaces[interface].interface_is_ok }}
    [Interface is OK: {{ to_parse_interfaces[interface].interface_is_ok }}]
```
{% endfor %}
```

Click the **Extensions** tab in VS Code and search for **Mermaid**, as shown in Figure 7-28.

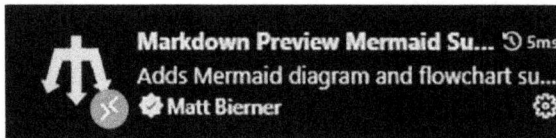

Figure 7-28 *Installing Mermaid Support for VS Code*

Then, on any valid markdown file (including our previous tabular markdown file), you can click the **Markmap** button in VS Code (see Figure 7-29) to preview the file as a mind map, as illustrated in Figure 7-30.

```mermaid
flowchart LR
    GigabitEthernet1[GigabitEthernet1]
    GigabitEthernet1 --> 10.10.20.48[IP Address: 10.10.20.48]
    GigabitEthernet1 --> up[Status: up]
    GigabitEthernet1 --> up[Protocol: up]
    GigabitEthernet1 --> other[Method: other]
    GigabitEthernet1 --> YES[Interface is OK: YES]
```

Figure 7-29   *Render as Mermaid Preview Button*

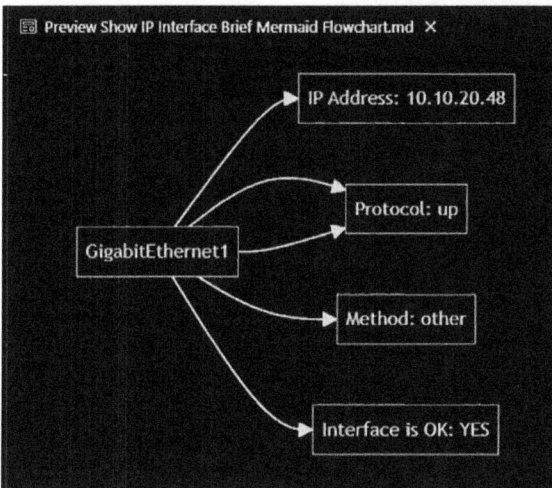

Figure 7-30   *show ip interface brief as Mermaid Flow Chart*

Mermaid flow charts are just one example of the powerful JavaScript-enabled Markdown format. Let's take a look at class diagrams, which are particularly useful for automated network documentation.

Markdown: Mermaid Class Diagrams

According to the article "Database Modelling in UML" by Geoffrey Sparks, "The class is the basic logical entity in the UML. It defines both the data and the behavior of a structural unit. A class is a template or model from which instances or objects are created at run time. When we develop a logical model such as a structural hierarchy in UML we explicitly deal with classes."

UML class diagrams can be automated from the pyATS JSON structured data using Jinja2 templates and Mermaid class diagrams, as demonstrated in Example 7-14.

Example 7-14   *Transform JSON Mermaid Class Diagram*

```
aetest.loop.mark(MD_Mermaid_Class, device_name=testbed.devices)
# Testcases - Markdown Mermaid Class File
class Parse_And_Save_Show_IP_Interface_Brief_to_MD_Mermaid_Class(aetest.Testcase):
    """Capture Show IP Interface Brief and transform
       to Markdown Mermaid Class documentation"""
    @aetest.test
    def save_markdown_mermaid_class_file(self,testbed,device_name):
        #Set current device in loop to self.device
        self.device = testbed.devices[device_name]

        #Parse show ip interface brief to JSON
        self.parsed_show_ip_interface_brief = \\
        self.device.parse("show ip interface brief")

        # Load the Jinja2 template
        md_mermaid_class_show_ip_interface_brief = \\
        self.device.api.load_jinja_template \\
        (path="", file="markdown_mermaid_class.j2", \\
         to_parse_interfaces= \\
         self.parsed_show_ip_interface_brief['interface'])

        #Save Markdown Table to file
        with open("Show IP Interface Brief Mermaid Class.md", "w") as md_file:
            md_file.write(md_mermaid_class_show_ip_interface_brief)
```

The class diagram is like the flowchart diagram, but it does have its own syntax, as shown in Example 7-15.

Example 7-15   *The markdown_mermaid_class.j2 Jinja2 Template*

```
{% for interface in to_parse_interfaces %}
```mermaid
classDiagram
    class Interface {
        +String name
        +String ipAddress
        +String status
        +String protocol
```
```

```
        +String method
        +bool interfaceIsOk
    }


    note for {{ interface }} "{{ interface }}\nIP Address:
    {{ to_parse_interfaces[interface].ip_address }}\n
    Status: {{ to_parse_interfaces[interface].status }}\n
    Protocol: {{ to_parse_interfaces[interface].protocol }}\n
    Method: {{ to_parse_interfaces[interface].method }}\n
    Interface is OK: {{ to_parse_interfaces[interface].interface_is_ok }}"
```
{% endfor %}

Figure 7-31 shows the rendered Mermaid markdown class diagram.

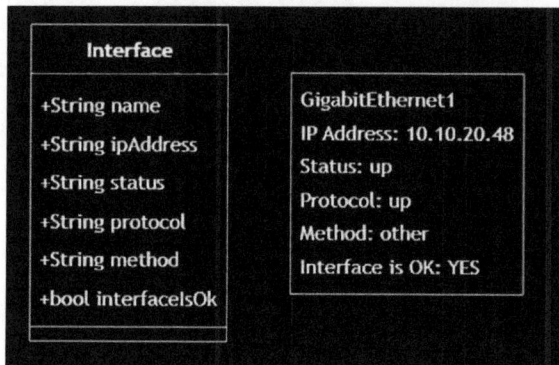

Figure 7-31   *show ip interface brief as Mermaid Class Diagram*

Using pyATS, Jinja2, and Mermaid, you could conceivably fully implement Unified Modeling Language (UML) for network documentation.

Markdown: Mermaid State Diagrams

In addition to flowcharts and class diagrams, Mermaid allows for state diagrams as well. Example 7-16 demonstrates how to create a new test that will use Jinja2 to generate a state diagram.

Example 7-16   *Transform JSON Mermaid State Diagram*

```
aetest.loop.mark(MD_Mermaid_State, device_name=testbed.devices)
# Testcases - Markdown Mermaid State File
class Parse_And_Save_Show_IP_Interface_Brief_to_MD_Mermaid_State(aetest.Testcase):
    """Capture Show IP Interface Brief and transform
        to Markdown Mermaid State documentation"""
    @aetest.test
    def save_markdown_mermaid_state_file(self,testbed,device_name):
        #Set current device in loop to self.device
        self.device = testbed.devices[device_name]

        #Parse show ip interface brief to JSON
        self.parsed_show_ip_interface_brief = \\
        self.device.parse("show ip interface brief")

        # Load the Jinja2 template
        md_mermaid_state_show_ip_interface_brief = \\
        self.device.api.load_jinja_template \\
        (path="", file="markdown_mermaid_state.j2",
         to_parse_interfaces= \\
         self.parsed_show_ip_interface_brief['interface'])

        #Save Markdown Table to file
        with open("Show IP Interface Brief Mermaid State.md", "w") as md_file:
            md_file.write(md_mermaid_state_show_ip_interface_brief)
```

The state diagram, again, has its own syntax, as shown in Example 7-17.

Example 7-17   *The markdown_mermaid_state.j2 Jinja2 Template*

```
{% for interface in to_parse_interfaces %}
```mermaid
stateDiagram-v2
        state {{ interface }} {
            [*] --> {{ to_parse_interfaces[interface].status }}

            state {{ to_parse_interfaces[interface].status }}

            {{ to_parse_interfaces[interface].status }}
            --> Up : Interface Up
            {{ to_parse_interfaces[interface].status }}
            --> Down : Interface Down
            {{ to_parse_interfaces[interface].status }}
            --> AdminDown : Admin Shutdown
        }
```
{% endfor %}
```

Figure 7-32 illustrates the rendered Mermaid markdown state diagram.

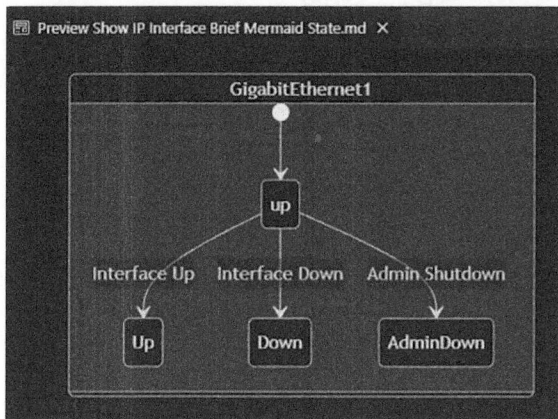

Figure 7-32   *show ip interface brief as Mermaid State Diagram*

Markdown: Mermaid Entity Relationship Diagrams

Along with flow charts, class diagrams, and state diagrams, we can also represent entity relationships in Mermaid diagrams. Peter Chin, in a paper titled "The Entry-Relationship Model—Toward a Unified View of Data," states the following:

> "An entity-relationship model (or ER model) describes interrelated things of interest in a specific domain of knowledge. A basic ER model is composed of entity types (which classify the things of interest) and specifies relationships that can exist between entities (instances of those entity types)."

Representing the network state as entity relationship diagrams is extremely powerful, and they are easy to make with Mermaid, as demonstrated in Example 7-18.

Example 7-18   *Transforming a JSON Mermaid Entity Relationship Diagram*

```
aetest.loop.mark(MD_Mermaid_Entity_Relationship, device_name=testbed.devices)
# Testcases - Markdown Mermaid Entity Relationship File
class MD_Mermaid_Entity_Relationship(aetest.Testcase):
    """Capture Show IP Interface Brief and transform
       to Markdown Mermaid Entity Relationship documentation"""
    @aetest.test
    def save_markdown_mermaid_entity_relationship_file(self,testbed,device_name):
        #Set current device in loop to self.device
        self.device = testbed.devices[device_name]
```

```
#Parse show ip interface brief to JSON
self.parsed_show_ip_interface_brief = \\
self.device.parse("show ip interface brief")

# Load the Jinja2 template
md_mermaid_entity_relationship_show_ip_interface_brief = \\
self.device.api.load_jinja_template\\
(path="", file="markdown_mermaid_entity_relationship.j2", \\
 to_parse_interfaces=self.parsed_show_ip_interface_brief['interface'])

#Save Markdown Table to file
with open\\
("Show IP Interface Brief Mermaid Entity Relationship.md", "w")\\
 as md_file:
    md_file.write(md_mermaid_entity_relationship_show_ip_interface_brief)
```

The entity relationship diagram, again, has its own syntax, as shown in Example 7-19.

Example 7-19   *The markdown_mermaid_entity_relationship.j2 Jinja2 Template*

```
{% for interface in to_parse_interfaces %}
```mermaid
erDiagram
    {{ interface }}
    {{ interface }}_IPAddress
    {{ interface }}_Status
    {{ interface }}_Protocol
    {{ interface }}_Method
    {{ interface }}_InterfaceIsOk

    {{ interface }} ||--o{ {{ interface }}_IPAddress : has
    {{ interface }} ||--o{ {{ interface }}_Status : has
    {{ interface }} ||--o{ {{ interface }}_Protocol : uses
    {{ interface }} ||--o{ {{ interface }}_Method : uses
    {{ interface }} ||--o{ {{ interface }}_InterfaceIsOk : status
```
{% endfor %}
```

Figure 7-33 shows the rendered Mermaid markdown entity relationship diagram.

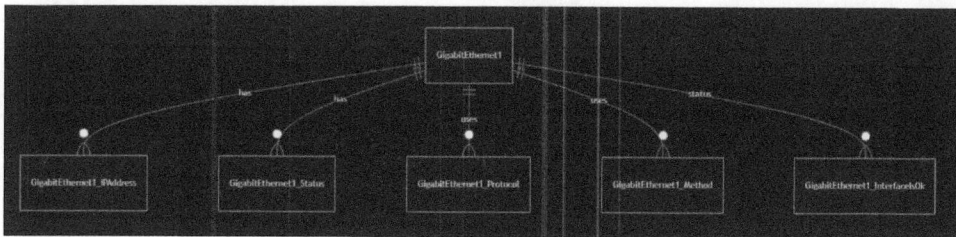

Figure 7-33 *show ip interface brief as Mermaid Entity Relationship Diagram*

Markdown: Mermaid Mind Maps

Similar to Markmap, Mermaid also supports its own interpretation of mind maps. It is not interactive, nor can it be exported to HTML, but it does render natively in GitHub and is a handy lightweight version that does not require the Markmap extension. Example 7-20 shows how to write the new pyATS testcase, whereas Example 7-21 shows the Jinja2 template format for Mermaid mind maps.

Example 7-20 *Transform JSON Mermaid Mind Map*

```
aetest.loop.mark(MD_Mermaid_Mind_Map, device_name=testbed.devices)
# Testcases - Markdown Mermaid Mind Map File
class MD_Mermaid_Mind_Map(aetest.Testcase):
    """Capture Show IP Interface Brief and transform
       to Markdown Mermaid Mind Map documentation"""
    @aetest.test
    def save_markdown_mermaid_mind_map_file(self,testbed,device_name):
        #Set current device in loop to self.device
        self.device = testbed.devices[device_name]

        #Parse show ip interface brief to JSON
        self.parsed_show_ip_interface_brief = \\
        self.device.parse("show ip interface brief")

        # Load the Jinja2 template
        md_mermaid_mind_map_show_ip_interface_brief = \\
        self.device.api.load_jinja_template\\
        (path="", file="markdown_mermaid_mind_map.j2", \\
        to_parse_interfaces=\\
        self.parsed_show_ip_interface_brief['interface'])

        #Save Markdown Table to file
        with open("Show IP Interface Brief Mermaid Mind Map.md", "w") as md_file:
            md_file.write(md_mermaid_mind_map_show_ip_interface_brief)
```

Example 7-21   *The markdown_mermaid_mindmap.j2 Jinja2 Template*

```
```mermaid
mindmap
  root((Network Interfaces))
{% for interface in to_parse_interfaces %}
    {{ interface }}
      IP Address
        {{ to_parse_interfaces[interface].ip_address }}
      Status
        {{ to_parse_interfaces[interface].status }}
      Protocol
        {{ to_parse_interfaces[interface].protocol }}
      Method
        {{ to_parse_interfaces[interface].method }}
      Interface is OK
        {{ to_parse_interfaces[interface].interface_is_ok }}
{% endfor %}
```
```

Figure 7-34 shows the rendered Mermaid mind map diagram.

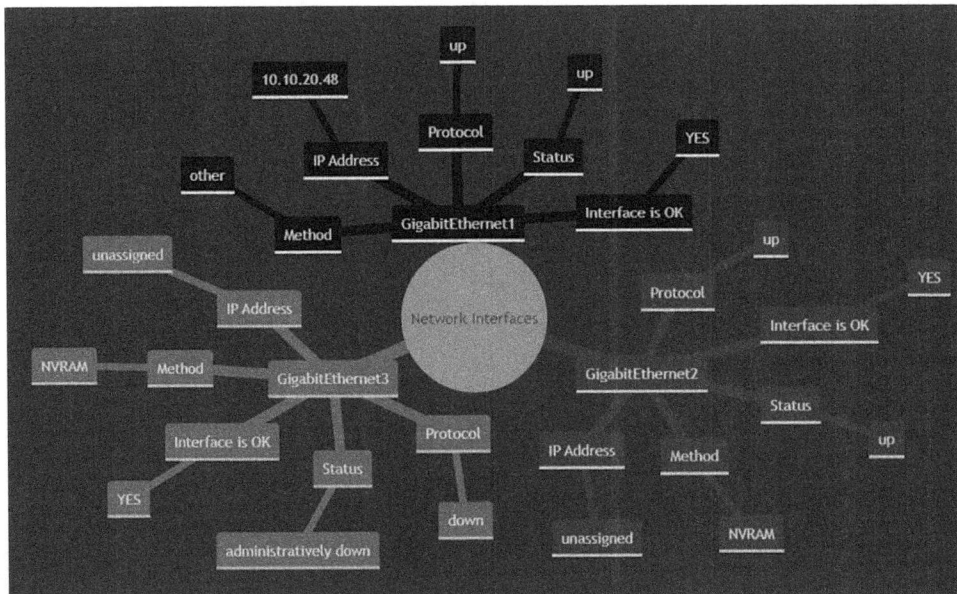

Figure 7-34   *show ip interface brief as Mermaid Mind Map*

As demonstrated, Mermaid adds an entirely new aspect to markdown—powerful business-ready visualizations. For a more interactive, tabular experience, basic Hypertext Markup Language (HTML) can be used to create tables. Once we have an HTML table, we can use a free, open-source set of tools from https://www.datatables.net to transform the basic table into an interactive experience.

HTML

HTML is the standard language used to create and design web pages. One of the fundamental elements in HTML is the table, which allows web developers to organize and display data in rows and columns. To create a table in HTML, specific tags such as <table>, <tr>, <td>, and <th> are used. The <table> tag initiates the table structure, <tr> defines a row, <td> represents a data cell, and <th> is used for table headers. By nesting these tags appropriately, developers can structure data in a tabular format, making it easier for users to read and understand, as demonstrated in Example 7-22.

Example 7-22   *Transform JSON to HTML*

```
aetest.loop.mark(HTML, device_name=testbed.devices)
# Testcases - Markdown Mermaid HTML File
class Parse_And_Save_Show_IP_Interface_Brief_to_HTML(aetest.Testcase):
    """Capture Show IP Interface Brief and transform to HTML documentation"""
    @aetest.test
    def save_html_file(self,testbed,device_name):
        #Set current device in loop to self.device
        self.device = testbed.devices[device_name]

        #Parse show ip interface brief to JSON
        self.parsed_show_ip_interface_brief = \\
        self.device.parse("show ip interface brief")

        # Load the Jinja2 template
        html_show_ip_interface_brief = \\
        self.device.api.load_jinja_template\\
        (path="", file="html.j2", \\
         to_parse_interfaces=\\
         self.parsed_show_ip_interface_brief['interface'])

        #Save Markdown Table to file
        with open("Show IP Interface Brief.html", "w") as md_file:
            md_file.write(html_mermaid_mind_map_show_ip_interface_brief)
```

HTML is very simple and straightforward syntax to make a basic table from the JSON. You can even use the CSV template as a base template and replace the commas with the appropriate opening and closing HTML tags, as demonstrated in Example 7-23.

Example 7-23   *The html.j2 Jinja2 Template*

```
<h1>Show IP Interface Brief</h1>
<table>
  <thead>
    <tr>
      <th>Interface</th>
      <th>IP Address</th>
      <th>Status</th>
      <th>Protocol</th>
      <th>Method</th>
      <th>Interface is OK</th>
    </tr>
  </thead>
  <tbody>
{%- for interface in to_parse_interfaces %}
<tr>
    <td>{{ interface }}</td>
    <td>{{ to_parse_interfaces[interface].ip_address }}</td>
    <td>{{ to_parse_interfaces[interface].status }}</td>
    <td>{{ to_parse_interfaces[interface].protocol }}</td>
    <td>{{ to_parse_interfaces[interface].method }}</td>
    <td>{{ to_parse_interfaces[interface].interface_is_ok }}</td>
  </tr>
{%- endfor %}
  </tbody>
</table>
```

Figure 7-35 shows a basic HTML page rendered by Jinja2.

Figure 7-35   *show ip interface brief as a Basic HTML Table*

This simple table can be enhanced in a few ways. First, we can add some simple logic and change the colors of cells to red or green based on their state, as demonstrated in Example 7-24.

Example 7-24 *Enhancing the HTML with Jinja2 Logic*

```
{% if to_parse_interfaces[interface].status == "up"%}
    <td style="color: green;">{{ to_parse_interfaces[interface].status }}</td>
{% else %}
    <td style="color: red;">{{ to_parse_interfaces[interface].status }}</td>
{% endif %}
    <td>{{ to_parse_interfaces[interface].protocol }}</td>
    <td>{{ to_parse_interfaces[interface].method }}</td>
{% if to_parse_interfaces[interface].status == "up"%}
    <td style="color: green;">
    {{ to_parse_interfaces[interface].interface_is_ok }}
    </td>
{% else %}
    <td style="color: red;">
    {{ to_parse_interfaces[interface].interface_is_ok }}
    </td>
{% endif %}
```

Figure 7-36 shows how we can enhance, with HTML tags, the appearance of the data, such as using colors like red and green to indicate the health of the interfaces.

Figure 7-36 *show ip interface brief as an HTML Table with Logic and Color*

Datatables

Datatables.net is a highly flexible and feature-rich jQuery plugin designed to enhance the functionality of standard HTML tables. By integrating Datatables.net with your web application, you can effortlessly transform a basic HTML table into a dynamic table with advanced features such as pagination, sorting, searching, and more. One of the standout features of Datatables.net is its ability to automatically detect table headers and footers. The plugin uses the headers and footers to generate controls such as sorting arrows and search fields. The visual appearance and interactivity are achieved through a combination of CSS and JavaScript provided by the Datatables.net library. This means that developers don't have to write extensive code to get a professional and functional table; instead, they can rely on the power of Datatables.net to handle the heavy lifting.

Create a new Jinja2 template called **datatable_headers.j2** (see Example 7-25) and another called **datatable_footers.j2** (see Example 7-26). We will *include* these templates inside our base html.j2 file. The headers file will contain links to Cascading Style Sheets (CSS) and JavaScript (JS), while the footer will contain an actual JavaScript. The only other modification to our HTML template is providing the table an ID.

Example 7-25 *datatable_header.j2 Example*

```
<html>
<head>
<script src="https://ajax.googleapis.com/ajax/libs/jquery/3.5.1/jquery.min.js">
</script>
<script src="https://cdn.datatables.net/1.11.4/js/jquery.dataTables.min.js">
</script>
<script src="https://cdn.datatables.net/buttons/2.0.0/js/dataTables.buttons.min.js">
</script>
<script src="https://cdnjs.cloudflare.com/ajax/libs/jszip/3.1.3/jszip.min.js">
</script>
 script src="https://cdnjs.cloudflare.com/ajax/libs/pdfmake/0.1.53/pdfmake.min.js">
</script>
<script src="https://cdnjs.cloudflare.com/ajax/libs/pdfmake/0.1.53/vfs_fonts.js">
</script>
 script src="https://cdn.datatables.net/buttons/2.0.0/js/buttons.html5.min.js">
</script>
 script src="https://cdn.datatables.net/buttons/2.0.0/js/buttons.print.min.js">
</script>
<script src="https://cdn.datatables.net/colreorder/1.5.4/js/dataTables.colReorder.
min.js"></script>
<script src="https://cdn.datatables.net/buttons/2.0.0/js/buttons.colVis.min.js">
</script>
<script src="https://cdn.datatables.net/keytable/2.6.4/js/dataTables.keyTable.min.js">
</script>
<script src="https://cdn.datatables.net/select/1.3.3/js/dataTables.select.min.js">
</script>
<script src="https://cdn.datatables.net/fixedheader/3.1.9/js/dataTables.fixedHeader.
min.js"></script>
<link rel="stylesheet" href="https://cdn.datatables.net/fixedheader/3.1.9/css/
fixedHeader.dataTables.min.css">
<link rel="stylesheet" href="https://cdn.datatables.net/select/1.3.3/css/select.
dataTables.min.css">
<link rel="stylesheet" href="https://cdn.datatables.net/keytable/2.6.4/css/keyTable.
dataTables.min.css">
<link rel="stylesheet" href="https://cdn.datatables.net/1.11.4/css/jquery.
dataTables.min.css">
</head>
```

Example 7-26   *datatable_footer.j2 Example*

```
<script type = "text/javascript">
  $(document).ready(function(){
    $('#datatable thead tr')
    .clone(true)
    .addClass('filters')
    .appendTo('#datatable thead');

    var table = $('#datatable').DataTable({
        keys: true,
        dom: 'Bfrtip',
        lengthMenu: [
            [ 10, 25, 50, -1 ],
            [ '10 rows', '25 rows', '50 rows', 'Show all' ]
        ],
        buttons: [
        'pageLength','colvis','copy', 'csv', 'excel', 'pdf', 'print'
        ],
        colReorder: true,
        select: true,
        orderCellsTop: true,
        fixedHeader: true,
        initComplete: function () {
        var api = this.api();

        // For each column
        api
            .columns()
            .eq(0)
            .each(function (colIdx) {
                // Set the header cell to contain the input element
                var cell = $('.filters th').eq(
                    $(api.column(colIdx).header()).index()
                );
                var title = $(cell).text();
                $(cell).html('<input type="text" placeholder="' + title + '" />');

                // On every keypress in this input
                $(
                    'input',
```

```
                    $('.filters th').eq($(api.column(colIdx).header()).index())
            )

            .off('keyup change')
            .on('keyup change', function (e) {
                e.stopPropagation();

                // Get the search value
                $(this).attr('title', $(this).val());
                var regexr = '({search})';
                //$(this).parents('th').find('select').val();

                var cursorPosition = this.selectionStart;
                // Search the column for that value
                api
                    .column(colIdx)
                    .search(
                        this.value != ''
                            ? regexr.replace('{search}',
                            '((( ' + this.value + ')))')
                            : '',
                        this.value != '',
                        this.value == ''
                    )
                    .draw();

                $(this)
                    .focus()[0]
                    .setSelectionRange(cursorPosition, cursorPosition);
            });
        });
    },
});
});
</script>
</body></html>
```

Example 7-27 shows the updated Jinaj2 template for the HTML page that includes
headers and footers to create a data table instead of a basic HTML page.

Example 7-27 *Updated html.j2*

```
{%- include 'datatable_header.j2' %}
<h1>Show IP Interface Brief</h1>
<table id="datatable">
  <thead>
    <tr>
      <th>Interface</th>
      <th>IP Address</th>
      <th>Status</th>
      <th>Protocol</th>
      <th>Method</th>
      <th>Interface is OK</th>
    </tr>
  </thead>
  <tbody>
{%- for interface in to_parse_interfaces %}
<tr>
    <td>{{ interface }}</td>
    <td>{{ to_parse_interfaces[interface].ip_address }}</td>
{% if to_parse_interfaces[interface].status == "up"%}
    <td style="color: green;">{{ to_parse_interfaces[interface].status }}</td>
{% else %}
    <td style="color: red;">{{ to_parse_interfaces[interface].status }}</td>
{% endif %}
    <td>{{ to_parse_interfaces[interface].protocol }}</td>
    <td>{{ to_parse_interfaces[interface].method }}</td>
{% if to_parse_interfaces[interface].status == "up"%}
    <td style="color: green;">
    {{ to_parse_interfaces[interface].interface_is_ok }}
    </td>
{% else %}
    <td style="color: red;">
    {{ to_parse_interfaces[interface].interface_is_ok }}
    </td>
{% endif %}
  </tr>
{%- endfor %}
  </tbody>
</table>
{%- include 'datatable_footer.j2' %}
```

Figure 7-37 shows the enhanced data table HTML page, complete with search, sort, filter, and a variety of other capabilities such as printing and exporting to other file types.

Figure 7-37 *show ip interface brief as an HTML Datatable*

Now we have pagination, sort, search, filter, print, and many more options such as reordering columns by simply including the datatable header and footer code with our basic HTML table.

Summary

In the realm of network automation, pyATS stands out as a robust tool that empowers network engineers to test networks *and* create automated documentation. This chapter delves into the intricacies of using pyATS to generate network documentation in various formats, including JSON, YAML, CSV, Markdown tables, Mermaid diagrams, and HTML integrated with Datatables. At its core, pyATS facilitates the extraction of network data, which can then be transformed and rendered into structured formats. The use of Jinja2 templates further enhances this process, allowing for the creation of customized CSV files, Markdown tables, and other formats tailored to specific needs. Particularly noteworthy is the integration with Mermaid, a popular tool for generating diagrams, and HTML combined with Datatables, which transforms basic tables into dynamic, interactive ones. The power of pyATS lies not just in its versatility but also in its ease of use and safety. For those embarking on their network automation journey, pyATS serves as an excellent starting point. Its intuitive nature ensures that even those new to automation can harness its capabilities without a steep learning curve. Moreover, because it runs in "read-only" mode, network operations are ensured to remain uninterrupted and secure. For enterprises, the value proposition of pyATS is undeniable. In an era where network complexities are everincreasing, having fully automated and up-to-date documentation is not just a luxury but a necessity. pyATS provides this, ensuring that network configurations, topologies, and other critical data are always at one's fingertips. This not only aids in troubleshooting and network optimization but also in compliance and auditing processes. In conclusion, pyATS is more than just a tool; it's a transformative solution for modern network management. Its combination of power, ease, and ability to safely gather network state makes it an invaluable asset for any enterprise aiming for efficient and automated network operations.

References

Geoffrey Sparks, "Database Modelling in UML," https://www.methodsandtools.com/archive/archive.php?id=9

Peter Chin, "The Entry-Relationship Model—Toward a Unified View of Data," http://faculty.ndhu.edu.tw/~wpyang/DatabaseTeachingCenter/File2AdvancedDB/4References/erd.pdf

Chapter 8

Automated Network Testing

In this chapter, we delve into the realm of automated network testing, leveraging the robust capabilities of Cisco's pyATS framework in conjunction with the test-driven development (TDD) methodology. The interaction of pyATS and TDD paves the way for a meticulous testing paradigm, enabling not only safe, read-only testing but also actionable testing that encompasses configuration management in the face of failed tests. Through a pragmatic lens, we will explore real-world use cases illustrating how these intertwined methodologies foster a resilient, self-healing network infrastructure. By engendering a proactive testing culture, we aim to significantly mitigate network vulnerabilities and ensure a higher standard of network reliability and performance. This chapter is set to equip you with the knowledge and practical insights to navigate the complex landscape of automated network testing and configuration management, showcasing the profound impact of a well-orchestrated testing strategy on network robustness and operational excellence. We can connect to our devices using traditional SSH or use modern interfaces such as RESTCONF to gather the network state. This connection approach is important in determining whether you will be using pyATS parsers or RESTCONF YANG endpoints.

This chapter covers the following topics:

- An approach to network testing
- Software version testing
- Interface testing
- Neighbor testing
- Reachability testing
- Intent-validation testing
- Feature testing

An Approach to Network Testing

Heading into the domain of network testing is akin to navigating a labyrinth, with a myriad of pathways unfolding with every step. A well-thought-out approach is our compass in this scenario, guiding us through the intricacies and ensuring that we emerge triumphant on the other side. Embracing the principles of TDD and adapting its iterative rhythm to the network's beat form the crux of our strategy. It's like having a friendly debate with the network—proposing a point (our test), seeing how the network responds, and then tweaking either the network or our stance to reach a consensus. Our primary metric for assessing the network's condition is a specific benchmark parameter. Alongside this benchmark, I've initialized a descriptively named variable. If this variable is set to True by the conclusion of our evaluation, it indicates that the network did not meet our expectations and the test is deemed unsuccessful.

It's pivotal to highlight how networks and the TDD approach are integral to the software development lifecycle (SDLC). In the realm of network testing, TDD is not merely a methodology; it embodies the essence of software testing, a core part of the SDLC. By following a test-driven approach, we preemptively address potential defects and ensure that each phase of the lifecycle meets the prescribed quality standards through continuous validation. This proactive stance not only enhances the reliability and efficiency of the network but also ensures a seamless integration of network functionalities within the broader software development process.

Additionally, I like to jazz up my pyATS logs using the Rich library. Rich is a Python library for rich text used to beautify console and HTML outputs, making logs a visual delight rather than a chore to sift through. With Rich, my pyATS logs transform into a colorful, easy-to-decipher narrative of the test journey, where red and green indicators instantly tell me if a test has passed or failed. It's like having a traffic light system for my test results, making it super-intuitive to interpret the outcomes and elevating my pyATS test jobs to a production-grade level of finesse. This marriage of TDD, pyATS, and Rich not only ensures a robust testing framework but also makes the process visually engaging and professional, bridging the gap between meticulous testing and user-friendly reporting.

In an ideal world, we present our argument, the network rebuffs, we fine-tune the network's stance, and, voilà, we are in agreement. But sometimes, it's our argument that needs a slight rephrasing, maybe changing a stringent equality check to a more flexible greater-than or less-than comparison. This iterative dance of adjustments is what fine-tunes our network, leading to a harmonious dialogue that ensures robust performance. Here's a glimpse into some general good practices that form the bedrock of our approach to network testing:

- Keep the unit tests small.

- Keep the structure as setup/execute/cleanup.

- Always test a well-known state.

- Limit, or eliminate, dependencies between tests.

- Complex is fine; complicated is not.

- Avoid "all-knowing" tests.

- Set up a threshold and test against that threshold.

- Evaluate at the end of the test to pass or fail the test.

This methodical approach to testing, blended with the flexibility to adapt, sets the stage for fostering a robust, reliable, and responsive network infrastructure.

Software Version Testing

Let's start with a straightforward example of automated network testing, beginning with the pyATS parsed **show version** JSON. There are two approaches to follow TDD:

- Set the threshold to the target software version; fail the test; upgrade the software; pass the test.

- "Mock" this by setting the threshold to be "greater than" the current known installed version; fail the test; change the threshold to be "equals."

Using the second approach, and the Cisco DevNet Always-On IOS XE Sandbox (which you can use this code against to follow along if you don't have an IOS XE device yourself), can be used to test this code for free and without risk. Inside a Python virtual environment with pyATS installed first, make the **testbed.yaml** file, as shown in Example 8-1.

Example 8-1 *Cisco DevNet Always-On IOS-XE Sandbox pyATS testbed.yaml*

```
---
devices:
  Cat8000v:
    alias: "Sandbox Router"
    type: "router"
    os: "iosxe"
    platform: Cat8000v
    credentials:
      default:
        username: developer
        password: C1sco12345
    connections:
      cli:
        protocol: ssh
        ip: devnetsandboxiosxe.cisco.com
        port: 22
        arguments:
          connection_timeout: 360
```

Next, we will create a pyATS job file, called **ios_xe_version_job.py**, that will load the preceding testbed and call the Python script **ios_xe_version.py**, as shown in Example 8-2.

Example 8-2   *ios_xe_version_job.py pyATS Job File*

```
import os
from genie.testbed import load

def main(runtime):

    # ----------------
    # Load the testbed
    # ----------------
    if not runtime.testbed:
        # If no testbed is provided, load the default one.
        # Load default location of Testbed
        testbedfile = os.path.join('testbed.yaml')
        testbed = load(testbedfile)
    else:
        # Use the one provided
        testbed = runtime.testbed

    # Find the location of the script in relation to the job file
    testscript = os.path.join(os.path.dirname(__file__), 'ios_xe_version.py')

    # run script
    runtime.tasks.run(testscript=testscript, testbed=testbed)
```

Now we need to write the pyATS testscript to test the **show version** command output and pass or fail the test based on a threshold. The following steps will help you write the pyATS testscript:

Step 1.   **Set up logging:** The script begins by importing the necessary modules and setting up logging using the standard logging module in Python. This allows the script to display informative messages as it runs.

Step 2.   **Common setup (common_setup class):** This class contains setup procedures that are common to the entire test process.

- **connect_to_devices:** This function uses the pyATS framework to establish connections to all devices defined in the testbed.

- **loop_mark:** This function is crucial for running the testcase for each device in the testbed.

Step 3.   **Testcase (Test_Cisco_IOS_XE_Version class):** This class contains the core logic for the version-checking process:

- **setup:** Prepares for the upcoming tests by setting the current device in the loop.

- **get_parsed_version:** Uses the pyATS framework to execute the **show version** command on the device and then parses the output. The parsed output is stored in JSON format.

- **create_file:** Writes the parsed JSON output to a file named based on the device alias.

- **test_version:** This is where the primary logic resides:

    The desired version (threshold) is set to **"17.4"**.

    A table is set up using the Rich library to display results.

    The version retrieved from the device is compared to the threshold.

    If the versions don't match, the test fails for that device, and the result is added to the table with a Failed status. Otherwise, it's marked Passed.

    The table is then displayed in the terminal.

Step 4.   **Common cleanup (CommonCleanup class):** Contains cleanup procedures, like disconnecting from the devices after the tests are complete.

Step 5.   **Execution block:** If the script is run as a standalone program, the **aetest. main()** function is called, initiating the entire test process.

Example 8-3 shows the pyATS testscript in its entirety.

Example 8-3   *ios_xe_version.py pyATS Test Script*

```
import json
import logging
from pyats import aetest
from rich.table import Table
from rich.console import Console
from genie.utils.diff import Diff
from pyats.log.utils import banner


# ----------------
# Get logger for script
# ----------------

log = logging.getLogger(__name__)
```

```python
# ----------------
# AE Test Setup
# ----------------
class common_setup(aetest.CommonSetup):
    """Common Setup section"""
# ----------------
# Connected to devices
# ----------------
    @aetest.subsection
    def connect_to_devices(self, testbed):
        """Connect to all the devices"""
        testbed.connect()
# ----------------
# Mark the loop for Show Version
# ----------------
    @aetest.subsection
    def loop_mark(self, testbed):
        aetest.loop.mark(Test_Cisco_IOS_XE_Version, device_name=testbed.devices)


# ----------------
# Test Case #1
# ----------------
class Test_Cisco_IOS_XE_Version(aetest.Testcase):
    """Parse pyATS show version and test it against a threshold"""

    @aetest.test
    def setup(self, testbed, device_name):
        """ Testcase Setup section """
        # Set current device in loop as self.device
        self.device = testbed.devices[device_name]

    @aetest.test
    def get_parsed_version(self):
        parsed_version = self.device.parse("show version")
        # Get the JSON payload
        self.parsed_json=parsed_version

    @aetest.test
    def create_file(self):
        # Create .JSON file
        with open(f'{self.device.alias}_Show_Version.json', 'w') as f:
            f.write(json.dumps(self.parsed_json, indent=4, sort_keys=True))
```

```python
    @aetest.test
    def test_version(self):
        # Test for version 17.4
        version_threshold = "17.4"
        self.failed_version = {}
        table = Table(title="Show Version")
        table.add_column("Device", style="cyan")
        table.add_column("Threshold Version", style="magenta")
        table.add_column("Running Version", style="magenta")
        table.add_column("Passed/Failed", style="green")
        version = self.parsed_json['version']['version']
        if version:
            if version != version_threshold:
                table.add_row\
                (self.device.alias,version_threshold,version,'Failed',style="red")
                self.failed_version = version
            else:
                table.add_row\
                (self.device.alias,version_threshold,version,'Passed',style="green")
        else:
            table.add_row(self.device.alias,'N/A','N/A','N/A',style="yellow")
        # display the table
        console = Console(record=True)
        with console.capture() as capture:
            console.print(table,justify="left")
        log.info(capture.get())

        if self.failed_version:
            self.failed()
        else:
            self.passed()

class CommonCleanup(aetest.CommonCleanup):
    @aetest.subsection
    def disconnect_from_devices(self, testbed):
        testbed.disconnect()

# for running as its own executable
if __name__ == '__main__':
    aetest.main()
```

Inside your virtual environment, run the **pyats run job ios_xe_version_job.py** command to run the script. We will be testing, and failing the test, IOS XE software version of the Cisco DevNet Always-On Sandbox. Note that the Rich Python library needs to be installed into the virtual environment before you execute this job.

```
(venv)$ pip install rich
(venv)$ pyats run job ios_xe_version_iob.py
```

Figure 8-1 through Figure 8-4 illustrate the pyATS console logs, including the Rich tabular output.

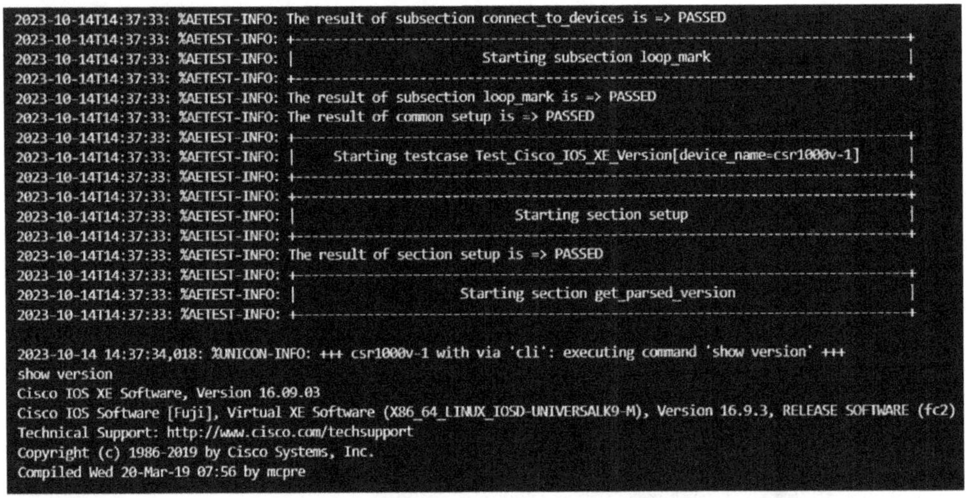

Figure 8-1 *Testing IOS XE Software Version*

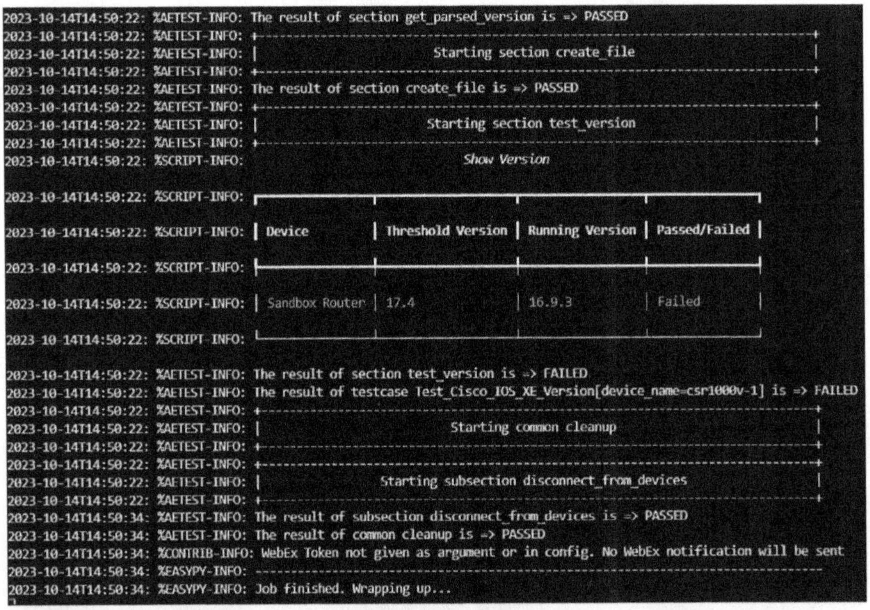

Figure 8-2 *Rich Table of pyATS Test Results Showing Initial Failed Test*

```
2023-10-14T14:37:46: %EASYPY-INFO: +--------------------------------------------------------+
2023-10-14T14:37:46: %EASYPY-INFO: |                   Task Result Summary                   |
2023-10-14T14:37:46: %EASYPY-INFO: +--------------------------------------------------------+
2023-10-14T14:37:46: %EASYPY-INFO: Task-1: ios_xe_version.common_setup                    PASSED
2023-10-14T14:37:46: %EASYPY-INFO: Task-1: ios_xe_version.Test_Cisco_IOS_XE_Version[device_name=csr100...  FAILED
2023-10-14T14:37:46: %EASYPY-INFO: Task-1: ios_xe_version.common_cleanup                  PASSED
2023-10-14T14:37:46: %EASYPY-INFO:
2023-10-14T14:37:46: %EASYPY-INFO: +--------------------------------------------------------+
2023-10-14T14:37:46: %EASYPY-INFO: |                   Task Result Details                   |
2023-10-14T14:37:46: %EASYPY-INFO: +--------------------------------------------------------+
2023-10-14T14:37:46: %EASYPY-INFO: Task-1: ios_xe_version
2023-10-14T14:37:46: %EASYPY-INFO: |-- common_setup                                        PASSED
2023-10-14T14:37:46: %EASYPY-INFO: |    |-- connect_to_devices                            PASSED
2023-10-14T14:37:46: %EASYPY-INFO: |    `-- loop_mark                                     PASSED
2023-10-14T14:37:46: %EASYPY-INFO: |-- Test_Cisco_IOS_XE_Version[device_name=csr1000v-1]  FAILED
2023-10-14T14:37:46: %EASYPY-INFO: |    |-- setup                                         PASSED
2023-10-14T14:37:46: %EASYPY-INFO: |    |-- get_parsed_version                            PASSED
2023-10-14T14:37:46: %EASYPY-INFO: |    |-- create_file                                   PASSED
2023-10-14T14:37:46: %EASYPY-INFO: |    `-- test_version                                  FAILED
2023-10-14T14:37:46: %EASYPY-INFO: `-- common_cleanup                                     PASSED
2023-10-14T14:37:46: %EASYPY-INFO:      `-- disconnect_from_devices                       PASSED
2023-10-14T14:37:46: %EASYPY-INFO: Sending report email...
2023-10-14T14:37:46: %EASYPY-INFO: Missing SMTP server configuration, or failed to reach/authenticate/send mail. Result notification email failed to send.
2023-10-14T14:37:46: %EASYPY-INFO: Done!

Pro Tip
-------
  Try the following command to view your logs:
      pyats logs view
```

Figure 8-3 *pyATS Test Results Summary*

Figure 8-4 *pyATS HTML Logs Respecting Rich Markup*

Now that we have failed our first automated network test, it's essential to delve into the JSON artifact that was generated during the process. This artifact contains valuable information, including the current software version running on our device. As of the writing of this book, the version on the csr1000v is identified as 16.9.3. To align our testing criteria with this newfound insight, we need to adjust our code. Specifically, within the ios_xe_version.py file, locate the **test_version** method of the **Test_Cisco_IOS_XE_Version** class. There, you will find a variable named **version_threshold**. Initially set to "**17.4**", this should now be updated to "**16.9.3**" to reflect the current software version of our network device. After making this adjustment, rerun the test. You should observe that the test now passes, a transition that is illustrated in Figure 8-5 through Figure 8-7.

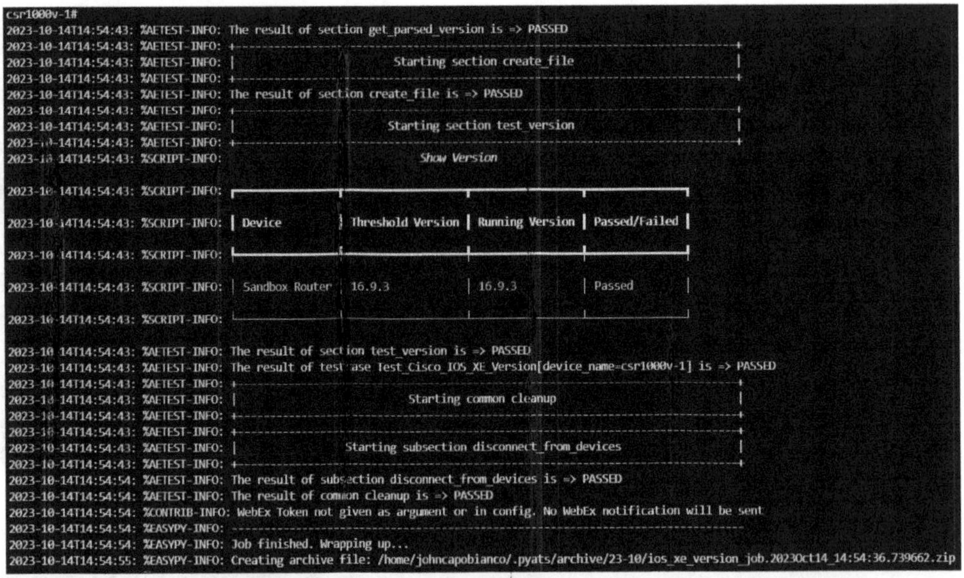

Figure 8-5 *Passing show version test as Rich Table*

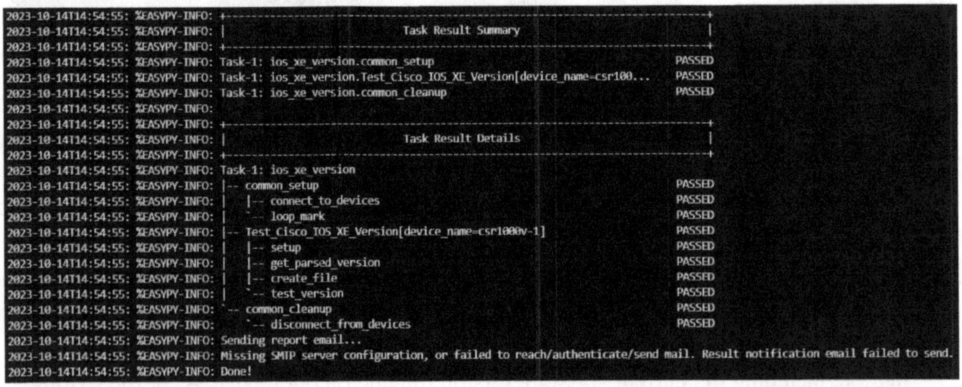

Figure 8-6 *Passing show version test pyATS Log Summary*

Figure 8-7 *Passing show version test in pyATS HTML Log Viewer*

In the preceding section, we embarked on a foundational journey into network test automation using the pyATS framework. We saw firsthand the benefits of automating repetitive tasks, such as version checking. Through the prism of TDD, we visualized how tests can first fail (highlighting a potential issue) and then pass once the necessary adjustments are made. This iterative process not only ensures consistency across your network but also offers a safety net against configuration drifts and discrepancies. In addition to streamlining network validation, running a pyATS job generates valuable artifacts that encapsulate the test outcomes. These job artifacts, specifically in XML and JSON formats, provide structured records of the results, facilitating easy aggregation, analysis, and long-term tracking of network health and behavior. The XML files are particularly useful for integration with other tools that utilize XML for data interchange, while the JSON files offer a lightweight, human-readable format that can be readily used for web applications and data processing pipelines. Together, they form a comprehensive audit trail for network changes and test verifications.

The artifacts produced—from the structured JSON outputs, to the visually appealing Rich tables and the comprehensive HTML logs—underscore the importance of clear and actionable feedback in network testing. These artifacts don't just serve as proof-of-test results; they facilitate seamless information exchange.

However, network testing is a vast domain, and version checking is just the tip of the iceberg. As our networks grow in complexity, so does the need for more intricate and comprehensive testing methodologies.

Interface Testing

Interfaces, of all kinds, can be easily learned and tested, at scale. In today's networked environments, ensuring the integrity and efficiency of interfaces, be they physical or virtual, is paramount. Many outages, latencies, and bottlenecks can be traced back to Layer 1/Layer 2 issues. Whether you're dealing with a vast enterprise network or a more confined setup, consistent testing against predefined thresholds is essential to maintaining smooth operations. With the advent of the pyATS framework, this rigorous testing is no longer a tedious manual process. Instead, it empowers network engineers to automate tests and analyze results in real time.

In the forthcoming sections, our focus shifts to leveraging pyATS, a comprehensive and versatile testing framework designed for network environments. This exploration aims to illustrate the utility of pyATS for evaluating the performance and compliance of network interfaces against predefined benchmarks. Utilizing a binary pass/fail indicator, we will systematically ascertain whether each interface aligns with our established criteria. The inclusion of iterative constructs, specifically *for loops*, facilitates the streamlined examination of numerous interfaces, enhancing the efficacy and scalability of our testing approach.

The outcome of these assessments will be meticulously organized and presented using Rich tables, offering an immediate visual appraisal of each interface's status. Interfaces satisfying the benchmarks will be marked in green, denoting compliance or performance that meets or surpasses our expectations. Conversely, interfaces failing to meet these benchmarks will be highlighted in red, signaling a need for further investigation or adjustment. This methodical approach not only expedites the evaluation process but also furnishes a concise overview of the network's operational health, underscoring the practical applications and significance of pyATS in contemporary network testing scenarios.

By adopting this structured and analytical methodology, we underscore the critical role of precise, automated testing in maintaining and optimizing network infrastructure. This section intends to equip you with the knowledge and tools necessary to implement effective testing strategies, thereby fostering a deeper understanding of network testing dynamics within the context of pyATS.

Communicating with Devices: The Role of SSH in Testing

As we delve into the mechanics of network testing using pyATS, it's crucial to acknowledge the underlying communication method that facilitates our interaction with network devices: Secure Shell (SSH). SSH serves as the foundational protocol for securely accessing and managing devices remotely. This method ensures a secure channel over an unsecured network, providing a robust means of executing commands and retrieving data. As we progress, we will also explore the integration of RESTCONF as an alternative communication method, offering a glimpse into the evolving landscape of network management protocols.

PyATS leverages its **.learn()** method to thoroughly inspect the state of interfaces on network devices. This powerful functionality generates structured JSON output, encapsulating a wealth of information about each interface's current status. For the sake of documentation and future reference, this output is saved locally as an artifact, serving as tangible evidence of the interface state at the time of testing. This data then forms the basis for our subsequent evaluation, employing the threshold and flag approach that proved effective in our software version assessment.

To streamline the process of testing interface states, it is advised that we create new files specifically for this task: **ios_xe_interfaces_job.py** and **ios_xe_interfaces.py**. A practical approach is to replicate the structure of the previously utilized ios_xe_version_job.py and ios_xe_version.py files. By copying, renaming, and then slightly refactoring this code, we can maintain consistency in our testing framework while minimizing the potential for errors. This strategy of code reuse not only accelerates the development of our interface test but also ensures a level of reliability and efficiency in our testing procedures.

By integrating SSH for secure communication and laying the groundwork for future RESTCONF exploration, this section aims to enhance your understanding of both the technical and strategic aspects of network testing. The practical application of pyATS's **.learn()** method and the strategic reuse of code demonstrate a methodical approach to verifying network interface states, highlighting the importance of adaptability and efficiency in network testing practices.

In the case of the job file, simply update the renamed file **ios_xe_interfaces_job.py**:

```
# Find the location of the script in relation to the job file
testscript = os.path.join(os.path.dirname(__file__), 'ios_xe_
interfaces.py')
```

For the actual testing script, in the renamed file there are a few modifications to make. First, in the Common Setup, update the following lines, the Python class testcase name, to reflect that we are testing interfaces now and on the device version (see Example 8-4).

Example 8-4 *ios_xe_interface.py pyATS Modifications*

```
# ----------------
# Mark the loop for Learn Interfaces
# ----------------
    @aetest.subsection
    def loop_mark(self, testbed):
        aetest.loop.mark(Test_Cisco_IOS_XE_Interfaces, device_name=testbed.devices)


# ----------------
# Test Case #1
# ----------------
class Test_Cisco_IOS_XE_Interfaces(aetest.Testcase):
    """Parse pyATS learn interface and test against thresholds"""
```

Next, within the testcase, modify **.parse()** to **.learn()** as follows:

```
@aetest.test
def get_parsed_version(self):
    parsed_version = self.device.learn("interface")
    # Get the JSON payload
    self.parsed_json=parsed_version.info
```

Now that we have the learned interface JSON inside the **self.parsed_json** variable, we can update the next step, which creates the artifact:

```
@aetest.test
def create_file(self):
    # Create .JSON file
    with open(f'{self.device.alias}_Learn_Interface.json', 'w') as f:
        f.write(json.dumps(self.parsed_json, indent=4, sort_
keys=True))
```

Let's stop here, run the test, and inspect what keypair values we can test against. First, comment out the **show version** test:

```
# @aetest.test
# def test_version(self):....
```

Now save the file and run the job:

```
(venv)$pyats run job ios_xe_interfaces_job.py
```

After the job completes, open the Sandbox_Router_Learn_Interface.json file, with the intention of finding important key values to test.

Figure 8-8 and Figure 8-9 show the structured output, as JSON, from the pyATS .learn(interface) command.

Figure 8-8 *pyATS Learn Interface JSON*

```
    "delay": 10,
    "description": "MANAGEMENT INTERFACE - DON'T TOUCH ME",
    "duplex_mode": "full",
    "enabled": true,
    "encapsulation": {
        "encapsulation": "arpa"
    },
    "flow_control": {
        "receive": false,
        "send": false
    },
    "ipv4": {
        "10.10.20.48/24": {
            "ip": "10.10.20.48",
            "prefix_length": "24",
            "secondary": false
        }
    },
    "mac_address": "0050.56bf.9379",
    "mtu": 1500,
    "oper_status": "up",
    "phys_address": "0050.56bf.9379",
    "port_channel": {
        "port_channel_member": false
    },
    "port_speed": "1000mbps",
    "switchport_enable": false,
    "type": "CSR vNIC"
},
"GigabitEthernet2": { ...
```

Figure 8-9 *pyATS Learn Interface JSON (Continued)*

Given the JSON output from pyATS for the interface GigabitEthernet1, there are several keys that can be of significance for automated network testing. Depending on the focus of the network tests and what you aim to measure or monitor, the following keys can be considered "important" or "valuable" for testing:

- **accounting:** This section can be useful for understanding the type and amount of traffic that an interface is handling.

- **arp:** The ARP traffic details can help in identifying any ARP-related anomalies.

- **ip:** IP traffic details can help in assessing the general load and kind of IP traffic.

- **auto_negotiate:** Whether the interface has auto-negotiation enabled can be crucial, especially in environments where there might be compatibility issues between connected devices.

- **bandwidth:** Monitoring the bandwidth of an interface ensures that it's not being overutilized or underutilized.

- **counters:** This section provides in-depth statistics about the traffic, and it contains important data that helps determine whether the interface is operating efficiently. For example, input/output errors and input/output rate counters can be found in this section.

■ **in_errors and out_errors:** Monitor these to ensure there aren't many errors, which might indicate issues.

■ **in_rate and out_rate:** These can help gauge the current traffic rate, which can be compared against expected values or thresholds.

■ **description:** Useful for ensuring that the interface's purpose aligns with its function. For example, if the description says, "MANAGEMENT INTERFACE – DON'T TOUCH ME," it should not have an exorbitant amount of non-management traffic.

■ **duplex_mode:** Ensuring that the duplex mode is as expected can prevent half-duplex-related issues.

■ **enabled:** Checking if the interface is active or not is crucial. A down interface might lead to network disruptions.

■ **ipv4:** Monitoring the IP address can be useful, especially if DHCP is in use or if there's a risk of IP conflicts.

■ **mac_address:** Useful for ensuring the integrity of the MAC address table and for security monitoring.

■ **mtu:** Ensuring that the MTU size is consistent can prevent fragmentation and communication issues.

■ **oper_status:** The operational status can quickly inform you if there's a problem with the interface.

■ **port_speed:** Monitoring this ensures that the interface is operating at its expected speed.

Each of the aforementioned keys can be used for different test scenarios. For example, for counters, you might want to set up an alert for when **in_errors** exceed a certain threshold. For IPv4, a test could ensure there are no IP conflicts in your network. Depending on the exact requirements and network setup, the tests might vary, but this list should give you a good starting point for automated network testing based on the pyATS Learn Interface JSON.

Try to reuse as much code as possible by refactoring the existing show version code unit test. Rewrite it as shown in Example 8-5 to create a single, small unit test for input errors.

Example 8-5 *ios_xe_interface.py pyATS Input Error Test*

```
@aetest.test
def test_input_errors(self):
    # Test for version interface input errors
    input_errors_threshold = 0
    self.failed_interface = {}
```

```
table = Table(title="pyATS Learn Interface")
table.add_column("Device", style="cyan")
table.add_column("Interface", style="cyan")
table.add_column("Input Error Threshold", style="green")
table.add_column("Input Errors", style="red")
table.add_column("Passed/Failed", style="green")
for intf,value in self.parsed_json.items():
    if 'counters' in value:
        counter = value['counters']['in_errors']
        if int(counter) > in_errors_threshold:
            table.add_row(self.device.alias,intf,str\\
            (in_errors_threshold),str(counter),'Failed',style="red")
            self.failed_version = int(counter)
        else:
            table.add_row(self.device.alias,intf,str\\
            (in_errors_threshold),str(counter),'Passed',style="green")
    else:
            table.add_row\\
            (self.device.alias,intf,in_errors_threshold,
             'N/A','Skipped',style="yellow")
# display the table
console = Console(record=True)
with console.capture() as capture:
    console.print(table,justify="left")
log.info(capture.get())

if self.failed_interface:
    self.failed()
else:
    self.passed()
```

The given test is designed to assess input errors on various interfaces of a network device. Let's look at how it works.

The test is initiated with a predefined threshold of input errors, which is set to zero. The function then proceeds to create a table with columns to display the device name, interface, input error threshold, actual input errors, and the result of the test (Passed/Failed). For each interface in the provided data (sourced from self.parsed_json), the function checks if there are any input errors. If the number of errors is greater than the threshold, the test is marked as Failed for that interface, and this failure is recorded. If no error counters are available for an interface, it's marked as Skipped. Once all interfaces are checked, the table is displayed on the console, showcasing the results. Finally, based on the presence or absence of failures, the entire test is marked as either Failed or Passed. The Rich library is used for creating and displaying this table, with colors indicating various statuses: green for Passed, red for Failed, and yellow for Skipped.

Save the updated file and run the test. There may or may not be input errors on the Sandbox interfaces, but remember the TDD approach: we want this test to fail at first. Therefore, you might have to adjust the threshold to be equal to zero to simulate the failed test. This example will show the subsequent passing example:

```
(venv)$ pyats run job ios_xe_interfaces_job.py
```

Figure 8-10 through Figure 8-12 show the passed test as a Rich table, which is rendered inside the HTML log viewer.

Figure 8-10   *pyATS Interface Input Error Test Rich Table Results – Passed*

Figure 8-11   *pyATS Interface Input Error Test Summary Results – Passed*

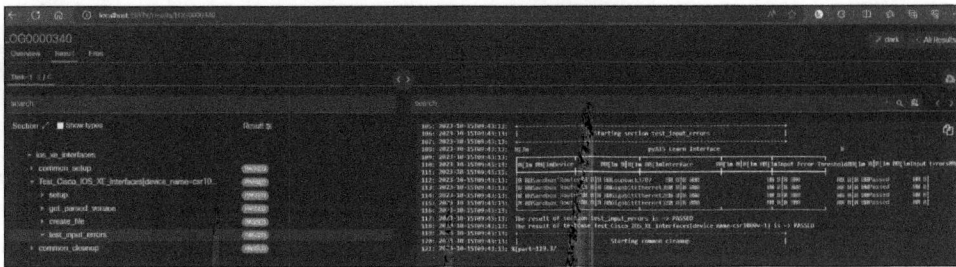

Figure 8-12 *pyATS Interface Input Error Test HTML Log Viewer – Passed*

Let's do one more example together—this time focusing on output errors. Perform a simple copy/paste operation and then perform a find/replace on the keyword "input," as shown in Figure 8-13. Replace all the instances of **input** with **output** and change the actual JSON key of **in_errors** to **out_errors**.

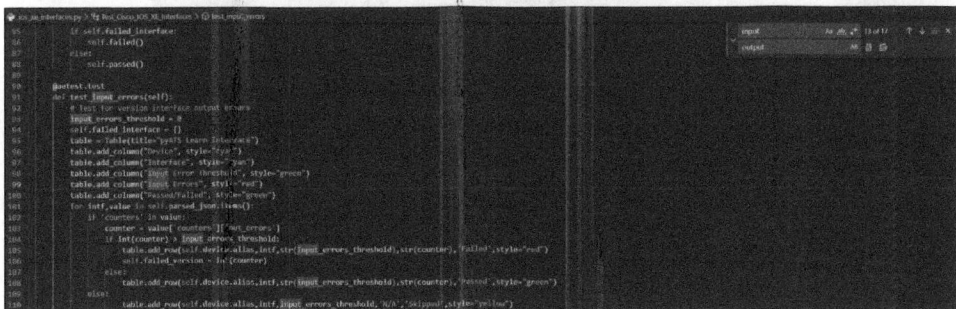

Figure 8-13 *Changing the pyATS Interface Input Error to Output Error Using Find/Replace*

Save the file and then rerun the job:

```
(venv)$ pyats run job ios_xe_interfaces_job.py
```

Figure 8-14 shows passed pyATS interface tests as Rich tables.

As you can see, it takes only seconds to scale interface tests from the original working test. Continue to add any tests you see that are relevant in order to practice. Try and test to see if the interface has a description, or to see the operational state, or to view more counters.

At a massive scale, there may be a better approach to using datafiles to define thresholds and pyATS parsing to gather network data while still utilizing the same approach with pyATS testing. So far, we have used SSH, which is great if your network does not support newer methods of connection such as RESTCONF/NETCONF.

```
+----------------------------------------------------------------+
|                 Starting section test_input_errors             |
+----------------------------------------------------------------+
                 pyATS Learn Interface Input Errors

  _____
| Device         | Interface      | Input Error Threshold | Input Errors | Passed/Failed |
|_____|_____|_____|_____|_____|

| Sandbox Router | Loopback3707   | 0                     | 0            | Passed        |
| Sandbox Router | GigabitEthernet3 | 0                   | 0            | Passed        |
| Sandbox Router | GigabitEthernet2 | 0                   | 0            | Passed        |
| Sandbox Router | GigabitEthernet1 | 0                   | 0            | Passed        |
|_____|_____|_____|_____|_____|

The result of section test_input_errors is => PASSED
+----------------------------------------------------------------+
|                 Starting section test_output_errors            |
+----------------------------------------------------------------+
                 pyATS Learn Interface Output Errors

  _____
| Device         | Interface      | Output Error Threshold | Output Errors | Passed/Failed |
|_____|_____|_____|_____|_____|

| Sandbox Router | Loopback3707   | 0                      | 0             | Passed        |
| Sandbox Router | GigabitEthernet3 | 0                    | 0             | Passed        |
| Sandbox Router | GigabitEthernet2 | 0                    | 0             | Passed        |
| Sandbox Router | GigabitEthernet1 | 0                    | 0             | Passed        |
|_____|_____|_____|_____|_____|
```

Figure 8-14   *pyATS Interface I/O Errors Testing*

Using RESTCONF

In the evolving landscape of network management, the shift from traditional mechanisms like SSH to more modern approaches such as RESTCONF is becoming increasingly evident. RESTCONF is a standard protocol designed by the Internet Engineering Task Force (IETF) to modify and retrieve the configuration of network devices using HTTP/HTTPS. At a massive scale, RESTCONF offers several distinct advantages over SSH:

- **Scalability:** By leveraging HTTP/HTTPS, RESTCONF can better handle multiple simultaneous connections, making it ideal for large network environments.

- **Structured data:** Unlike SSH, which often returns unstructured text that needs parsing, RESTCONF returns data in a structured format based on a YANG model, typically JSON or XML. This structured data format, based on a YANG model, ensures consistency and eases the process of data extraction and interpretation. This means we do not need a parser; instead, we can consume the JSON directly.

- **Stateless nature:** Each RESTCONF request is independent, ensuring no session maintenance, thus reducing overhead.

- **Standardized methods:** RESTCONF uses standard HTTP methods like GET, POST, DELETE, and PUT. This standardization simplifies interactions and ensures a uniform way of dealing with different network devices.

Integrating RESTCONF with pyATS is seamless, especially with the help of the REST Connector. This connector is purpose-built to facilitate communication with devices that support REST-based APIs. By using pyATS with the REST Connector, network engineers can extract data more efficiently and run tests more effectively, making the most of both worlds—the modernity and scalability of RESTCONF, and the robustness and flexibility of pyATS.

First, we need to ensure RESTCONF is enabled on the device. This is an unnecessary step for the DevNet Always On Sandbox since RESTCONF is already enabled, but for your own IOS XE devices, you will need to add the following lines of configuration to the device (this is also a validation step on the sandboxes to confirm the features have been enabled):

```
switch> enable
switch# conf t
switch(conf)# ip http secure-server
switch(conf)# restconf
switch(conf)# username cisco privilege 15
```

Inside your Python virtual environment, you will also need to install the pyATS REST Connector:

```
(venv)$ pip install rest.connector
```

We can reuse either of the pyATS job files we have created, calling the job **ios_xe_rest_interfaces_job.py**, updating the testbed to a new testbed, **rest_testbed.yaml**, and running a new script **ios_xe_rest_interfaces.py**.

The job file is relatively unchanged; we just update the testbed file name as well as the script name, as follows:

```
testbedfile = os.path.join('rest_testbed.yaml')
```

and

```
testscript = os.path.join(os.path.dirname(__file__), 'ios_xe_rest_
interfaces.py')
```

Next, we update the rest_testbed.yaml file to use the REST Connector, as demonstrated in Example 8-6.

Example 8-6   *rest_testbed.yaml*

```
---
devices:
    csr1000v-1:
        alias: 'csr1000v-1'
        type: 'router'
        os: 'iosxe'
        platform: csr1000v
        connections:
            rest:
                # Rest connector class
                class: rest.connector.Rest
                ip: sandbox-iosxe-recomm-1.cisco.com
                port: 443
                credentials:
                    rest:
                        username: developer
                        password: lastorangerestoreball8876
```

Finally, let's write a testcase using the REST Connector *method* of **.get()**, as demonstrated in Example 8-7 and Example 8-8.

Example 8-7   *ios_xe_rest_interfaces_job.py Sample*

```
import os
from genie.testbed import load

def main(runtime):

    # ----------------
    # Load the testbed
    # ----------------
    if not runtime.testbed:
        # If no testbed is provided, load the default one.
        # Load default location of Testbed
        testbedfile = os.path.join('rest_testbed.yaml')
        testbed = load(testbedfile)
    else:
        # Use the one provided
        testbed = runtime.testbed
```

```
    # Find the location of the script in relation to the job file
    testscript = os.path.join(os.path.dirname(__file__), 'ios_xe_rest_interfaces.
py')

    # run script
    runtime.tasks.run(testscript=testscript, testbed=testbed)
```

Example 8-8 *ios_xe_rest_interfaces.py Sample*

```python
import json
import logging
from pyats import aetest
from rich.table import Table
from rich.console import Console
from genie.utils.diff import Diff
from pyats.log.utils import banner


# ----------------
# Get logger for script
# ----------------

log = logging.getLogger(__name__)


# ----------------
# AE Test Setup
# ----------------
class common_setup(aetest.CommonSetup):
    """Common Setup section"""
# ----------------
# Connected to devices
# ----------------

    @aetest.subsection
    def connect_to_devices(self, testbed):
        """Connect to all the devices"""
        testbed.connect()
# ----------------
# Mark the loop for Learn Interfaces
# ----------------

    @aetest.subsection
    def loop_mark(self, testbed):
        aetest.loop.mark(Test_Cisco_IOS_XE_REST_Interfaces, device_name=testbed.
devices)
```

```
# ----------------
# Test Case #1
# ----------------
class Test_Cisco_IOS_XE_REST_Interfaces(aetest.Testcase):
    """Parse pyATS learn interface and test against thresholds"""

    @aetest.test
    def setup(self, testbed, device_name):
        """ Testcase Setup section """
        # Set current device in loop as self.device
        self.device = testbed.devices[device_name]

    @aetest.test
    def get_pre_test_yang_data(self):
        # Use the RESTCONF OpenConfig YANG Model
        parsed_openconfig_interfaces = self.device.rest.get("/restconf/data/opencon-
fig-interfaces:interfaces")
        # Get the JSON payload
        self.parsed_json=parsed_openconfig_interfaces.json()

    @aetest.test
    def create_pre_test_files(self):
        # Create .JSON file
        with open(f'{self.device.alias}_OpenConfig_Interfaces.json', 'w') as f:
            f.write(json.dumps(self.parsed_json, indent=4, sort_keys=True))

    @aetest.test
    def test_input_errors(self):
        # Test for version interface input errors
        input_errors_threshold = 0
        self.failed_interface = {}
        table = Table(title="OpenConfig YANG Interface Input Errors Test")
        table.add_column("Device", style="cyan")
        table.add_column("Interface", style="cyan")
        table.add_column("Input Error Threshold", style="green")
        table.add_column("Input Errors", style="red")
        table.add_column("Passed/Failed", style="green")
        for intf in self.parsed_json['openconfig-interfaces:interfaces']['inter-
face']:
            counter = intf['state']['counters']['in-errors']
            if int(counter) > input_errors_threshold:
                table.add_row(self.device.alias,str(intf['name']),str(input_errors_
threshold),str(counter),'Failed',style="red")
```

```
                self.failed_version = int(counter)
            else:
    table.add_row(self.device.alias,str(intf['name']),str(input_errors_threshold),str(
    counter),'Passed',style="green")
        # display the table
        console = Console(record=True)
        with console.capture() as capture:
            console.print(table,justify="left")
        log.info(capture.get())

        if self.failed_interface:
            self.failed()
        else:
            self.passed()

class CommonCleanup(aetest.CommonCleanup):
    @aetest.subsection
    def disconnect_from_devices(self, testbed):
        testbed.disconnect()

# for running as its own executable
if __name__ == '__main__':
    aetest.main()
```

Figure 8-15 shows the successful pyATS interface errors test in a Rich table.

Device	Interface	Input Error Threshold	Input Errors	Passed/Failed
cat8000v	GigabitEthernet1	0	0	Passed
cat8000v	GigabitEthernet2	0	0	Passed
cat8000v	GigabitEthernet3	0	0	Passed
cat8000v	Loopback0	0	0	Passed
cat8000v	Loopback10	0	0	Passed
cat8000v	Loopback109	0	0	Passed
cat8000v	VirtualPortGroup0	0	2	Failed

OpenConfig YANG Interface Input Errors Test

Figure 8-15 *pyATS Interface Input Error Testing with YANG Models and REST Connector*

Now use the remaining YANG modules to test the other interface counters and scale this testcase (see Figure 8-16). What you might have noticed is the dramatic performance improvements and speed of using REST Connector to get the JSON directly instead of using SSH and parsing **show** commands.

```
{
    "openconfig-interfaces:interfaces": {
        "interface": [
            {
                "config": {
                    "description": "MANAGEMENT INTERFACE - DON'T TOUCH ME",
                    "enabled": true,
                    "name": "GigabitEthernet1",
                    "type": "iana-if-type:ethernetCsmacd"
                },
                "name": "GigabitEthernet1",
                "openconfig-if-ethernet:ethernet": {
                    "config": {
                        "auto-negotiate": true,
                        "enable-flow-control": true,
                        "mac-address": "00:50:56:bf:78:8f"
                    },
                    "state": {
                        "auto-negotiate": true,
                        "counters": {
                            "in-8021q-frames": "0",
                            "in-crc-errors": "0",
                            "in-fragment-frames": "0",
                            "in-jabber-frames": "0",
                            "in-mac-control-frames": "0",
                            "in-mac-pause-frames": "0",
                            "in-oversize-frames": "0",
                            "out-8021q-frames": "0",
                            "out-mac-control-frames": "0",
                            "out-mac-pause-frames": "0"
                        },
                        "enable-flow-control": false,
                        "hw-mac-address": "00:50:56:bf:78:8f",
                        "mac-address": "00:50:56:bf:78:8f",
                        "negotiated-duplex-mode": "FULL",
                        "negotiated-port-speed": "openconfig-if-ethernet:SPEED_1GB",
                        "port-speed": "openconfig-if-ethernet:SPEED_1GB"
                    }
                },
                "state": {
                    "admin-status": "UP",
                    "counters": {
                        "in-broadcast-pkts": "0",
                        "in-discards": "0",
                        "in-errors": "0",
                        "in-fcs-errors": "0",
                        "in-multicast-pkts": "0",
```

Figure 8-16 *RESTCONF Sample Output from IOS XE*

Neighbor Testing

Testing network devices often goes beyond basic device software version and interface validation; understanding and verifying neighbor relationships is a critical component. Protocols such as BGP (Border Gateway Protocol), OSPF (Open Shortest Path First), CDP (Cisco Discovery Protocol), LLDP (Link Layer Discovery Protocol), and others play pivotal roles in ensuring network connectivity and topology awareness:

- Testing Border Gateway Protocol (BGP) and Open Shortest Path First (OSPF) neighbor relationships is paramount to ensuring that routing protocols are functioning as intended within a network. For instance, one can craft tests to ascertain whether the expected number of BGP peers is actively established or if OSPF adjacencies are correctly formed on anticipated interfaces. Should the count of established neighbors not meet or exceed a certain threshold, such a scenario would be marked as a failure. This criterion is vital because discrepancies in the expected and actual number of neighbors can signal potential misconfigurations or underlying network issues, impacting the stability and reliability of network operations.

- In addition to the fundamental tests, it's crucial to consider the broader implications of these neighbor relationships, especially in more complex networking scenarios such as peering at an Internet exchange point (IXP). A sample scenario to be wary of is when a peering partner inadvertently sends its entire routing table. Although mechanisms like setting a maximum number of receivable routes (max-receive limit) can mitigate the risk of overwhelming your router, this underscores the importance of comprehensive testing. Such tests should not only verify the presence and correctness of neighbor relationships but also ensure that safeguards are in place to prevent and handle potential misconfigurations or errors from peers. This added layer of testing provides a safety net against operational surprises that could otherwise disrupt network performance and connectivity.

- CDP and LLDP are discovery protocols that help in identifying directly connected devices. Tests can be designed to ensure that specific neighbors are discovered on certain interfaces. If a particular neighbor is missing, or an unexpected neighbor appears, it might point to a cabling error, a device malfunction, or a security concern, and the test would fail.

- Setting thresholds for the minimum number of expected neighbors adds a layer of robustness to the testing procedure. If the number of neighbors drops below this threshold, it's a clear indication that something's amiss. Similarly, tests can be tailored for specific critical neighbors. For instance, if a core router is expected on a particular interface but is not discovered, it's a serious concern that needs immediate attention.

Incorporating neighbor testing into network validation processes provides a holistic view of the network's health. By setting thresholds and specifically checking for critical neighbors, network administrators can swiftly identify and rectify potential issues, ensuring the stability and security of the entire infrastructure.

In Example 8-9, we first enable CDP globally and per interface, and then we test the state of CDP. We have added a bit of logic to illustrate how pyATS can work with IOS XE or NXOS, but the command differs slightly based on platform. We can add logic to handle the appropriate command on the appropriate platform, but a universal script can be used to push configuration for CDP neighbors.

Example 8-9 *Enable CDP with pyATS*

```
@aetest.setup
def enable_CDP_globally(self):
    if self.device.os == "iosxe":
        self.device.configure("cdp run")
    elif self.device.os == "nxos":
        self.device.configure("cdp enable")


@aetest.test
def enable_CDP_interfaces(self):
    interfaces = self.device.parse("show ip interface brief")
    for interface in interfaces['interface'].items():
        if self.device.os == "iosxe":
            if "Gigabit" in interface:
                self.device.configure(f""" interface { interface }
                cdp enable
                """)
        elif self.device.os == "nxos":
            if "Ethernet" in interface:
                self.device.configure(f""" interface { interface }
                cdp enable
                """)
    time.sleep(15)
```

After waiting 15 seconds for CDP to come up and establish neighbor relationships, we can now test for the number of neighbors, the presence of CDP, or even the IP reachability of each CDP neighbor with a ping test, as demonstrated in Example 8-10.

Example 8-10 *Testing CDP with pyATS*

```
@aetest.test
def gather_CDP_Neighbors(self):
    number_of_pings = 5
    self.ping_cdp_neighbors_results = {}
    self.failed_neighbors = {}
```

```
    # Setting up the Rich table
    table = Table(title="pyATS Ping CDP Neighbors Results")
    table.add_column("Device", style="cyan")
    table.add_column("Neighbor", style="cyan")
    table.add_column("Local Interface", style="cyan")
    table.add_column("Ping Result", style="green")
    table.add_column("Passed/Failed", style="green")

    # Check OS type and execute corresponding actions
    if self.device.os == "iosxe":
        cdp_cmd = "show cdp neighbors detail"
        address_key = 'entry_addresses'
    elif self.device.os == "nxos":
        cdp_cmd = "show cdp neighbors detail"
        address_key = 'interface_addresses'
    else:
        log.info("Unsupported OS type.")
        return

    self.cdp_neighbor_list = self.device.parse(cdp_cmd)

@aetest.test
def ping_CDP_Neighbors(self):     # Loop through neighbors and attempt pings
    for neighbor, value in self.cdp_neighbor_list['index'].items():
        for entry in value[address_key]:
            try:
                parsed_ping = \\
                Self.device.parse\\
                (f"ping { entry } source { value['local_interface'] }
                repeat { number_of_pings }")
                table.add_row\\
                (self.device.alias, neighbor, value['local_interface'],
                "Success", style="green")
                self.ping_cdp_neighbors_results[self.device.alias][neighbor]=\\
                parsed_ping
            except:
                table.add_row\\
                (self.device.alias, neighbor, value['local_interface'],
                 "Invalid PING", style="red")
                self.failed_neighbors[neighbor] = entry
                   Log.info\\
```

```
                    (f"Sorry but we cannot ping { entry }
                from { value['local_interface'] } ")

# Display the table
console = Console(record=True)
with console.capture() as capture:
    console.print(table, justify="left")
log.info(capture.get())

# Check for failed neighbors
if self.failed_neighbors:
    self.failed()
else:
    self.passed()
```

Testing neighbor relationships in a network lays the foundation for understanding device interconnectivity and ensuring protocol operations. By examining protocols like BGP, OSPF, CDP, and LLDP, we obtain insights into network topology and identify any potential misconfigurations or unexpected relationships. This understanding of the immediate network neighborhood forms the cornerstone of more extensive network diagnostics.

However, verifying neighbor relationships is just the beginning. Once we ascertain that devices recognize each other, the next critical step is to determine if they can effectively communicate. This is where reachability testing, such as pinging, comes into play. Ping is a simple yet powerful tool that provides immediate feedback on the health of network paths, checking if ICMP packets can successfully traverse from one device to another and then return. By integrating ping tests, we not only confirm that devices are aware of each other, but we also ensure that they can exchange data without hindrances.

In the upcoming section, we will delve deeper into the nuances of ping-based reachability tests, expanding on the initial example to cover a broader range of scenarios and further solidify our understanding of the network's operational status.

Reachability Testing

As seen with the CDP neighbor example, pyATS has the ability to parse a variety of ping commands. Figure 8-17 illustrates the various available ping parsers from pyATS.

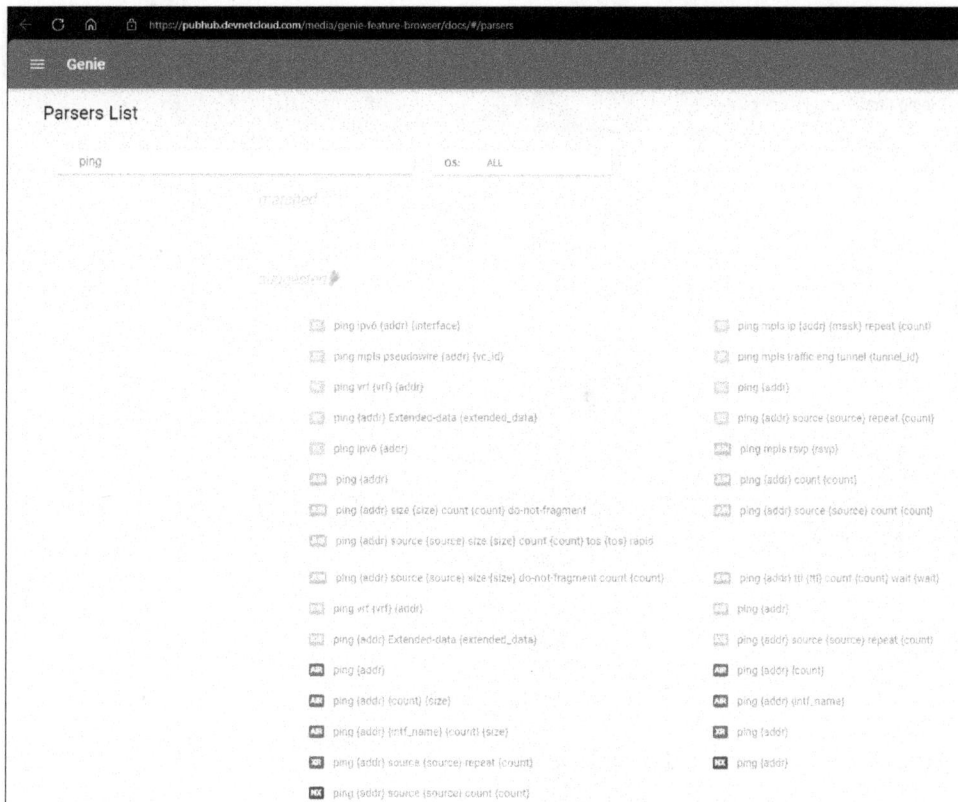

Figure 8-17   *Available pyATS Ping Parsers*

The actual schema from the parsed JSON is quite remarkable, with multiple options for testing, including the following:

- Success rate percentage
- Minimum millisecond response
- Average millisecond response
- Maximum millisecond response

Figures 8-18 and 8-19 illustrate the ping {addr} schema.

Figure 8-18 *pyATS Ping Parser Schema*

Figure 8-19 *pyATS Ping Parser Schema (Continued)*

The testcase in Example 8-11 expands on the CDP neighbor ping test by testing for success rate percentage.

Example 8-11 *Testing CDP Neighbor Ping Success Rate with pyATS*

```
@aetest.test
def test_CDP_Neighbor_ping_success_rate_percentage(self):
    # Test for ping success rate
    self.failed_success_rate={}
    self.no_ping_response_CDP_Neighbor = False
    table = Table(title="CDP Neighbor Ping Success Rate")
    table.add_column("Device")
    table.add_column("Neighbor")
    table.add_column("Destination")
    table.add_column("Success Rate Percentage")
    table.add_column("Passed/Failed")

    for interface, value in \\
    self.ping_cdp_neighbors_results[self.device.alias].items():
        success_rate = value['ping']['statistics']['success_rate_percent']
        if success_rate == 0:
            self.no_ping_response_CDP_Neighbor = True
        status = "Passed" if success_rate == 100.0 else "Failed"
        color = "green" if status == "Passed" else "red"
        table.add_row\\
        (self.device.alias, interface, value['ping']['address'],
        str(success_rate), status, style=color)

        if status == "Failed":
            self.failed_success_rate = success_rate

    console = Console()
    console.print(table)

    if self.failed_success_rate:
        Self.failed\\
        (f'{ self.device } Has An Interface That
        Had Less Than 100% PING Against CDP Neighbor')
    else:
        Self.passed\\
        (f'All Interfaces on { self.device } Had 100%
        PING Success Against CDP Neighbor')
```

For security purposes, you might want to test if a public IP address is reachable from your management interface, as demonstrated in Example 8-12. In this case, the logic might be inverted to *pass* the pyATS test in the event you *cannot* reach the public IP address; otherwise, it means your device does in fact have access to the public Internet, which might not be desirable.

Example 8-12   *Testing Public Internet Access from Your Device*

```
@aetest.test
def test_8_8_8_8_success_rate_percentage(self):
    # Test for ping success rate
    self.failed_8_8_8_8_success_rate={}
    self.no_ping_response_8_8_8_8 = False

    table = Table(title="Ping 8.8.8.8 Success Rate")
    table.add_column("Device")
    table.add_column("Interface")
    table.add_column("Success Rate Percentage")
    table.add_column("Passed/Failed")

    for interface, value in self.ping_8_8_8_8_results[self.device.alias].items():
        success_rate = value['ping']['statistics']['success_rate_percent']
        if success_rate == 0:
            self.no_ping_response_8_8_8_8 = True
            status = "Passed"
            color = "green"
        elif success_rate > 0:
            status = "Failed"
            color = "red"
            self.failed_8_8_8_8_success_rate = "Failed"

        table.add_row\\
        (self.device.alias, interface,
         str(success_rate), status, style=color)

    console = Console()
    console.print(table)

    if self.no_ping_response_8_8_8_8:
        self.passed(f'{ self.device } Can NOT PING 8.8.8.8')
    else:
        self.failed(f'{ self.device } *CAN* PING 8.8.8.8')
```

Having conducted basic reachability tests, we've gained valuable insights into the health and connectivity of our network infrastructure. Pinging, as one of the most rudimentary yet essential diagnostics, has allowed us to confirm that our devices not only recognize each other but can also effectively communicate. However, in an era defined by dynamic and complex network configurations, merely being "reachable" isn't enough. We must guarantee that our network behaves precisely as we intend it to. This introduces the paradigm of *network intent verification*.

Network intent verification goes beyond basic reachability; it seeks to align our network's actual state with our desired or intended state. By extending our testbed.yaml to include "intent," we can define our network's expected behavior in clear, actionable terms. This might encompass everything from routing preferences, to security postures, to quality of service (QoS) expectations. Once this is defined, we can then programmatically test if our live network configuration aligns with this intent, ensuring not just functionality but optimized and intended operation. In the following sections, we'll delve into the nuances of defining intent and, more importantly, verifying that our live network indeed reflects our strategic goals and objectives.

Intent Validation Testing

pyATS allows for "intent-based networking" via extension of the testbed.yaml file. Our original testbed file can be *extended* to include additional, custom fields—fields we will use to drive configuration and validate that the state of our network matches our *intended* configuration. First, let's extend the original testbed file. Create a new YAML file, called **intent_testbed.yaml**. As demonstrated in Example 8-13, the first line of code, after the YAML header delimiter, will be **extends: testbed.yaml**. This will preserve the original testbed and extend it to include our customized intent. Our new file will still include the devices and must match the original testbed file's device definition. We can then use a **custom** key and define our intent. It should be noted that our new extended testbed has a direct reference to the original testbed.yaml file.

Example 8-13   *intent_testbed.yaml*

```
---
extends: testbed.yaml
devices:
    csr1000v-1:
        custom:
            domain_name: "lab.devnetsandbox.local"
            interfaces:
                GigabitEthernet1:
                    type: ethernet
                    description: "MANAGEMENT INTERFACE - DON'T TOUCH ME"
                GigabitEthernet2:
                    type: ethernet
                    description: "Network Interface"
                GigabitEthernet3:
                    type: ethernet
                    description: "Intent Testing"
```

Our intent is to have a domain name of "lab.devnetsandbox.local" globally and to have three physical interfaces, each with specific descriptions. First, let's test and confirm that our domain name matches our intent; if it does not, we will *update the domain name dynamically* and then retest (see Example 8-14).

Example 8-14   *Testing Whether the Configured Domain Name Matches the Intended Domain Name*

```
    @aetest.test
    def get_parsed_config(self):
        parsed_config = self.device.learn("config")
        # Get the JSON payload
        self.parsed_json=parsed_config


@aetest.test
def test_ip_domain_name(self):
    # Test for domain name
    self.failed_domain_name = {}

    table = Table(title="Domain Name Verification")
    table.add_column("Device")
    table.add_column("Intent Domain Name")
    table.add_column("Configured Domain Name")
    table.add_column("Passed/Failed")

    intended_domain = self.device.custom.domain_name
    actual_domain = f"ip domain name { intended_domain }"

    if actual_domain in self.parsed_json:
        status = "Passed"
        color = "green"
    else:
        status = "Failed"
        color = "red"
        actual_domain = "Different Domain Name Configured"
        self.failed_domain_name = "Error"

    table.add_row(self.device.alias, intended_domain, actual_domain, status,
style=color)

    console = Console()
    console.print(table)
```

```
    if self.failed_domain_name:
        self.failed('Device Has An Incorrect Domain Name')
    else:
        self.passed('Device Has The Correct Domain Name')


    @aetest.test
    def update_domain_name(self):
        if self.failed_domain_name:
            self.pre_change_parsed_json = self.parsed_json
            self.device.configure(f"ip domain name { self.device.custom.domain_name
}")
        else:
            self.skipped('No domain name mismatches skipping test')


    @aetest.test
    def get_post_test_yang_data(self):
        if self.failed_domain_name:
            post_parsed_native = self.device.learn("config")
            self.post_parsed_json=post_parsed_native
        else:
            self.skipped('No domain name mismatches skipping test')


    @aetest.test
    def retest(self):
        if self.failed_domain_name:
            self.get_parsed_config()
            self.test_ip_domain_name()
        else:
            self.skipped('No domain name mismatches skipping test')
```

Now, using nested loops (one looping over the actual interfaces and the other looping over the intended interfaces), we will validate that the interface descriptions match. If they do not, we will update them accordingly and then retest to validate (see Example 8-15).

Example 8-15 *Testing Whether the Configured Interface Descriptions Match the Intent*

```
@aetest.test
def test_interface_description_matches_intent(self):
    # Test for interface description
    self.failed_interfaces = {}
```

```
    table = Table(title="Interface Description Verification")
    table.add_column("Device")
    table.add_column("Interface")
    table.add_column("Intent Desc")
    table.add_column("Actual Desc")
    table.add_column("Passed/Failed")

    for self.intf, value in self.parsed_interfaces.info.items():
        description_present = 'description' in value
        for interface, intent_value in self.device.custom.interfaces.items():
            if self.intf == interface:
                self.intended_desc = intent_value['description']
                actual_desc = value['description'] if description_present else ""
                status = "Passed" if actual_desc == \\
                self.intended_desc else "Failed"
                color = "green" if status == "Passed" else "red"

                table.add_row\\
                (intent_value['description'], self.intf,
                 self.intended_desc, actual_desc, status, style=color)

                if status == "Failed":
                    self.failed_interfaces[self.intf] = self.intended_desc
                    self.interface_name = self.intf
                    self.error_counter = self.failed_interfaces[self.intf]
                    if not description_present:
                        self.update_interface_description()

    console = Console()
    console.print(table)

     # should we pass or fail?
    if self.failed_interfaces:
        self.failed('Some interfaces intent / actual descriptions mismatch')
    else:
        self.passed('All interfaces intent / actual descriptions match')

def update_interface_description(self):
    if self.failed_interfaces:
        self.device.configure(f'''interface { self.intf }
                                  description { self.intended_desc }
                            ''')
```

```
        else:
            self.skipped('No description mismatches skipping test')


@aetest.test
def get_post_test_interface_data(self):
    if self.failed_interfaces:
        self.pre_change_parsed_json = self.parsed_interfaces
        self.post_parsed_interfaces = self.device.learn("interface")
    else:
        self.pre_change_parsed_json = self.parsed_interfaces
        self.skipped('No description mismatches skipping test')


@aetest.test
def retest(self):
    if self.failed_interfaces:
        self.get_pre_test_interface_data()
        self.test_interface_description_matches_intent()
    else:
        self.skipped('No description mismatches skipping test')
```

In the digital age, network management has evolved from manual configuration and troubleshooting to a more dynamic, intent-driven approach. The fundamental premise is simple, yet powerful: rather than focusing solely on individual configurations, we articulate our broader operational goals and then leverage automation to ensure our network aligns with these objectives. By defining our "intent," we set the North Star for our network's desired state. The two cases in point, interface descriptions and domain names, might seem rudimentary, but they exemplify the immense potential of this methodology. Whether it's a single device attribute or a complex orchestration of multiple network elements, the intent-driven paradigm allows us to automatically configure, validate, and remediate any discrepancies in real time. This not only enhances operational efficiency but also helps to ensure consistency and compliance across the network landscape. Moreover, the scalability of this approach is capable of serving very large topologies. Regardless of the size and complexity of our infrastructure, intent-driven tests and configurations can be seamlessly extrapolated to manage every device, every attribute, and every interaction.

Finally, and possibly most important, we will focus on feature testing with BGP.

Feature Testing

In the vast landscape of network testing, automated feature testing stands out as a crucial and evolving component. One of the essential features often tested is the Border Gateway Protocol (BGP), which plays a pivotal role in the global Internet routing infrastructure.

BGP is responsible for managing how packets get routed across the Internet through the exchange of routing and reachability information among edge routers. As such, ensuring the correct and efficient functioning of BGP is paramount.

Feature testing using BGP primarily involves assessing the establishment of BGP sessions, the exchange of routing updates, and the handling of various BGP attributes and policies. Through automation, network engineers can simulate various scenarios, verify configurations, and ensure that the BGP implementation aligns with the expected standards and behaviors. The script in Example 8-16 serves as an example of how one might automate BGP feature testing.

The provided script showcases a structured approach to BGP feature testing. Utilizing Python libraries like pyATS and Genie, the script automates the connection to network devices, extracts and parses BGP-related information, and performs specific tests to ensure the network's BGP functionality is as expected. The tests include verifying the establishment of BGP neighbors, simulating a BGP change by shutting down interfaces, checking routing tables, reverting the change, and ensuring BGP is in the same, expected state as before the change was implemented.

Example 8-16   *End-to-End BGP Feature Testing*

```
import logging
from genie.utils import Dq
from pyats import aetest
from unicon.core.errors import ConnectionError
import time


logger = logging.getLogger(__name__)
logger.setLevel("INFO")


class CommonSetup(aetest.CommonSetup):
    @aetest.subsection
    def connect_to_devices(self, testbed):
        """Connect to all testbed devices"""
        try:
            testbed.connect()
        except ConnectionError:
            self.failed(f"Could not connect to all devices in {testbed.name}")
        # Print log message confirming all devices are in a 'connected' state
        logger.info(f"Connected to all devices in {testbed.name}")


class BGPTestcase(aetest.Testcase):
    @aetest.test
```

```
def check_bgp_neighbors(self, testbed):
    """Check number of established BGP neighbors on each device"""
    # Parse BGP neighbor command on each device
    r1_bgp_neighs = testbed.devices["cat8k-rt1"].parse("show bgp neighbors")
    r2_bgp_neighs = testbed.devices["cat8k-rt2"].parse("show bgp neighbors")


    # Capture 'established' neighbors from each device
    # Set class variables for future comparison
    self.r1_pre_estab_neighbors = \\
    Dq(r1_bgp_neighs).contains("Established").get_values("neighbor")
    self.r2_pre_estab_neighbors = \\
    Dq(r2_bgp_neighs).contains("Established").get_values("neighbor")


    # Confirm 'established' BGP neighbors were found
    if self.r1_pre_estab_neighbors and self.r2_pre_estab_neighbors:
        self.passed(f"Established neighbors were found for each router.")
    else:
        self.failed("One of the routers has 0 established BGP neighbors.")

@aetest.test
def shutdown_bgp(self, testbed):
    """Shutdown BGP neighbors by shutting interfaces"""
    testbed.devices["cat8k-rt1"].configure("interface g4\r shut")
    testbed.devices["cat8k-rt2"].configure("interface g4\r shut")

@aetest.test
def check_routing(self, testbed):
    """Check routing tables for received BGP routes"""
    # Parse routing tables on each device
    rt1_bgp_routes = testbed.devices["cat8k-rt1"].parse("show ip route")
    rt2_bgp_routes = testbed.devices["cat8k-rt2"].parse("show ip route")


    # Drill down to IPv4 AF routes
    rt1_ipv4_routes = \\
    rt1_bgp_routes["vrf"]["default"]["address_family"]["ipv4"]["routes"]

    rt2_ipv4_routes = \\
    rt2_bgp_routes["vrf"]["default"]["address_family"]["ipv4"]["routes"]


    # Look for specific received BGP routes from each router
    rt1_bgp_route = rt1_ipv4_routes.get("10.60.0.0/24")
    rt2_bgp_route = rt2_ipv4_routes.get("10.50.0.0/24")
```

```
        # Determine whether routes are found and
        # pass/fail the test based on the result
        if rt1_bgp_route is not None:
            Self.failed\\
            (f"BGP route for 10.60.0.0/24
             is still present in cat8k-rt1's route table")
        elif rt2_bgp_route is not None:
            Self.failed\\
            (f"BGP route for 10.50.0.0/24 is
             still present in cat8k-rt2's route table")
        else:
            self.passed("BGP routes are not found in either routing table.")

@aetest.test
def unshut_bgp(self, testbed):
    """Reactivate BGP by unshutting WAN interfaces on each device"""
    testbed.devices["cat8k-rt1"].configure("interface g4\r no shut")
    testbed.devices["cat8k-rt2"].configure("interface g4\r no shut")
    # Allow time for BGP to re-establish
    logger.info("Allow time for BGP to re-establish...")
    time.sleep(20)

@aetest.test
def check_post_routing(self, testbed):
    """Check routing tables for received BGP routes"""
    # Parse routing tables on each device
    rt1_bgp_routes = testbed.devices["cat8k-rt1"].parse("show ip route")
    rt2_bgp_routes = testbed.devices["cat8k-rt2"].parse("show ip route")

    # Drill down to IPv4 AF routes
    rt1_ipv4_routes = \\
    rt1_bgp_routes["vrf"]
                  ["default"]
                  ["address_family"]
                  ["ipv4"]
                  ["routes"]

    rt2_ipv4_routes = \\
    rt2_bgp_routes["vrf"]
                  ["default"]
                  ["address_family"]
                  ["ipv4"]
                  ["routes"]
```

```
        # Look for received BGP routes from each router
        rt1_bgp_route = rt1_ipv4_routes.get("10.60.0.0/24")
        rt2_bgp_route = rt2_ipv4_routes.get("10.50.0.0/24")


        # Determine whether routes are found
        # Pass/fail the test based on the result
        if rt1_bgp_route is None:
            Self.failed\\
            (f"BGP route for 10.60.0.0/24
             was not found in cat8k-rt1's route table")
        elif rt2_bgp_route is None:
            Self.failed\\
            (f"BGP route for 10.50.0.0/24
            was not found in cat8k-rt2's route table")
        else:
            self.passed("BGP routes are found in both routing tables.")

@aetest.test
def check_post_bgp_neighbors(self, testbed):
    """Check number of established BGP neighbors on each device"""
    # Parse BGP neighbor command on each device
    r1_bgp_neighs = testbed.devices["cat8k-rt1"].parse("show bgp neighbors")
    r2_bgp_neighs = testbed.devices["cat8k-rt2"].parse("show bgp neighbors")


    # Capture 'established' neighbors from each device
    r1_post_estab_neighbors = \\
    Dq(r1_bgp_neighs).contains("Established").get_values("neighbor")


    r2_post_estab_neighbors = \\
    Dq(r2_bgp_neighs).contains("Established").get_values("neighbor")


    # Compare the list of 'established' neighbors
    # with the list of 'established' neighbors found pre-testing

    if self.r1_pre_estab_neighbors == \\
    r1_post_estab_neighbors and \\
    self.r2_pre_estab_neighbors == r2_post_estab_neighbors:
        Self.passed\\
        (f"The same number of established
        neighbors were found on each router.")
    else:
        Self.failed\\
```

```
                  ("One of the routers have a different
                  number of established BGP neighbors after testing.")

class CommonCleanup(aetest.CommonCleanup):
    @aetest.subsection
    def disconnect_from_devices(self, testbed):
        """Disconnect from all devices"""
        testbed.disconnect()
        logger.info(f"Disconnected from all devices in {testbed.name}")



# Standalone execution
if __name__ == '__main__':
    from pyats.topology import loader
    from dotenv import load_dotenv

    # Load environment variables
    load_dotenv()

    # Load the testbed file
    cml_tb = loader.load("cml_testbed.yaml")

    # and pass all arguments to aetest.main() as kwargs
    aetest.main(testbed=cml_tb)

    # To run standalone execution:
    # python bgp_testscript.py
```

Automated network testing, as demonstrated, not only increases efficiency but also reduces human error, ensuring a more robust and reliable network by verifying network operational state before and after changes. As networks continue to evolve and grow in complexity, the role of automation in feature testing, especially with protocols as vital as BGP, will undoubtedly become even more significant.

Summary

Diving into the world of automated network testing reveals the synergy between Cisco's pyATS framework and test-driven development (TDD). This union crafts a meticulous testing culture, offering both passive, read-only tests and actionable ones that integrate configuration management in response to anomalies.

The core highlights include the following:

- **Integrated methodology:** The union of pyATS and TDD results in a transformative approach to network testing, paving the way for a resilient and self-repairing network infrastructure.

- **Navigating network testing:** Network testing, likened to a labyrinth, demands a sound strategy. TDD principles, when synchronized with the network's pulse, create a dynamic dialogue characterized by iterative challenges and responses until alignment is achieved.

- **Threshold flag and variable evaluation:** These two pivotal tools form the testing backbone and set the network health benchmark.

- **Enhanced logging with Rich:** The Rich library transforms mundane pyATS logs into vibrant, color-coded narratives, simplifying result interpretation and adding professional flair to the testing process.

- **Foundational practices for network testing:** Emphasis is placed on structured testing, clarity in tests, and avoiding overly complex or omniscient tests.

- **Broad testing dimensions:**

  - **Software version testing:** Validates that the network components operate on the latest and most compatible software versions.

  - **Interface counter testing:** Observes interface operations, highlighting any discrepancies in data flow.

  - **Neighbor testing:** Ensures the health and active status of neighboring devices within the network.

  - **Reachability testing:** Assesses the efficacy of communication between various network nodes.

  - **Feature testing:** Guarantees that specific network features function as intended. This can be very useful when you're testing a feature before and after a change occurs in the network.

The synthesis of these methodologies and tools fortifies network resilience and drives operational excellence, emphasizing the paramount importance of strategic testing in network robustness.

pyATS Triggers and Verifications

Automated network tests face challenges in achieving futureproofing due to the dynamic nature of networks and their evolving requirements. The pyATS library (Genie) provides the Genie Harness to execute network tests with dynamic and robust capabilities. The Genie Harness is part of the pyATS library (Genie), which is built on the foundation of pyATS. It introduces the ability for engineers to create dynamic, event-driven tests with the use of processors, triggers, and verifications. In this chapter, the following topics are going to be covered:

- Genie objects

- Genie Harness

- Triggers

- Verifications

The Genie Harness can be daunting for new users of the pyATS library (Genie), as there are many configuration options and parameters. The focus of this chapter will be to provide an overview of Genie Harness and its features and wrap up with a focus on triggers and verifications. By the end of the chapter, you'll understand the powerful capabilities of Genie Harness and how to write and execute triggers and verifications.

Genie Objects

One of the hardest parts of network automation is figuring out how to normalize and structure the data returned from multiple network device types. For example, running commands to gather operational data (CPU, memory, interface statistics, routing state, and so on) from a Cisco IOS XE switch, IOS XR router, and an ASA firewall may have different, but similar, output. How do we account for the miniscule differences across the different outputs? The pyATS library (Genie) abstracts the details of the parsed data returned from devices by creating network OS-agnostic data models. Two types of Genie objects represent these models: Ops and Conf. In the following sections, you'll dive into the details of each object.

Genie Ops

As you may be able to guess from the name, the Genie Ops object learns everything about the operational state of a device. The operational state data is represented as a structured data model, or feature. A *feature* is a network OS-agnostic data model that is created when pyATS "learns" about a device. Multiple commands are executed on a device and the output is parsed and normalized to create the structure of the feature. You can access the learned feature data by accessing **<feature>.info**. To learn more information about the available models, check out the Genie models documentation (https://pubhub.devnetcloud.com/media/genie-feature-browser/docs/#/models). Example 9-1 shows how to instantiate an Ops object for BGP and learn the BGP feature on a Cat8000v device. Figure 9-1 shows the associated output.

Example 9-1 *Genie Ops Object*

```
from genie import testbed
from genie.libs.ops.bgp.iosxe.bgp import Bgp
import pprint

# Load Genie testbed
testbed = testbed.load(testbed="testbed.yaml")

# Find the device using hostname or alias
uut = testbed.devices["cat8k-rt2"]
uut.connect()

# Instantiate the Ops object
bgp_obj = Bgp(device=uut)

# This will send many show commands to learn the operational state
# of BGP on this device
bgp_obj.learn()

pprint.pprint(bgp_obj.info)
```

```
{'instance': {'default': {'bgp_id': 65001,
                          'vrf': {'default': {'cluster_id': '10.254.1.2',
                                              'neighbor': {'172.16.1.1': {'address_family': {'ipv4 unicast': {'bgp_table_version': 2,
                                                                                                             'path': {'memory_usage': 136,
                                                                                                                      'total_entries': 1},
                                                                                                             'prefixes': {'memory_usage': 248,
                                                                                                                          'total_entries': 1},
                                                                                                             'routing_table_version': 2,
                                                                                                             'total_memory': 680}},
                                                                          'bgp_neighbor_counters': {'messages': {'received': {'keepalives': 0,
                                                                                                                              'notifications': 0,
                                                                                                                              'opens': 0,
                                                                                                                              'updates': 0},
                                                                                                                 'sent': {'keepalives': 0,
                                                                                                                          'notifications': 0,
                                                                                                                          'opens': 0,
                                                                                                                          'updates': 0}}},
                                                                          'bgp_session_transport': {'connection': {'last_reset': 'never',
                                                                                                                    'state': 'Active'}},
                                                                          'bgp_version': 4,
                                                                          'remote_as': 65000,
                                                                          'session_state': 'Active',
                                                                          'shutdown': False}}}}}}
```

Figure 9-1   *Genie Ops Object Output*

Genie Conf

The Genie Conf object allows you to take advantage of Python's object-oriented programming (OOP) approach by building logical representations of devices and generating configurations based on the logical device and its attributes. Just like Genie Ops, the structure of Genie Conf objects is based on the feature models. The best way to describe the Conf object is by example. Example 9-2 shows how to build a simple network interface object for an IOS XE device. You'll notice that the code is very comprehensive, specifically for a network engineer, and will even generate the configuration!

Example 9-2   *Genie Conf Object*

```
from genie import testbed
from genie.libs.conf.interface.iosxe.interface import Interface
import pprint


# Load Genie testbed
testbed = testbed.load(testbed="testbed.yaml")

# Find the device using hostname or alias
uut = testbed.devices["cat8k-rt2"]

# Instantiate the Conf object
interface_obj = Interface(device=uut, name="Loopback100")

# Add attributes to the Interface object
interface_obj.description = "Managed by pyATS"
interface_obj.ipv4 = "1.1.1.1"
interface_obj.ipv4.netmask = "255.255.255.255"
interface_obj.shutdown = False
```

```
# Build the configuration for the interface
print(interface_obj.build_config(apply=False))




! Code Execution Output
interface Loopback100
 description Managed by pyATS
 ip address 1.1.1.1 255.255.255.255
 no shutdown
 exit
```

You may notice the last line of code actually builds the necessary configuration for you! To go one step further, you can add the interface object to a testbed device and push the configuration to a device. The Conf object allows you to drive the configuration of network devices with Python, which takes you one step further into the world of automation.

Genie Harness

The Genie Harness is structured much like a pyATS testscript. There are three main sections: Common Setup, Triggers and Verifications, and Common Cleanup. The Common Setup section will connect to your devices, take a snapshot of current system state, and optionally configure the devices, if necessary. The Triggers and Verifications section, which you will see later in the chapter, will execute the triggers and verifications to perform tests on the devices. This is where the action happens! The Common Cleanup section confirms that the state of the devices is the same as their state in the Common Setup section by taking another snapshot of the current system state and comparing it with the one captured in the Common Setup section.

gRun

The first step to running jobs with Genie Harness is creating a job file. Job files are set up much like a pyATS job file. Within the job file there's a main function that is used as an entry point to run the job(s). However, instead of using **run()** within the main function to run the job, you must use **gRun()**, or the "Genie Run" function. This function is used to execute pyATS testscripts with additional arguments that provide robust and dynamic testing. Datafiles are passed in as arguments, which allows you to run specific triggers and verifications. Example 9-3 shows a Genie job file from the documentation that runs one trigger and one verification.

Example 9-3   *Genie Harness – Job file*

```
from genie.harness.main import gRun

def main():
        # Using built-in triggers and verifications
        gRun(trigger_uids=["TriggerSleep"],
        verification_uids=["Verify_IpInterfaceBrief"])
```

The trigger and verification names in Example 9-3 are self-explanatory: sleep for a specified amount of time and parse and verify the output of **show ip interface brief** command. Each trigger and verification is part of the pyATS library. If these were custom-built triggers or if a testbed device did not have the name or alias of "uut" in your testbed file, you would need to create a mapping datafile. A mapping datafile is used to create a relationship, or mapping, between the devices in a testbed file and Genie. It's required if you want to control multiple connections and connection types (CLI, XML, YANG) to testbed devices. By default, Genie Harness will only connect to the device with the name or alias of "uut," which represents "unit under testing." The uut name/alias requirement allows Genie Harness to properly load the correct default trigger and verification datafiles. Otherwise, you must include a list of testbed devices to the triggers/verifications in the respective datafile. If this doesn't make sense, don't worry; we will touch on datafiles and provide examples further in the chapter that should help provide clarity.

To wrap up the Genie Harness example, the Genie job is run with the same Easypy runtime environment used to run pyATS jobs. To run the Genie job from the command line, enter the following:

```
(.venv)dan@linux-pc# pyats run job {job file}.py --testbed-file
{/path/to/testbed}
```

Remember to have your Python virtual environment activated! In the next section, we will jump into datafiles and how they are defined.

Datafiles

The purpose of a datafile is to provide additional information about the Genie features (triggers, verifications, and so on) you would like Genie Harness to run during a job. For example, the trigger datafile may specify which testbed devices a trigger should run on in the job. There are many datafiles available in Genie Harness. However, many of them are optional and are only needed if you're planning to modify the default datafiles provided. The default datafiles can be found at the following path:

```
$VIRTUAL_ENV/lib/python<version>/site-packages/genie/libs/sdk/genie_
yamls
```

Within the genie_yamls directory, you'll find default datafiles that apply to all operating systems (OSs) and others that are OS-specific. These default datafiles are only implicitly activated when a testbed device has either a name or alias of uut. If there isn't a testbed device with that name or alias, the default datafile will not be implicitly passed to that job. I would highly recommend checking out (not editing) the default datafiles. If you'd like to edit one, you may create a new datafile in your local directory and extend the default one—but don't jump too far ahead yet! This topic will be covered later in the chapter. Here's a list of the different datafiles that can be passed to **gRun**:

- Testing datafile

- Mapping datafile

- Verification datafile

- Trigger datafile

- Subsection datafile

- Configuration datafile

- PTS datafile

Each datafile serves a purpose to a specific Genie Harness feature, but one only needs to be specified if you are deviating from the provided default datafile. For example, the Profile the System (PTS) default datafile only specifies to run on the testbed device with the alias uut. If you would like it to run on more devices, you'll need to create a **pts_data-file.yaml** file that maps devices to the device features you want profiled by PTS and include the **pts_datafile** argument to **gRun**.

Device Configuration

Applying device configuration is always a hot topic when it comes to network automation. In the context of pyATS and the pyATS library (Genie), the focus is on applying (and reverting) the configuration during testing. In many cases, you'll want to test a feature by configuring it on a device, testing its functionality, and removing the configuration before the end of testing. The pyATS library (Genie) provides many ways to apply configuration to devices. Here are some of the options you have to configure a network device during testing:

- Manual configuration before testing begins (not recommended).

- Automatically apply the configuration to the device in the Common Setup and Common Cleanup sections with TFTP/SCP/FTP. A config.yaml file can be provided to the **config_datafile** argument of **gRun**, which specifies the configuration to apply.

- Automatically apply the configuration to the device in the Common Setup and Common Cleanup sections using Jinja2 template rendering. The Jinja2 template filename will be passed to **gRun** using the **jinja2_config** argument and the device variables will be passed as key-value pairs using the **jinja2_arguments** argument.

You will dive into Jinja2 templates further and how to use them to generate configurations in Chapter 10, "Automated Configuration Management." For now, just understand that you can standardize the configuration being pushed to the network devices under testing using configuration templates and a template rendering engine (Jinja2) to render the templates with device variables, resulting in complete configuration files.

All configurations should be built using the **show running** style, which means you create your configuration files how they would appear when you view a device's configuration using the **show running-config** command. This differs from how you would configure a device interactively via SSH using the **configure terminal** approach.

After the devices under testing have been configured, the pyATS library learns the configuration of the devices via the check_config subsection. The check_config subsection runs twice: once during Common Setup and another time during Common Cleanup. It collects the **running-config** of each device in the topology and compares the two configuration "snapshots" to ensure the configuration remains the same before and after testing.

PTS (Profile the System)

Earlier in this chapter, you saw examples of device features that can be learned (for example, BGP) during testing. This is made possible by the network OS-agnostic models built into pyATS. These models create the foundation for building reliable data structures and provide the ability to parse data from the network.

PTS provides the ability to "profile" the network during testing. PTS creates profile snapshots of each feature in the Common Setup and Common Cleanup sections. PTS can learn about all the device features, or a specific list of device features can be provided as a list to the **pts_features** argument of **gRun**. Example 9-4 shows how **gRun** is called in a job file with a list of features passed to the **pts_features** argument.

Example 9-4   *PTS Feature*

```
from genie.harness.main import gRun


def main():
        # Profiling built-in features (models) w/o the PTS datafile
        gRun(pts_features=["bgp", "interface"])
```

Along with having the ability to profile a subset of features/device commands, PTS, by default, will only run on the device with the device alias uut. To have more devices profiled by PTS, you'll need to supply a pts_datafile.yaml file. The datafile can provide a list of devices to profile and describe specific attributes to ignore in the output when comparing snapshots (such as timers, uptime, and dates). Example 9-5 shows a PTS datafile, and Example 9-6 shows the updated **gRun** call, with the **pts_datafile** argument included.

Example 9-5 *PTS Datafile – ex0906_pts_datafile.yaml*

```
extends: "%CALLABLE{genie.libs.sdk.genie_yamls.datafile(pts)}"

bgp:
  devices: ["cat8k-rt1", "cat8k-rt2"]
  exclude:
    - up_time

interface:
  devices: ["cat8k-rt1", "cat8k-rt2"]
```

Example 9-6 *PTS Datafile Argument*

```
from genie.harness.main import gRun

def main():
        # Profiling built-in features (models) w/ the PTS datafile
        gRun(pts_features=["bgp", "interface"],
        pts_datafile="ex0906_pts_datafile.yaml")
```

PTS Golden Config

PTS profiles the operational state of the network devices under testing, but how do we know the state is what we expect? PTS provides a "golden config" snapshot feature that compares the profiles learned by PTS to what is considered the golden snapshot. Each job run generates a file named pts that is saved to the pyATS archive directory of the job. Any PTS file can be moved to a fixed location and used as the golden snapshot. Like the **pts_datafile** argument, the **pts_golden_config** argument can be passed to **gRun**, which points to the golden PTS snapshot used to compare against the current test run snapshots.

There's a lot to digest with Genie Harness as well as a lot of different options, and an understanding of these features and when to use them is critical. In the following sections, we will turn our attention to triggers and verifications. Let's take a look at verifications first, as triggers rely on them to perform properly.

Verifications

A verification runs one or multiple commands to retrieve the current state of the device. The main purpose of a verification is to capture and compare the operational state before and after a trigger, or set of triggers, performs an action on a device. The state of the device can be retrieved via the multiple connection types offered by the pyATS library (CLI, YANG, and so on). Verifications typically run in conjunction with triggers to verify the trigger action did what it was supposed to do and to check for unexpected results, such as changes to a feature you didn't initiate.

Verification Types

Verifications can be broken down into two types: global and local. The difference between the two is related to scoping and when each verification type runs within a script.

- **Global verifications:** Global verifications are used to capture a snapshot of a device before a trigger is executed. Global verifications run immediately after the Common Setup section and before a trigger in a script. If more than one trigger is executed, subsequent snapshots are captured before and after each trigger using the same set of verifications.

- **Local verifications:** Local verifications are independent of global verifications and run as subsections within a trigger. More specifically, a set of snapshots is taken before a trigger action and a subsequent set is taken after to compare to the first one. Local verifications confirm the trigger action did what it was supposed to do (configure/unconfigure, shut/no shut, and so on).

Figure 9-2 shows where and when the different verification types run in a Genie job.

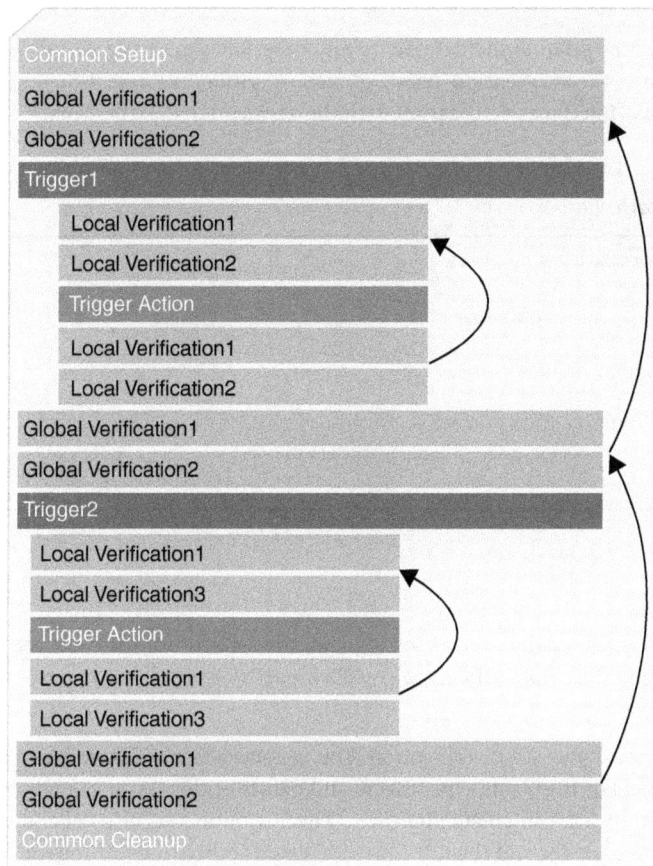

Figure 9-2 *Verification Execution*

Verification Datafile

A verification datafile is used to customize the execution of built-in or custom verifications. Like other datafiles, verification_datafile.yaml must be provided to **gRun** using the **verification_datafile** argument. Example 9-7 shows a verification datafile that extends the default verification datafile (via the **extends:** key) and overrides the default setting of connecting to only the "uut" device (via the **devices:** key) for the **Verify_Interfaces** verification. If you wanted to change the list of devices to connect to for another verification, you would need to add that verification in the datafile.

Example 9-7   *Verification Datafile – ex0908_verification_datafile.yaml*

```
# Extend default verification datafile to inherit the required keys
# (class, source, etc.) per the verification datafile schema
extends: "%CALLABLE{genie.libs.sdk.genie_yamls.datafile(verification)}"

Verify_Interfaces:
  devices: ["cat8k-rt1", "cat8k-rt2"]
```

In order to run the verifications listed in the datafile, or any other built-in verifications, you'll need to include them in the **verification_uids** argument to **gRun**. Example 9-8 shows how to run the Verify_Interfaces verification with the verification datafile from Example 9-7.

Example 9-8   *gRun – Verifications*

```
from genie.harness.main import gRun


def main():
        gRun(
        verification_uids=["Verify_Interfaces"],
        verification_datafile="ex0908_verification_datafile.yaml"
        )
```

Writing a Verification

The process in which a feature is verified during testing may be different for different use cases. If a built-in verification does not suffice, the pyATS library (Genie) allows you to create your own verification.

There are several ways to create your own verification. You can use a Genie Ops feature (model), a parser, or callable. For the Genie Ops feature and parser options, you can use an existing model or parser, or you can create your own. The last option, using a callable, is discouraged, as it isn't OS-agnostic and does not provide extensibility to use different management interfaces (CLI, YANG, and so on). Example 9-9 shows a custom verification built with the **show bgp all** parser. Take note of the list of excluded values from the parsed data (found under the **exclude:** key). The reason is because many of these values are dynamic (such as timers, counters, and so on) and are almost guaranteed to be different between snapshots. Remember, if a parsed value is different between snapshots, the verification will fail.

Example 9-9 *Custom Verification – ex0910_verification_datafile.yaml*

```
# Local verification datafile that already extends the default datafile
extends: verification_datafile.yaml

Verify_Bgp:
    cmd:
        class: show_bgp.ShowBgpAll
        pkg: genie.libs.parser
    context: cli
    source:
        class: genie.harness.base.Template
    devices: ["cat8k-rt1", "cat8k-rt2"]
    iteration:
        attempt: 5
        interval: 10
    exclude:
    - if_handle
    - keepalives
    - last_reset
    - reset_reason
    - foreign_port
    - local_port
    - msg_rcvd
    - msg_sent
    - up_down
    - bgp_table_version
    - routing_table_version
    - tbl_ver
    - table_version
    - memory_usage
    - updates
    - mss
    - total
    - total_bytes
    - up_time
```

```
    - bgp_negotiated_keepalive_timers
    - hold_time
    - keepalive_interval
    - sent
    - received
    - status_codes
    - holdtime
    - router_id
    - connections_dropped
    - connections_established
    - advertised
    - prefixes
    - routes
    - state_pfxrcd
```

To run the custom verification, you follow the same process as running any other verification. Pass the verification datafile with the custom verification to **gRun** via the **verification_datafile** argument and add the custom verification name to the list of verifications in the **verification_uids** argument. Example 9-10 shows the updated **gRun** call with the custom verification name and datafile. Remember, the custom verification datafile (Example 9-9) extends the original verification datafile (Example 9-7), which essentially inherits all the built-in verifications from the pyATS library (Genie).

Example 9-10 *gRun – Custom Verification*

```
from genie.harness.main import gRun


def main():
        gRun(
        verification_uids=["Verify_Interfaces", "Verify_Bgp"],
        verification_datafile="ex0910_verification_datafile.yaml"
        )
```

Triggers

Triggers perform a specific action, or a sequence of actions, on a device to alter its state and/or configuration. As examples, actions may include adding/removing parts of a configuration, flapping protocols/interfaces, or performing high availability (HA) events such as rebooting a device. The important part to understand is that triggers are what alter the device during testing.

The pyATS library (Genie) has many prebuilt triggers available for Cisco IOS/IOS XE, NX-OS, and IOS XR. All prebuilt triggers are documented, describing what happens when the trigger is initiated and what keys/values to include in the trigger datafile specifically for that trigger. For example, the **TriggerShutNoShutBgpNeighbors** trigger performs the following workflow:

1. Learn BGP Ops object and verify it has "established" neighbors. If there aren't any "established" neighbors, skip the trigger.

2. Shut the BGP neighbor that was learned from step 1 with the BGP Conf object.

3. Verify the state of the learned neighbor(s) in step 2 is "down."

4. Unshut the BGP neighbor(s).

5. Learn BGP Ops again and verify it is the same as the BGP Ops snapshot in step 1.

As you might recall from earlier in this chapter, the Genie Ops object represents a device/ feature's operational state via a Python object, and the Genie Conf object represents a feature, as a Python object, that can be configured on a device. The focus of the Conf object is *what* feature you want to apply on the device, not *how* to apply it per device (OS) platform. This allows a network engineer to focus on the network features being tested and not on the low-level details of how the configuration is applied.

Now that there's a general understanding of what triggers do, let's check out how they can be configured using a trigger datafile.

Trigger Datafile

As with other features of the pyATS library (Genie), to run triggers, there needs to be a datafile—more specifically, a trigger datafile (trigger_datafile.yaml). The pyATS library provides a default trigger datafile found in the same location as all the other default data-files, discussed earlier in the chapter. However, if you want to customize any specific trigger settings, such as what devices or group of devices to run on during testing (any device besides uut), or to run a custom trigger, you'll need to create your own trigger datafile. A complete example can be found at the end of the chapter that includes both triggers and verifications, but let's focus now on just triggers in a brief example. Example 9-11 shows a custom trigger file that flaps the OSPF process on the targeted devices (iosv-0 and iosv01). Example 9-12 shows how to include the appropriate trigger and trigger datafile in the list of arguments to **gRun**.

Example 9-11   *Trigger Datafile – ex0912_trigger_datafile.yaml*

```
extends: "%CALLABLE{genie.libs.sdk.genie_yamls.datafile(trigger)}"


# Custom trigger - created in Example 9-14
TriggerShutNoShutOspf:
# source imports the custom trigger, just as you would any other Python class
  source:
    class: ex0915_custom_trigger.ShutNoShutOspf
  devices: ["iosv-0", "iosv-1"]
```

Example 9-12 *gRun – Triggers and Trigger Datafile*

```
from genie.harness.main import gRun


def main():
        gRun(
        trigger_uids=["TriggerShutNoShutOspf"],
        trigger_datafile="ex0912_trigger_datafile.yaml"
        )
```

Trigger Cluster

The last neat trigger feature to cover is the ability to execute a group of multiple triggers and verifications in one cluster trigger. First, a trigger datafile must be created with the list of triggers and verifications, the order in which to run them, and a list of testbed devices to run them against. Example 9-13 shows a trigger datafile configured for a trigger cluster and the accompanying test results if it was run.

Example 9-13 *Trigger Cluster*

```
TriggerCombined:
    sub_verifications: ['Verify_BgpVrfAllAll']
    sub_triggers: [ 'TriggerSleep', 'TriggerShutNoShutBgp']
    sub_order: ['TriggerSleep', 'Verify_BgpVrfAllAll',
    'TriggerSleep','TriggerShutNoShutBgp','Verify_BgpVrfAllAll']
    devices: ['uut']


-- TriggerCombined.uut                                              PASSED
  |-- TriggerSleep_sleep.1                                          PASSED
  |-- TestcaseVerificationOps_verify.2                              PASSED
  |-- TriggerSleep_sleep.3                                          PASSED
  |-- TriggerShutNoShutBgp_verify_prerequisite.4                    PASSED
  |   |-- Step 1: Learning 'Bgp' Ops                                PASSED
  |   |-- Step 2: Verifying requirements                            PASSED
  |   '-- Step 3: Merge requirements                                PASSED
  |-- TriggerShutNoShutBgp_shut.5                                   PASSED
  |   '-- Step 1: Configuring 'Bgp'                                 PASSED
  |-- TriggerShutNoShutBgp_verify_shut.6                            PASSED
  |   '-- Step 1: Verifying 'Bgp' state with ops.bgp.bgp.Bgp        PASSED
  |-- TriggerShutNoShutBgp_unshut.7                                 PASSED
  |   '-- Step 1: Unconfiguring 'Bgp'                               PASSED
  |-- TriggerShutNoShutBgp_verify_initial_state.8                   PASSED
  |   '-- Step 1: Verifying ops 'Bgp' is back to original state     PASSED
  '-- TestcaseVerificationOps_verify.9                              PASSED
```

You may notice that the triggers have accompanying local verifications that run before and after the trigger is run to ensure the action was actually taken against the device. This is the true power of triggers. One of the biggest reasons people are skeptical about network automation is due to the lack of trust. Did this automation script/test really do what it's supposed to do? Triggers provide that verification out of the box through global and local verifications.

What if we wanted to build our own trigger with verifications? In the next section, you'll see how to do just that!

Writing a Trigger

The pyATS library (Genie) provides the ability to write your own triggers. A trigger is simply a Python class that has multiple tests in it that either configure, verify, or unconfigure the configuration or device feature you're trying to test.

To begin, your custom trigger must inherit from a base **Trigger** class. This base class contains common setup and cleanup tasks that help identify any unexpected changes to testbed devices not currently under testing (for example, a device rebooting). For our custom trigger, we are going to shut and unshut OSPF. Yes, this trigger already exists in the library, but it serves as a great example when you're beginning to create custom triggers. The workflow is going to look like this:

1. Check that OSPF is configured and running.

2. Shut down the OSPF process.

3. Verify that OSPF is shut down.

4. Unshut the OSPF process.

5. Verify OSPF is up and running.

In Examples 9-14 and 9-15, you'll see the code to create the custom OSPF trigger and the associated job file, running it with **gRun**. To run the job file, you'll need the following files:

- ex0915_custom_trigger.py

- ex0916_custom_trigger_job.py

- ex0915_custom_trigger_datafile.yaml

- testbed2.yaml

The testbed2.yaml file has two IOSv routers, named "iosv-0" and "iosv-1," running OSPF. The file ex0915_custom_trigger_datafile.yaml is used to map the custom OSPF triggers and the testbed devices:

```
# ex0915_custom_trigger_datafile.yaml

extends: "%CALLABLE{genie.libs.sdk.genie_yamls.datafile(trigger)}"

# Custom trigger
TriggerMyShutNoShutOspf:
    # source imports the custom trigger
    source:
        class: ex0915_custom_trigger.MyShutNoShutOspf
    devices: ["iosv-0", "iosv-1"]
```

Example 9-14   *Custom Trigger and Job File*

```
import time
import logging
from pyats import aetest

from genie.harness.base import Trigger
from genie.metaparser.util.exceptions import SchemaEmptyParserError

log = logging.getLogger()

class MyShutNoShutOspf(Trigger):
"""Shut and unshut OSPF process. Verify both actions."""

                @aetest.setup
                def prerequisites(self, uut):
                """Check whether OSPF is configured and running."""

                # Checks if OSPF is configured. If not, skip this trigger
                try:
                                output = uut.parse("show ip ospf")
                except SchemaEmptyParserError:
                    self.failed(f"OSPF is not configured on device {uut.name}")

                # Extract the OSPF process ID
                self.ospf_id = list(output["vrf"]["default"]["address_family"] \
                ["ipv4"]["instance"].keys())[0]
```

```python
# Checks if the OSPF process is enabled
ospf_enabled = output["vrf"]["default"]["address_family"] \
["ipv4"]["instance"][self.ospf_id]["enable"]

if not ospf_enabled:
        self.skipped(f"OSPF is not enabled on device {uut.name}")


@aetest.test
def ShutOspf(self, uut):
"""Shutdown the OSPF process"""

uut.configure(f"router ospf {self.ospf_id}\n shutdown")
time.sleep(5)


@aetest.test
def verify_ShutOspf(self, uut):
"""Verify ShutOspf worked"""

output = uut.parse("show ip ospf")

ospf_enabled = output["vrf"]["default"]["address_family"] \
["ipv4"]["instance"][self.ospf_id]["enable"]

if ospf_enabled:
            self.failed(f"OSPF is enabled on device {uut.name}")

@aetest.test
def NoShutOspf(self, uut):
"""Unshut the OSPF process"""

uut.configure(f"router ospf {self.ospf_id}\n no shutdown")

@aetest.test
def verify_NoShutOspf(self, uut):
"""Verify NoShutOspf worked"""

output = uut.parse("show ip ospf")

ospf_enabled = output["vrf"]["default"]["address_family"] \
["ipv4"]["instance"][self.ospf_id]["enable"]

if not ospf_enabled:
            self.failed(f"OSPF is enabled on device {uut.name}")
```

Example 9-15 *Running a Custom Trigger – ex0915_custom_trigger_job.py*

```
from genie.harness.main import gRun

def main():
        gRun(
        trigger_uids=["TriggerMyShutNoShutOspf"],
        trigger_datafile="ex0915_custom_trigger_datafile.yaml",
        )




# Running the job using 'pyats run job' command
# pyats run job ex0916_custom_trigger_job.py --testbed-file testbed2.yaml
```

Figure 9-3 shows some sample job output.

```
+----------------------------------------------------------+
|                    Task Result Details                   |
+----------------------------------------------------------+
Task-1: genie_testscript
|-- common_setup                                     PASSED
|   |-- connect                                      PASSED
|   |-- configure                                    SKIPPED
|   |-- configuration_snapshot                       PASSED
|   |-- save_bootvar                                 PASSED
|   |-- learn_system_defaults                        PASSED
|   |-- initialize_traffic                           SKIPPED
|   `-- PostProcessor-1                              PASSED
|-- TriggerMyShutNoShutOspf.iosv-0                   PASSED
|   |-- prerequisites                                PASSED
|   |-- ShutOspf                                     PASSED
|   |-- verify_ShutOspf                              PASSED
|   |-- NoShutOspf                                   PASSED
|   `-- verify_NoShutOspf                            PASSED
|-- TriggerMyShutNoShutOspf.iosv-1                   PASSED
|   |-- prerequisites                                PASSED
|   |-- ShutOspf                                     PASSED
|   |-- verify_ShutOspf                              PASSED
|   |-- NoShutOspf                                   PASSED
|   `-- verify_NoShutOspf                            PASSED
`-- common_cleanup                                   PASSED
    |-- verify_configuration_snapshot                PASSED
    |-- stop_traffic                                 SKIPPED
    `-- PostProcessor-1                              PASSED
```

Figure 9-3 *Job Results*

Trigger and Verification Example

Now it's time to combine the triggers and verifications into one complete example. In the example, we will build on previous examples where we flap (shut/unshut) the OSPF process on two IOSv routers. The trigger has local verifications that will confirm OSPF is indeed shut down and confirm that it comes up after being unshut. In addition to the local verifications defined in the trigger, the job will also introduce a global verification to check the router link states (LSA Type 1) in the OSPF LSDB (link state database). This global verification will run before and after the trigger. This allows us to confirm that LSA Type 1 packets are being exchanged before and after testing and that the router link types have not changed during testing.

Let's see how this example can be implemented and executed using Genie Harness and also within a pyATS testscript.

Genie Harness (gRun)

This whole chapter has been focused on Genie Harness, so let's not rehash the details. Example 9-16 shows an example of the job file, trigger datafile, and verification datafile used to execute the job. Figure 9-4 shows the associated test results.

Example 9-16   *Trigger and Verification Example – ex0917_complete_example.py*

```
# Job file - ex0917_complete_example.py
from genie.harness.main import gRun

def main():
gRun(
        trigger_uids=["TriggerShutNoShutOspf"],
        trigger_datafile="ex0917_trigger_datafile.yaml",
        verification_uids=["Verify_IpOspfDatabaseRouter"],
        verification_datafile="ex0917_verification_datafile.yaml"
        )




# Trigger datafile - ex0917_trigger_datafile.yaml
extends: "%CALLABLE{genie.libs.sdk.genie_yamls.datafile(trigger)}"


# Custom trigger - created before Example 9-14
TriggerShutNoShutOspf:
# source imports the custom trigger, just as you would any other Python class
  source:
    class: ex0915_custom_trigger.MyShutNoShutOspf
  devices: ["iosv-0", "iosv-1"]
```

```
# Verification datafile - ex0917_verification_datafile.yaml
extends: "%CALLABLE{genie.libs.sdk.genie_yamls.datafile(verification)}"



Verify_IpOspfDatabaseRouter:
  devices: ["iosv-0", "iosv-1"]
```

```
+-------------------------------------------------------------------+
|                       Task Result Summary                         |
+-------------------------------------------------------------------+
Task-1: genie_testscript.common_setup                        PASSED
Task-1: genie_testscript.Verifications.TriggerShutNoShutOspf.iosv-0  PASSED
Task-1: genie_testscript.TriggerShutNoShutOspf.iosv-0        PASSED
Task-1: genie_testscript.Verifications.TriggerShutNoShutOspf.iosv-1  PASSED
Task-1: genie_testscript.TriggerShutNoShutOspf.iosv-1        PASSED
Task-1: genie_testscript.Verifications.post                  PASSED
Task-1: genie_testscript.common_cleanup                      PASSED


+-------------------------------------------------------------------+
|                       Task Result Details                         |
+-------------------------------------------------------------------+
Task-1: genie_testscript
|-- common_setup                                             PASSED
|   |-- connect                                              PASSED
|   |-- configure                                            SKIPPED
|   |-- configuration_snapshot                               PASSED
|   |-- save_bootvar                                         PASSED
|   |-- learn_system_defaults                                PASSED
|   |-- initialize_traffic                                   SKIPPED
|   `-- PostProcessor-1                                      PASSED
|-- Verifications.TriggerShutNoShutOspf.iosv-0               PASSED
|   |-- Verify_IpOspfDatabaseRouter.iosv-0.1                 PASSED
|   |   `-- verify                                           PASSED
|   `-- Verify_IpOspfDatabaseRouter.iosv-1.1                 PASSED
|       `-- verify                                           PASSED
|-- TriggerShutNoShutOspf.iosv-0                             PASSED
|   |-- prerequisites                                        PASSED
|   |-- ShutOspf                                             PASSED
|   |-- verify_ShutOspf                                      PASSED
|   |-- NoShutOspf                                           PASSED
|   `-- verify_NoShutOspf                                    PASSED
|-- Verifications.TriggerShutNoShutOspf.iosv-1               PASSED
|   |-- Verify_IpOspfDatabaseRouter.iosv-0.2                 PASSED
|   |   `-- verify                                           PASSED
|   `-- Verify_IpOspfDatabaseRouter.iosv-1.2                 PASSED
|       `-- verify                                           PASSED
|-- TriggerShutNoShutOspf.iosv-1                             PASSED
|   |-- prerequisites                                        PASSED
|   |-- ShutOspf                                             PASSED
|   |-- verify_ShutOspf                                      PASSED
|   |-- NoShutOspf                                           PASSED
|   `-- verify_NoShutOspf                                    PASSED
|-- Verifications.post                                       PASSED
|   |-- Verify_IpOspfDatabaseRouter.iosv-0.3                 PASSED
|   |   `-- verify                                           PASSED
|   `-- Verify_IpOspfDatabaseRouter.iosv-1.3                 PASSED
|       `-- verify                                           PASSED
`-- common_cleanup                                           PASSED
    |-- verify_configuration_snapshot                        PASSED
    |-- stop_traffic                                         SKIPPED
    `-- PostProcessor-1                                      PASSED
```

Figure 9-4 *Trigger and Verification Example Results*

pyATS

Triggers and verifications can be run in a pyATS testscript as a testcase or within a test section. Custom trigger and verification datafiles can be provided using the **--trigger-datafile** and **--verification-datafile** arguments when calling a pyATS job.

To run it as its own testcase, you'll just need to create a class that inherits from **GenieStandalone**, which inherits from the pyATS **Testcase** class. The inherited class you create will provide a list of triggers and verifications. The same trigger and verification datafiles will be included as options when running the pyATS job via the command line.

The second way to include triggers and verifications in a pyATS testscript is by including them in an individual test section, as part of a pyATS testcase. The **run_genie_sdk** function allows you to run triggers or verifications as steps within a section.

Example 9-17 shows how to include triggers and verifications as their own testcase and within an existing subsection in a pyATS testscript. To run the testscript, you'll need to add the **--trigger-datafile** and **--verification-datafile** arguments with the appropriate data-files to map the custom trigger and verifications to the additional devices. These datafiles are not included in the example. Figure 9-5 shows the testscript results.

Example 9-17 *Triggers and Verifications in pyATS Testscript*

```
from pyats import aetest
import genie
from genie.harness.standalone import GenieStandalone, run_genie_sdk


class CommonSetup(aetest.CommonSetup):
""" Common Setup section """

        @aetest.subsection
        def connect(self, testbed):
        """"Connect to each device in the testbed."""

                genie_testbed = genie.testbed.load(testbed)
                self.parent.parameters["testbed"] = genie_testbed
                genie_testbed.connect()


# Call Triggers and Verifications as independent pyATS testcase
class GenieOspfTriggerVerification(GenieStandalone):
"""Shut/unshut the OSPF process and verify LSA Type 1 packets are still being
exchanged before and after testing."""
```

```
            # Must specify 'uut'
            # If other devices are included in the datafile(s), they will be tested
            uut = "iosv-0"
            triggers = ["TriggerShutNoShutOspf"]
            verifications = ["Verify_IpOspfDatabaseRouter"]




# Calling Triggers and Verifications within a pyATS section
class tc_pyats_genie(aetest.Testcase):
        """Testcase with triggers and verifications."""
        # First test section
        @aetest.test
        def simple_test_1(self, steps):
        """Sample test section."""

            # Run Genie triggers and verifications
            # Note that you must specify the order of each trigger and verification
            run_genie_sdk(self,
                            steps,
                            ["Verify_IpOspfDatabaseRouter", \
                             "TriggerShutNoShutOspf", \
                             "Verify_IpOspfDatabaseRouter"],
                            uut="iosv-0"
                            )



class CommonCleanup(aetest.CommonCleanup):
    """Common Cleanup section"""

    @aetest.subsection
    def disconnect_from_devices(self, testbed):
            """Disconnect from each device in the testbed."""
            testbed.disconnect()
```

```
+---------------------------------------------------------------------+
|                        Task Result Summary                          |
+---------------------------------------------------------------------+
Task-1: ex0918_pyats_testscript.common_setup                   PASSED
Task-1: ex0918_pyats_testscript.GenieOspfTriggerVerification   PASSED
Task-1: ex0918_pyats_testscript.tc_pyats_genie                 PASSED
Task-1: ex0918_pyats_testscript.common_cleanup                 PASSED


      +---------------------------------------------------------+
      |                  Task Result Details                    |
      +---------------------------------------------------------+
Task-1: ex0918_pyats_testscript
|-- common_setup                                               PASSED
|    `-- connect                                               PASSED
|-- GenieOspfTriggerVerification                               PASSED
|    |-- Verify_IpOspfDatabaseRouter.iosv-0_verify.1           PASSED
|    |-- Verify_IpOspfDatabaseRouter.iosv-1_verify.2           PASSED
|    |-- TriggerShutNoShutOspf.iosv-0_prerequisites.3          PASSED
|    |-- TriggerShutNoShutOspf.iosv-0_ShutOspf.4               PASSED
|    |-- TriggerShutNoShutOspf.iosv-0_verify_ShutOspf.5        PASSED
|    |-- TriggerShutNoShutOspf.iosv-0_NoShutOspf.6             PASSED
|    |-- TriggerShutNoShutOspf.iosv-0_verify_NoShutOspf.7      PASSED
|    |-- TriggerShutNoShutOspf.iosv-1_prerequisites.8          PASSED
|    |-- TriggerShutNoShutOspf.iosv-1_ShutOspf.9               PASSED
|    |-- TriggerShutNoShutOspf.iosv-1_verify_ShutOspf.10       PASSED
|    |-- TriggerShutNoShutOspf.iosv-1_NoShutOspf.11            PASSED
|    |-- TriggerShutNoShutOspf.iosv-1_verify_NoShutOspf.12     PASSED
|    |-- Verify_IpOspfDatabaseRouter.iosv-0_verify.13          PASSED
|    `-- Verify_IpOspfDatabaseRouter.iosv-1_verify.14          PASSED
|-- tc_pyats_genie                                             PASSED
|    `-- simple_test_1                                         PASSED
|        |-- STEP 1: Verify_IpOspfDatabaseRouter.iosv-0        PASSED
|        |-- STEP 2: TriggerShutNoShutOspf.iosv-0             PASSED
|        `-- STEP 3: Verify_IpOspfDatabaseRouter.iosv-0        PASSED
`-- common_cleanup                                             PASSED
     `-- disconnect_from_devices                               PASSED
```

Figure 9-5 *Triggers and Verifications in pyATS Testscript Results*

Summary

This chapter covered a lot of information about the pyATS library (Genie), the Genie Harness, with its many different features, along with triggers and verifications. The Genie Harness allows you to take advantage of the pyATS infrastructure without having to dive-deep into code. The goal of the pyATS library (Genie) is to be modular and robust. Triggers and verifications are a perfect example. They make it easy to quickly build dynamic testcases that change with your network requirements. I highly recommend taking a closer at the code examples in this chapter and trying them out for yourself. Within minutes, you'll see how quickly you can test a network feature with speed and accuracy!

Chapter 10

Automated Configuration Management

Configuration management has always been a struggle for network engineers. Whether it be lack of standard enforcement or constant firefighting, untracked configuration is in the network. Untracked configuration in this context refers to the ad-hoc changes that may be required to fix an issue or even a legacy, stale configuration that was never cleaned up. Regardless, configuration management is a top use case for many engineers when they begin their network automation journey. In this chapter, we are going to take a look at how configuration can be managed declaratively, with the concept of intent-based networking, and pushed to devices using pyATS and the pyATS library (Genie). The following topics will be covered in this chapter:

- Intent-based network configuration
- Generating configurations with pyATS
- Configuring devices with pyATS

Intent-Based Network Configuration

How many times have you been asked to roll out a simple change, such as updating the logging hosts or NTP servers on your network device, and before you write your implementation plan, you wonder what's even configured there in the first place? Intent-based networking is an emerging concept where "intent" on how a network should be configured and operated is declared, and with the help of telemetry data, analysis, and a sprinkle of artificial intelligence (AI), you can confirm the network intent is operating as expected.

The next question asked is, "Wow, that's something I'm interested in, but how could I get started?" The best part about pyATS is that it's a testing framework first, with the ability to configure devices. In no way is pyATS a full-blown intent-based networking solution, but it can help with the low-hanging fruit and get you started down the right path. In the following section, we will explore how to generate device configuration using the popular Python templating engine Jinja2 and with pyATS library (Genie) Conf objects. Wrapping up the chapter, we will discuss the different ways to push configuration to entire testbeds or specific devices.

Generating Configurations with pyATS

Configurations cannot be generated without declaring network intent. It could simply be that you want 10.1.1.1 and 10.1.1.2 to be configured as logging hosts on every device in your network—that's intent! Network intent can be declared using data models. Data models allow you to produce a standardized and abstract definition of data, which provides a common and predictable structure for data extraction. Data modeling is useful when you have *a lot* of data from multiple, potentially untrusted sources, and it needs to be organized and validated. In the case of configuration management, the data required for network intent can be collected from a source of truth (SoT), such as NetBox, or something much simpler, such as an Excel spreadsheet. Regardless of the data source, for any automation effort, external data should be modeled and validated into a data model. In the case of using a source of truth, this may seem redundant, but it does provide a level of certainty that the data is what you expect it to be. Once data is modeled, we need to extract the key values from the data models and implement them into low-level details, such as a config script.

Data Modeling and Validation

Whether data is coming from an external system, such as a source of truth like NetBox, or from a local spreadsheet saved on your laptop, it's paramount that data is modeled and validated before being used for further processing. The purpose is to confirm the data is what you expect it is before consuming it. The worst thing that can happen is you use the wrong value in your automation workflow and cause a major outage. Try explaining to your manager or the CTO that you took down the network due to an unexpected value that was not validated in your automation workflow—not a good day.

In network automation, YANG is the de facto choice when it comes to data modeling. YANG models network state and configuration data that is stored on a network device. However, Python and other programming languages offer many data modeling and validation libraries. A popular data modeling library in Python is Pydantic (https://docs.pydantic.dev/latest/). Pydantic is a data validation library that allows you to build models that represent your data and validate "untrusted" data against the defined models. There are some additional features in Pydantic, such as data coercion and serialization, but I recommend checking out the documentation for more details. Another popular Python data validation library is Cerebrus (https://docs.python-cerberus.org/). Cerebrus is more lightweight than Pydantic and quickly allows you to validate data against a schema that you define. Data modeling and validation can be a book itself, but the goal of it, as it relates to network automation, is to provide an interface to define the structure and validate the data you're working with, whether that's configuration or operational state data.

Data Templates

Once data is modeled and validated against a predefined schema, it's time to consume the data. To generate *dynamic* configuration files for different types of devices, operating systems, and so on, you must use a templating engine that allows for variable substitution and conditional logic. Python has a popular templating engine called Jinja2. The Jinja2 templating engine allows you to pass in data and provides the ability to write programming logic (conditional statements, loops, and so on) directly in the template. Example 10-1 shows a simple Jinja2 template that renders an interface configuration.

Example 10-1 *Jinja2 Template – Single Interface*

```
interface {{ interface.name }}
  description {{ interface.description }}
  ip address {{ interface.ip_address }} {{ interface.mask }}
```

You can see the template variables identified by the double curly braces. This template works great for one interface, but what if we wanted to render configurations for all interfaces on a device? Let's add some additional logic to the template and introduce looping with a conditional to accomplish this goal. Example 10-2 shows how we can expand the previous template to accommodate for more interfaces and determine whether to explicitly activate the port.

Example 10-2 *Jinja2 Template – All Interfaces*

```
{% for interface in interfaces %}
interface {{ interface.name }}
  description {{ interface.description }}
  ip address {{ interface.ip_address }} {{ interface.mask }}
  {% if interface.active %}
  no shutdown
  {% else %}
  shutdown
  {% endif %}
{% endfor %}
```

By adding the **for** loop, you can provide any number of interfaces in a list to the template to be rendered. Another addition is the **if** conditional to determine whether to activate or shut down the port. This logic can be helpful for a device with many interfaces, such as a switch, where it can be tough to manage which interfaces should be active.

Jinja2 allows you to pass data, or what the library calls *context*, into the template as keyword arguments, a Python list, or a Python dictionary. In the two previous examples, variables are passed in as a dictionary (Example 10-1) and a list (Example 10-2). You can tell by the dot notation used to reference each interface's attribute (that is, ***interface.description***). Example 10-1 references a single dictionary that was passed in with the keyword **interface**, which is why the word "interface" is referenced in the template. Example 10-2 references a list of dictionaries, with each dictionary referencing an interface. The **interface** key in this scenario references each individual list item in the **for** loop. Example 10-3 shows what an interface dictionary may look like being passed into either template.

Example 10-3   *Jinja2 Context – Interface Dictionary*

```
"interface:" {
  "name": "GigabitEthernet1",
  "description": "WAN Link",
  "ip_address": "10.100.1.10",
  "mask": "255.255.255.252",
  "active": True
}
```

A Python dictionary is easy enough to read, but not every network engineer is a programmer, so how do we define an intended configuration to generate without having to introduce code? YAML can be our friend in this situation. YAML is a human-readable data serialization language that makes it easy to define and consume data. A YAML document can be used to define intended configuration parameters by all engineers, programmers, and non-programmers. The best part about YAML is that it's a data serialization language, which means it can translate data into a data structure such as a Python dictionary. There is wide support for YAML across many programming languages, which allows you to translate a YAML document to different data structures across different programming languages. Example 10-4 shows how the Python dictionary in Example 10-3 can be represented in a YAML document.

Example 10-4   *YAML Document – Interface*

```
---
interfaces:
    - name: GigabitEthernet1
      description: WAN Link
      ip_address: 10.100.1.10
      mask: 255.255.255.252
      active: True
```

The Python dictionary and YAML document look similar, but the YAML document can easily be edited and stored outside of the code. Technically, Python dictionaries can be stored outside of the main code module and used as data parameters, but conventionally this doesn't make sense. YAML provides flexibility and extensibility to store configuration parameters without locking you into a particular programming language. To wrap up how Jinja2 templates can be used to generate device configurations, Example 10-5 shows the rendered configuration for interface GigabitEthernet1 using the previously defined Jinja2 template in Example 10-2 based on the data context provided in Example 10-4.

Example 10-5   *Jinja2 Template – Rendered Configuration*

```
interface GigabitEthernet1
  description WAN Link
  ip address 10.100.1.10 255.255.255.252
  no shutdown
```

The pyATS library has a Device API called **load_jinja_template** that allows you to render a device configuration using a Jinja2 template and keyword arguments. It's really a helper function to easily generate a config without having to worry about the details of creating a Jinja2 environment and other Jinja2 configuration details. You simply need to pass the path of the directory containing Jinja2 templates, the Jinja2 template name, and any keyword arguments you'd like to use as variables when rendering the template. The rendered template string will be returned, which will be the generated configuration. Example 10-6 shows how to use the **load_jinja_template** API to render a short logging configuration. You'll notice that a list of logging hosts is passed in with the **hosts** keyword argument. This illustrates the flexibility of using a templating engine to render configurations by passing in arbitrary values that serve as variables in the template.

Example 10-6   *Device API – load_jinja_template*

```
from pyats.topology.loader import load

# Load testbed and select "router1" device
my_testbed = load("testbed.yaml")
xe_router = my_testbed.devices["cat8k-rt1"]

# Render config with load_jinja_template() device API
logging_config = xe_router.api.load_jinja_template(
    path="config_templates", file="ios_logging.j2", \
    hosts=["10.1.1.1", "10.1.1.2"]
)

# Print rendered config
print(logging_config)
```

Genie Conf Objects

In Chapter 9, "pyATS Triggers and Verifications," you learned about Genie Conf and how it takes advantage of Python's object-oriented (OOP) approach to building logical representations of real-world network objects. The beauty of Genie Conf is that it focuses on what you want to do, instead of how to do it. This removes the complexity of figuring out how to implement the same network feature across different device platforms. For example, Genie Conf allows you to build an object that represents a network interface and attach it to a testbed device. Example 10-7 depicts how easy it is to build a network interface for a Cisco IOS-XE device.

Example 10-7   *Genie Conf – Cisco IOS-XE Interface Object*

```
from pyats.topology.loader import load
from genie.conf.base import Interface


# Load testbed and select "router1" device
my_testbed = load("testbed.yaml")
xe_router = my_testbed.devices["cat8k-rt1"]


# Create "GigabitEthernet4" Interface object for router1 device
xe_interface = Interface(device=xe_router, name="GigabitEthernet5")


# Set interface attributes
xe_interface.ipv4 = "10.1.1.1"
xe_interface.ipv4.netmask = "255.255.255.0"
xe_interface.shutdown = False



# To find out more about the different Interface attributes:
# print(dir(xe_interface))
# Check out the Interface object source code on GitHub:
# https://github.com/CiscoTestAutomation/genielibs/blob/master/pkgs/# conf-pkg/src/
  genie/libs/conf/interface/__init__.py#L187
```

Once the object is built and saved as a variable, you simply need to call on the **build_config()** method to build the configuration and configure the testbed device. Example 10-8 shows how easy this is to do.

Example 10-8   *Genie Conf – Build Config*

```
# Assumes the Interface object from Example 10-7 is created


# Build config (without pushing configuration to the device)
print(xe_interface.build_config(apply=False))
```

```
# Build and push config to device
xe_interface.build_config()

# Remove the config from the device
xe_interface.build_unconfig()
```

Simple as that! These examples can be extended to other network features, such as BGP, OSPF, NTP, LLDP, and many more! For a full list of available features, check out the pyATS library (Genie) models here: https://pubhub.devnetcloud.com/media/genie-feature-browser/docs/#/models. All the available models have an associated Conf object that can be used as a reference to configure the corresponding object and build the appropriate device config.

Configuring Devices with pyATS

The most nerve-racking part of automation is pushing configuration to network devices. As the saying goes, "Automation just increases the blast radius." Although the author is unknown, the quote holds true. If your automation script is accurate and executes without error, perfect! However, if you have a bug in your automation script (sometimes something small as a typo), your error just propagates at a faster rate than if you made that error manually. Luckily, there are many automation platforms and libraries that can help avoid these problems by providing boilerplate code/templates to push configuration to network devices. In the following sections, we will look at different methods to configure network devices using pyATS. I'll preface this by saying that pyATS is not a configuration management tool. Its primary focus is network testing. However, given that the pyATS framework offers the ability to push configuration, we can very easily manage configuration using pyATS, without needing to install more tooling or libraries.

File Transfer

PyATS provides the ability to transfer files to or from a remote host. You can accomplish file transfers by using the FileUtils utility module or the embedded pyATS file transfer server.

FileUtils Module

FileUtils is a multivendor, OS- and protocol-agnostic file transfer utility that abstracts the complexities of transferring files between local and remote hosts. The module also provides the ability, through a plugin, to manipulate files relative to remote hosts versus local execution. Protocols supported for copying files include FTP, SFTP, TFTP, and SCP. Other file operations, such as deleting a file, renaming a file, and checking whether a file exists are only supported when using FTP and SFTP. In order to use the SFTP and SCP connection types, you must install the paramiko package, and additionally for SCP, the scp package. Example 10-9 shows how to specify a remote server in the testbed file and the different use cases for using the FileUtils module.

Example 10-9   *FileUtils Module*

```
# Contents of testbed YAML file testbed.yaml
testbed:
  servers:
    server_alias:
      server: myserver.domain.com
      address: 1.1.1.1
      credentials:
        default:
          username: my_username
          password: my_password


# -------------------------------------------------------------------------
# Examples
# --------
#
from pyats.utils.fileutils import FileUtils
from pyats.topology.loader import load

tb = load("testbed.yaml")

# This with statement ensures that any sessions are automatically closed
# if something goes wrong.
with FileUtils(testbed=tb) as futils:
    # Copy local file to remote location
    # Note the two ways of specifying server the name: FQDN or testbed alias
    futils.copyfile(
        source = '/local/path/to/file',
        destination = 'ftp://server_alias/remote/path/to/file')

    futils.copyfile(
        source = 'file:///local/path/to/file',
        destination = 'tftp://myserver.domain.com/remote/path/to/file',
        timeout_seconds=80)


    # Copy remote file to local location, specifying the server via its address:
    futils.copyfile(
        source = 'scp://1.1.1.1/path/to/file',
        destination = '/local/path/to/file')
```

You can see the many different use cases for the FileUtils module, but you may be thinking, "Is it too much for simple configuration management?" Depending on the specifics, you may be correct, and pyATS has another option available.

Embedded pyATS File Transfer Server

The embedded file transfer server in pyATS allows you to launch a file server automatically with any pyATS job run. The embedded file server uses the **FileServer** class and supports FTP, HTTP, TFTP, and SCP. The server will figure out the local interface IP to use by looking at the device.spawn.id process ID. You can also provide a **subnet:** key in the testbed to identify which interface can reach the testbed devices, in the case there are multiple connected network interfaces. Example 10-10 shows how to specify an embedded file transfer server in a testbed and how to copy a file to a device using the pyATS library (Genie) API.

Example 10-10 *Embedded File Transfer Server*

```
testbed:
  servers:
    myftpserver:
      dynamic: true
      protocol: ftp
      subnet: 10.0.0.0/8  # Optional
      path: /path/to/root/dir


-------------------------------------------------------------------------


from pyats.topology.loader import load


# Load testbed and select "router1" device
my_testbed = load("testbed.yaml")
xe_router = my_testbed.devices["cat8k-rt1"]

# Copy myimage.bin to flash:/ on router via FTP
xe_router.api.copy_to_device(
    protocol="ftp",
    server="myftpserver",
    remote_path="myimage.bin",
    local_path="flash:/",
)
```

In the testbed, the **dynamic** key must be set to true, as it tells the file transfer server plugin that a server should be started. The **protocol** key identifies which file transfer protocol to use. The **subnet** key is optional and is an alternative way to identify which network interface is connected to the network with access to the testbed devices. The **path** key is used to identify the root directory of the FTP/TFTP server. The default root directory is /.

Once the server is defined in the testbed, we can use it to copy files to a device from the server. The **copy_to_device** device API allows us to easily transfer files by providing arguments to specify the protocol, server, remote path, and local path of the file. Since we are copying to the device from the device's perspective, the **remote_path** argument refers to the file path on the server and the local path refers to the file path on the device. It might seem confusing at first, but just keep in mind that we are using the device's API to copy the file, so the arguments are from the device's perspective. Alternatively, if you don't want to define the server in the testbed, you can use the **FileServer** class as a context manager. Example 10-11 shows the **FileServer** class creating a file transfer server on the fly within the script and copying a file to a device. The example is identical to Example 10-10 but shows another way to execute the file transfer.

Example 10-11   *FileServer Class – Context Manager*

```
from genie.libs.filetransferutils import FileServer
from pyats.topology.loader import load

# Load testbed and select "router1" device
my_testbed = load("testbed.yaml")
xe_router = my_testbed.devices["cat8k-rt1"]

# Instantiate file server in script instead of defining in testbed
with FileServer(
    protocol="ftp", path="/path/to/root/dir", \
    testbed=my_testbed, name="mycontextserver"
) as fs:
    xe_router.api.copy_to_device(
        protocol="ftp",
        server="mycontextserver",
        remote_path="myimage.bin",
        local_path="flash:/",
    )
```

If you want to stick with a more traditional approach where you specify the **copy:** command on the network device, you can use the copy service provided by the Unicon library. As a reminder, Unicon is the connection library used by pyATS to abstract the low-level interactions required to control a device. Unicon services are APIs used to perform common tasks (such as execute commands, configure, reload, and so on) on devices in a platform-agnostic way. The copy service provides an interface for the various Cisco

IOS flavors to copy software images, configs, and other files in and out of a device's flash memory. Example 10-12 shows how to copy the **running-config** to the **startup-config** and copy a file via TFTP to the bootflash of a device.

Example 10-12   *Unicon Copy Service*

```
from pyats.topology.loader import load


# Load testbed and select "router1" device
my_testbed = load("testbed.yaml")
xe_router = my_testbed.devices["cat8k-rt1"]


# Saving running-config to startup-config on router
out = xe_router.copy(source="running-conf", dest="startup-config")


# Copy 'copy-test.txt' file from TFTP server to bootflash on router
out = xe_router.copy(
    source="tftp:",
    dest="bootflash:",
    source_file="copy-test.txt",
    dest_file="copy-test.txt",
    server="10.105.33.158",
)
```

Bringing it back to configuration management, the different file transfer server options allow you to copy configurations to and from testbed devices. In some cases, this can be more efficient than executing the **show running-config** or **show startup-config** command, which requires pagination and may not be scraped properly if the terminal dimensions are not set properly. Remember, this is just one option to manage configuration. In the next section, we will look at configuring entire testbeds and individual devices directly using the available pyATS library (Genie) APIs.

pyATS Library API

The pyATS library (Genie) provides thousands of device APIs to view, analyze, and configure testbed devices. The APIs are essentially functions built to provide convenience when you're configuring a specific feature on a device (such as BGP, NTP, logging, and so on) or for quickly viewing device information. Example 10-13 shows a few examples of how to use the device APIs to configure and view device information.

Example 10-13   *pyATS Library – Device API*

```
from pyats.topology.loader import load

# Load testbed and select cat8k-rt1 device
my_testbed = load("testbed.yaml")
xe_router = my_testbed.devices["cat8k-rt1"]
# Connect to cat8k-rt1
xe_router.connect()

# Shut Loopback0 interface
xe_router.api.shut_interface(interface="Loopback0")
# Check the config of Loopback0 interface
xe_router.api.get_interface_config(interface="Loopback0")
# Unshut Loopback0 interface
xe_router.api.unshut_interface(interface="Loopback0")

# Disconnect from the device
xe_router.disconnect()
```

The example uses a simple example of flapping (shut/no shut) a loopback interface, but it shows off the abstraction the device API functions provide to the developer. You no longer need to know the specific commands or configurations that need to be applied to perform specific actions. The API function names provide enough context to understand what actions are being taken on the device. This is great and all, but what if there's not an API available for your specific use case? You have two choices: either you can create a new API function that performs your action or you can build a custom configuration using one of the methods shown in the "Generating Configuration with pyATS" section. In the following sections, you'll see how to configure all devices in a testbed using the configure method and how to configure a specific device using the configure service from the Unicon library.

Testbed-Wide Configuration

The pyATS topology package, the package that allows you to define Interface, Device, Link, and Testbed objects in pyATS, provides a **configure** method in the Testbed class that configures testbed devices in parallel with a given list or string of commands. If you don't want to configure every device in the testbed, the **configure** method allows you to provide a list of device objects. Example 10-14 shows how to configure the same logging hosts on every testbed device.

Example 10-14   *Testbed Configure*

```
from pyats.topology.loader import load

tb = load("testbed.yaml")
# Connect to all testbed devices
tb.connect()
# Configure logging host on all testbed devices
tb.configure("logging host 10.1.1.1")
# Disconnect from all testbed devices
tb.disconnect()
```

Device Configuration

The Unicon library provides a configure service that allows you to configure a device
with a list or string of commands. This may sound similar to testbed configuration,
because it is. The major difference is the testbed configuration option allows you to
configure all or a select number of testbed devices in parallel. The configure service
has many arguments that allow you to customize the behavior of the configuration.
Table 10-1 shows all the available arguments for the configure service.

Table 10-1   *Unicon Configure Service Arguments*

Argument	Type	Description
timeout	int	Timeout value for the duration of command execution.
error_pattern	list	List of RegEx strings to check output for errors.
append_error_pattern	list	List of RegEx strings append to error_pattern.
reply	Dialog	Additional dialog.
command	list	List of commands to configure.
prompt_recovery	bool (default False)	Enable/disable prompt recovery feature.
force	bool (default False)	For XR, runs commit force at end of config.
replace	bool (default False)	For XR, runs commit replace at end of config.
lock_retries	int (default 0)	Retry times if config mode is locked.
lock_retry_sleep	int (default 2 sec)	Sleep between lock retries.

Argument	Type	Description
target	str (default "active")	Target RP where service is executed. For DualRp only.
bulk	bool (default False)	If False, send all commands in one send-line. If True, send commands in chunked mode.
bulk_chunk_lines	int (default 50)	Maximum number of commands to send per chunk; 0 means to send all commands in a single chunk.
bulk_chunk_sleep	float (default 0.5 sec)	Sleep between sending command chunks.

As you can see, there are many customization options when it comes to using the configure service. Example 10-15 shows how to configure a hostname and logging hosts on a testbed device using a string and a list of commands.

Example 10-15　*Device Configure Service*

```
from pyats.topology.loader import load

# Load testbed and select cat8k-rt1 device
my_testbed = load("testbed.yaml")
xe_router = my_testbed.devices["cat8k-rt1"]
xe_router.connect()
# Configure hostname and disable console logging with a multiline string
xe_router.configure(
    """
    hostname cat8k-1
    no logging console
    """
)


# Configure logging hosts with a list
cmd = ["logging host 10.1.1.1", "logging host 10.1.1.2"]
xe_router.configure(cmd, timeout=30)
xe_router.disconnect()
```

The example is great if you're trying to send small, static config snippets, but what if we wanted to push a more complex configuration that requires variables, conditionals, and other logic? In the next section, you'll see how to do that with the available Jinja2 device APIs.

Jinja2 Configuration

Earlier in the chapter, you saw how device configurations can be generated using Jinja2 templates. The **configure_by_jinja2** API is another device API that generates configuration and configures a device with that rendered configuration. The configure_by_jinja2 device API uses two other device APIs under the hood to generate and push configuration to a device. The **get_jinja_template** and **change_configuration_using_jinja_templates** APIs are used to get the Jinja2 template, render it, and push each line of rendered configuration using the Unicon configure service mentioned in the previous section. Remember, the device APIs are meant to make it easier to consume the pyATS library with easy-to-read API method names. The pyATS library (Genie) parsers and APIs are open source and can be found on GitHub, so the code can be reviewed to confirm there is no hidden magic. Example 10-16 shows how to use the **configure_by_jinja2** device API to generate and push the rendered configuration to a device.

Example 10-16   *Device API – configure_by_jinja2*

```
from genie import testbed

# Load testbed and select "router1" device
my_testbed = testbed.load("testbed.yaml")
xe_router = my_testbed.devices["cat8k-rt1"]

# Configuration is rendered and pushed to "router1" device
logging_config = xe_router.api.configure_by_jinja2(
    path="config_templates", file="ios_logging.j2", \
    hosts=["10.1.1.1", "10.1.1.2"]
)
```

Genie Harness

Genie Harness allows you to create event-driven testscripts using triggers and verifications. Triggers and verifications are covered heavily in Chapter 9, but we are going to focus on the configuration options provided by Genie Harness. Once a testbed and mapping file are created, you may begin using Genie Harness. In the following sections, you'll see how to use a config datafile to apply configuration to testbed devices and check the configurations in the Common Setup and Common Cleanup sections during test execution.

Config Datafile

Genie Harness allows you to pass a config datafile to **gRun**, which is a YAML file that contains the configuration that should be applied to each testbed device, and in what order. The **config_datafile** argument expects a path to a configs.yaml file that contains the configuration that should be applied to each device. In configs.yaml, configurations are specified in the order they should be applied to a device, the type of configuration, and whether to apply it in the setup or cleanup section. The configuration can be a typical configuration file or a Jinja2 template with keyword arguments. For more information on the available keys in the configs.yaml file, I highly recommend checking out the Genie Harness datafile schema documentation:

> https://pubhub.devnetcloud.com/media/genie-docs/docs/userguide/harness/user/datafile.html?highlight=config%20datafile#configuration-datafile

Example 10-17 shows a configs.yaml file that can be provided to **gRun** as the **config_datafile** argument. You'll notice that a typical configuration file is applied in the first step and a Jinja2 template is rendered to apply the second configuration. Remember, the configurations are applied in the order they are numbered.

Example 10-17   *Genie Harness – Config Datafile*

```
devices:
    uut:
        1:
            config: /path/to/my/configuration
            sleep: 3
            invalid: ['overlaps', '(.*inval.*)']
        2:
            jinja2_config: routing.j2
            jinja2_arguments:
                lstrip_blocks: true
                trim_blocks: true
                bgp_data:
                    bgp_asn: 100
                    neighbor_ips: [
                        '1.1.1.1', '2.2.2.2'
                    ]
            configure_arguments:
                bulk: True
                timeout: 180
```

In the example, you'll see two configurations that will be applied to the "uut" testbed device. If there were multiple devices, the configurations would be applied in parallel using multiprocessing across all testbed devices. The first configuration, identified by the **1** key, is a typical configuration file that can be read from a file and applied directly to the device. The **sleep** key provides a timer, in seconds, to wait until moving on to the next configuration. This allows time for the device configuration to stabilize before continuing. The **config** key provides a path to the configuration file you want to apply to the device. The **invalid** keyword contains phrases and/or RegEx patterns that will match any warnings or errors that may arise when applying the configuration. If there is a match on any phrase or RegEx pattern, the configuration section will fail.

The second configuration, identified by the **2** key, is applied after the first configuration and requires a Jinja2 template, provided by the **jinja2_config** key, to render a configuration using the keyword **arguments** under the **jinja2_arguments** key. The **lstrip_blocks** and **trim_blocks** keys are two arguments that handle whitespace. The **lstrip_blocks** argument determines whether to trim any leading whitespace. The **trim_blocks** argument determines whether to trim any newlines. The **bgp_data** key will be provided to the Jinja2 template as a dictionary, so all nested variables will be accessible via dot notation, which means all variable names in the template must be prepended with **bgp_data**. For example, if we wanted to access the **bgp_asn** argument from the YAML file in the Jinja2 template, we would use the following syntax:

```
{{ bgp_data.bgp_asn }}
```

The **configure_arguments** key provides keyword arguments to the Unicon configure service, which were presented in Table 10-1 earlier in the chapter.

To recap, there are two configurations applied to the "uut" testbed device in the numbered order shown. The first configuration uses a typical configuration file that is read and applied to the device. The second configuration uses a Jinja2 template and keyword arguments to provide template variables to render a configuration. Once a configuration is rendered, it is applied to the device. Because it's the second configuration applied, any configuration that was the same in the first configuration will be overwritten.

Config Check

After the testbed devices are configured, you may want to check the configuration before and after testing to ensure it is the same. Genie learns the configuration of every testbed device in the check_config subsection, which executes in the Common Setup and Common Cleanup sections during testing. In the Common Setup section, a configuration snapshot is created for every device by executing the **show running-config** command. After testing, in the Common Cleanup section, a second configuration snapshot is created by collecting the **running-config** again. Afterwards, the two configuration snapshots are compared to ensure nothing has changed. Certain values may change during testing, and you may want those values to be excluded during the comparison. To exclude specific configuration values, you must specify an **exclude_config_check** key in configs. yaml, which accepts a string or RegEx pattern to match values to exclude. Example 10-18 shows a configs.yaml file that with the **exclude_config_check** key excludes interface descriptions from comparison.

Example 10-18 *Check Config – Exclude Values*

```
devices:
    uut:
        1:
            config: <full path to config file>
            sleep: 5
            invalid : ['(.*ERROR.*)']
exclude_config_check: ['(.*description.*)']
```

Summary

In this chapter, we reviewed managing device configuration using intent-based configuration as well as generating configurations using data models, Jinja2 templates, and Genie Conf objects. Lastly, we reviewed the many ways to push configuration using a file transfer server, Genie Conf objects, and pyATS device APIs. It's important to remember that pyATS is a testing framework, and all the methods discussed in this chapter are meant to be used during testing. For true configuration management, pyATS should be complemented by other tooling that focuses on configuration management, such as Ansible. The beauty of network automation and software engineering is that it's up to the engineer to find the best tool for the job, not necessarily using the same tool for every job.

## Chapter 11

Network Snapshots

One of the worst calls you can receive after a change window is from the network operating center (NOC) team because they received alerts that there are issues in the network. Regardless of a network change's risk, any change to the network has a chance to cause catastrophic failure. Some failures are due to poor preparation, such as not simulating the change in a lab environment; some may be innocent failures, such as triggering an unknown software bug. Wouldn't it be great if there was an "easy button" for capturing what changed in the network during a change window, instead of anxiously waiting for a phone call after the window hoping you didn't break anything? Well, this chapter may help you get there!

The purpose of this chapter is to show how you can better prepare for network issues and outages by creating snapshots of the network before and after changes. Once snapshots are created of the network's state, we will look at how we can compare each network state, pre- and post-change, to determine what exactly changed using a unified diff. The following topics will be covered in this chapter:

- Network profiling
- Comparing network state
- Polling expected state
- The Robot framework with Genie

Network Profiling

The concept of profiling, in terms of software development, refers to the analysis of a program's behavior. Much like software, a network has its own behavior in the form of packet forwarding. Many factors play into the network's behavior to forward packets, including network hardware, network operating system (NOS), the device configuration, all the way down to the protocol level. If you've been in the networking field for a while, you most likely have a story about how one of these factors caused an outage for you and your team—whether it was a hardware failure, software bug, or forgetting the **add** keyword when adding a VLAN to an existing trunk port (it's okay; we've all been there). All the different factors that go into a network's behavior shouldn't surprise you as a network engineer, but how many network engineers truly know what is going on in their network at a moment's notice? Network engineers build networks with an intended design to cause certain behavior, but over time, how can we confirm that behavior is still the same? In the following sections, we will see how the pyATS library (Genie) can help us learn and compare the state of the network with network features.

What Is a Network Snapshot?

In the context of networking, a *snapshot* is a moment-in-time capture of the current behavior and operating state of the network. The operating state of the network may include operational state data such as environmental values (CPU, memory, temperature, and so on), physical state (interface status, stack cable connectivity, and so on), or logical state (routing tables/policies, access lists, spanning tree, LLDP, and so on) and may include device configurations. Device configurations don't necessarily provide operational state data about the network, but it is beneficial to capture with operational state data in the case of unexpected behavior. The pyATS library (Genie) provides the ability to "learn" about, or take snapshots of, the network using its available features (models). In the following section, we'll explain what network features are and how to use the pyATS library (Genie) to learn them.

Network Features

The pyATS library (Genie) provides an Ops module that contains different network features that can be learned about a network device. A network feature is typically a network protocol, such as BGP, OSPF, VTP, or NTP, but can also refer to hardware, such as an interface or device platform information. Network features are represented as Python objects with attributes that represent the features' configuration. For example, an interface object has attributes for IP address, mask, and interface state (shutdown/not shutdown).

When a network feature is learned about a network device, multiple **show** commands are executed and parsed to form a unified data structure that can be compared in the future. To improve performance, a connection pool can be created to send the multiple commands in parallel to a network device. You may wonder why not just execute and parse the individual **show** commands you're interested in versus running a series of commands? In some scenarios, it may make sense to only parse the **show** command(s) you're interested in, but the main reason is for data consistency. When you're learning device features, different commands are executed, depending on the device type and OS, with each command having a different output structure. The learn functionality abstracts the details about which commands to execute to gather the appropriate data, per device platform, and provides a consistent data output for each device feature.

One or more network features can be learned about a network device at a time. Features can be learned using the pyATS CLI or using pyATS Python API. Example 11-1 shows how to use the pyATS CLI to learn the BGP feature from a Cisco IOS-XE testbed device named cat8kv-rt1, with the associated output.

Example 11-1 *Learn BGP Feature – CLI*

```
(.venv)ch11_network_snapshots$ pyats learn bgp --testbed-file ../testbed.yaml
 --devices cat8k-rt1 --output bgp_state

! Output
Learning '['bgp']' on devices '['cat8k-rt1']'
100%|██████████████████████████████████████████████| 1/1
  [00:17<00:00, 17.40s/it]
+=============================================================================+
| Genie Learn Summary for device cat8k-rt1                                    |
+=============================================================================+
|  Connected to cat8k-rt1                                                     |
|  -   Log: bgp_state/connection_cat8k-rt1.txt                               |
|----------------------------------------------------------------------------|
|  Learnt feature 'bgp'                                                       |
|  -  Ops structure:  bgp_state/bgp_iosxe_cat8k-rt1_ops.txt                   |
|  -  Device Console: bgp_state/bgp_iosxe_cat8k-rt1_console.txt               |
|=============================================================================|
```

You'll notice the output provides a path to the "Log," "Ops structure," and "Device Console." The Log is the connection log, which contains connection information for attempts to connect to the network device. The Ops structure is the unified data model of the parsed BGP data collected from running multiple commands. The Device Console file contains the raw command outputs before they are parsed by the pyATS library (Genie). These outputs are what you would see if you logged in to the device yourself and ran the associated commands in a terminal session.

Using the pyATS CLI is the recommended approach for learning device features because it's faster and allows you to store and view the learned data quickly. However, you can also use the pyATS API to programmatically learn devices features. Example 11-2 shows how to programmatically learn the BGP feature from the same Cisco IOS-XE device.

Example 11-2   *Learn BGP Feature – API*

```
from pyats.topology.loader import load

# Initialize and load testbed
testbed = load("../testbed.yaml")
# Identify device from testbed
xe_router = testbed.devices["cat8k-rt1"]

# Connect to device, learn BGP feature, and disconnect from the device
xe_router.connect()
bgp_state = xe_router.learn("bgp")
xe_router.disconnect()

# Print BGP feature output
print(bgp_state.info)
```

Now that you understand how to capture network snapshots using the available features/
models provided by the pyATS library (Genie), let's see how we can operationalize the
process and use it to compare network state before and after a change in the network.

Comparing Network State

One of the toughest, and sometimes tedious, tasks to perform before and after a change
is gathering the state of the network. Beyond running the necessary **show** commands and
gathering device logs, the process can become worse if the change fails and you must
compare the before and after captured states. Who wants to read line by line and look for
that one interface counter or that one route that changed? To add to it, even if a change
fails, it's common for some counters and miscellaneous data points to be different due to
the nature of change. For example, in the case of a BGP routing change that causes neighbors to flap (disconnect and re-establish right away), the age of routes or the uptime of
routing neighbors or peers will change, regardless of whether the change was successful
or had to be reverted. This can create a lot of noise for the network admin or engineer on
the maintenance window that is trying to validate the network is operating as expected
after a change. In the following sections, you'll see how we can verify the state of the network before and after a change by using **genie diff** to compare data structures created by
the **genie learn** and **genie parse** commands. Along with using the commands, you'll also
see how to use Python code to learn and run the differentials.

Pre-change Snapshots

Before making any changes to the network, it's a good idea, if not a requirement, to capture a known-good state of the network. This known-good state could include relevant
show command output that captures the current behavior of the network and device
configurations.

Beginning in this section, we are going to create a scenario where we simulate a small change to two routers using OSPF to exchange routes. In the process of the change, we are going to break the OSPF adjacency. To capture the network state before and after changes, we are going to learn the OSPF and Interface features using the pyATS library (Genie). Let's start by learning the OSPF and Interface features on both routers and capturing the output into a directory called "pre_change". Example 11-3 shows how we can learn the OSPF feature using the **genie learn** Genie command-line command with the associated output.

Example 11-3   *Genie Learn OSPF – Pre-Change*

```
(.venv)ch11_network_snapshots$ genie learn ospf interface --testbed-file ../testbed.
 yaml --output pre_change

! Output
Learning '['ospf', 'interface']' on devices '['cat8k-rt2', 'cat8k-rt1']'
100%|███████████████████████████████████████████████████
       ██████| 2/2 [00:22<00:00, 11.01s/it]
+=============================================================================+
| Genie Learn Summary for device cat8k-rt2                                    |
+=============================================================================+
|  Connected to cat8k-rt2                                                     |
|  -   Log: pre_change/connection_cat8k-rt2.txt                               |
|-----------------------------------------------------------------------------|
|  Learnt feature 'ospf'                                                      |
|  -  Ops structure:  pre_change/ospf_iosxe_cat8k-rt2_ops.txt                 |
|  -  Device Console: pre_change/ospf_iosxe_cat8k-rt2_console.txt             |
|-----------------------------------------------------------------------------|
|  Learnt feature 'interface'                                                 |
|  -  Ops structure:  pre_change/interface_iosxe_cat8k-rt2_ops.txt            |
|  -  Device Console: pre_change/interface_iosxe_cat8k-rt2_console.txt        |
|=============================================================================|


+=============================================================================+
| Genie Learn Summary for device cat8k-rt1                                    |
+=============================================================================+
|  Connected to cat8k-rt1                                                     |
|  -   Log: pre_change/connection_cat8k-rt1.txt                               |
|-----------------------------------------------------------------------------|
|  Learnt feature 'ospf'                                                      |
|  -  Ops structure:  pre_change/ospf_iosxe_cat8k-rt1_ops.txt                 |
|  -  Device Console: pre_change/ospf_iosxe_cat8k-rt1_console.txt             |
|-----------------------------------------------------------------------------|
|  Learnt feature 'interface'                                                 |
|  -  Ops structure:  pre_change/interface_iosxe_cat8k-rt1_ops.txt            |
|  -  Device Console: pre_change/interface_iosxe_cat8k-rt1_console.txt        |
|=============================================================================|
```

You should now see a directory named pre_change in your current directory, which contains the different files shown in the output. Feel free to take a look at each generated file, but the files we are mostly concerned with are the Ops structure files, which contain the parsed JSON output based on the respective Genie model. These are the files that will be used in the differential to find what changed in the network. Once the pre-change snapshot is taken, the intended network changes will be deployed. After the changes are deployed, we will capture a post-change snapshot.

Post-change Snapshots

The operation to create a post-change network snapshot is basically identical to the pre-change snapshot, in that you capture all the same network features, with the only difference being that you must save the snapshot to a separate directory. Example 11-4 shows how to learn the OSPF and Interface features using the **genie learn** command again, but this time, saving the output to a directory called "post_change".

Example 11-4   *Genie Learn OSPF – Post-Change*

```
(.venv)ch11_network_snapshots$ genie learn ospf interface --testbed-file ../testbed.
yaml --output post_change

! Output
Learning '['ospf', 'interface']' on devices '['cat8k-rt2', 'cat8k-rt1']'
100%|█████████████████████████████████████████████████████████████████
       █████| 2/2 [00:22<00:00, 11.24s/it]
+=============================================================================+
| Genie Learn Summary for device cat8k-rt2                                    |
+=============================================================================+
|  Connected to cat8k-rt2                                                     |
|  -  Log: post_change/connection_cat8k-rt2.txt                               |
|-----------------------------------------------------------------------------|
|  Learnt feature 'ospf'                                                      |
|  -  Ops structure:  post_change/ospf_iosxe_cat8k-rt2_ops.txt                |
|  -  Device Console: post_change/ospf_iosxe_cat8k-rt2_console.txt            |
|-----------------------------------------------------------------------------|
|  Learnt feature 'interface'                                                 |
|  -  Ops structure:  post_change/interface_iosxe_cat8k-rt2_ops.txt           |
|  -  Device Console: post_change/interface_iosxe_cat8k-rt2_console.txt       |
|=============================================================================|


+=============================================================================+
| Genie Learn Summary for device cat8k-rt1                                    |
+=============================================================================+
|  Connected to cat8k-rt1                                                     |
|  -   Log: post_change/connection_cat8k-rt1.txt                              |
```

```
| --------------------------------------------------------------------- |
|   Learnt feature 'ospf'                                               |
|   -  Ops structure:   post_change/ospf_iosxe_cat8k-rt1_ops.txt        |
|   -  Device Console:  post_change/ospf_iosxe_cat8k-rt1_console.txt    |
| --------------------------------------------------------------------- |
|   Learnt feature 'interface'                                          |
|   -  Ops structure:   post_change/interface_iosxe_cat8k-rt1_ops.txt   |
|   -  Device Console:  post_change/interface_iosxe_cat8k-rt1_console.txt|
| ===================================================================== |
```

To recap, we now have two directories, pre_change and post_change, which contain the raw (*_console.txt files) and parsed JSON (*_ops.txt files) outputs for the learned features on each testbed device. Now that we have the data, we will begin the adventure of figuring out what's changed in the network!

Snapshot Differentials

Capturing the operational state of the network is great, but now we are stuck with an abundance of data without obvious direction. If you think back to the original problem of figuring out what's changed in the network, we are kind of still at the starting line of having to read and compare **show** command output before and after a network change or event. However, the main difference is that we now have data that can be consumed programmatically—this is key! On top of that, the data is in a structure based on a unified data model that is vendor-agnostic, so no matter what type of network device it is, the data will look the same based on the feature's data model.

The ability to assume a data structure is foundational to building robust automation with confidence, as it allows you to easily extract interesting data points without having to unknowingly iterate and waste computing cycles searching through a data structure, reducing efficiency and error-proneness. Because the data structures are identical, we can easily compare the differences between the two snapshots using the pyATS library (Genie). In the following sections, we are going to see how we can use the Genie CLI, specifically the **genie diff** command, to compare learned feature output to see what exactly changed once the network change occurred.

Genie Diff

The **genie diff** Genie CLI command allows you to compare two outputs of learned network features. The outputs are directories that were created as a result of the **genie learn** command, so we simply need to reference the directory names. Example 11-5 builds on the previous example of figuring out what broke the OSPF adjacency between the two routers by comparing the "pre_change" directory of learned output with the "post_change" directory of learned output. The differential result is in a unified diff format, which indicates what content was included in one of the outputs and not the other. The minus sign (-) indicates what was in the first output but missing in the second output. The plus sign (+) indicates what was included in the second output but was not in the first output.

Example 11-5   *Genie Diff*

```
(.venv)ch11_network_snapshots$ genie diff pre_change post_change -output
change_diff

# Output
+==============================================================================+
| Genie Diff Summary between directories pre_change/ and post_change/           |
+==============================================================================+
|   File: ospf_iosxe_cat8k-rt1_ops.txt                                         |
|    - Diff can be found at change_diff/diff_ospf_iosxe_cat8k-rt1_ops.txt       |
|------------------------------------------------------------------------------|
|   File: ospf_iosxe_cat8k-rt2_ops.txt                                         |
|    - Identical                                                               |
|------------------------------------------------------------------------------|
|   File: interface_iosxe_cat8k-rt1_ops.txt                                    |
|    - Identical                                                               |
|------------------------------------------------------------------------------|
|   File: interface_iosxe_cat8k-rt2_ops.txt                                    |
|    - Diff can be found at change_diff/diff_interface_iosxe_cat8k-rt2_ops.txt  |
|------------------------------------------------------------------------------|
```

The **genie diff** output clearly identifies which feature outputs are identical and which ones are different. If there's a difference, the unified diff is stored in the output directory specified in the **genie diff** command. In the case of our example, the "change_diff" directory will store all diffs. Based on the output from Example 11-5, there should be two files in the change_diff directory: one text file that identifies differences with the OSPF feature on the cat8k-rt1 device (diff_ospf_iosxe_cat8k-rt1_ops.txt) and another that identifies differences with the Interfaces feature on the cat8k-rt2 device (diff_interface_iosxe_cat8k-rt2_ops.txt) before and after the network change. Example 11-6 show the content of each file.

Example 11-6   *Genie Diff – Unified Diff*

```
(.venv)ch11_network_snapshots$ genie diff pre_change post_change -output
change_diff

# diff_ospf_iosxe_cat8k-rt1_ops.txt output

--- pre_change/ospf_iosxe_cat8k-rt1_ops.txt
+++ post_change/ospf_iosxe_cat8k-rt1_ops.txt
```

```
info:
  vrf:
    default:
      address_family:
        ipv4:
          instance:
            10:
              areas:
                0.0.0.0:
                  database:
                    lsa_types:
-                     2:
-                       lsa_type: 2
-                       lsas:
-                         172.16.1.1 10.254.1.1:
-                           adv_router: 10.254.1.1
-                           lsa_id: 172.16.1.1
-                           ospfv2:
-                             body:
-                               network:
-                                 attached_routers:
-                                   10.254.1.1:
-                                   10.254.1.2:
-                                 network_mask: 255.255.255.0
-                             header:
-                               adv_router: 10.254.1.1
-                               age: 72
-                               checksum: 0x7CD9
-                               length: 32
-                               lsa_id: 172.16.1.1
-                               option: None
-                               seq_num: 80000004
-                               type: 2


--------------------------------------------------------------------------------


# diff_interface_iosxe_cat8k-rt2_ops.txt output

--- pre_change/interface_iosxe_cat8k-rt2_ops.txt
+++ post_change/interface_iosxe_cat8k-rt2_ops.txt
info:
  GigabitEthernet4:
-  mtu: 1500
+  mtu: 1530
```

Sometimes the unified diff can lead to an obvious answer to why the network broke. Let's dive into the diff output provided in Example 11-6. Starting with the changes in the OSPF snapshot on the cat8k-rt1 device, it looks like we are missing an LSA type 2 that was present in the pre-change snapshot. I'd like to take a moment and recognize that this analysis requires a network engineer and their knowledge of the OSPF routing protocol. I mention this due to the fact that when many network engineers hear the phrase "network automation," they jump to the conclusion you must convert to being a full-time programmer. Programming is just part of the job; it got us to this point. However, when it comes to troubleshooting what's wrong with the network by analyzing the data at hand, you must fall back on network engineering knowledge. Being able to read the data is one thing, but understanding what's wrong is another...but I digress.

Back to the issue at hand, let's remind ourselves about the OSPF routing protocol and link state advertisements (LSAs). An LSA is used to communicate information between OSPF routers and is stored in the OSPF link state database (LSDB). There are 11 different LSA types used to communicate different types of information in the same or between areas. A type 2 LSA is an OSPF network LSA that is used in a multi-access network where a designated router (DR) and backup designated router (BDR) election occurs and advertises a list of routers connected to the network within the same OSPF area. Theoretically, the direct link between the two routers could be configured as point-to-point, but for example purposes it is not. Since the type 2 LSA is now missing in the post-change snapshot, it makes sense that the OSPF routes were missing and the adjacency was no longer up between the two routers. The cat8k-rt1 device doesn't even know about the other router (cat8k-rt2) because it did not receive a type 2 network LSA from cat8k-rt2. Let's move on to the second differential found on the interfaces feature for cat8k-rt2.

The second differential discovered was between the interfaces feature captured on the cat8k-rt2 device. Referring to Example 11-6, you'll notice that the MTU configured on interface GigabitEthernet4 changed from 1500 to 1530. This may seem like a miniscule change, but let's put our network engineer hats on and think about this.

Let's review the problem: once a change was put in place, the two routers were no longer OSPF neighbors and lost their OSPF routes learned from one another. That is apparent in the first differential found on cat8k-rt1, where the type 2 LSA is no longer in the OSPF database, which tells us it lost its adjacency to cat8k-rt2, but not exactly *how* it did. That leaves us with the second differential found on cat8k-rt2, where the interface MTU was changed. This, again, is a moment where it helps to be a network engineer. For two routers to become OSPF neighbors, several criteria must match between the router interfaces forming the adjacency:[1]

- Area number
- Area type (that is, stub or NSSA)
- Authentication (if configured)
- Network and subnet mask
- Hello and dead timers
- MTU

The last bullet point is interesting, considering the MTU did change on the adjacent inter-face (GigabitEthernet4) on cat8k-rt2. Once we remove the MTU configuration from the interface and set it back to the default of 1500 bytes, the adjacency comes back up! Both routers see each other as OSPF neighbors in a FULL state. We fixed the issue!

In case you are trying to replicate this same issue in your own lab, you should note that these routers are running Cisco IOS XE 17.6.1 in CML, and Per-Port MTU configuration was enabled starting with Cisco IOS XE 17.1.1. If you are running on an older IOS-XE software version, you would need to configure the MTU size at the system level, which could lead to further problems.[2]

Data Exclusions

When you're comparing network data, there may be some data points that are irrelevant for comparison. For example, packet counters, timestamps, and uptime are data points that naturally change between snapshots but do not provide value when comparing data. To avoid these values from being identified in the differential, they must be excluded from being compared in the differential. The pyATS library (Genie) has different methods to exclude certain data points when comparing datasets. In the following sections, we will review the built-in mechanisms and ways to add or customize the data points to exclude in a differential.

Default Exclusions

The pyATS library (Genie) has a pre-created list of keys that are ignored when compar-ing two data structures that are a result of the **genie parse** or **genie learn** command. The purpose of ignoring these keys is to reduce the amount of noise you must fight through while comparing data. For example, data points like counters, timestamps, and uptime are all irrelevant when it comes to differentiating data, as you expect these values to change in time. For reference, the pyATS library (Genie) looks in the following locations for the default list of excluded keys:

- **genie learn:** $VIRTUAL_ENV/lib/python{version}/site-packages/genie/libs/sdk/genie_yamls/pts_datafile.yaml

- **genie diff:** $VIRTUAL_ENV/lib/python{version}/site-packages/genie/libs/sdk/genie_yamls/{os}/verification_datafile.yaml

The PTS datafiles are referenced when learning a device feature using the **genie learn** command. The PTS datafile will have each device feature with exclude keys that are lists of keys to exclude in the learned output. The verification datafiles are referenced when parsing command output using the **genie parse** command. The verification datafiles, found under each OS, contain **show** commands that have exclude keys that list the keys to exclude in the parsed output. The one difference between the two features is that **genie diff** can reference OS-specific verification datafiles that are found in the genie_yamls/{os} directories. For example, there's a verification datafile for different Cisco network operating systems, including IOS, IOS-XE, NX-OS, and IOS-XR. These OS-specific datafiles inherit from the "base" datafile found in the genie_yamls/ directory.

Additional Exclusions

Additional keys can be provided to Genie Diff by specifying them with the **--exclude** argument. To add the key **'state'** to be excluded from a differential, you would enter the following:

```
(.venv)ch11_network_snapshots$ genie diff pre_change post_change -exclude 'state' --output change_diff
```

When you add the **'state'** key to the **--exclude** argument, the value of the **'state'** key will not be compared between the two snapshots. You can add more keys to be excluded by separating them with a space. You can add more keys to the excluded list by entering the following:

```
(.venv)ch11_network_snapshots$ genie diff pre_change post_change -exclude 'state' 'uptime' 'msg_sent' --output change_diff
```

Custom Exclusions

You also have the ability to remove the default exclusions by specifying the **--no-default-exclusion** argument to **genie diff**. Expanding on the previous example, you can provide your own custom list of exclusions using the **--exclude** argument. The following code melds these two ideas together by removing the default exclusions and only excluding the **'state'**, **'uptime'**, and **'msg_sent'** keys from being compared in the differential:

```
(.venv)ch11_network_snapshots$ genie diff pre_change post_change -no-default-exclusion -exclude 'state' 'uptime' 'msg_sent' --output change_diff
```

Polling Expected State

The pyATS library (Genie) provides a polling mechanism to verify if a feature is in an expected state. It couples with the **genie learn** functionality and is useful if a feature was recently configured and is still stabilizing while being learned. To verify a feature's operational state, you must create a **verify** function that determines whether the feature was learned successfully. The **verify** function will raise an exception if it's determined a feature was not learned. If a feature was not learned, the function will wait for a period of time and try to verify again. A feature can be polled using the **learn_poll()** function.

The **learn_poll()** function accepts three arguments:

- **sleep:** An integer representing the amount of time (in seconds) to wait between polls

- **attempt:** An integer representing how many times to try to verify a feature

- **verify:** A callable that will receive the feature as an argument and confirm whether the feature is in an expected state. If no exception is raised by the callable, the feature "passes" and is considered in an expected state.

Example 11-7 shows how to poll the OSPF feature and verify the proper instance number was learned.

Example 11-7   *Polling OSPF Feature*

```
from pyats.topology.loader import load

def verify_ospf_state(obj, process_id, router_id):
  # Obj is the feature object which is being learnt
  # Confirm learned process ID and router ID are correct
  assert (
    obj.info["vrf"]["default"]["address_family"]["ipv4"]["instance"][process_id][
    "router_id"] == router_id
  )

# Load testbed and connect to 'cat8k-rt1' device
testbed = load("../testbed.yaml")
xe_router = testbed.devices["cat8k-rt1"]
xe_router.connect()

# Learn OSPF feature from device
ospf = xe_router.learn("ospf")

# Learn OSPF and confirm the process ID and corresponding router ID are correct
# using the verify_ospf_state function. It will try up to 5 times,
# and sleep 10 seconds between each attempt.
ospf.learn_poll(
  verify=verify_ospf_state,
  sleep=10,
  attempt=5,
  process_id="10",
  router_id="10.254.1.1",
)

print("Feature verified - Polling was a success!")

# Disconnect from device
xe_router.disconnect()
```

Robot Framework with Genie

The Robot Framework is a generic open-source automation framework built on Python. It uses keywords with human-readable syntax to perform specific tasks. The abstraction created by utilizing provided keywords makes it a powerful and flexible tool to hand off to other non-programmers. Each library that makes up the pyATS framework—Unicon, pyATS, and the pyATS library (Genie)—have Robot keywords published. Chapter 22, "Robot Framework," is dedicated to just the Robot Framework, but for now let's focus on how it can help us learn device features.

Before we begin, we must install the optional robot subpackage for pyATS and the pyATS library (Genie). To do this using pip, enter the following:.

```
(.venv)ch11_network_snapshots$ pip install genie.libs.robot pyats.
robot
```

Once the Robot Framework and pyATS library (Genie) robot subpackage are installed, we are ready to explore! As mentioned, the Robot Framework relies on keywords to drive its automation. The pyATS library (Genie) has many keywords to running triggers, verifications, learning device features, custom verify functions, Genie device API support, and even support for the helpful Dq library!

Example 11-8 shows the Robot Framework test suite file used to learn the OSPF feature from the cat8k-rt1 testbed device, and Example 11-9 shows the associated output. Take note of how easy it is to follow along!

Example 11-8   *Robot Framework – Learn OSPF*

```
# learn_ospf.robot
*** Settings ***
# Importing test libraries, resource files and variable files.
Library pyats.robot.pyATSRobot
Library genie.libs.robot.GenieRobot


*** Variables ***
${testbed} testbed.yaml


*** Test Cases ***

Initialize
  # Initializes the pyATS/Genie Testbed
  use genie testbed "${testbed}"

  # Connect to 'cat8k-rt1' device
  connect to device "cat8k-rt1"

Learn OSPF
  # Learn OSPF feature from 'cat8k-rt1' device and assign to 'output' variable
  ${output}= learn "ospf" on device "cat8k-rt1"
```

Once we have the test suite file created, we only need one command to run the tests:

```
(.venv)ch11_network_snapshots$ robot learn_ospf.robot
```

Example 11-9 *Robot Framework – Learn OSPF Output*

```
================================================================================
Learn Ospf
================================================================================
Initialize                                                            | PASS |
--------------------------------------------------------------------------------
Learn OSPF                                                            | PASS |
--------------------------------------------------------------------------------
Learn Ospf                                                            | PASS |
2 tests, 2 passed, 0 failed
================================================================================
Output:   /Users/danwade/temp/pyats-book-
   examples/ch11_network_snapshots/robot_framework/output.xml
Log:      /Users/danwade/temp/pyats-book-
   examples/ch11_network_snapshots/robot_framework/log.html
Report:   /Users/danwade/temp/pyats-book-
   examples/ch11_network_snapshots/robot_framework/report.html
```

You may think it's a joke, but this is how easy it is to write a Robot Framework test suite file! The only work you must do is familiarize yourself with the keywords used for each library you're interested in using. You may be wondering where all the magic happens. It's all defined in the libraries specified in the *** **Settings** *** section. The libraries reference Python modules that map the robot keywords to code logic. We will discuss the Robot Framework in more detail in Chapter 22. The purpose in this chapter is to show off the ease of use the Robot Framework provides for users who want to get started with pyATS and learn about the network without having much or any programming experience.

Summary

Having a snapshot of your network can be crucial when troubleshooting a network issue, or just learning about your network and how it's operating. In this chapter, we looked at the importance of network profiling, how to create network snapshots, and the multiple ways to compare network snapshots using **genie diff** and the Robot Framework. Using network snapshots is one of the easiest ways to get started with network automation, as it has minimal risk (all data is gathered using read-only actions) and huge benefits with understanding how your network is actually running.

References

1. https://www.cisco.com/c/en/us/support/docs/ip/open-shortest-path-first-ospf/13699-29.html

2. https://www.cisco.com/c/en/us/td/docs/switches/lan/catalyst9300/software/release/17-1/configuration_guide/int_hw/b_171_int_and_hw_9300_cg/configuring_per_port_mtu.pdf

## Chapter 12

Recordings, Playbacks, and Mock Devices

A lesser-known feature of Cisco's pyATS is the ability to make recordings of jobs, which opens up a realm of possibilities for network testing and simulation. By recording interactions with real devices, engineers can replay these sequences to verify the consistency of test results across different runs. More intriguingly, these recordings can serve as the foundation for creating mock devices, allowing for extensive testing without the need for physical hardware. This feature not only streamlines the test automation process but also provides a cost-effective solution for simulating complex network environments and troubleshooting without impacting live systems. This chapter covers the following topics:

- Recording pyATS jobs
- Playback recordings
- Creating mock devices
- Mock device CLI

Recording pyATS jobs

The ability to record pyATS jobs allows you to enhance several key operational areas of your network automation, adding immediate value to the evolving test-driven approach:

- **Change management artifacts:** Within the scope of change management, the ability to record pyATS jobs is a powerful tool for ensuring transparency and accountability. Before a network change is applied, a recording can be created to capture the pre-change state of the network. This serves as an immutable reference point, a source of truth, which can be invaluable in complex environments where changes have a domino effect on interconnected systems. After the change is made, another recording can be captured, and the two can be compared to ensure that the intended changes have been applied correctly. These recordings become integral parts of the change documentation, providing a detailed narrative of the network's evolution. They also play a significant role in post-implementation reviews, offering insights that can refine future change processes and help in avoiding similar issues. This can be extremely helpful for troubleshooting problems as well as ensuring that compliance with polices and procedures is being met.

■ **Troubleshooting:** The role of pyATS job recordings in troubleshooting cannot be overstated. When network issues occur, engineers are often under pressure to resolve them quickly. Having a library of recordings depicting the network in its various operational states can dramatically reduce the time it takes to identify the root cause of a problem. By examining the differences between a healthy network state and the current state during an outage or performance degradation, engineers can direct their efforts toward the areas that show deviations. This methodical approach not only speeds up the resolution process but also helps in developing a more profound understanding of the network's dynamics, leading to improved designs and configurations that are resilient to similar issues in the future.

■ **Training and simulations:** For training new network engineers, pyATS job recordings are akin to flight simulators used in pilot training—they provide a realistic, risk-free environment to learn and make mistakes. Trainees can replay various recorded scenarios to see how the network should perform, allowing them to recognize when something is amiss. These recordings can also be used to simulate network failures or attacks, providing practical experience that textbook learning cannot match. In a simulation, trainees can practice troubleshooting, implement changes, and even test disaster recovery procedures. This experiential learning accelerates the development of their skills and builds confidence in their abilities to manage real-world network challenges.

■ **Forensics and security analysis:** In the context of network forensics, the value of pyATS job recordings lies in their ability to serve as a digital ledger, documenting the state of the network at discrete intervals over time. While pyATS is not designed for continuous monitoring, its snapshots provide critical insights in the wake of a security incident. Analysts can leverage these recordings to piece together the events leading up to a security breach, offering a detailed retrospective examination rather than real-time surveillance.

This analytical process might include identifying the entry points an intruder used or mapping out the spread of malicious software within the network. Although pyATS does not track every moment of network activity, its recorded states offer a series of detailed cross-sections. By examining these snapshots, investigators can discern patterns, trace the evolution of the network's configuration, and detect anomalies. This retrospective analysis is instrumental in pinpointing vulnerabilities that were exploited during an attack and in detecting potential threats that might have initially gone unnoticed.

Furthermore, the insights gleaned from this post-incident analysis are invaluable for bolstering network security. They contribute to the formulation of more robust security policies, the design of fortified network architectures, and the enhancement of incident response strategies. In sum, while pyATS may not provide the continuous monitoring traditionally associated with threat vector tracking, its capability to capture and archive the network's state at various points makes it a potent tool for security analysis and forensics, aiding in the aftermath of security breaches by illuminating vulnerabilities and guiding the reinforcement of network defenses.

■ **Other uses:** Beyond these specific scenarios, the versatility of pyATS job recordings extends into various other domains. For capacity planning, recordings can be used to model how the network will behave under projected growth scenarios, allowing for informed decision-making regarding upgrades and expansions. In a development context, recordings enable software teams to test their applications against a "live" network without the need to actually deploy on one, thus identifying potential issues early in the development cycle. Recordings can also be an asset during vendor evaluations, providing a consistent baseline against which to measure the performance of new equipment or software. This can be particularly beneficial when conducting proof of concept (PoC) testing, as it ensures that all vendors are assessed under the same network conditions, ensuring fairness and accuracy in the evaluation process.

Recording pyATS jobs is as simple as adding the **--record** flag and the name of the recording. Any existing pyATS job can be recorded as follows:

```
(venv)$ pyats run job learn_all_job.py --record learn_all_recording
```

You may notice this job is titled "learn_all_job.py". Let's write the job file and script that can use any existing SSH-based testbed file and use the pyATS models, all of them, to take a complete snapshot of a device. When we get to the mock devices later in this chapter, we can use this complete snapshot as the foundation for our mock device. Example 12-1 shows the basic pyATS job file while Example 12-2 shows the pyATS script to "learn all" features from a device.

Example 12-1 *pyATS Job File – learn_all_job.py*

```
import os
from genie.testbed import load

def main(runtime):

    # ----------------
    # Load the testbed
    # ----------------
    if not runtime.testbed:
        # If no testbed is provided, load the default one.
        # Load default location of Testbed
        testbedfile = os.path.join('testbed.yaml')
        testbed = load(testbedfile)
    else:
        # Use the one provided
        testbed = runtime.testbed

    # Find the location of the script in relation to the job file
    testscript = os.path.join(os.path.dirname(__file__), 'learn_all.py')

    # run script
    runtime.tasks.run(testscript=testscript, testbed=testbed)
```

Example 12-2   *pyATS Testscript – learn_all.py*

```python
import json
import logging
from pyats import aetest
from rich.table import Table
from rich.console import Console
from pyats.log.utils import banner


# ----------------
# Get logger for script
# ----------------

log = logging.getLogger(__name__)


# ----------------
# AE Test Setup
# ----------------
class common_setup(aetest.CommonSetup):
    """Common Setup section"""
# ----------------
# Connected to devices
# ----------------

    @aetest.subsection
    def connect_to_devices(self, testbed):
        """Connect to all the devices"""
        testbed.connect()
# ----------------
# Mark the loop for Show Version
# ----------------

    @aetest.subsection
    def loop_mark(self, testbed):
        aetest.loop.mark(Learn_All, device_name=testbed.devices)


# ----------------
# Test Case #1
# ----------------
class Learn_All(aetest.Testcase):
    """Learn all pyATS models"""

    @aetest.test
    def setup(self, testbed, device_name):
        """ Testcase Setup section """
        # Set current device in loop as self.device
        self.device = testbed.devices[device_name]
```

```
    @aetest.test
    def learn_all(self):
        self.learn_all = self.device.learn("all")


    @aetest.test
    def create_file(self):
        # Create .JSON file
        with open(f'{self.device.alias}_Learn_All.json', 'w') as f:
            f.write(json.dumps(self.learn_all.info, indent=4, sort_keys=True))

class CommonCleanup(aetest.CommonCleanup):
    @aetest.subsection
    def disconnect_from_devices(self, testbed):
        testbed.disconnect()


# for running as its own executable
if __name__ == '__main__':
    aetest.main()
```

Playback Recordings

Now that we have recorded our pyATS "learn all" job, we can play back the recording. Using the same pyATS **run job** command used to run and record, the job is simply updated with the **--replay** flag and the name of the recording you want to replay:

```
(venv)$ pyats run job learn_all_job.py --replay learn_all_recording
```

If you want to adjust the speed of the playback, you can also include the **--speed** flag followed by the replay speed (such as 0.5 for half-speed or 2.0 for double speed). To replay the original recording at quarter speed, you can use the following command:

```
(venv)$ pyats run job learn_all_job.py --replay learn_all_recording
--speed 0.25
```

The playback functionality in Cisco's pyATS is a testament to the platform's flexibility and user-centric design. This capability allows network engineers to replay previously recorded jobs with precision, providing a dynamic way to analyze and understand network behavior over time or under different conditions. The command line simplicity, as demonstrated by the **--replay** flag, belies the underlying sophistication of being able to view network interactions as they occurred. This can be particularly useful when trying to replicate transient issues that are difficult to diagnose, as the playback can be paused, rewound, or sped up to scrutinize specific events in greater detail.

Adjusting the speed of playback with the **--speed** flag adds another layer of versatility. Running a playback at a slower speed can be invaluable when training personnel, as it gives them more time to process the information and understand the sequence of events. Conversely, speeding up the playback can be beneficial when trying to evaluate the network's performance over extended periods, condensing hours of activity into a shorter, more manageable timeframe.

The ability to replay network behavior exactly as it occurred is not just a convenience; it's a potent tool for validation and verification of network configurations and behaviors. This level of control and repeatability is what makes pyATS a robust option for network engineers looking to enhance their testing and troubleshooting processes.

While the ability to play back recordings is indeed remarkable, pyATS doesn't stop there. The true potential of these recordings unfolds when we leverage them to create mock devices. In the upcoming section, we'll explore how pyATS takes a recorded job—a faithful representation of a device's behavior in the network—and uses it to construct a mock device complete with a working command-line interface. This capability elevates the recordings from being mere snapshots of the past to active components of a simulated network environment. By doing so, engineers can rehearse configurations, experiment with responses to hypothetical scenarios, and test the impact of changes in a controlled, risk-free manner. Stay tuned as we delve into the transformative process of turning recordings into interactive, virtual devices that respond just like their physical counterparts.

Mock Devices

In the ever-evolving landscape of network engineering, the ability to model and simulate network components is not just beneficial—it's essential. Cisco's pyATS framework elevates this capability through its mock device feature, which allows engineers to create and interact with virtual network devices that behave like real hardware. These mock devices are not mere static models; they possess a working command-line interface (CLI) that responds to commands as a physical device would. This technology enables a hands-on experience without the associated costs or risks of manipulating actual network hardware. It should be noted that mock devices *support only* **show** *commands*; they do not have a configuration mode and are not intended to be a replacement for full device simulation tools such as CML, GNS3, EVE-NG, and containerlab.

Use Cases for Mock Devices

A variety of use cases and possibilities exist for mock devices, including the following:

- **Testing:** Mock devices serve as a cornerstone for rigorous testing protocols. Developers and network engineers can use these virtual replicas to test new software, updates, or configurations before they're rolled out to the production network. This approach helps in identifying potential issues early, ensuring that only thoroughly vetted changes make their way into the live environment.

- **Change management:** Change management processes thrive on predictability and control. With mock devices, engineers can simulate the impact of changes in a sandbox environment, allowing for detailed planning and analysis. This helps in foreseeing the consequences of a change, which is invaluable for minimizing downtime and avoiding costly mistakes during actual implementation.

- **Troubleshooting and training:** For troubleshooting, mock devices can replicate an environment where an issue has been reported, allowing for a safe exploration of solutions. Similarly, they are an excellent resource for training, providing a practical and interactive platform for network professionals to hone their skills and learn to resolve a wide array of simulated network issues.

■ **Audits and compliance:** Audits and compliance checks often require demonstration of network behavior under certain conditions. Mock devices can be configured to replicate those conditions, providing auditors with the necessary evidence of compliance without the need to interact with the live environment, thus maintaining network integrity.

■ **Prototyping and proof of concept:** In the realm of network design and innovation, the utilization of mock devices plays a pivotal role during the initial phases of prototyping and proof of concept (PoC) testing. These virtual or simulated devices enable teams to conceptualize and evaluate new network architectures or the integration of emerging solutions without the immediate need for physical hardware. This approach facilitates a swift and flexible examination of theoretical scenarios and design principles, significantly expediting the development process.

However, it's important to note that while mock devices are invaluable for the early stages of design and testing, the transition to employing physical hardware becomes necessary to validate the network's operational performance in a real-world environment. After completing the PoC phase and establishing the viability of the proposed design, incorporating actual hardware devices is crucial for recording authentic outputs and conducting rigorous, in-depth analyses. This dual-phase approach—starting with mock devices for initial exploration and moving to physical hardware for final validation—ensures a comprehensive evaluation of network designs. It allows for iterative refinement based on both theoretical exploration and tangible performance metrics. Consequently, this methodology not only accelerates the design and testing cycles but also enhances the thoroughness of network planning discussions, ensuring that the eventual implementation is both robust and aligned with the project's objectives.

Additional Use Cases

Beyond these applications, mock devices open up opportunities for disaster recovery planning, allowing engineers to view changes in the network before and after a disaster recovery procedure has been initiated. They can also be used for security training, such as simulating network breaches to train security teams on detection and response. In sales and marketing, mock devices can demonstrate the functionality of network products in a controlled environment, facilitating demonstrations and evaluations.

Unicon, an integral part of Cisco's pyATS ecosystem, introduces an advanced capability to create mock devices that simulate real network hardware. This feature is particularly powerful because it uses a YAML file to define the device behavior. Engineers have the option to either manually craft this YAML file or generate it dynamically from an existing recording. For instance, the creation of a mock device YAML from a recorded session is as straightforward as running the following command:

```
(venv)$ python -m unicon.playback.mock --recorded-data \\
    learn_all_recording --output data/mock_data.yaml
```

Once this YAML file is in place, it becomes the blueprint for a mock device. Launching a mock device that mimics a Cisco IOS XE device is done through a simple command:

```
(venv)$ python -m unicon.mock.mock_device --os iosxe \\
        --mock_data_dir data --state connect


python -m unicon.mock.mock_device --os iosxe \\
        --mock_data_dir data --state connect


/usr/lib/python3.10/runpy.py:126: RuntimeWarning: 'unicon.mock.mock_
device'
found in sys.modules after import of package 'unicon.mock',
but prior to execution of 'unicon.mock.mock_device';
this may result in unpredictable behaviour  warn(RuntimeWarning(msg))
/home/testuser/code/pyatsbook/.venv/bin/python:
No code object available for unicon.mock.mock_device
```

This command spins up a virtual device that behaves like the specified operating system, complete with a CLI that can be interacted with just as if it were a real device. This simulated environment is ideal for various testing scenarios and allows network engineers to script, test, and validate network configurations in a controlled manner.

Mock Device CLI

Now that you are connected to the CLI of your mock device, try some commands! Enter enable mode. Run **show** commands. Fully explore the mock device and recognize it is a complete stateful clone of the IOS XE router by using the pyATS "learn all" models to create the recording. Also, it should be noted that the mock device integrates seamlessly into testscripts by defining the device connection within a script's configuration. This is exemplified in the connection setup, where a mock device is treated like any other device within the testscript, as demonstrated in Example 12-3.

Example 12-3 *Using a Mock Device in a Testbed*

```
---
connections:
  defaults:
    class: 'unicon.Unicon'
  a:
    command: mock_device_cli --os iosxe --mock_data_dir data --state connect
    protocol: unknown
```

This integration into automated testscripts means that the mock device can be used as part of a CI/CD pipeline, ensuring that any network changes are validated against a controlled environment that accurately reflects the behavior of the production network.

By default, the mock device stores only the first output for each command. However, Unicon provides the flexibility to record every instance of a command's output. This is essential for testing scenarios where a command may produce different outputs on subsequent executions. The `--allow-repeated-commands` flag is used to enable this feature:

```
(venv)$ python -m unicon.playback.mock --recorded-data \\
        learn_all_recording --output data/mock_data.yaml
--allow-repeated-commands
```

In the YAML file, commands are structured in a way that allows for a circular response type. This means that for a command with multiple recorded outputs, the mock device will cycle through these outputs, providing a new response each time the command is executed. Example 12-3 demonstrates a sample structure from a YAML file:

Example 12-4   *Mock Device Cycling Through Responses*

```
---
execute:
  commands:
    show interfaces GigabitEthernet1:
      response:
        - "GigabitEthernet1 is up, line protocol is up..."
        - "GigabitEthernet1 is up, line protocol is up..."
      response_type: circular
```

This structure ensures that the mock device will never "run out" of responses, allowing for more comprehensive testing scenarios where the command's output may change over time or due to different network conditions.

Summary

In this chapter, we explored the robust capabilities of Cisco's pyATS for enhancing network testing, change management, and troubleshooting through recordings and mock devices. We began by uncovering the utility of pyATS job recordings, a feature that allows network engineers to capture and replay the intricacies of network behavior with precision. This functionality not only aids in consistent testing but also serves as a valuable asset for change management, providing a verifiable baseline that is crucial for ensuring system integrity.

We then delved into the process of replaying these recordings, detailing how engineers can use the `--replay` and `--speed` flags to replay job recordings at varying speeds. This feature is instrumental in troubleshooting scenarios, training sessions, and audit compliances, offering a controlled environment to revisit and analyze network events as they unfolded.

Further, we examined the creation and application of mock devices. By transforming recorded data into dynamic, virtual devices, pyATS opens up a realm of possibilities for network simulation. We highlighted the use of the Unicon plugin to create these mock devices, which can be interacted with through a fully functional CLI, providing an authentic experience without the need for physical hardware. The mock devices are not only perfect for testing and validating network changes in a risk-free environment but also offer a platform for rigorous training and development.

The following is a recap of the steps involved:

Step 1.   Run pyATS job with --**record** flag.

Step 2.   Create a YAML file for the mock device using **unicon.playback.mock** Python module.

Step 3.   Use the **mock_device_cli** command to create and connect to the mock device using the created YAML file in step 2.

In addition, the chapter discussed how mock devices, with their circular response capability, ensure a comprehensive testing approach by allowing repeated command execution with varied outputs. This feature is particularly useful for simulating and testing different network states and behaviors, further enhancing the network engineer's toolkit.

In summary, pyATS's recording and mock device capabilities are indispensable tools in the modern network engineer's arsenal. They provide an unprecedented level of control and insight into network operations, paving the way for more resilient, efficient, and secure network infrastructures.

## Chapter 13

Working with Application Programming Interfaces (API)

Particularly highlighting a selection of APIs and the REST Connector, this chapter aims to take a deeper dive into these powerful tools. We've already seen how pyATS, with its extensive range of built-in device APIs, enhances the network automation landscape. In this chapter, we'll explore in greater detail the REST Connector, an integral component that enables pyATS to directly interact with sophisticated systems like Cisco Catalyst Center and Cisco APIC for ACI. This exploration is crucial for understanding how pyATS not only simplifies but also revolutionizes network management. Additionally, we'll delve into the nuances of the YANG connector and the gNMI capabilities within pyATS. This deep dive is particularly relevant for network engineers and NetDevOps practitioners who are transitioning from traditional CLI and SSH parsing to more advanced and scalable methods like REST APIs. By the end of this chapter, you'll have a comprehensive understanding of how pyATS is empowering this shift and paving the way for the future of network automation.

This chapter covers the following topics:

- pyATS APIs
- REST connector
- YANG connector
- gNMI

pyATS APIs

The pyATS Library's Device API functions for a range of tasks on a device, including modifying configurations, checking their states, and retrieving current system or configuration data. The pyATS Library facilitates interactions with devices using specific Device API functions. This process is comparable to parsing a device, enabling you to carry out particular tasks by employing the relevant **device.api.function_name()** method. Additionally, we will demonstrate how to conduct basic operations on a device using these API functions, offering practical insights into their usage.

In this simple example, we first need to load our testbed, create the testbed and device objects, and finally connect to the device, as demonstrated in Example 13-1.

Example 13-1 *Load Testbed and Connect To Device*

```
# Load testbed and connect to the device
>>>   from genie.testbad import load
>>>   testbed = load('mock.yaml')
>>>   device = testbed.devices['csr1000v-1']
>>>   device.connect()
```

The system connects to the device and displays the connection details. Once you're connected, you can perform operations on the device. Now you are able to perform the Device API functions (operations). For example, to get all the routes on the device, enter the following to yield the results in Example 13-2:

```
>>> routes = device.api.get_routes()
```

Example 13-2 *Raw Non-parsed Cisco **show ip route** Output at the CLI*

```
[2019-12-03  13:02:28,738]  +++ csr1000v-1: executing command 'show ip route'
show ip route
Codes: L - local, C - connected, S - static, R - RIP, M - mobile, B = BGP
       D - EIGRP, EX - EIGRP external, O - OSPF, IA - OSPF inter area
       N1 - OSPF NSSA external type 1, N2 - OSPF NSSA external type 2
       E1 - OSPF external type 1, E2 - external type 2
       i - IS-IS, su - IS-IS summary, L1 - IS-IS level 1, L2 - IS-IS level-2
       ia - IS-IS inter area, * - candidate default, U - per user static route
       o - ODR, P - periodic downloaded static route, H - NHRP, l - LISP
       a - application route
       + - replicated route, %% - next hop override, p - overrides from PfR

Gateway of last resort is 10.255.0.1 to network 0.0.0.0

S*    0.0.0.0/0 [254/0] via 10.255.0.1
      10.0.0.0/8 is variably subnetted, 10 subnets, 3 masks
C        10.0.1.0/24 is directly connected, GigabitEthernet2
L        10.0.1.1/32 is directly connected, GigabitEthernet2
C        10.0.2.0/24 is directly connected, GigabitEthernet3
L        10.0.2.1/32 is directly connected, GigabitEthernet3
C        10.1.1.1/32 is directly connected, Loopback0
```

```
O        10.2.2.2/32 [110/2] via 10.0.2.2, 6d01h, GigabitEthernet3
                     [110/2] via 10.0.1.2, 00:02:40, GigabitEthernet2
C        10.11.11.11/32 is directly connected, Looback1
B        10.22.22.22/32 [200/0] via 10.2.2.2, 6d01h
C        10.255.0.0/16 is directly connected, GigabitEthernet1
L        10.255.8.19/32 is directly connected, GigabitEthernet1
switch#
>>>
>>> print(routes)
['0.0.0.0/0', '10.0.1.0/24', '10.0.1.1/32', '10.0.2.0/24', '10.0.2.1/32',
>>>
```

Another example might be to shut down an interface on the device. Here we are shutting down interface GigabithEthernet3:

```
>>> device.api.shut_interface(interface = 'GigabitEthernet3')
```

For a complete list of operations (Device APIs) that the pyATS Library can perform, you can either visit the APIs page (https://pubhub.devnetcloud.com/media/genie-feature-browser/docs/#/apis) or from the Python prompt, enter the following to yield the results in Example 13-3:

```
>>> dir(device.api)
```

Example 13-3 *Python dir() Output Inspecting device.api*

```
analyze_rate
analyze_udp_in_mpls_packets
bits_to_netmask
change_configuration_using_jinja_templates
change_hostname
check_traffic_drop_count
check_traffic_expected_rate
check_traffic_transmitted_rate
clear_bgp_neighbors_soft
clear_interface_config
clear_interfade_counters
clear_ip_bgp_vrf_af_soft
clear_packet_buffer
compare_clear_archive_config_dicts
compare_config_dicts
config_acl_on_interface
config_extended_acl
config_interface_carrier_delay
config_interface_mtu
...
```

Similar to the functionality described in the chapter on parsers, the API page also features the capability to search and filter based on supported operating systems, a process illustrated in Figure 13-1. This feature enables users to efficiently navigate through the available options, ensuring they can quickly find the information relevant to their specific operating system requirements.

Figure 13-1   *pyATS API Explorer Example*

As we delve deeper into the functionalities of the pyATS Library's Device API, it's crucial to understand the extensive capabilities these APIs bring to network automation. These APIs go beyond mere interactions with individual devices; they are designed for the programmatically driven management of entire network topologies. This distinction is important to note: while the term "individual device interactions" might suggest the possibility of manual API calls, in reality, the pyATS Library APIs are meant for automated, script-driven engagements. This automation framework empowers network engineers to execute complex tasks across their network infrastructure with precision and efficiency.

The utilization of pyATS Library APIs significantly reduces the need for manual intervention in routine network operations, such as configuration, testing, and troubleshooting. For instance, automated scripts can swiftly configure multiple devices, deploy consistent updates across the network, or perform health checks to ensure network reliability and performance. These programmatic capabilities not only accelerate operational workflows but also minimize the risk of human error, leading to more reliable network management practices.

In summary, the pyATS Library's Device API offers a powerful suite of tools for automating network management tasks. By harnessing these APIs, network professionals can achieve remarkable efficiencies, transforming complex, time-consuming processes into streamlined, accurate, and highly scalable operations.

A key advantage of using pyATS APIs lies in their ability to abstract complexity. Whether you are dealing with a single device or an intricate network of devices, the APIs provide a consistent and simplified interface for executing a wide range of operations. This uniformity is crucial in reducing the learning curve and increasing efficiency, especially when managing diverse network components.

Take, for example, the **get_software_version()** IOS XE pyATS Device API, as demonstrated in Figure 13-2.

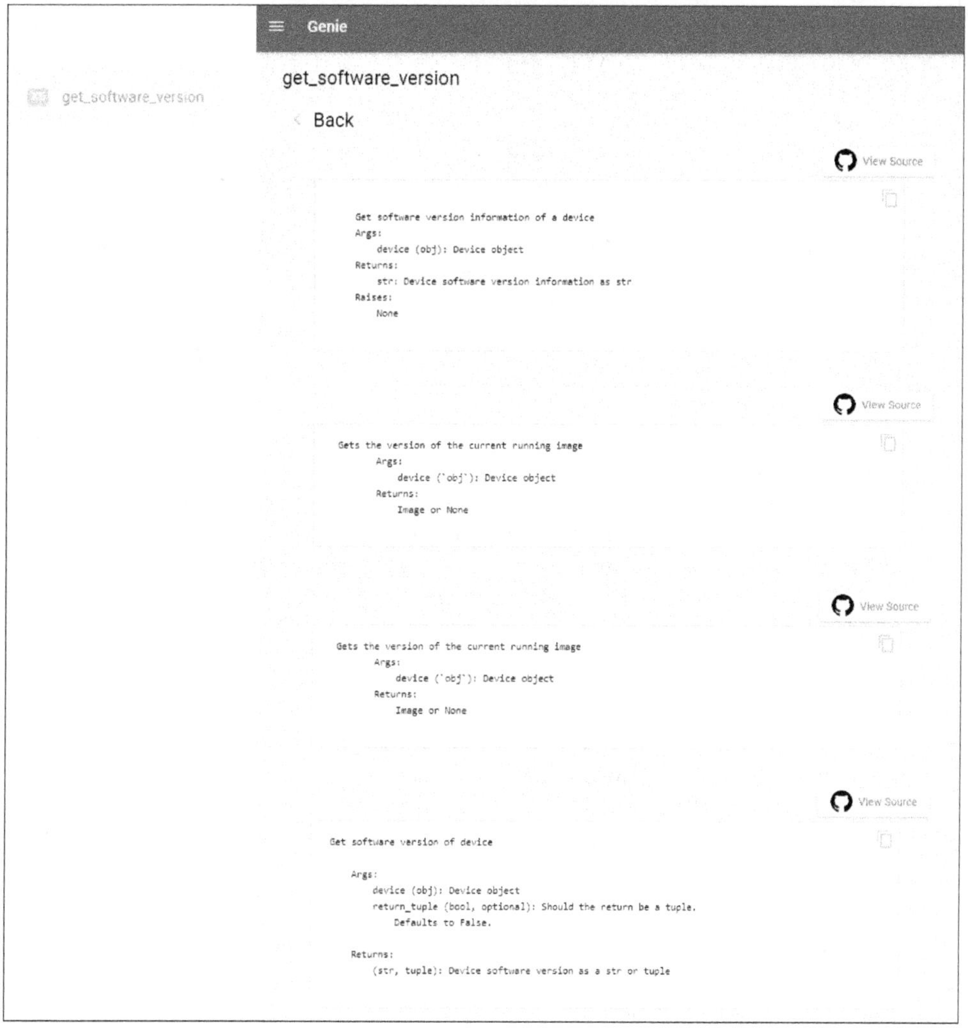

Figure 13-2 *The pyATS get_software_version API GitHub Source Code References*

In a pyATS job, you could use the code in Example 13-4 to log the software version (or write tests against the returned software version).

Example 13-4   *get_software_version() Device API*

```
class Test_Cisco_IOS_XE_Version(aetest.Testcase):
    """Parse pyATS show version and test against a threshold"""


    @aetest.test
    def setup(self, testbed, device_name):
        """ Testcase Setup section """
        # Set current device in loop as self.device
        self.device = testbed.devices[device_name]


    @aetest.test
    def get_software_version(self):
        software_version = self.device.api.get_software_version()
        log.info(f"The software version is: { software_version }")


2023-12-03T10:31:37: %SCRIPT-INFO: The software version is: Cisco IOS Software [Ben-
   galuru], Catalyst L3 Switch Software (CAT9K_IOSXE), Experimental Version 17.6.4
   RELEASE SOFTWARE (fc1)
2023-12-03T10:31:37: %SCRIPT-INFO: Copyright (c) 1986-2022 by Cisco Systems, Inc.
2023-12-03T10:31:37: %SCRIPT-INFO: Compiled Sun 14-Aug-22 08:58 by mcpre
```

The pyATS APIs are designed to provide a high degree of flexibility and control over network management tasks. Through these APIs, network engineers have the capability to configure devices individually or by network segments, monitor states dynamically, and analyze trends to better understand network behavior. This control ranges from executing straightforward tasks, such as modifying device configurations or toggling interface states, to orchestrating complex multidevice workflows and embedding custom logic for proactive network maintenance.

It's important to clarify that while the APIs enable extensive data collection and analysis, suggesting they facilitate predictive analysis might overstate their capabilities. The reality is that pyATS primarily offers robust tools for data gathering and interrogation. The sophisticated task of predictive analysis, which involves forecasting future network behaviors and potential issues based on complex algorithms and models, typically requires additional layers of analytical tools and expertise beyond what pyATS directly provides.

Furthermore, the phrase "the tool consumes itself" merits further explanation to avoid confusion. This expression intends to convey that pyATS leverages its own set of tools and libraries, such as a variety of parsers, to simplify and streamline development processes. Essentially, pyATS uses its components to abstract and handle the complexity of various operations, making development workflows more intuitive and less cumbersome for users. This self-referential design philosophy enables developers to focus on higher-level automation and network management tasks without getting bogged down with the intricacies of individual device commands and responses.

In essence, the pyATS framework empowers network engineers with comprehensive tools for managing and analyzing network infrastructure efficiently. By automating routine tasks and providing detailed insights into network operations, pyATS enhances the ability to maintain robust, high-performing networks.

Moreover, the integration of pyATS APIs with existing network management tools and processes is seamless, enhancing the capabilities of network professionals. This integration not only saves time but also reduces the potential for errors, resulting in more reliable network operations.

In conclusion, the pyATS APIs stand as a testament to the evolution of network management tools, encapsulating a profound blend of simplicity, control, and versatility that caters to the modern network professional's needs. By elegantly abstracting the complexities inherent in network operations and offering a uniform methodology for managing a wide array of network tasks, these APIs serve as a linchpin for enhancing operational efficiency. Network engineers are equipped with the means to seamlessly manage not just isolated devices but entire network topologies, paving the way for a more proactive and less reactive network management paradigm.

The true power of the pyATS APIs lies in their capacity to transform the network management landscape. They not only streamline and automate day-to-day tasks but also furnish network professionals with the insights and agility needed to anticipate and mitigate network issues before they escalate. This strategic advantage ensures that networks are not only robust and reliable but also adaptable to the fast-paced changes and challenges of today's digital world.

Thus, the pyATS APIs do more than just simplify network management—they redefine it. By empowering network professionals with advanced tools that foster innovation, efficiency, and strategic foresight, the pyATS framework is instrumental in driving forward the future of network operations. The impact of these APIs extends beyond mere technical capabilities, marking a significant leap forward in how networks are designed, managed, and optimized for the demands of tomorrow.

REST Connector

Before we delve deeper into the capabilities and applications of the pyATS REST (Representational State Transfer) Connector, it's beneficial to establish a basic understanding of the RESTful verbs, which are fundamental to how REST operates. These verbs, or methods, define the action types that can be performed on the resources identified by URIs (Uniform Resource Identifiers). The following are the most commonly used RESTful verbs:

- **GET:** Retrieves data from a specified resource
- **POST:** Submits data to be processed to a specified resource, often resulting in a change in state or side effects on the server
- **PUT:** Replaces all current representations of the target resource with the uploaded content

- **DELETE:** Removes all current representations of the target resource given by a URI

- **PATCH:** Applies partial modifications to a resource

Understanding these verbs is crucial, as they form the backbone of interactions within RESTful services, allowing for a standardized approach to performing Create, Read, Update, Delete (CRUD) operations across various network platforms and technologies.

With this primer in mind, the pyATS REST Connector emerges as a powerful tool in the realm of network automation and management. Its versatility is highlighted by its application across a diverse range of network environments, not limited to IOS XE using RESTCONF. The REST Connector facilitates seamless integration and automation across various platforms, including but not limited to NXOS, NSO, DNAC Catalyst Center, and APIC for ACI, as well as major network management systems like BIG-IP, vManage for SD-WAN, DCNM, and Nexus Dashboard. This adaptability makes the pyATS REST Connector a key component in modern network management strategies, enabling users to efficiently manage and automate network operations across a wide spectrum of technologies. Through the use of RESTful APIs and the specific verbs that dictate the nature of each request, network professionals can streamline processes, ensuring robust, reliable network performance with the flexibility required for today's dynamic network environments.

The REST package can be installed via pip in the same virtual environment where pyATS has been installed:

```
pip install rest.connector
```

To specify a REST connection for your device, add the **rest.connector.Rest** class to the connection for your devices in the testbed YAML file, as demonstrated in Example 13-5.

Example 13-5 *REST Connector Class in testbed.yaml*

```
---
devices:
    cat8000v:
        alias: 'cat8000v'
        type: 'router'
        os: 'iosxe'
        platform: csr1000v
        connections:
            rest:
                # Rest connector class
                class: rest.connector.Rest
                ip: 10.10.20.48
                port: 443
                credentials:
                    rest:
                        username: developer
                        password: C1sco12345
```

NXOS

The REST Connector's support for key HTTP methods—GET, POST, DELETE, PATCH, and PUT—within the context of the NXOS NX-API opens a gateway to high-efficiency network automation and operational management for Cisco Nexus data centers. By leveraging the syntax **device.rest.<*method*>(*url*)**, network engineers can directly interact with the NX-API, facilitating structured JSON responses. This interaction is not just about executing commands; it's about streamlining the data exchange between network devices and management systems, thereby enabling rapid, real-time decision-making and operational adjustments.

This capability is particularly significant for environments that have not transitioned to a software-defined networking (SDN) framework with Cisco ACI and APIC. For these traditional or hybrid network setups, the REST Connector serves as a critical automation tool, offering a bridge to more advanced network automation without the need for a full SDN overhaul. The ability to programmatically manage network configurations, monitor device states, and implement changes across the Nexus data center via RESTful APIs means that even non-SDN environments can achieve a degree of agility and efficiency typically associated with more modern network architectures.

Furthermore, this approach empowers network teams to operationalize their data centers with greater precision and control. The direct and structured interaction with devices via the NX-API reduces the likelihood of human error, enhances the speed of network adjustments, and supports a more proactive network management strategy. For organizations navigating the transition to SDN or evaluating their future network infrastructure plans, the REST connector and its support for NXOS NX-API present a powerful mechanism to maximize current network capabilities while laying the groundwork for future advancements.

In essence, the REST Connector not only facilitates immediate operational improvements for Nexus data centers but also represents a strategic investment in the network's adaptability and readiness for evolving technologies. By harnessing these capabilities, network professionals can ensure robust, reliable network performance today, while positioning themselves favorably for tomorrow's technological shifts.

The GET API method is fundamental in the realm of RESTful web services, serving a crucial role in retrieving resources from a server. When applied within the context of network device management via pyATS, the GET command is executed to fetch data or state information from the network device.

GET API

Purpose: Sends a GET command to the device

Arguments:

- **dn:** Unique distinguished name (mandatory)
- **headers:** Headers for the GET command (default: None)

- **timeout:** Maximum disconnection time (default: 30 seconds)

- **expected_return_code:** Expected return code (default: None)

- **url** = '/api/mo/sys/bgp/inst/dom-default/af-ipv4-mvpn.json'

- **output** = device.rest.get(url)

POST API

Purpose: Sends a POST command to the device

Arguments:

- **dn:** Unique distinguished name (mandatory)

- **headers:** Headers for the POST command (default: None)

- **timeout:** Maximum disconnection time (default: 30 seconds)

- **expected_return_code:** Expected return code (default: None)

```
payload =  payload = """
{
  "bgpInst": {
    "attributes": {
      "isolate": "disabled",
      "adminSt": "enabled",
      "fabricSoo": "unknown:unknown:0:0",
      "ctrl": "fastExtFallover",
      "medDampIntvl": "0",
      "affGrpActv": "0",
      "disPolBatch": "disabled",
      "flushRoutes": "disabled"
    },
    "children": [
      {
        "bgpDom": {
          "attributes": {
            "name": "default",
            "pfxPeerTimeout": "30",
            "pfxPeerWaitTime": "90",
            "clusterId": "",
            "maxAsLimit": "0",
            "reConnIntvl": "60",
            "rtrId": "0.0.0.0"
          },
          "children": [
```

```
{
    "bgpRtCtrl": {
      "attributes": {
        "logNeighborChanges": "disabled",
        "enforceFirstAs": "enabled",
        "fibAccelerate": "disabled",
        "supprRt": "enabled"
      }
    }
  },
  {
    "bgpPathCtrl": {
      "attributes": {
        "alwaysCompMed": "disabled",
        "asPathMultipathRelax": "disabled",
        "compNbrId": "disabled",
        "compRtrId": "disabled",
        "costCommunityIgnore": "disabled",
        "medConfed": "disabled",
        "medMissingAsWorst": "disabled",
        "medNonDeter": "disabled"
      }
    }
  }
            ]
        }
    }
  ]
}
}
"""
```

- **url** = 'api/mo/sys/bgp/inst.json'
- **output =** device.rest.post(url, payload)

DELETE API

Purpose: Sends a DELETE command to the device

Arguments:

- **dn:** Unique distinguished name (mandatory)
- **headers:** Headers for the DELETE command (default: None)
- **timeout:** Maximum disconnection time (default: 30 seconds)

- **expected_return_code:** Expected return code (default: None)

- **url** = '/api/mo/sys/bgp/inst/dom-default/af-ipv4-mvpn.json'

- **output** = device.rest.delete(url)

A delete is only expected to return a successful status code, such as 200 or 201, but will not return any payload to the user.

PATCH API

Purpose: Sends a PATCH command to the device

Arguments:

- **dn:** Unique distinguished name (mandatory)

- **payload:** Payload for the PATCH command (mandatory)

- **headers:** Headers for the PATCH command (default: None)

- **timeout:** Maximum disconnection time (default: 30 seconds)

- **expected_return_code:** Expected return code (default: None)

```
payload = """{
    "intf-items": {
      "phys-items": {
        "PhysIf-list": [
          {
            "adminSt": "down",
            "id": "eth1/2",
            "userCfgdFlags": "admin_layer,admin_state"
          }
        ]
      }
    }
}
"""
```

- **url** = '/api/mo/sys/bgp/inst/dom-default/af-ipv4-mvpn.json'

- **output** = device.rest.patch(url, payload)

PUT API

Purpose: Sends a PUT command to the device

Arguments:

- **dn:** Unique distinguished name (mandatory)

- **payload:** Payload for the PUT command (mandatory)

- **headers:** Headers for the PUT command (default: None)

- **timeout:** Maximum disconnection time (default: 30 seconds)

- **expected_return_code:** Expected return code (default: None)

```
payload = """{
    "intf-items": {
      "phys-items": {
        "PhysIf-list": [
          {
            "adminSt": "down",
            "id": "eth1/2",
            "userCfgdFlags": "admin_layer,admin_state"
          }
        ]
      }
    }
}
"""
```

- **url** = '/api/mo/sys/bgp/inst/dom-default/af-ipv4-mvpn.json'

- **output** = device.rest.put(url, payload)

NSO

Cisco Network Services Orchestrator (NSO) is a powerful network automation and orchestration platform designed for complex, multivendor networks. At its core, NSO enables network operators and engineers to automate the provisioning and management of network services, significantly reducing manual efforts and the potential for errors. Cisco NSO offers northbound APIs that enable external applications and systems to interact with it. These APIs are crucial for integrating NSO into larger IT and network systems, allowing for automated workflows and enhanced functionality.

NSO provides a RESTful API, enabling easy integration with web-based systems and custom applications. This API is used for CRUD operations on network services and devices managed by NSO. In addition to REST, NSO supports NETCONF and RESTCONF protocols, providing additional options for network management and automation. When integrated with NSO, pyATS can be a powerful tool for test-driven network automation:

Key features of Cisco NSO include:

- **Multivendor support:** NSO is designed to work with a wide range of network devices and services across different vendors, offering a unified approach to network automation.

- **Service abstraction:** NSO abstracts network services from the underlying hardware, allowing network services to be defined and deployed independently of the specific devices.

■ **Model-driven architecture:** NSO uses YANG modeling to define network services and devices, enabling a high level of customization and flexibility.

■ **Scalability and reliability:** NSO is built to scale for large enterprise and service provider networks, ensuring high performance and reliability.

■ **Network change automation:** NSO automates the process of network change management, ensuring consistent and error-free network updates.

■ **Automated testing:** pyATS can be used to automatically test network configurations and services provisioned by NSO. This helps in validating network changes and ensuring compliance with desired network states.

■ **REST Connector:** pyATS's REST Connector enables it to interact with NSO's REST API, allowing for automated and programmatic control over network configurations and services.

■ **Continuous integration/continuous deployment (CI/CD):** Together, NSO and pyATS can be part of a CI/CD pipeline for networks, where changes are tested and deployed automatically, increasing agility and reducing the time to deploy new services or updates.

■ **Proactive monitoring and validation:** With pyATS, network engineers can proactively monitor and validate the state of the network managed by NSO, ensuring that the network is functioning as intended and rapidly identifying any issues.

The combination of Cisco NSO's network orchestration capabilities with pyATS's automated testing and validation framework forms a powerful duo for modern network automation. This integration allows organizations to rapidly deploy and validate network changes, ensuring high levels of network performance, reliability, and compliance with business requirements.

Connect API

Purpose: Establishes a connection to the device

Features:

■ Supports specifying a port for connection

■ Allows specifying a username and password in the testbed YAML file

Default Settings:

■ **Port:** 8080 (if not specified)

■ **Username/password:** admin/admin (if not specified)

```
testbed:
    name: myTestbed
```

```
devices:
    PE1:
        custom:
            abstraction:
                order: [os]
        connections:
            rest:
                class: rest.connector.Rest
                ip: 1.2.3.4
                port: 8080
                credentials:
                    rest:
                        username: admin
                        password: admin
```

GET API

Purpose: Sends a GET request to the device

Arguments:

- **api_url:** API URL string (required)

- **content_type:** Return content type (json/xml) (optional, default: json)

- **headers:** Dictionary of headers (optional)

- **expected_status_codes:** Expected status codes (optional, default: 200)

- **timeout:** Timeout in seconds (optional, default: 30)

- **url** = '/api/running/devices'

- **output** = device.rest.get(url)

POST API

Purpose: Sends a POST request with an optional payload to the device

Arguments:

- **api_url:** API URL string (required)

- **payload:** Payload to send (string/dict; optional)

- content_type, headers, expected_status_codes, timeout (as in GET API)

Note   The script detects XML/JSON based on payload or passed content type.

- **url** = '/api/running/devices/device/R1/_operations/check-sync'

- **output** = device.rest.post(url)

PATCH API

Purpose: Sends a PATCH request with payload to the device

Arguments: Similar to POST API, with payload being required

```
config_routes = """
{
  "tailf-ned-cisco-ios:route": {
    "ip-route-forwarding-list": [
      {
        "prefix": "10.6.1.0",
        "mask": "255.255.255.0",
        "forwarding-address": "10.2.2.2"
      }
    ]
  }
}
"""
```

- **output** = device.rest.patch("/api/running/devices/device/R1/config/ios:ip/route", payload=config_routes)

PUT API

Purpose: Sends a PUT request with payload to the device

Arguments: Same as POST and PATCH APIs

```
config_routes = """
{
  "tailf-ned-cisco-ios:route": {
    "ip-route-forwarding-list": [
      {
        "prefix": "10.1.1.0",
        "mask": "255.255.255.0",
        "forwarding-address": "10.2.2.2"
      },
      {
        "prefix": "10.2.1.0",
```

```
         "mask": "255.255.255.0",
         "forwarding-address": "10.2.2.2"
      }
    ]
  }
}
"""
```

- output = device.rest.put("/api/running/devices/device/R1/config/ios:ip/route", payload=config_routes)

DELETE API

Purpose: Sends a DELETE request to the device

Arguments: Same as GET API, without **api_url**

- **ouput** = device.rest.delete('/api/running/devices/device/R1/config/ios:ip/route')

Catalyst Center

The Cisco Catalyst Center stands as a pivotal network control and management hub. It streamlines and centralizes network operations, facilitating easier management and hastening digital transformation.

Key features of Cisco Catalyst Center include the following:

- **Automated network management:** Catalyst Center offers sophisticated automation for network tasks like provisioning, policy management, and monitoring, significantly reducing manual intervention.

- **Intent-based networking:** Catalyst Center translates business objectives into network policies, ensuring the network's alignment with business needs.

- **API integration and openness:** The platform supports extensive API integrations, allowing for a flexible, customizable network ecosystem.

- **AI-powered insights:** Catalyst Center provides AI-driven analytics for network optimization and proactive troubleshooting.

- **Enhanced security:** Catalyst Center strengthens network security with advanced analytics and automated policy enforcement.

Catalyst Center includes northbound APIs, enabling external systems and applications to communicate with it. These are RESTful APIs that facilitate various functions, including network configuration and monitoring, essential for automation and integration with other systems. The APIs allow for seamless integration with network automation tools like pyATS, via the REST Connector, enhancing operational capabilities. The integration of Catalyst Center with pyATS, Cisco's automated test system, offers a powerful solution for network management and automation. pyATS can automatically test and validate network configurations managed by Catalyst Center, ensuring accuracy and consistency. pyATS's REST Connector can interface with Catalyst Center's APIs. This capability allows for the automated control and monitoring of network configurations and operations. The integration allows for continuous testing and validation of the network, ensuring it meets specified requirements and functions optimally. This setup can be integrated into CI/CD pipelines, where network changes are continuously tested and validated, leading to improved network agility and reliability.

Cisco Catalyst Center and pyATS REST Connector create a robust environment for network automation and management. This integration empowers organizations to efficiently manage their networks, automate crucial testing and validation processes, and ensure network operations align with strategic business goals. This results in enhanced operational efficiency and network reliability, essential for modern network environments.

Connect API

Purpose: Establishes a connection to the device using Cisco Catalyst Center's REST implementation

Features:

- Supports specifying a custom port for connection
- Allows setting a username and password in the testbed YAML file
- Option to skip SSL certificate verification

Default Settings:

- **Port:** 443 (if not specified)
- **SSL Verification:** Enabled (set verify: False to disable)

```
testbed:
    name: myTestbed

devices:
    dnac:
        custom:
            abstraction:
                order: [os]
        connections:
            rest:
                class: rest.connector.Rest
                ip: 1.2.3.4
                port: 8080
```

```
                verify: False
                credentials:
                    rest:
                        username: admin
                        password: admin
```

GET API

Purpose: Sends a GET request to the device

Arguments:

- **api_url**: API URL string (required)

- **timeout**: Timeout in seconds (optional, default: 30)

- **url** = '/dna/intent/api/v1/interface'

- **output** = device.rest.get(url)

Disconnect API

Purpose: Disconnects from the device

Implementation: Follows the generic implementation for disconnecting

```
# Example
# -------
#
#   loading & using REST testbed yaml file in pyATS

# import the topology module
from pyats import topology

# load the above testbed file containing REST device
testbed = topology.loader.load('/path/to/rest/testbed.yaml')

# get device by name
device = testbed.devices['PE1']

# connect to it
device.connect(via='rest')

# disconnect rest connection
device.rest.disconnect()
```

IOS XE

Cisco IOS XE is an operating system optimized for programmability, scalability, and maintainability at the network edge. It brings together the best of Cisco's IOS technologies and modern software principles. One of the key features of IOS XE is its support for RESTCONF, a network management protocol based on HTTP/HTTPS for accessing data defined in YANG, using the datastores defined in NETCONF. RESTCONF in IOS XE allows for device management and configuration using standard HTTP methods (GET, POST, PUT, PATCH, and DELETE) applied to a conceptual datastore. RESTCONF primarily uses JSON for data encoding, making it straightforward to integrate with modern software tools and languages that natively handle JSON. For the REST Connector to function with IOS XE devices, RESTCONF must be enabled on the device. There are some direct benefits of using RESTCONF, not limited to massive performance and speed improvements over SSH parsing:

- **Direct JSON data access:** Unlike traditional methods that might require SSH and custom parsers, RESTCONF provides direct access to a device's YANG data models using JSON as the data format. This simplifies data retrieval and parsing, making it more efficient and less error-prone.

- **Standard HTTP methods:** Utilizing standard HTTP methods for configuration and management tasks aligns well with modern web technologies and practices.

- **Avoiding SSH parsers:** By using RESTCONF, there's no need for complex SSH command scripts and parsers to retrieve data. Instead, structured/modeled data can be directly accessed and manipulated using RESTful APIs.

In a network automation context, the REST Connector can be used to interface with IOS XE devices over RESTCONF. This allows network automation tools to directly interact with the devices using standardized APIs, streamlining configuration and management processes, and ensuring compatibility with modern network management practices. IOS XE's support for RESTCONF represents a significant step toward more dynamic and programmable network management. It facilitates easier integration with automation tools, ensuring more efficient and reliable network operations. This approach aligns well with the current trends in network programmability and automation, making it a valuable feature for network administrators and engineers.

Connect API

Purpose: Establishes a REST connection to an IOS-XE device

Features:

- Option to specify a custom port
- User credentials set in the testbed YAML file
- Optional proxy settings

Default Settings:

- **Port:** 443 (if not specified)

- **Username/password:** admin/admin (if not specified)

```
testbed:
    name: myTestbed

devices:
    eWLC:
        os: iosxe
        type: router
        connections:
            rest:
                class: rest.connector.Rest
                ip: 1.2.3.4
                port: 443
                credentials:
                    rest:
                        username: admin
                        password: admin
```

GET API

Purpose: Sends a GET request to the IOS-XE device

Arguments:

- **api_url:** Required API URL string

- **default_content_type:** JSON/XML (default: json)

- **timeout:** Timeout in seconds (optional, default: 30)

- **url** = '/restconf/data/site-cfg-data/'

- **output** = device.rest.get(url)

POST API

Purpose: Sends a POST request with an optional payload to the IOS-XE device

Arguments:

- **api_url:** Required API URL string

- **payload:** String or dict payload (optional)

■ content_type, headers, expected_status_codes, timeout (as in GET API)

■ **url** = '/restconf/data/site-cfg-data/ap-cfg-profiles/'

■ **output** = device.rest.post(url)

PATCH API

Purpose: Sends a PATCH request with payload to the IOS-XE device

Arguments: Similar to POST API, with payload being required

```
payload = """
"Cisco-IOS-XE-wireless-site-cfg:ap-cfg-profile": {
    "hyperlocation": {
        "hyperlocation-enable": true,
    }
}
"""
```

■ **output** = device.rest.patch("/restconf/data/site-cfg-data/ap-cfg-profiles/ap-cfg-profile=default-ap-profile", payload=payload)

PUT API

Purpose: Sends a PUT request with payload to the IOS-XE device

Arguments: Same as POST and PATCH APIs

```
payload = """
"Cisco-IOS-XE-wireless-site-cfg:ap-cfg-profile": {
    "hyperlocation": {
        "hyperlocation-enable": true,
    }
}
"""
```

■ **output** = device.rest.put("/restconf/data/site-cfg-data/ap-cfg-profiles/ap-cfg-profile=default-ap-profile", payload=payload)

DELETE API

Purpose: Sends a DELETE request to the IOS-XE device

Arguments: Same as GET API, without api_url

■ **output** = device.rest.delete('/restconf/data/site-cfg-data/ap-cfg-profiles/ap-cfg-profile=test-profile')

Cisco ACI APIC

Software-defined networking (SDN) has revolutionized network management, emphasizing flexibility, programmability, and dynamic control. Cisco's Application Centric
Infrastructure (ACI) is at the forefront of this evolution, representing a sophisticated SDN
solution that fuses software and hardware to create a highly responsive and efficient
network ecosystem. Cisco ACI leverages the principles of SDN to offer a centralized,
policy-driven environment that significantly automates network provisioning and management. By abstracting the underlying network infrastructure, ACI enables network
engineers to concentrate on network performance and behavior rather than on individual
component configurations.

Key Features of Cisco ACI:

- **Centralized policy management:** ACI enables the definition and application of network policies centrally, ensuring consistent enforcement across the network fabric.

- **Application-focused networking:** The network configurations and policies in ACI
  are designed to support specific applications, thereby enhancing both performance
  and security.

- **Seamless integration and automation:** ACI integrates with a variety of automation
  tools and cloud management solutions, facilitating a more agile network that can
  swiftly respond to changing demands.

At the heart of Cisco ACI is the Application Policy Infrastructure Controller (APIC),
which serves as the primary entity for managing the ACI fabric. APIC offers a centralized
point of automation and management, crucial for maintaining the efficiency and consistency of the ACI architecture.

Key Aspects of Cisco APIC:

- **Management Information Tree (MIT):** APIC utilizes the MIT, a hierarchical structure that represents all elements and policies within the network. Each component in
  the ACI fabric is represented as a node with a unique, hierarchical path.

- **API exposure for every object:** APIC ensures that every object within the MIT is
  accessible via a REST API, enabling comprehensive programmatic access and manipulation. This capability is integral to custom development and automation within
  ACI.

The integration of Cisco ACI APIC with the pyATS REST Connector creates a robust
combination for network automation and testing. The pyATS REST Connector allows
pyATS to interact directly with APIC's REST APIs, enabling the automation of tasks
and validation of configurations within the ACI environment. There are three operation
modes with pyATS REST Connector for Cisco ACI APIC:

- **REST APIs:** Direct interaction with APIC using RESTful APIs for network automation tasks.

- **APIC SDK (Cobra):** An alternative approach using the Cobra SDK, which provides a Python library for more advanced operations with APIC, simplifying the complexity of REST API interactions.

- **Test-driven automation:** By integrating pyATS, network engineers can adopt a test-driven approach to network changes, ensuring that these changes meet desired outcomes and adhere to defined policies. The combination of APIC's policy-driven architecture with pyATS's automated testing capabilities leads to more efficient, reliable, and agile network operations, aligning with modern network management practices. The synergy between Cisco ACI, APIC, and pyATS represents a powerful and dynamic approach to network management in the SDN era. This combination enhances network automation, accuracy, and efficiency, meeting the critical needs of contemporary and future network environments.

GET API

Purpose: Sends a GET request to retrieve data from APIC

Arguments:

- **dn:** Unique distinguished name of the object (mandatory)

- **query_target:** Scope of the query ('self', 'children', 'subtree'; default: 'self')

- **rsp_subtree:** Level of child objects in the response ('no', 'children', 'full'; default: 'no')

- **rsp_prop_include:** Properties to include ('all', 'naming-only', 'config-only'; default: 'all')

- **rsp_subtree_include:** Additional contained objects or options (default: None)

- **rsp_subtree_class, query_target_filter, target_subtree_class, order_by:** Additional filters and sorting options (default: None)

- **expected_status_code:** Expected HTTP status code (default: None)

- **timeout:** Maximum time for the operation (default: 30 seconds)

- **url** = 'api/node/class/fvTenant.json'

- **output** = device.get(url, query_target='self', rsp_subtree='no', query_target_filter=", rsp_prop_include='all')

POST API

Purpose: Sends a POST request to create or update data in APIC

Arguments:

- **dn:** Unique distinguished name of the object (mandatory)

- **payload:** Data to send (dict or string; mandatory)

- **xml_payload:** Set to True if payload is XML (default: False)
- **expected_status_code:** Expected result (default: 200)
- **timeout:** Maximum time for the operation (default: 30 seconds)

```
payload = """
{
  "fvTenant": {
    "attributes": {
      "dn": "uni/tn-test",
      "name": "test",
      "rn": "tn-test",
      "status": "created"
    },
    "children": []
  }
}
"""
```

- **url** = 'api/node/mo/uni/tn-test.json'
- **output =** device.rest.post(url, payload)

DELETE API

Purpose: Sends a DELETE request to remove data from APIC

Arguments:

- **dn:** Unique distinguished name of the object (mandatory)
- **expected_status_code:** Expected HTTP status code (default: 200)
- **timeout:** Maximum time for the operation (default: 30 seconds)
- **url** = 'api/v1/schema/583c7c482501002501061985'
- **output** = device.delete(url)

Query

Purpose: Mimics the query function from the **MoDirectory** class in the Cobra SDK

Arguments:

- **queryObject:** A query object (Mandatory).

Additional Information: Refer to the Cisco APIC Python API for more details.

Purpose: Mimics the commit function from the MoDirectory class in the Cobra SDK

Config

Purpose: Mimics the configuration function from the Cobra SDK

Arguments:

- **configObject:** The configuration request to commit (mandatory)

Additional Information: Refer to the Cisco APIC Python API for more details.

LookupByDn

Purpose: Mimics the **lookupByDn** function from the **MoDirectory** class in the Cobra SDK

Arguments:

- **dnStrOrDn:** DN (Distinguished Name) of the object to look up (mandatory)
- **queryParams:** Additional filters (optional)

Additional Information: Refer to the Cisco APIC Python API for more details.

LookupByClass

Purpose: Mimics the **lookupByClass** function from the **MoDirectory** class in the Cobra SDK

Arguments:

- **classNames:** Class name or list of classes (mandatory)
- **parentDn:** DN of the parent object (optional)
- **kwargs:** Additional filters (optional)

Additional Information: Refer to the Cisco APIC Python API for more details.

Exists

Purpose: Mimics the exists function from the **MoDirectory** class in the Cobra SDK.

Arguments:

- **dnStrOrDn:** DN of the object to check (mandatory)

Additional Information: Refer to the Cisco APIC Python API for more details.

Get_Model

Purpose: Automatically imports the required library and returns the model class

Arguments:

- **model:** Unique identifier of the module and class (for example, fv.Tenant; mandatory).

- **tenant_class** = device.cobra.get_model(model='fv.Tenant')

Create

Purpose: Automatically imports the required library and instantiates the model object

Arguments:

- **model:** Unique identifier of the module and class (e.g., fv.Tenant; mandatory)

- **parent_mo_or_dn:** The parent Managed Object (MO) or DN (mandatory)

- **extra_parms:** Additional attributes for the object (optional)

- **tenant** = device.cobra.create(model='fv.Tenant', parent_mo_or_dn='uni', name='test')

Config_And_Commit

Purpose: Adds a Managed Object (MO) to **ConfigRequest** and pushes it to the device

Arguments:

- **mo:** Object to be committed (mandatory)

- **expected_status_code:** Expected result (default: 200)

  (Assuming **'tenant'** object is created using the **create** function.)

- **tenant** = device.cobra.config_and_commit(mo=tenant)

BIG-IP

F5 BIG-IP is a widely recognized product suite, primarily known for its advanced load-balancing capabilities, which are integral in modern network architectures. BIG-IP extends beyond traditional load balancing to deliver a range of network functions, including traffic management, security, access control, and performance optimization. In today's complex network environments, particularly those incorporating Cisco technologies, F5 BIG-IP plays a crucial role in ensuring efficient traffic distribution, application delivery, and network reliability.

One of the strengths of F5 BIG-IP lies in its compatibility with multiplatform API integrations, making it a versatile choice in diverse network setups. This adaptability is particularly beneficial when integrating with automation and testing frameworks like pyATS, a test automation framework developed by Cisco. pyATS is not exclusive to Cisco devices and supports a variety of multiplatform APIs, including those of F5 BIG-IP.

The integration of F5 BIG-IP with pyATS and its REST Connector underscores a strategic move toward unified network management and automation. By leveraging the RESTful APIs of BIG-IP, network engineers can utilize pyATS to automate, test, and develop network configurations and operations across their entire infrastructure, not just Cisco devices. This unified approach simplifies the management of multivendor environments, a common scenario in modern networks, and enhances the ability to maintain consistent and reliable network services.

The use of pyATS in conjunction with F5 BIG-IP's APIs facilitates a more streamlined and efficient workflow for network operations. It enables automated testing and deployment of configurations, performance monitoring, and quick adaptation to changing network demands. This compatibility demonstrates the flexibility and power of pyATS as a tool for comprehensive network automation and testing, capable of bridging different platforms and technologies within a network.

The integration of F5 BIG-IP with pyATS and its REST Connector represents an important step toward more cohesive, automated, and efficient network management practices. By embracing this combination, organizations can effectively manage their F5 load balancers alongside other network components, ensuring high performance, security, and reliability across their entire network infrastructure.

Connect API

Purpose: Establishes a connection to an F5 BIG-IP device

Features:

- Supports custom port configuration
- Username and password specified in the testbed YAML file
- Utilizes token authentication for communication with BIG-IP

Default Settings:

- **Port:** 443 (if not specified)
- **SSL Verification:** Enabled if **verify** is True
- **Protocol:** HTTP (if not specified)

```
testbed:
    name: myTestbed
```

```
devices:
    bigip01.lab.local:
        alias: 'bigip01'
        type: 'bigip'
        os: 'bigip'
        custom:
            abstraction:
                order: [os]
        connections:
            rest:
                class: rest.connector.Rest
                ip: 1.2.3.4
                port: 443
                protocol: http
                credentials:
                    rest:
                        username: admin
                        password: admin

device = testbed.devices['bigip01']
device.connect(alias='rest', via='rest')
```

Disconnect API

Purpose: Deletes the current API token in use

GET API

Purpose: Retrieves data from the F5 BIG-IP device

Arguments:

- **api_url**: API URL string (mandatory)

- **timeout**: Timeout in seconds (optional, default: 30)

- **url** = '/mgmt/tm/ltm/node'

- **nodes** = device.rest.get(url)

POST API

Purpose: Sends data to the F5 BIG-IP device

Arguments:

- **api_url**: API URL string (mandatory)

- **payload**: Data payload (mandatory)

- **timeout:** Timeout in seconds (optional, default: 30)

- **url** = '/mgmt/tm/ltm/node/'

- **data** = {"name": "wa12", "partition": "Common", "address": "119.119.192.193"}

- **node** = device.rest.post(url, data)

PATCH API

Purpose: Updates specific fields of data on the F5 BIG-IP device

Arguments:

- **api_url:** API URL string (mandatory)

- **payload:** Data payload (mandatory)

- **timeout:** Timeout in seconds (optional, default: 30)

- **url** = '/mgmt/tm/ltm/node/~Common~wa12'

- **data** = {"session": "user-disabled"}

- **node** = device.rest.patch(url, data)

PUT API

Purpose: Replaces data on the F5 BIG-IP device

Arguments:

- **api_url:** API URL string (mandatory)

- **payload:** Data payload (mandatory)

- **timeout:** Timeout in seconds (optional, default: 30)

- **url** = '/mgmt/tm/ltm/pool/wa12'

- **data** = {"members": "wa13:80"}

- **pool** = device.rest.put(url, data)

DELETE API

Purpose: Removes data from the F5 BIG-IP device

Arguments:

- **api_url:** API URL string (mandatory)

- **timeout:** Timeout in seconds (optional, default: 30)

- **url** = '/mgmt/tm/ltm/node/wa12'

- **node** = device.rest.delete(url)

SD-WAN vManage

In the realm of advanced networking solutions, Cisco SD-WAN emerges as a transformative force, reshaping how wide area networks (WANs) are managed and optimized. This innovative approach is central to Cisco's intent-based networking, which aims to create adaptive networks that align seamlessly with evolving business needs. The cornerstone of Cisco SD-WAN's functionality and user experience is vManage, a sophisticated network management system designed for ease of use in configuring, monitoring, and maintaining SD-WAN networks.

vManage offers a centralized platform that simplifies the management of the SD-WAN network. Its graphical user interface is intuitive, making it easier for network administrators to oversee the entire network infrastructure, including routers and controllers. This centralized management is crucial in enhancing the operational efficiency of SD-WAN deployments.

Cisco SD-WAN is not just about simplified management; it also brings robust security features to the table, ensuring secure and reliable connectivity across various network types. The system is designed to optimize network performance, employing intelligent path control and application-aware routing to enhance the user experience. Moreover, its scalability and seamless cloud integration capabilities make Cisco SD-WAN a versatile solution for modern network environments.

The true power of vManage in the context of Cisco SD-WAN lies in its API capabilities. These RESTful APIs are designed to be user-friendly, promoting ease of integration and use. They open up vast opportunities for automation, enabling network administrators to streamline operations and adapt quickly to network changes.

The integration of vManage APIs with the pyATS REST Connector marks a significant advancement in network automation. pyATS, known for its robust testing and automation capabilities, extends its proficiency to Cisco SD-WAN through this integration. This fusion allows network engineers to interact programmatically with vManage, bringing a new level of automation and efficiency to network operations.

Through the pyATS REST Connector, tasks that were once manual and time-consuming can now be automated. This includes everything from deploying configuration changes to conducting in-depth performance analytics and troubleshooting. The REST Connector acts as a bridge between pyATS and vManage, ensuring that the benefits of automation are fully realized in the SD-WAN environment.

The collaborative use of Cisco SD-WAN's vManage and pyATS REST Connector exemplifies a forward-thinking approach to network management. This integration not only simplifies the complexities associated with managing modern networks but also ensures that networks are agile, secure, and aligned with business objectives. By leveraging the combined strengths of vManage APIs and the pyATS REST Connector, network professionals are equipped to manage, automate, and optimize their networks effectively, paving the way for a more dynamic and resilient network infrastructure.

Connect API

Purpose: Establishes a connection to a vManage device

Features:

- Supports specifying a custom port for connection

- Allows setting a username and password in the testbed YAML file

- Uses token-based authentication

Default Settings:

- **Port:** 8443 (if not specified)

- **SSL Verification:** Disabled if **verify** is set to False

```
testbed:
    name: myTestbed

devices:
    vmanage:
        os: vmanage
        type: vmanage
        custom:
            abstraction:
                order: [os]
        connections:
            rest:
                class: rest.connector.Rest
                ip: "2.3.4.5"
                port: 8443
                verify: False
                credentials:
                    rest:
                        username: admin
                        password: admin

device = testbed.devices['vmanage']
device.connect()
```

GET API

Purpose: Retrieves data from the vManage device

Arguments:

- **mount_point:** API URL string (required)
- **headers:** Additional headers (optional)
- **timeout:** Timeout in seconds (optional, default: 30 seconds)
- **mount_point** = 'dataservice/device'
- **output** = device.get(mount_point=mount_point)

POST API

Purpose: Sends data to the vManage device

Arguments:

- **mount_point:** API URL string (required)
- **payload:** Data payload (required)
- **headers:** Additional headers (optional)
- **timeout:** Timeout in seconds (optional, default: 30 seconds)

```
mount_point = "dataservice/template/feature/"
payload =   {"templateName":"cli-add-stp",
            "templateDescription":"cli-add-stp",
            "templateType":"cli-template",
            "deviceType":["vedge-ISR-4451-X"],
            "templateMinVersion":"15.0.0",
            "templateDefinition":{"config":{"vipObjectType":"object",
                                            "vipType":"constant",
                                            "vipValue":"spanning-tree
                                                mode rapid-pvst"}},
            "factoryDefault":false}
device.rest.post(mount_point=mount_point, payload=payload)
```

PUT API

Purpose: Updates data on the vManage device

Arguments:

- **mount_point:** API URL string (required)
- **payload:** Data payload (required)

■ **headers:** Additional headers (optional)

■ **timeout:** Timeout in seconds (optional, default: 30 seconds)

```
mount_point = "dataservice/template/feature/{oid}"
payload =   {"templateName":"cli-add-stp",
            "templateDescription":"cli-add-stp",
            "templateType":"cli-template",
            "deviceType":["vedge-ISR-4451-X"],
            "templateMinVersion":"15.0.0",
            "templateDefinition":{"config":{"vipObjectType":"object",
                                            "vipType":"constant",
                                            "vipValue":"spanning-tree
                                                mode rapid-pvst"}},
            "factoryDefault":false}
device.rest.put(mount_point=mount_point, payload=payload)
```

DELETE API

Purpose: Removes data from the vManage device

Arguments:

■ **mount_point:** API URL string (required)

■ **headers:** Additional headers (optional)

■ **timeout:** Timeout in seconds (optional, default: 30 seconds)

```
mount_point = "dataservice/device/unreachable/"
device.rest.delete(mount_point=mount_point)
```

DCNM

Cisco Data Center Network Manager (DCNM) is a comprehensive management solution designed specifically for the data centers' network infrastructure. As an integral component of Cisco's data center solutions, DCNM provides centralized management for Cisco Nexus data centers. It simplifies, automates, and enhances the operational management of Cisco-based data center networks, ensuring they are efficient, scalable, and secure.

The core strength of Cisco DCNM lies in its capability to manage a wide array of network configurations and operations, ranging from LAN fabric to SAN. This includes automation of routine tasks, efficient monitoring and troubleshooting, as well as providing deep insights into the network performance. DCNM facilitates a streamlined approach to managing data center components, which is crucial in today's complex and dynamic IT environments.

GET API

Purpose: Sends a GET request to retrieve information from the device

Arguments:

- **api_url:** Unique distinguished name that describes the API endpoint (mandatory)
- **headers:** Additional headers for the GET command (optional, default: None)
- **timeout:** Maximum time for the operation (default: 30 seconds)
- **expected_return_code:** Expected HTTP status code (optional, default: None)
- **url** = '/rest/top-down/fabrics/single_leaf_demo/vrfs'
- **output** = device.rest.get(url)

POST API

Purpose: Sends a POST request to create or update resources on the device

Arguments:

- **api_url:** Unique distinguished name that describes the API endpoint (mandatory)
- **payload:** Data payload to send (mandatory)
- **headers:** Additional headers for the command (optional, default: None)
- **timeout:** Maximum time for the operation (default: 30 seconds)
- **expected_return_code:** Expected HTTP status code (optional, default: None)

```
payload = """
{
    "fabric": "single_leaf_demo",
    "networkName": "MyNetwork_10000",
    "serialNumber": "FDO22230J8W",
    "vlan": "1000",
    "dot1QVlan": 1,
    "untagged": False,
    "detachSwitchPorts": "Ethernet1/5",
    "switchPorts": "",
    "deployment": False
}
"""
```

- **url** = '/rest/top-down/fabrics/single_leaf_demo/networks/MyNetwork_10000/attachments/'
- **output** = device.rest.post(url, payload)

DELETE API

Purpose: Sends a DELETE request to remove resources from the device

Arguments:

- **api_url:** Unique distinguished name that describes the API endpoint (mandatory)
- **headers:** Additional headers for the command (optional, default: None)
- **timeout:** Maximum time for the operation (default: 30 seconds)
- **expected_return_code:** Expected HTTP status code (optional, default: None)
- **url** = '/rest/top-down/fabrics/single_leaf_demo/networks/MyNetwork_10000'
- **output** = device.rest.delete(url)

PATCH API

Purpose: Sends a PATCH request to update specific fields of data on the device

Arguments:

- **api_url:** Unique distinguished name that describes the API endpoint (mandatory)
- **payload:** Data payload to send (mandatory)
- **headers:** Additional headers for the command (optional, default: None)
- **timeout:** Maximum time for the operation (default: 30 seconds)
- **expected_return_code:** Expected HTTP status code (optional, default: None)

```
payload = """
{
    "fabric": "single_leaf_demo",
    "networkName": "MyNetwork_10000",
    "serialNumber": "FDO22230J8W",
    "vlan": "1000",
    "dot1QVlan": 1,
    "untagged": False,
    "detachSwitchPorts": "Ethernet1/6",
    "switchPorts": "",
    "deployment": False
}
"""
```

- **url** = '/rest/top-down/fabrics/single_leaf_demo/networks/MyNetwork_10000'
- **output** = device.rest.patch(url)

PUT API

Purpose: Sends a PUT request to replace data on the device

Arguments:

- **api_url:** Unique distinguished name that describes the API endpoint (mandatory)

- **payload:** Data payload to send (mandatory)

- **headers:** Additional headers for the command (optional, default: None)

- **timeout:** Maximum time for the operation (default: 30 seconds)

- **expected_return_code:** Expected HTTP status code (optional, default: None)

```
payload = """
{
    "fabric": "single_leaf_demo",
    "networkName": "MyNetwork_10000",
    "serialNumber": "FDO22230J8W",
    "vlan": "1000",
    "dot1QVlan": 1,
    "untagged": False,
    "detachSwitchPorts": "Ethernet1/7",
    "switchPorts": "",
    "deployment": False
}
"""
```

- **url** = '/rest/top-down/fabrics/single_leaf_demo/networks/MyNetwork_10000'

- **output** = device.rest.put(url)

Nexus Dashboard

The Cisco Nexus Dashboard stands at the forefront of data center networking innovation, offering a centralized and scalable platform for managing multicloud networks. As a hub for network operations, it provides a unified interface that brings together various Cisco data center products like Nexus switches, ACI, and cloud services. This central point of control is crucial for enhancing visibility, simplifying network operations, and optimizing the performance of intricate network infrastructures.

GET Command

Purpose: Retrieves information from the device

Arguments:

- **api_url:** Subdirectory part of the URL (mandatory)

- **expected_status_code:** Expected HTTP status code (default: 200)

- **timeout:** Maximum time for the operation (default: 30 seconds)
- **retries:** Number of retries in case of transmission error (default: 3 times)
- **retry_wait:** Time to wait between retries (default: 10 seconds)
- **url** = '/api/config/class/localusers/'
- **output** = device.get(url)

POST Command

Purpose: Sends data to the device

Arguments:

- **api_url:** Subdirectory part of the URL (mandatory)
- **payload:** Dictionary containing the information to send (mandatory)
- **expected_status_code:** Expected HTTP status code (default: 200)
- **content_type:** Format of the request data (json/xml/form; default: json)
- **timeout:** Maximum time for the operation (default: 30 seconds)
- **retries:** Number of retries (default: 3 times)
- **retry_wait:** Time between retries (default: 10 seconds)

```
payload = """
{
  "loginID": "test",
  "loginPasswd: "cisco!123"
}
"""
```

- **url** = 'api/config/localusers/test'
- **output =** device.rest.post(url, payload)

PUT Command

Purpose: Updates data on the device

Arguments:

- **api_url:** Subdirectory part of the URL (mandatory)
- **payload:** Dictionary containing the information to send (optional)
- **expected_status_code:** Expected HTTP status code (default: 200)
- **content_type:** Format of the request data (json/xml/form; default: json)

- **timeout:** Maximum time for the operation (default: 30 seconds)
- **retries:** Number of retries (default: 3 times)
- **retry_wait:** Time between retries (default: 10 seconds)

DELETE Command

Purpose: Removes data from the device

Arguments:

- **api_url:** Subdirectory part of the URL (mandatory)
- **expected_status_code:** Expected HTTP status code (default: 200)
- **timeout:** Maximum time for the operation (default: 30 seconds)
- **retries:** Number of retries (default: 3 times)
- **retry_wait:** Time between retries (default: 10 seconds)
- **url** = 'api/config/localusers/test'
- device.delete(url)

YANG Connector

The pyATS YANG Connector is an innovative module designed to interface with data model interfaces (DMIs), specifically focusing on implementations of the NETCONF client and gNMI client. It is a part of the broader pyATS framework, a Cisco-developed automated testing ecosystem. This module stands out for its integration of a popular open-source package ncclient, starting with version 2.0.0. Consequently, the **yang.connector.Netconf** class in this module supports all high-level APIs of ncclient, enhancing its capabilities and user experience.

This YANG Connector module is compliant with various standards like NETCONF versions 1.0 and 1.1 (RFC 6241) and supports NETCONF over SSH (RFC 6242), including the Chunked Framing Mechanism. It also aligns with gNMI version 0.8.0 and is pyATS compliant. A significant aspect of this module is its seamless integration into the pyATS ecosystem. The introduction of pyATS Connection Manager facilitates the establishment of NETCONF connections in the pyATS topology model, thereby streamlining network automation and testing processes.

The yang.connector module is easy to install and can be obtained from the popular Python package index PyPi. To install it, users can simply use the command **pip install yang.connector**. For those who wish to upgrade to the latest version, the command **pip install --upgrade yang.connector** can be used.

One of the key features of the pyATS YANG Connector is its capacity to act as a NETCONF client within the pyATS framework. This Pythonic approach allows for more intuitive and efficient handling of network configurations and operations, especially for users familiar with Python. The flexibility offered by this module, including the ability to send almost any RPC requests (even those with syntax errors for negative test cases), makes it an invaluable tool for comprehensive network testing and data model validation.

In essence, the pyATS YANG Connector is a crucial component for network engineers and developers working within Cisco's ecosystem, particularly for those focused on data model testing and automation. It not only facilitates a more Pythonic and streamlined approach to using NETCONF clients but also ensures compatibility and integration within the extensive pyATS testing framework.

The NETCONF client in pyATS provides extensive capabilities for managing network devices using the NETCONF protocol, particularly effective with YANG data models. Here is a comprehensive summary of its functionalities, including examples of how to use them in pyATS.

Topology YAML Configuration

Understanding the topology YAML configuration is a foundational step in leveraging NETCONF with pyATS for network automation. The YAML file, as exemplified in Example 13-6, meticulously outlines the device specifics and the parameters necessary for establishing a NETCONF connection. Let's delve into what's essential in this configuration and why it matters.

The essential elements of topology YAML configuration are as follows:

- **Device details:** This includes the name, type, and any other identifiers for the device. Specifying each device accurately is crucial for ensuring that the automation framework can correctly recognize and interact with it.

- **Connection parameters:** These define how the NETCONF session should be initiated. Essential parameters often include the device's IP address, port number, username, and password. This information is critical for establishing a secure and reliable connection to the device.

- **NETCONF-specific settings:** Certain settings are unique to NETCONF, such as netconf_port and netconf_yang. These settings enable pyATS to understand how to communicate with the device using NETCONF, including which port to connect to and how to interpret the device's YANG models.

The topology YAML configuration serves as a blueprint for the automation framework, providing it with all the necessary details to establish connections and interact with network devices via NETCONF. Its importance lies in several key areas:

- **Automation precision:** By accurately describing each device and its connection details, the YAML configuration ensures that scripts and automation tasks target the correct devices with the appropriate parameters, reducing errors and improving efficiency.

■ **Scalability:** A well-defined YAML configuration makes it easier to scale network automation efforts. As new devices are added to the network, they can be seamlessly incorporated into automation routines by adding their details to the YAML file.

■ **Flexibility and customization:** The configuration allows for customization of connection parameters and device interactions. This flexibility is essential for adapting automation tasks to the specific requirements and constraints of different network environments.

In summary, the topology YAML configuration is not just a requirement; it's a cornerstone of using NETCONF with pyATS for network automation. It encapsulates the critical information and settings that enable automated tools to accurately and efficiently manage network devices, laying the groundwork for sophisticated automation strategies that can adapt to the complexities of modern networks.

Example 13-6 *Sample YAML Configuration*

```
devices:
  asr22:
    type: 'ASR'
    credentials:
      default:
        username: admin
        password: admin
    connections:
      netconf:
        class: yang.connector.Netconf
        ip: "2.3.4.5"
        port: 830
        username: admin
        password: admin
        credentials:
          netconf:
            username: ncadmin
            password: ncpw
```

Connect Function

Purpose: Establishes a NETCONF session with the device

```
from pyats.topology import loader
testbed = loader.load('/path/to/asr22.yaml')
device = testbed.devices['asr21']
device.connect(alias='nc', via='netconf')
```

Connected Property

Purpose: Checks if there is an active connection to the NETCONF server

Usage: **device.nc.connected** returns True if a connection is established

Server Capabilities:

- **Purpose:** Retrieves the capabilities advertised by the NETCONF server
- **Usage:** Loop through **device.nc.server_capabilities** to list capabilities

Timeout Setting:

- **Purpose:** Configures the timeout for synchronous RPC requests
- **Usage:** Adjust **device.nc.timeout** as needed

Get Function

- Purpose: Retrieves running configuration and device state information
- Usage Example: Implement with either a subtree filter or an XPATH filter

Get Config Function

- Purpose: Retrieves specific configurations from the device
- Usage Example: Use subtree or XPATH filters to specify configuration data to retrieve

Edit Config Function

- Purpose: Loads configurations to the target datastore on the device
- Usage Example: Use XML strings or ElementTree format for specifying configurations

Request Function

- Purpose: Sends custom RPC requests in string format
- Usage Example: Define the RPC request in string format and send using the **request()** method

Get Schema Function

- Purpose: Retrieves schema information from the device

- Usage Example: Use **device.nc.get_schema('ietf-interfaces')** to fetch schema details

Disconnect and Close Session Functions

- Purpose: Closes the NETCONF session or transport connection

- Usage: Utilize **device.nc.disconnect()** for disconnecting or **device.nc.close_session()** for terminating the session

This summary of the NETCONF client in pyATS illustrates the flexibility and power of this tool for network device management, offering a wide range of capabilities for configuration and state retrieval, alongside effective session management.

gNMI

The introduction of the gNMI (gRPC Network Management Interface) client within the pyATS framework marks a significant advancement in network device management and configuration. Leveraging the efficiencies of gRPC (Google Remote Procedure Call) alongside the structured data models defined by YANG (Yet Another Next Generation), gNMI offers a unified and modern approach to interacting with network devices. This is especially beneficial for devices that are designed to support YANG models through gRPC protocols, facilitating a more streamlined and programmable interface for network operations.

gNMI is designed to simplify the process of device management and configuration by providing a consistent and powerful interface that covers a wide range of network management tasks. Through the use of gRPC, gNMI enables high-performance, bi-directional streaming of configuration and telemetry data between devices and management applications. This makes it possible to not only push configuration changes to devices but also to subscribe to and receive real-time updates about the network state, enhancing monitoring and automation capabilities.

Setting Up and Connecting with gNMI Client

To connect to a device using gNMI, you need to define the device in the YAML topology file with gNMI-specific configurations:

```
devices:
  asr22:
    type: 'ASR'
    tacacs:
      login_prompt: "login:"
```

```
        password_prompt: "Password:"
        username: "admin"
    passwords:
      tacacs: admin
      enable: admin
      line: admin
    connections:
      gnmi:
        class: yang.connector.gNMI
        ip: "2.3.4.5"
        port: 50052
        timeout: 10
```

```
device.connect(alias='gnmi', via='gnmi')
```

Before utilizing the gNMI functionalities, you need to set up the environment and establish a connection to the device:

```
from pyats.topology import loader
from yang.connector.gnmi import Gnmi

# Load the testbed file
testbed = loader.load('testbed.static.yaml')

# Access the device from the testbed
device = testbed.devices['uut']

# Connect to the device using gNMI
device.connect(alias='gnmi', via='yang2')
```

Fetching Device Capabilities

One of the primary uses of the gNMI client is to retrieve the capabilities of the network device, such as supported YANG models and gNMI version:

```
# Fetching the capabilities of the device
resp = device.capabilities()

# Accessing the gNMI version from the response
gnmi_version = resp.gNMI_version
print("gNMI Version:", gnmi_version)
```

- **Capabilities:** Query the device for its capabilities, like supported YANG models, gNMI version, and so on.

- **Get requests:** Retrieve data from the device using YANG models.

- **Set requests:** Modify configurations on the device using YANG models.

- **Subscribe requests:** Subscribe to specified paths for notifications on changes.

- **Additional functionalities:**

  - **Session management:** Establish and manage gNMI sessions with network devices.

  - **Error handling:** Handle errors and exceptions gracefully during gNMI transactions.

  - **Logging and output:** Log and output device responses for debugging and analysis.

The gNMI client in pyATS provides advanced functionalities for network device management using the gRPC protocol and YANG models. Following is a detailed overview of additional gNMI capabilities, including **SetRequest**, **GetRequest**, and connecting to gNMI.

gNMI SetRequest

Purpose: **SetRequest** in gNMI can perform transactions similar to a Netconf **edit-config**. It allows for the modification of device configurations.

Usage Example:

To use **SetRequest**, first inspect it to see if there's an instance of **ConfigDelta**:

```
print(delta.gnmi)
```

Send the **SetRequest**:

```
reply = device.gnmi.set(delta.gnmi)
print(reply)
```

gNMI GetRequest

Purpose: Collect data from the device.

Building **GetRequest**:

Use the gnmi_pb2 module to build the **GetRequest**.

Example:

```
from yang.ncdiff import gnmi_pb2
# Additional steps to build the request
request = gnmi_pb2.GetRequest(...)
print(request)
```

Send the GetRequest and print the reply.

```
reply = device.gnmi.get(request)
print(reply)
```

Creating an Instance of Config

Purpose: Instantiate a **Config** object to manage configurations.

Instantiate a **Config** object with the reply from a gNMI operation.

The JSON content in the gNMI reply is converted to XML.

```
from yang.ncdiff import Config
config = Config(device.nc, reply)
print(config)
```

This extended summary provides a deeper insight into the gNMI client's capabilities in pyATS, highlighting how it can be used for advanced device configuration and data retrieval tasks. The gNMI client, with its support for **SetRequest** and **GetRequest**, offers a powerful tool for modern network management, particularly in environments where gRPC and YANG models are prevalent.

The gNMI client in pyATS offers advanced functionalities for managing network devices, particularly through the use of **Config** and **ConfigDelta** classes. These classes provide a versatile and powerful way to handle device configurations.

Creating Config Objects

From XML String: Config objects can be instantiated using an XML string representing the device's configuration:

```
xml = """ <nc:config xmlns:nc="urn:ietf:params:xml:ns:netconf:b
ase:1.0">
    <device-config xmlns="http://example.com/ns/device-config">
        <hostname>MyRouter</hostname>
        <interfaces>
            <interface>
                <name>GigabitEthernet0/0</name>
                <enabled>true</enabled>
                <ipv4>
                    <address>
                        <ip>192.168.1.1</ip>
                        <netmask>255.255.255.0</netmask>
                    </address>
                </ipv4>
            </interface>
```

```
        </interfaces>
      </device-config>
    </nc:config>
    """
    from yang.ncdiff import Config
    config = Config(device.nc, xml)
```

Storing Operational Data: Config objects can also hold operational data:

```
device.nc.timeout = 120
device.nc.load_model('openconfig-interfaces')
reply = device.nc.get(models='openconfig-interfaces')
state = Config(device.nc, reply)
```

Support for XPath: Config instances support XPath for querying specific config data:

```
ret = config.xpath('/nc:config/oc-if:interfaces/oc-if:interface
[oc-if:name:GigabitEthernet0/1')
```

Partial Config with Filter: You can retrieve a part of the configuration using filters:

```
from lxml import etree

# Creating an XML filter for NTP settings
f = etree.Element("filter")
ios_native = etree.SubElement(f, "native", xmlns="http://cisco.com/ns/
yang/Cisco-IOS-XE-native")
etree.SubElement(ios_native, "ntp")

# Use the filter in a get_config call
reply = device.nc.get_config(source='running', filter=f)

# Extract the config from the reply
c1 = device.nc.extract_config(reply)

# Assuming 'config' is a Config object that contains the full
configuration
# Apply a filter to extract only the NTP settings from 'config'
c2 = config.filter('.//ios:native/ios:ntp')
```

Comparing Configs

"Less Than or Equal To" Logic: It is defined as all nodes in one config existing in another config.

c1 > c2 # True if c1 has more configuration nodes than c2.

ConfigDelta represents the difference between two Config instances or a single Config instance and an **edit-config.** It's used for calculating diffs similar to NETCONF **edit-config,** RESTCONF Requests, or gNMI **SetRequest.**

From Two **Config** Objects: Created by subtracting two Config objects.

From **Config** Object and **edit-config:** Can be instantiated using a **Config** object and an XML string representing an **edit-config.**

```
from yang.ncdiff import ConfigDelta
delta = ConfigDelta(config1, delta=edit_config_xml)
```

Predicting and Verifying Config Changes: The result of applying a **ConfigDelta** can be predicted, sent to the device, and then verified.

```
config2 = config1 + delta
reply = device.nc.edit_config(target='running', config=delta.nc)
config3 = device.nc.extract_config(reply)
assert config2 == config3
```

Creating ConfigDelta Objects with Special Requirements

Custom NETCONF Operations: You can specify how the **ConfigDelta** should handle NETCONF operations like 'create', 'replace', and 'delete':

```
delta = ConfigDelta(config_src=config1, config_dst=config2,
                    preferred_create='create',
                    preferred_replace='replace',
                    preferred_delete='remove')
```

These functionalities of **Config** and **ConfigDelta** in the gNMI client allow for sophisticated handling of device configurations, making them essential tools in network automation and testing processes. They offer the flexibility to manage device configurations precisely, predict changes, and verify the results, thereby ensuring the desired state of network devices.

In pyATS, the **RPCReply** class, originally from the ncclient package, is enhanced by yang. ncdiff to support additional functionalities, notably XPath queries. This enhancement allows users to perform more complex and precise queries on the data returned in an **RPCReply** object.

XPath Queries

Purpose: **RPCReply** supports the **xpath()** method, enabling users to execute XPath queries on the returned data.

■ **reply** = device.nc.get(models='openconfig-network-instance')

- ret = reply.xpath('/nc:rpc-reply/nc:data/oc-netinst:network-instances/oc-netinst:network-instance/oc-netinst:interfaces/oc-netinst:interface/oc-netinst:id/text()')

- assert(set(ret) == {'GigabitEthernet0/0'})

Facilitating XPath Queries with ns_help()

Purpose: To assist with XPath queries, **ns_help()** can be used to view the mapping between prefixes and URLs, which is crucial for formulating correct XPath expressions.

Usage in Error Handling: In scenarios where an **rpc-error** is received, **ns_help()** is particularly useful. It lists all the namespaces, even those not declared in the model schema, and generates prefixes for them.

```
reply = device.nc.edit_config(delta, target='running')
reply.ok   # Returns False in case of an error
reply.ns_help()   # Outputs namespace mappings
```

Important Note on RPCReply and XPath

The integration of XPath with **RPCReply** in pyATS enhances the usability and effectiveness of handling device responses. Users can now execute more complex queries and extract specific pieces of data from the responses efficiently. The addition of **ns_help()** further aids in this process by resolving any namespace-related queries or issues, especially in cases where errors are encountered. This advanced functionality makes pyATS a more powerful tool for network engineers, allowing for detailed and precise analysis of network device configurations and states.

Summary

In the realm of network automation, pyATS stands as a robust and versatile tool, offering a plethora of built-in APIs that cater to an extensive range of operations. One of the key strengths of these APIs is their sheer number—with thousands available, they offer a near guarantee that there's an API tailored for any specific task a network engineer might need to perform. This abundance transforms network automation into an "easy mode," simplifying complex operations and enhancing efficiency.

A pivotal component of pyATS is the REST Connector, a feature that expands the system's capabilities by enabling seamless integration with various platforms. These platforms encompass a wide array of technologies such as NXOS, NSO, DNAC, IOS XE RESTCONF, APIC, BIG-IP, vManage, DCNM, and Nexus Dashboard. The REST Connector acts as a bridge, allowing pyATS to function as a unified platform for managing an organization's entire network infrastructure. This integration is vital for organizations looking to streamline their network operations and ensure smooth, cohesive management across different platforms and devices.

Moreover, pyATS's versatility is further enhanced by its YANG connector, which serves as a NETCONF client, enabling the system to interact efficiently with network configurations and operations that utilize the NETCONF protocol. This feature extends the system's reach into more intricate aspects of network management, allowing for more refined and detailed control over network devices and configurations.

Finally, the integration of gNMI (gRPC Network Management Interface) with pyATS marks a significant leap forward. gNMI is a modern protocol designed for streaming telemetry and managing network configurations. This integration allows for more advanced, real-time monitoring and management capabilities, further solidifying pyATS as a comprehensive solution for network automation and management.

In summary, Cisco's pyATS, with its extensive range of APIs, REST Connector, and additional tools like the YANG connector and gNMI, presents itself as a powerful, all-encompassing platform for network automation. It offers a simplified yet effective approach for managing diverse network environments, making it an indispensable tool for organizations aiming to streamline their network operations and embrace modern automation technologies.

## Chapter 14

Parallel Call (pcall)

When running tests against the network, you may find it slow to connect to each device and collect output one by one. You may wonder if there's a quicker way to connect to multiple devices at the same time. In this chapter, we are going to cover the topic of asynchronous programming, applying parallelism and multiprocessing to pyATS, and wrap up with a performance comparison when connecting to multiple testbed devices, running a command, and returning the parsed output.

This chapter covers the following topics:

- Scaling performance

- Parallel call (pcall)

- Performance comparison

Scaling Performance

When considering performance-scaling measures, many think it's as simple as increasing the number of resources (CPU, memory, and so on). While that may be the case for scaling systems, there are a few different options when it comes to software development. When designing software, we must consider the type of operations our software is performing. The types of operations can include CPU bound, I/O bound, and even GPU bound. In networking, many of the time-consuming software operations are I/O (input/ output) bound. Waiting for the connection to the device, interacting with the device, and closing the device connection is considered an I/O operation. In the following sections, we are going to look at the different methods of performing asynchronous operations to optimize and scale the performance of our pyATS automation scripts.

Asynchronous Programming

In Python, the standard asynchronous library is asyncio, which provides the ability to write Python code that runs *concurrently*. Concurrency is the ability to run multiple tasks in an overlapping timeframe. To avoid confusion, let's take a common networking task as an example—configuring multiple devices. As mentioned previously, connecting to and interacting with a network device is considered an I/O operation, as the program is waiting for the device interaction to complete before proceeding with further instructions. This causes the execution of your code to be much longer since each device connection is a blocking I/O operation. Concurrency allows your code to cooperate and take turns executing on the same thread. To extend the example of configuring devices, this time using concurrency, once you connect to your first device and begin waiting for the device interaction to complete, your async code can "park" that function and begin interacting with the second network device. Once the second device interaction starts and the waiting period begins, your async code can jump back to the first device interaction to check if it's finished and/or move on to the next device connection. In summary, concurrency optimizes the use of a single thread by executing non-blocking code using the asyncio library.

To provide a more complete context, let's dive into a couple concepts of the asyncio library. Asyncio has the concepts of coroutines and event loops. Coroutines are functions that can be suspended and resumed.

In a networking use case, coroutines are functions that contain the code to connect and execute commands on network devices. Coroutines are often referred to as *awaitables*, as coroutines can pause their execution and wait for other coroutines. The coroutine execution is managed and scheduled by the event loop. Event loops are the core of asyncio and run all the asynchronous tasks in the current thread. They can be started, stopped, or run forever. Event loops provide a solid foundation for event-driven automation since they are non-blocking and can execute a callback function when data is received or an event occurs.

Asyncio and concurrency are popular with I/O-bound operations, such as connecting to a network device, as they allow multiple coroutines to cooperatively execute on a single thread. Async programming in Python is a subject of its own, and there are many books and blog series that dive deeper into the topic, but it's good to understand the difference between concurrency and parallelism (described further in the chapter). To learn more about async programming in Python, check out PEP 492, which is about coroutines with async and await syntax (https://peps.python.org/pep-0492/).

Threading

Threads are lightweight in nature and share the same memory within the same process. In contrast to asyncio, threads are created and execute independently. Asyncio utilizes only one thread to manage and schedule code execution. Threads are more user-friendly to create and manage, but there are a few downsides:

- They are not interruptible.

- Thread safety—all shared objects must use lock acquire/release methods.

- They are prone to race conditions—the program depends on a specific execution order.

One of the biggest concerns of threading is that threads are not interruptible. This can lead to threads hanging forever, which can cause unforeseen circumstances when dealing with tasks that can alter network state and connectivity. Additionally, there's the concept of thread safety, where a library is "thread-safe" if its implementation avoids race conditions. Race conditions occur when multiple threads are trying to access the same shared data or resource, or when the order of execution of multiple threads must be specific. If shared resources need to be accessed, a lock must be acquired during usage and released when usage completes. The lock acquire/release mechanism allows multiple threads to share resources. It should be noted that the pyATS library is not thread-safe. Threading is lightweight and can be very scalable when managed correctly.

Multiprocessing

The final asynchronous method is using Python's multiprocessing package, which allows you to create child processes to execute your code, similar to how you can create threads with the threading module. However, creating new child processes, also known as forking, is considered more expensive in terms of resources, as spawning processes have a larger memory footprint. The larger memory footprint is due to each child process having its own memory space, which can be advantageous when managing independent state. In contrast to threading, multiprocessing has the following advantages:

- Separate memory space for each child process—no race conditions.

- Child processes are easily interruptible.

- No global interpreter lock (GIL), which allows you to take advantage of multiple CPU cores.

Because all instances of code are executed in their own processes, in separate memory space, you can run all your code in parallel without worrying about execution order or whether the code is awaitable. For processes to communicate with one another, a pipe must be created to pass messages between two processes. Queues may also be created, which will allow for multiple producers and consumers, but we will focus on pipes. To pass messages in a pipe between processes, the sending process must serialize Python objects to a byte stream using Python's pickle module, and the receiving process will then deserialize the byte stream back to the Python object using the same pickle module. The serialization and deserialization process, plus the amount of data being transferred, can cause an increase in overhead. This is not to scare you; rather, it's a word of caution when you're transferring data between processes.

The concept of running tasks independently, at the same time, is known as *parallelism*. This allows your code to run at scale with the least complexity. The only tradeoff is that process-based parallelism requires more compute resources than concurrency. However, parallelism in Python allows you to take advantage of all the hardware resources (CPU, memory, and so on). The pyATS library utilizes the multiprocessing package to enable parallelism with the **Pcall** class, which will be discussed in detail later in the chapter.

Tip   To better understand the differences between concurrency and parallelism, check out the "Concurrency and Burgers" story found in the documentation of the Python web framework FastAPI: https://fastapi.tiangolo.com/async/#concurrency-and-burgers.

Easypy

Easypy uses multiprocessing to fork child processes for each task in a pyATS job file. Each job file task will run in its own memory space and pipe back its results back to the job file to create an aggregated result. Multiprocessing is also used to auto-create a new log file for each forked process, enable auto-reconnect to the Easypy Reporter server, and allow the pdb debugger to work with task processes when the **pdb=True** flag is used.

Logging and Reporting

The logging module in the Python standard library is thread-safe but is not process-aware. Normally, the user would have to figure out how to generate and manage logs for each forked process. However, Easypy takes care of this for us by creating a TaskLogHandler log handler and attaching it to the pyATS logger so that one TaskLog is created per Task (forked process).

For reporting purposes, Easypy uses the Reporter package, which is a client/server model that uses Unix sockets to communicate Task results. Each Task has a ReportClient client connection object that communicates with the Reporter server. The Reporter aggregates Task result reports when all Tasks are completed. Easypy automatically configures and handles the ReportClient client connection for each Task (forked process).

Parallel Call (pcall)

Parallel call, also known as pcall, is an API built on top of the multiprocessing package to enable parallelism in pyATS. The pcall API is included in the pyats.async_ module. Pcall supports calling any function or method in parallel, as long as the return value of the function or method is a pickleable object. As mentioned previously, the pickle module in the Python standard library serializes and deserializes Python object structures to bytes. Only specific Python object types can be pickled. A few examples of pickleable objects include integers, strings, bytes, and tuples. Lists as well as dictionaries that contain pickleable objects can also be pickled. Pickling is an interesting word to use for serializing and deserializing data, but it's a crucial concept to understand in Python. For more information about pickling and the pickle module, check out the Python standard library documentation at https://docs.python.org/3/library/pickle.html.

Usage

Pcall is a simple abstraction of the Python multiprocessing package that covers a majority of use cases of parallel processing. For more advanced use cases, it's recommended to use the multiprocessing package directly.

The pcall API comes with the following built-in features:

- Builds arguments for the child processes

- Creates and gracefully terminates the child processes

- Returns the target (function) results in their called order

- Re-raises child process errors in the parent process

Example 14-1 shows a simple example of multiplying two numbers that are provided as arguments (**num1** and **num2**).

Example 14-1 *Pcall Usage – Simple Math*

```
from pyats.async_ import pcall

# multiply function that is run in parallel
def multiply(num1, num2):

  return num1*num2
result = pcall(multiply, num1=(1,2,3), num2=(7,8,9))
print(result)
# (7, 16, 27)
```

Pcall zips the two arguments together to create multiple calls to the given target (function). The **zip()** function in Python works by taking two iterables and aggregates their values from left to right. For example, the arguments **num1=1** and **num2=7** would be paired and passed in together since they are the first values in each argument. Each pair of zipped arguments is used to call the target(s) and run in parallel. You may be wondering what happens if the length of the iterables doesn't match? What if **num1** only had two values instead of three? The iteration stops when the shortest argument iterable is reached—no exception is raised. That means that the first two argument values in the **num1** argument iterable would be zipped with the **num2** argument values, but the third value in the **num2** argument iterable (**9**) would essentially be dropped. Table 14-1 shows the available arguments to pcall.

Table 14-1   *Pcall Arguments*

Argument	Description
targets	A single or list of callable targets to be invoked in parallel
timeout	Max runtime of any single child process, in seconds
cargs	Positional arguments common to all target calls
iargs	List of specific positional arguments for each target
ckwargs	Keyword arguments common to all target calls
ikwargs	List of specific keyword arguments for each target
varkwargs	Any other variable keyword arguments and value lists for each target

Not all the available arguments are used, with many being optional. The goal of Table 14-1 is to show the different types of arguments that can be provided to pcall targets. You can provide arguments that are common and provided to all targets (**cargs** and **ckwargs**) and other arguments that are for each individual target (**iargs** and **ikwargs**). Now that you understand the different types of arguments, let's take a deeper look into targets and how arguments are passed into them.

Targets

Functions passed to pcall to run in parallel are called *targets*. There can be more than one target provided to pcall. The following two sections discuss scenarios for when a single target is provided and when multiple targets are provided, respectively.

Single Target

If a single target is provided to pcall, the number of parallel calls is determined by the number of arguments. If positional and keyword arguments are used together, their number of arguments must match. Otherwise, any extra argument values are dropped, which matches the behavior of the **zip()** function described earlier. Example 14-2 shows how arguments dictate the number of parallel calls to a single target.

Example 14-2   *Single Target*

```
from pyats.async_ import pcall

# pcall target function
def func(*args, **kwargs):
  """Function returns all arguments back"""
  return {"args": args, "kwargs": kwargs}

def print_pcall_results(results: tuple):
  """Loop through pcall results and print each child process' results"""
  for i in range(len(results)):
    print(f"Child process {results.index(results[i]) + 1}: {results[i]}")
```

```
# Positional argument building example
parg_results = pcall(func, cargs=(1, 2, 3), iargs=((4,), (5,), (6,)))
print("Positional args:")
print_pcall_results(parg_results)


# Keyword argument building example
kwarg_results = pcall(
  func, ckwargs={"a": 1, "b": 2}, ikwargs=[{"c": 3}, {"c": 4}, {"c": 5}]
)
print(f"\nKeyword args:")
print_pcall_results(kwarg_results)


# Variable keyword argument building example
varkwargs_results = pcall(func, x=(1, 2, 3), y=(4, 5, 6), z=(7, 8, 9))
print(f"\nVariable Keyword args:")
print_pcall_results(varkwargs_results)
```

Multiple Targets

Multiple targets are provided to pcall as a list of function names. Each target is run in its own child process with positional and keyword arguments corresponding to each target as they appear. Example 14-3 shows how arguments are provided to multiple targets.

Example 14-3 *Multiple Targets*

```
from pyats.async_ import pcall

# pcall target functions
def func1(*args, **kwargs):
  return dict(name="func1", arg=args, kwargs=kwargs)


def func2(*args, **kwargs):
  return dict(name="func2", arg=args, kwargs=kwargs)


def func3(*args, **kwargs):
  return dict(name="func3", arg=args, kwargs=kwargs)

def print_pcall_results(results: tuple):
  """Loop through pcall results and print each child process' results"""
  for i in range(len(results)):
    print(f"Child process {results.index(results[i]) + 1}: {results[i]}")


# Positional argument building example
```

```
parg_results = pcall(
  [func1, func2, func3], cargs=[1, 2, 3], iargs=[[4, 5, 6], [7, 8, 9], [9, 10, 11]]
)
print("Positional args:")
print_pcall_results(parg_results)


# Keyword argument building example
kwarg_results = pcall(
  [func1, func2, func3],
  ckwargs={"a": 1, "b": 2},
  ikwargs=[{"c": 3}, {"c": 4}, {"c": 5}],
)
print(f"\nKeyword args:")
print_pcall_results(kwarg_results)
```

You'll notice common arguments (**cargs** and **ckwargs**) are included in each target run, while individual arguments (**iargs** and **ikwargs**) are provided to each target in the order in which they appear. It's important to think about and organize the arguments you're providing to pcall targets.

Error Handling

Exceptions that occur in the child processes invoked by pcall are re-raised to the parent process as ChildProcessException exceptions. The details and traceback of the original exception that occurred in the child process is included in the __cause__ attribute of the ChildProcessException exception. When child processes exceed the timeout value, a SIGTERM is sent to the offending child process, which kills the process, and a TimeoutError exception is raised. You may run into a ChildProcessException exception if connecting to a network device fails when trying to connect to multiple network devices in parallel. The child process trying to connect to the network device will raise an exception to the parent process, which will result in a code failure during execution. However, the exception can be caught in Python using a **try...except** block. To further understand how to catch exceptions in Python, check out the Python standard library documentation on exceptions and errors: https://docs.python.org/3/tutorial/errors.html#handling-exceptions. If the exception is caught appropriately, we can log and react to the exception being raised. In the case of connecting to network devices, we can log a message such as "Failed to connect to {hostname}." Error handling is important to building resilient software and should be considered, especially with I/O tasks that are prone to failures.

Logging

Each pcall process will automatically create its own TaskLog file. This logging functionality is very similar to Easypy logging, which attaches a TaskLogHandler log handler to automatically create a TaskLog file for each Task (child process). Once pcall finishes, the log files from each Task (child process) are automatically joined into a main TaskLog, with each Task having its own section.

Pcall Object

This section dives deeper into the **Pcall** class in the pyats.async_ module. Interestingly, the pcall API is actually a class method of the **Pcall** class. The **Pcall** class is meant to be like **multiprocessing.Pool**, where you can create a pool of processes to execute tasks. There are only a couple of differences between the two. One difference is that **multiprocessing.Pool** will create a fixed number of workers to process task(s), while the **Pcall** class will dynamically create processes based on user inputs and only run the function (target) once per process. Another difference is that the **Pcall** class will build arguments for each function (target), while **multiprocessing.Pool** expects the user to provide the exact arguments needed. Table 14-2 shows the available methods and properties of the Pcall class.

Table 14-2 *Pcall Methods/Properties*

Attribute	Description
__init__	Takes in the exact same arguments as pcall.
pids	Tuple of all child processes.
living	Tuple of all currently alive child processes.
results	Tuple of results from all child processes. None if no results are currently available.
start()	Starts all child worker processes.
join()	Blocks the current process and waits for all children to finish, or until timeout is reached.
terminate()	Terminates all children processes with SIGTERM.

You can instantiate your own instances of the **Pcall** class to poll and control the forked parallel processes. Processes are started with the **start()** method, and all process results are collected with the **join()** method. The **results** property will return None if the processes have not been started or joined. Example 14-4 shows a basic example of instantiating an instance of the **Pcall** class and manually starting and collecting the results of the child processes.

Example 14-4 *A Pcall Class Instance*

```
from pyats.async_ import Pcall

# define a function to be called in
def add(x, y):
    return x + y

# create a Pcall object
p = Pcall(add, x = (1, 2, 3), y = (4, 5, 6))

# start all child processes
```

```
p.start()

# wait for everything to finish
p.join()

# collect results
results = p.results

print(results)
# (5, 7, 9)
```

Performance Comparison

Up to this point, we've talked about theory and provided simple examples. How about we test how pcall can help improve performance of a real-world network task? In this section, we are going to look at a common scenario of executing a command and parsing the output from multiple testbed devices. The testbed will consist of five IOSvL2 devices in Cisco Modeling Labs (CML). Figure 14-1 shows the topology in CML.

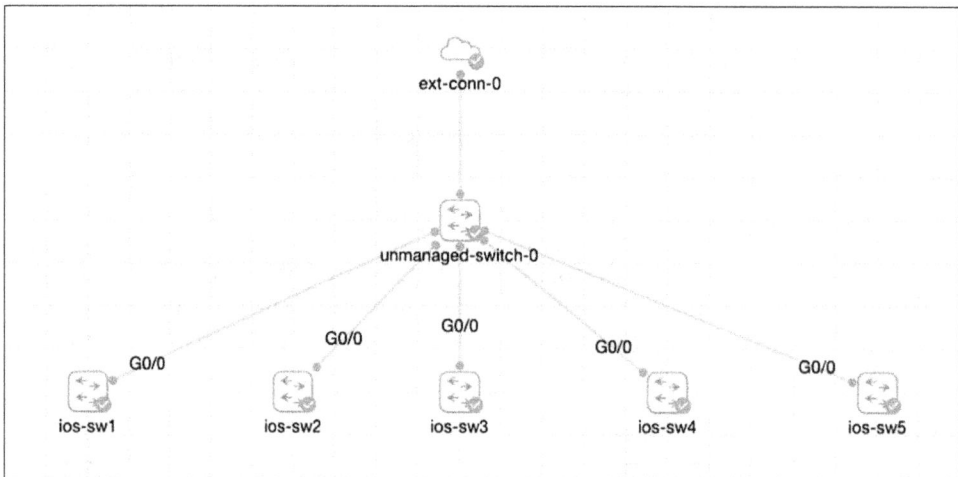

Figure 14-1 *CML Topology*

The networking task is very simple—connect to each testbed device, run the **show version** command, parse the output, and disconnect from each device. All default settings and arguments are used during testing—including a graceful disconnect delay when disconnecting from each testbed device. This is a default connection setting in the Unicon library meant to prevent connection issues, so this task can be further optimized to execute even more quickly! However, the purpose of the task is to show the performance difference between serial and parallel execution of the same task. Example 14-5 shows the function that will execute the described task.

Example 14-5 *The show version Function*

```
def get_version(device):
    # Connect to device
    device.connect()
    # Parse "show version" output
    sh_ver = device.parse("show version")
    # Disconnect from device
    device.disconnect()
    # Print and return output
    print(sh_ver)
    return sh_ver
```

The function takes a pyATS Device object as an input argument. The function code is commented to describe each step of the task, including connecting to the device, running the command, disconnecting from the device, and printing/returning the parsed output.

Now that we have the function we are going to use to test execution performance, let's look at how we are going to execute the code. For serial execution, we will simply loop through the testbed devices, passing in each individual testbed device to the function. Example 14-6 shows the code to execute the function with each testbed device, one by one.

Example 14-6 *Serial Execution*

```
from pyats.topology.loader import load

def main():
    # Load testbed
    testbed = load("iosv_testbed.yaml")

    # Connect to each device one by one and get version
    for dev in testbed.devices.values():
        get_version(device=dev)
```

First, we load the pyATS testbed file and then loop through the testbed devices to execute the task function with each device. Now, let's look at running the same task in parallel using the pcall API. Example 14-7 shows the slightly modified code to run the same task in parallel, across multiple devices at once.

Example 14-7   *Parallel Execution with pcall*

```
from pyats.topology.loader import load
from pyats.async_ import pcall

def main():
    # Load testbed
    testbed = load("iosv_testbed.yaml")

    # List of all device objects from testbed
    tb_devices = testbed.devices.values()
    # [<Device ios-sw1 at 0x1142c3640>, <Device ios-sw2 at 0x1142c3610>,
    # <Device ios-sw3 at 0x1142ba820>, <Device ios-sw4 at 0x1142bab50>,
    # <Device ios-sw5 at 0x114633430>]

    # Connect and get version of each testbed device in parallel
    # Results must be a pickleable object. In this case, we are returning a dict
type
    version_results = pcall(get_version, device=tb_devices)

    # All results are stored in a tuple
    print(f"SHOW VERSION OUTPUT:\n {version_results}")
```

The **get_version** function is the target of pcall, with a list of testbed Device objects used as arguments. Since a single target is used, the number of arguments determines the number of parallel processes. There is a list of five Device objects provided to the **device** argument of the target, resulting in five total child processes running in parallel.

Now's the time you've been waiting for—what was the difference in performance between serial and parallel execution? Example 14-8 shows the different execution times captured by the Linux **time** command, which tracks how long a command takes to run.

Example 14-8   *Execution Times*

```
Serial Execution time: time python perf_testing_serial.py  3.81s user 9.10s
system 13% cpu 1:34.78 total

Parallel Execution time: time python perf_testing_pcall.py  9.74s user 12.54s
system 70% cpu 31.799 total
```

The serial execution took 1 minute and 34 seconds to execute, while the parallel execution using pcall only took 31 seconds. The parallel execution time using pcall was three times faster than the serial execution time! As I mentioned before, the execution could further be optimized by changing connection settings, but the goal is to show how much quicker execution times are when enabling parallelism.

One other number to point out from Example 14-8 is the CPU utilization percentage. You'll notice that the CPU was running at 70% utilization when executing the parallel tasks versus 13% utilization when executing the serial tasks. This is expected due to the nature of pcall spawning child processes for each parallel task. In this case, five processes were created for each Device object argument.

To wrap up the performance comparison, pcall was used to connect to, collect/parse **show** command output, and disconnect from five testbed devices in parallel. The parallelism provided by pcall led to the execution time being about one-third the time when executing the same instructions serially. With that being said, the tradeoff to running the tasks in parallel is an increase in CPU utilization due to the overhead required to run multiple child processes.

Summary

In this chapter, we dove into the interesting world of asynchronous (async) programming in Python. We started with looking at the different methods to achieve async programming in Python using asyncio, threading, and parallelism with the multiprocessing package. We then dove right into the Parallel Call (pcall) API found within the pyats.async_ module. Pcall is essentially an API built on the multiprocessing package from the Python standard library. We covered pcall usage, targets, error handling, logging, and the intricacies of the **Pcall** object itself. Lastly, we tested and compared the performance of running tasks in parallel versus serially. The results showed that parallelism achieves significantly quicker execution (three times faster than serial execution), with the tradeoff of requiring more CPU and memory resources.

pyATS Clean

Have you ever begun testing a new feature in a lab network topology, whether in your personal lab or at your workplace, and wished every device was in a stable state with a standardized base configuration and software image? Network testing can be a process when you must initialize the lab environment, including all the network devices, before you even begin testing. PyATS Clean is a feature in the pyATS library (Genie) that integrates the process of wiping a device, upgrading to a golden software image, and/ or applying a base configuration into your network testing. The "cleaning" process helps automate the exercise of initializing a network device and preparing it for testing. In this chapter, we are going to look at the process of device cleaning using pyATS Clean by reviewing the following topics:

- Getting started
- Clean YAML
- Clean execution
- Developing clean stages

Getting Started

Before we dive into pyATS Clean, let's start with why device cleaning is important and check what network device operating systems and platforms are supported by pyATS Clean. The genie.libs.clean module (pyATS Clean) is installed as part of the pyATS library (Genie) package, so if you have the latest pyATS and pyATS library (Genie) versions installed, you are good to go. If you would like access to the latest Genie libraries (parsers, APIs, triggers, and verifications), you should upgrade to the latest versions of pyATS using pip or the **pyats version update** command inside the Python virtual environment.

Once pyATS Clean is installed, you must have a testbed file. This should be a given, but in addition, you must have the operating system (OS) and platform defined for each device. The OS and platform values are used to determine whether the device is supported by pyATS Clean and which cleaners are applicable to the device (more on that later). Example 15-1 shows how a device should be defined in a testbed file for pyATS Clean.

Example 15-1   *Clean Testbed Devices*

```
devices:
    cat8k-rt1:
        type: router
        os: iosxe
platform: c8kv
        custom:
            abstraction:
                order: [os, platform]

        credentials:
            default:
                username: cisco
                password: cisco
        connections:
            defaults:
                class: unicon.Unicon
```

Device Cleaning

Device cleaning is the concept of initializing a device with a specific software image and base configuration, regardless of the device's current state or configuration. The purpose of the pyATS Clean framework is to essentially reset a device, regardless of its current state, and start with a "clean slate." Cleaning a device can be helpful in many scenarios, including the following:

- Automated regression/sanity testing during software upgrades

- To recover devices and bring them back to an operational state

- To re-initialize a device with a standard base configuration and/or software image

- To remove unwanted files on a device

The pyATS Clean framework is broken down into stages. Clean stages are used to organize cleaning instructions, such as connecting a device, manipulating device configuration, or rebooting a device. Dividing the workflow into smaller stages allows for easier debugging when failures occur and modularity to allow easier modification of any step/stage for different OS/platform types. Clean stages will be further broken down in the Clean YAML section of the chapter.

To summarize, pyATS Clean is meant to help automatically reset network devices that may not be in an ideal operating state. It's not meant to assist with configuration drift or check for configuration compliance. It's simply meant to reset network devices and create a foundation for additional test automation. Now that we understand the purpose of device cleaning, let's see what types of devices are supported by pyATS Clean.

Supported Devices

Two main types of devices are supported by pyATS Clean: network devices and power cyclers. Power cyclers are out-of-band (OOB) devices that can power on/off and reboot network devices, such as power distribution units (PDUs) and even VMWare ESXi for virtual machines! In the following sections, we are going to break down which device types and platforms are supported by pyATS Clean.

Network Platforms

To use pyATS Clean, testbed devices must be a supported product, which is a combination of a supported OS and platform. Table 15-1 shows the currently supported OS/platforms as of version 23.11.

Table 15-1 *Supported OS/Platforms (v23.11)*

Product	OS	Platform
ASR 1000	iosxe	
ASR 1000v	iosxe	
ISR	iosxe	
Catalyst 9000	iosxe	cat9k
Catalyst 8000v	iosxe	c8kv
Nexus 7000	nxos	n7k
Nexus 9000	nxos	n9k
Nexus 9000v	nxos	n9k
Nexus 9000 (aci mode)	nxos	aci
NCS 5500	iosxr	
ASR 9000 x64	iosxr	
ASR 9000 px	iosxr	
APIC	apic	
Catalyst WS-C3560CX	ios	cat3k

Power Cyclers

Along with specific network products supported, a list of power cyclers is supported by pyATS Clean. Power cyclers are essentially your common OOB, PDU, and power management devices. Supported vendors include Raritan, APC, Dualcomm, Cyberswitching, and even ESXi for virtual machines! Power cyclers allow a device to auto-recover from a hung state or when connectivity to the device is lost. This is a huge feature! The device recovery process is automatically activated when connecting to a device fails or an exception is raised while a device is being upgraded. The device recovery process includes the following steps:

1. Connect to the power cycler and reboot the device.

2. Connect to the device and break the boot sequence to bring up the ROMMON prompt.

3. Boot the device with the "golden image" provided as part of the Clean YAML file.

4. Pass control back to pyATS Clean and continue the cleaning process.

To use a power cycler during the device cleaning process, you must define the power cycler in your testbed file. Table 15-2 shows a list of the supported power cycler types and their associated testbed schemas.

Table 15-2 *Supported Power Cyclers*

Power Cycler	Testbed Schema/Example
raritan-px	Testbed Schema

```
-------------

devices:
  <device>:
    peripherals:
      power_cycler:
        type: raritan-px
        connection_type: snmp
        host (str): Ip address for Powercycler.
        outlets (list): Power ports associated with your device.
        read_community (str, optional): 'private' or 'public'.
            Defaults to 'public'.
        write_community (str, optional): 'private' or 'public'.
            Defaults to 'private'.
```

Testbed Example

```
-------------

devices:
  PE1:
    peripherals:
      power_cycler:
        - type: raritan-px
          connection_type: snmp
          host: 127.0.0.1
          outlets: [20]
```

Power Cycler	Testbed Schema/Example

raritan-px2 (snmp)	**Testbed Schema**
	
	`-------------`
	
	```devices:```
	``` <device>:```
	``` peripherals:```
	``` power_cycler:```
	``` - type: raritan-px2```
	``` connection_type: snmp```
	``` host (str): Ip address for Powercycler.```
	``` outlets (list): Power ports associated with your```
	```device.```
	``` read_community (str, optional): 'private' or 'public'.```
	``` Defaults to 'public'.```
	``` write_community (str, optional): 'private' or```
	```'public'.```
	``` Defaults to 'private'.```
	
	Testbed Example
	
	`--------------`
	
	```devices:```
	``` PE1:```
	``` peripherals:```
	``` power_cycler:```
	``` - type: raritan-px2```
	``` connection_type: snmp```
	``` host: 127.0.0.1```
	``` outlets: [20]```

raritan-px2 (snmpv3)	**Testbed Schema**
	
	`-------------`
	
	```devices:```
	``` <device>:```
	``` peripherals:```
	``` power_cycler:```
	``` - type: raritan-px2```
	``` connection_type: snmpv3```
	``` host (str): Ip address for Powercycler.```
	``` outlets (list): Power ports associated with your```
	```device.```

Power Cycler	Testbed Schema/Example

```
                    username (str): username for Powercycler.
                    auth_key (str): authentication password.
                    auth_protocol (str): authentication protocol.
                    priv_key (str): private protocol password.
                    priv_protocol (str): private protocol type.
                    security_level (str): Different security levels.

              Snmpv3 supports three security levels:
                1. AuthPriv (Authentication and privacy)
                2. AuthNoPriv (Authentication)
                3. NoAuthNoPriv (None)

              Snmpv3 supported authentication protocols:
                'md5', 'sha', 'sha224', 'sha256, 'sha384', 'sha512'

              Snmpv3 supported private protocols:
                'des', '3des', 'aes128', 'aes192', 'aes256'


          Testbed Example
          --------------

          Type 1: (AuthPriv)

          devices:
              PE1:
                peripherals:
                  power_cycler:
                      type: raritan-px2
                      connection_type: snmpv3
                      host: pdu_host
                      outlets: [15]
                      username: test_user
                      auth_key: ****
                      auth_protocol: md5
                      priv_key: ****
                      priv_protocol: aes128
                      security_level: authpriv

          Type 2: (AuthNoPriv)

          devices:
            PE1:
              peripherals:
                power_cycler:
```

Power Cycler	Testbed Schema/Example
| | ```
type: raritan-px2
connection_type: snmpv3
host: pdu_host
outlets: [15]
username: test_user
auth_key: ****
auth_protocol: md5
security_level: authnopriv


Type 3: (NoAuthNoPriv)


devices:
  PE1:
    peripherals:
      power_cycler:
          type: raritan-px2
          connection_type: snmpv3
          host: pdu_host
          outlets: [15]
          username: test_user
          security_level: noauthnopriv
``` |
| generic-cli | Testbed Schema

```
devices:
  <device>:
    peripherals:
      power_cycler:
        - type: generic-cli
          host (str): Ip address for Powercycler.
          connection_type: ssh
          outlets (list, optional): Power ports associated with
your device.
          commands (dict):
              power_on (str): Command to power on the
Powercycler
              power_off (str): Command to power off the
Powercycler
```


Description

The **commands** argument takes in any **power_on** and **power_off** commands, which are mandatory. |
```

Power Cycler	Testbed Schema/Example
	Example: 1 (if outlets are used)

These commands should have outlet string on it, if the power cycle is based on outlet. Example commands:

```
power_on: "power outlets {outlet} on"
power_off: "power outlets {outlet} off"
```

It is mandatory to specify {outlet} as this string format.

Example: 2 (if device names are used)

If the device name is used to powercycle, refer to the following example commands:

```
power_on: "power-tool %{self} on"
power_off: "power-tool %{self} off"
```

Here, %{self} takes the device name from the testbed.

Testbed Example

```
devices:
  PE1:
    peripherals:
      power_cycler:
        - type: generic-cli
          host: 127.0.0.1
          connection_type: ssh
          outlets: [6]
          commands:
              power_on: "power outlets {outlet} on"
              power_off: "power outlets {outlet} off"
```

Raritan	Testbed Schema

```
devices:
  <device>:
    peripherals:
      power_cycler:
        - type: Raritan
          host (str): Ip address for Powercycler.
          connection_type: ssh
          outlets (list): Power ports associated with your
device.
```

Power Cycler	**Testbed Schema/Example**
	Description

Description

The **power_on** and **power_off** commands for Raritan are added by default. The user needs to pass the outlets.

Testbed Example

```
devices:
  PE1:
    peripherals:
      power_cycler:
          - type: Raritan
            host: 127.0.0.1
            connection_type: telnet
            outlets: [7]
```

APC

Testbed Schema

```
devices:
  <device>:
    peripherals:
      power_cycler:
          - type: apc
            connection_type: snmp
            host (str): Ip address for Powercycler.
            outlets (list): Power ports associated with your
device.
            read_community (str, optional): 'private' or 'public'.
                Defaults to 'public'.
            write_community (str, optional): 'private' or 'public'.
                Defaults to 'private'.
```

Testbed Example

```
devices:
  PE1:
    peripherals:
      power_cycler:
          - type: apc
            connection_type: snmp
            host: 127.0.0.1
            outlets: [20]
```

Power Cycler	Testbed Schema/Example

apc-rpdu	**Testbed Schema**
	
	`--------------`
	
	``
	devices:
	  <device>:
	    peripherals:
	      power_cycler:
	        - type: apc-rpdu
	          connection_type: snmp
	          host (str): Ip address for Powercycler.
	          outlets (list): Power ports associated with your
	device.
	          read_community (str, optional): 'private' or 'public'.
	              Defaults to 'public'.
	          write_community (str, optional): 'private' or
	'public'.
	              Defaults to 'private'.
	``
	
	Testbed Example
	
	`---------------`
	
	``
	devices:
	  PE1:
	    peripherals:
	      power_cycler:
	        - type: apc-rpdu
	          connection_type: snmp
	          host: 127.0.0.1
	          outlets: [20]
	``

Dualcomm	**Testbed Schema**
	
	`--------------`
	
	``
	devices:
	  <device>:
	    peripherals:
	      power_cycler:
	        - type: dualcomm
	          connection_type: snmp
	          host (str): Ip address for Powercycler.
	          outlets (list): Power ports associated with your
	device.
	          read_community (str, optional): 'private' or 'public'.
	              Defaults to 'public'.
	``

Power Cycler	Testbed Schema/Example

```
                write_community (str, optional): 'private' or
'public'.
                    Defaults to 'private'.


Testbed Example

--------------

devices:
  PE1:
    peripherals:
      power_cycler:
        - type: dualcomm
          connection_type: snmp
          host: 127.0.0.1
          outlets: [20]
```

Cyberswitching

```
Testbed Schema

-------------

devices:
  <device>:
    peripherals:
      power_cycler:
        - type: cyberswitching
          connection_type: telnet
          host (str): Cyberswitching device from Testbed YAML.
          outlets (list): Lines associated with your device.


Testbed Example

--------------

devices:
  PE1:
   peripherals:
      power_cycler:
        - type: cyberswitching
          connection_type: telnet
          host: my-cyberswitching
          outlets: [20]
  my-cyberswitching:
    # Fill out the rest of this device as normal
    # such as connection info, credentials, etc.
```

Power Cycler	Testbed Schema/Example
ESXi	Testbed Schema

```
--------------

devices:
  <device>:
    peripherals:
      power_cycler:
        - type: esxi
          connection_type: ssh
          host (str): ESXi device from Testbed YAML.
          outlets (list): VM IDs associated with your device.
```

Testbed Example

```
---------------

devices:
  PE1:
    peripherals:
      power_cycler:
        - type: esxi
          connection_type: ssh
          host: my-esxi
          outlets: [20]

  my-esxi:
    # Fill out the rest of this device as normal
    # such as connection info, credentials, etc.
```

Given the list of supported power cyclers, let's define an APC rack PDU (apc-rpdu) as a power cycler that's connected to the cat8k-rt1 testbed device from Example 15-1. Example 15-2 shows an updated testbed file that includes a power cycler with the cat8-rt1k device.

Example 15-2 *Testbed with Power Cycler*

```
devices:
    cat8k-rt1:
        peripherals:
            power_cycler:
                - type: apc-rpdu
                  connect_type: snmp
                  host: 10.1.1.1
                  outlets: [20]
```

You may have more than one power cycler connected to your device (for example, two power supplies in your network device). Example 15-3 shows how two APC RPDU power cyclers are connected to the cat8k-rt1 device.

Example 15-3   *Testbed with Two Power Cyclers*

```
devices:
    cat8k-rt1:
        peripherals:
            power_cycler:
                - type: apc-rpdu
                  connect_type: snmp
                  host: 10.1.1.1
                  outlets: [20]
                - type: apc-rpdu
                  connect_type: snmp
                  host: 10.1.1.2
                  outlets: [21]
```

Now that you understand the device types supported and how they need to be defined in a testbed file, let's see how we can define a pyATS Clean YAML file that provides instructions to pyATS Clean.

Clean YAML

The pyATS Clean framework is built around a modular design that is driven by YAML. The goal of pyATS Clean is to break down the device cleaning process into smaller stages. The Clean YAML file has multiple components, including devices and clean stages, which we will dive deeper into later in this section. To introduce the Clean YAML file, let's look at how we would start one. We must first define where the cleaner module and the cleaner class are located. You can think of this section as an **import** statement in Python. Example 15-4 shows how we can define the cleaner module and class to begin writing the Clean YAML file.

Example 15-4   *Clean YAML – Cleaner Module and Class*

```
cleaners:
    # This means to use the cleaner class `PyatsDeviceClean`
    PyatsDeviceClean:
        # The module is where the cleaner class above can be found
        module: genie.libs.clean
```

This pyATS cleaner class is used to discover and run the clean stages within a Clean YAML file. You can find more information about the **PyatsDeviceClean** class by reviewing the source code in the genie.libs.clean module in the genielibs library package on GitHub:

https://github.com/CiscoTestAutomation/genielibs/blob/master/pkgs/clean-pkg/src/genie/libs/clean/clean.py

Devices

The next section in the Clean YAML file is the **devices** block. The **devices** block is optional, but it's commonly used because it allows you to customize the cleaning on a per-device basis. The **devices** block is responsible for declaring what images to load, what clean stages to run, and the order to run the clean stages on a per-device basis. This may be a given, but the devices must exist in your testbed file. There are two locations to define the devices you would like to clean. You must define a list of devices in the **cleaners** block, from the previous section, and in a new top-level key named **devices** that contains the devices listed under the **cleaners** block. Example 15-5 shows how to add two devices in the Clean YAML file.

Example 15-5 *Clean YAML – Devices*

```
cleaners:
    PyatsDeviceClean:
        module: genie.libs.clean
        devices: [PE1, PE2]

devices:
    PE1:

    PE2:
```

By defining the devices in the **cleaners** block, we are instructing pyATS Clean as to which devices need to be cleaned. The details on how to clean each device are under the device name key in the **devices** block. The **devices** block is used to provide the clean stages to run and the order in which to run them. Currently, each device is blank without clean stages or an execution order. Let's change that and move on to discuss a core concept of pyATS Clean—clean stages.

Clean Stages

Clean stages are the foundation to pyATS Clean and provide a modular structure in which stages are organized by OS and platform. You can think of clean stages as small steps in the cleaning process. The pyATS library (Genie) comes with predefined clean stages, which can be viewed in the online Clean Stages browser (https://pubhub.devnetcloud.com/media/genie-feature-browser/docs/#/clean). The Clean Stages browser is the same web interface browser that provides available device APIs, parsers, models, triggers, and verifications in the pyATS library (Genie). Figure 15-1 shows the Clean Stages browser and some of the available clean stages.

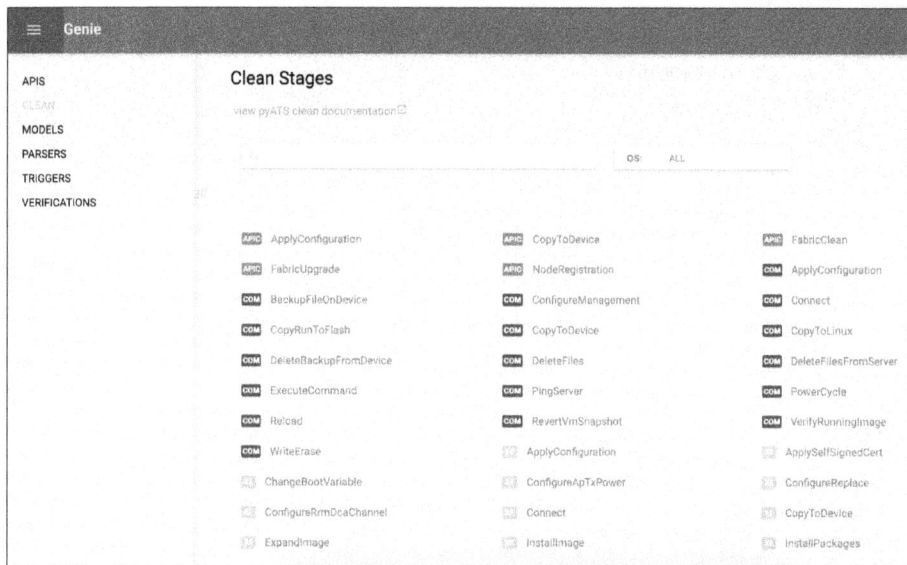

Figure 15-1 *Clean Stages Browser*

In the Clean Stages browser, you can filter down the available clean stages to a specific OS. The list of OSs is much smaller than the OS list for parsers, but there are clean stages for most Cisco products, including APIC, IOS XE, IOS-XR, NX-OS, and even AIRE-OS! There is also one category in the OS list called "COM," which stands for "common" and applies to all device OSs and platforms. For example, the "connect" clean stage, which is used to connect to a device, is considered a common (COM) clean stage. Many of the common clean stages are made possible due to using the Unicon library on the backend for all device interactions. The complexity of the different connection options, prompt handling, and so on is abstracted from the user. Remember, Unicon uses information available in the testbed file to determine how to connect to a device, using the defined connection types under each device.

Once you find a clean stage you're interested in using in your clean process, you can select it in the Clean Stages browser to see the different arguments available. Figure 15-2 shows the common "ExecuteCommand" clean stage, which allows you to execute an arbitrary list of commands.

You can see there are four arguments with this clean stage: **via**, **commands**, **execute_ timeout**, and **sleep_time**. The **via** argument allows you to specify the connection to use from the testbed to connect to the device. The default connection will be used if a connection is not specified in the **via** argument. The **commands** argument accepts a list of commands to run on the device. The **execute_timeout** argument is the maximum time (in seconds) allowed for each command to execute. The default timeout is 60 seconds. The last argument, **sleep_time**, is the amount of time (in seconds) to sleep after running each command. The default sleep time is 10 seconds. Adding to our previous Clean YAML file example, we will add two stages: Connect and ExecuteCommand (see Example 15-6). Ultimately, we will connect to each device and run the **show version** command.

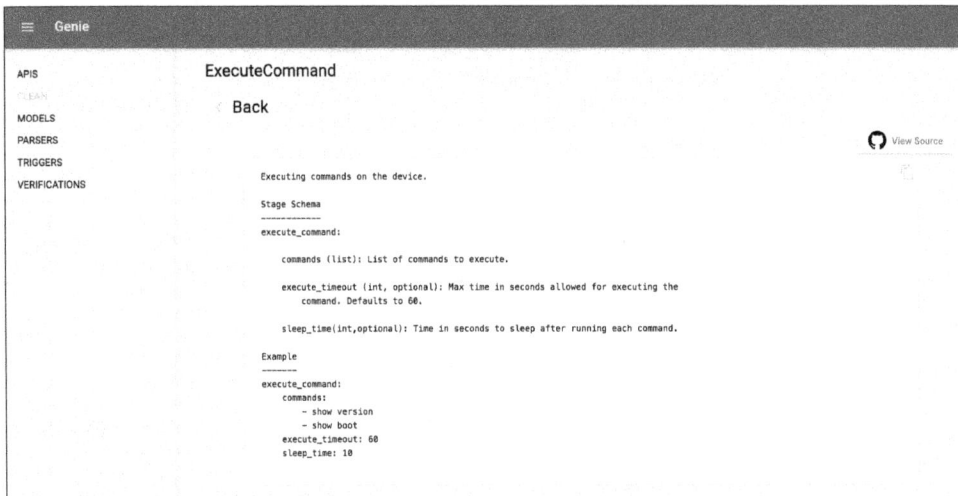

Figure 15-2 *ExecuteCommand Clean Stage*

Example 15-6 *Clean YAML – Clean Stages*

```
cleaners:
    PyatsDeviceClean:
        module: genie.libs.clean
        devices: [PE1, PE2]


devices:
    PE1:
        connect:
            timeout: 30

        execute_command:
            commands:
                - show version
            execute_timeout: 60
            sleep_time: 10


    PE2:
        connect:
            timeout: 30

        execute_command:
            commands:
                - show version
            execute_timeout: 60
            sleep_time: 10
```

Once we specify the clean stages and the argument values for each stage under each device, we need to specify the order in which the clean stages execute. We can do this by specifying an **order** key at the end of each **device** block. Example 15-7 adds an **order** key at the same level as the stages.

Example 15-7   *Clean YAML – Stage Execution Order*

```
cleaners:
    PyatsDeviceClean:
        module: genie.libs.clean
        devices: [PE1, PE2]

devices:
    PE1:
        connect:
            timeout: 30

        execute_command:
            commands:
                - show version
            execute_timeout: 60
            sleep_time: 10

        order:
        - connect
        - execute_command

    PE2:
        connect:
            timeout: 30

        execute_command:
            commands:
                - show version
            execute_timeout: 60
            sleep_time: 10

        order:
        - connect
        - execute_command
```

We've now defined devices, clean stages for each device, and the order in which each stage must execute. In the next section we will look at how we can perform clean operations on a group of devices, instead of having to specify each device separately.

Device Groups

Device groups allow you to group devices and perform the same clean operations against them. The purpose of device groups is to remove duplicate clean stages and arguments across multiple devices and define all the clean operations for a set of similar devices under a device group. Device groups are defined under the **groups** block. The **groups** block is optional in your Clean YAML file. Devices can be grouped by OS, platform, or an arbitrary list of devices that you deem similar. Only one group method is supported by device groups. Example 15-8 shows each of the different device grouping methods: by OS, platform, and an arbitrary list of devices.

Example 15-8 *Clean YAML – Device Groups*

```
cleaners:
    PyatsDeviceClean:
        module: genie.libs.clean
        # Like devices, device groups must be defined in the cleaners block
        # to be cleaned. Any group not defined here will not be cleaned.
        groups: [group_by_device, group_by_platform, group_by_os]

groups:
    group_by_device:
        # Arbitrary list of testbed devices
        # They do not need to be the same platform or have the same OS
        devices:
            - PE1
            - PE2

    group_by_platform:
        # More than one platform can be added to a platform group
        # We are including all Cat8kv and Cat9k testbed devices in this group
        platforms:
            - c8kv  # Only cat8kv devices
            - cat9k # Only cat9k devices

    group_by_os:
        # More than one OS can be added to a platform group
        # We are only including iosxe testbed devices in this group
        os:
            - iosxe
```

Now that we have three different device groups defined using each of the available grouping methods, let's see how we can add clean stages to execute against each device group. Much like with devices, you add each clean stage as a key at the same level as the list of devices is defined. Along with defining each clean stage and its associated arguments, the **order** key must also be defined to instruct how the clean stages should be executed. Example 15-9 shows two clean stages, connect and execute_command, and the execution order added to each device group.

Example 15-9   *Clean YAML – Device Group Clean Stages*

```
cleaners:
   PyatsDeviceClean:
      module: genie.libs.clean
 # Like devices, device groups must be defined in the cleaners block
 # to be cleaned. Any group not defined here will not be cleaned.
      groups: [group_by_device, group_by_platform, group_by_os]

groups:
   group_by_device:
      # Arbitrary list of testbed devices
      # They do not need to be the same platform or have the same OS
      devices:
         - PE1
         - PE2

      connect:

      execute_command:
         commands:
            - show version

      order:
         - connect
         - execute_command

   group_by_platform:
      # More than one platform can be added to a platform group
      # We are including all Cat8kv and Cat9k testbed devices in this group
      platforms:
         - c8kv  # Only cat8kv devices
         - cat9k # Only cat9k devices

      execute_command:
         commands:
            - show version

      order:
         - connect
         - execute_command
```

```
    group_by_os:
        # More than one OS can be added to a platform group
        # We are only including iosxe testbed devices in this group
        os:
            - iosxe

        execute_command:
            commands:
                - show version

        order:
            - connect
            - execute_command
```

The last key piece of information for device groups is the ability to override any clean stage operation for a specific device. To override a clean stage for a specific device in a device group, define the device with the clean stage under the **devices** block. Example 15-10 shows how we can override the command to execute on the PE1 device in the execute_command clean stage.

Example 15-10   *Clean YAML – Device Group Override*

```
cleaners:
    PyatsDeviceClean:
        module: genie.libs.clean
    # Other groups removed for brevity
        groups: [group_by_device]

groups:
    group_by_device:
        # Arbitrary list of testbed devices
        # They do not need to be the same platform or have the same OS
        devices:
            - PE1
            - PE2

        connect:

        execute_command:
            commands:
                - show version

        order:
            - connect
            - execute_command
```

```
devices:
    PE1:
        execute_command:
            commands:
                - show running-config
```

Instead of executing the **show version** command on PE1, we are going to execute the **show running-config** command. You may notice that the **order** key was not specified. This is due to it already being defined under the device group, and we are not overriding the order of execution, only the execute_command clean stage.

Useful Features

Up to this point, we've gone over the concepts of pyATS Clean and the different components that make up the pyATS Clean YAML file. In this section, we are going to take a deeper look into two useful features that enhance the pyATS Clean experience. The first feature we are going to dive into is software image management. One of the top use cases for network testing is validating that a certain version of software runs and operates as expected. The goal of testing software images is to ensure there are no bugs that pop up when running certain features or performing certain operations on a device. It's in no way foolproof, but it gives you a sense of confidence if you can confirm all the features running in your network operate as expected when testing a new version of software. The second useful feature is device recovery. Device recovery is specific to testing new features or functionality on network devices. During testing, you may run the risk of a device malfunctioning or losing connectivity to a device during testing. The device recovery process helps automate the recovery of the device, including applying a golden image and base configuration. Let's dive into each feature!

Software Image Management

During testing, you may want to declare a single software image to be used across all clean stages or try a different software version across different clean stages. The pyATS clean framework has made it easy to define a software image in one location and use it throughout the Clean YAML file. Once a software image is defined, you have the ability to override the software image in a specific clean stage. This can be helpful if you want to use multiple software images in the same Clean YAML file.

Instead of defining the image in each clean stage that requires it, we only need to reference the image once with the **images** key under the device. Example 15-11 shows how we can reference one software image for the PE1 device, which can be used across all the clean stages executed on the PE1 device.

Example 15-11   *Clean YAML – Software Image Management*

```
cleaners:
    PyatsDeviceClean:
        module: genie.libs.clean
        devices: [PE1]

devices:

    PE1:

        images:
        - /path/to/image.bin

        connect:

        copy_to_linux:
            destination:
                directory: /tftp-server
                hostname: 127.0.0.1
            unique_number: 12345

        copy_to_device:
            origin:
                hostname: 127.0.0.1
            destination:
                directory: 'bootflash:'

        change_boot_variable:

        reload:

        verify_running_image:

        order:
        - connect
        - copy_to_linux
        - copy_to_device
        - change_boot_variable
        - reload
        - verify_running_image
```

The advantage of using the **images** key is that you no longer need to specify the image name as an argument to any clean stage that requires it. For example, the last clean stage in the Clean YAML file, verify_running_image, requires an **images** argument that accepts a list of image(s) that should be running on the device. Since we provided the separate **images** key, we no longer need to provide that **image** argument to the verify_running_image clean stage. There are a few other clean stages that require an image path as an argument, but the concept stays the same.

Each platform requires a slightly different structure to define the software images under the **images** key. Refer to the pyATS library documentation for the latest **images** key structure for the appropriate OS and platform.

In the event the image needs to be overridden in a specific clean stage, you need to add a few extra keys. By default, if an **images** key is defined and an **image** argument provided in a clean stage, then the image provided by the **images** key will be used. It's a little backwards from what's expected. Many may expect the **image** argument provided directly to the clean stage would override the "global" **images** key, but that's not the case. In order for the clean stage argument to override the **images** key, you must add a new key under the device called **image_management** with a child key called **override_stage_images** set to False. Example 15-12 shows how a Clean YAML file would look with the override settings.

Example 15-12   *Clean YAML – Image Override*

```
cleaners:
    PyatsDeviceClean:
        module: genie.libs.clean
        devices: [PE1]

devices:

    PE1:

        image_management:
          override_stage_images: False

        images:
        - /path/to/image_56789.bin

        change_boot_variable:
            images:
            - bootflash:/image_12345.bin
```

In Example 15-12, image_12345.bin would be used instead of image_56789.bin since the **image_management** configuration settings are provided. It's good to note that the software image override configuration is only available for IOS XE devices (as of version 23.11).

Device Recovery

Have you ever begun network testing and made a change that cut off connectivity to your device? Or have you ever had a software upgrade fail? These scenarios are more common when cleaning a device due to the number of changes being reverted to reset the device back to a baseline state. The pyATS Clean framework accounts for these device malfunctions and failures and can assist in recovering devices.

The device recovery feature can clear the terminal/console line to a device and use a power cycler to reboot the device. In addition, once the device is back up, a golden image will be loaded on the device and a base configuration can optionally be applied. To summarize, the device recovery feature can recover a device and reset it to a usable state, regardless of its current state.

You may wonder, "how can this device recovery feature be invoked if we don't know when the device failure occurs?" The device recovery feature is automatically invoked after each clean stage is executed and will determine whether to continue or reset a device. Figure 15-3 shows the flow of device recovery and when it's invoked.

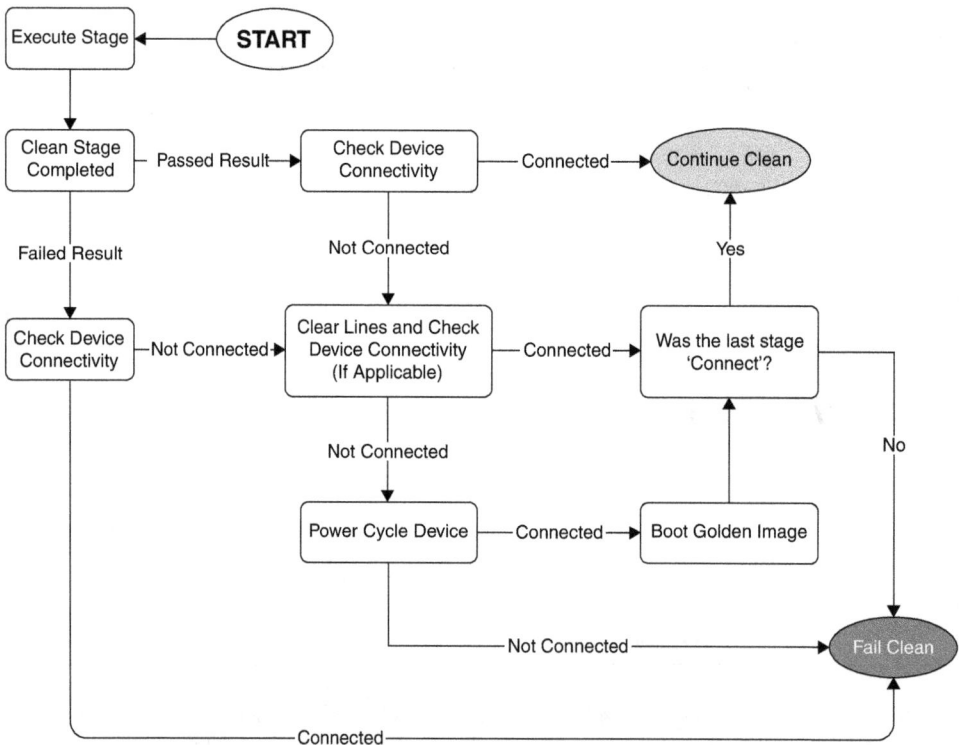

Figure 15-3 *Device Recovery Diagram*

Device recovery requires a golden image be configured for each device in your Clean YAML file. The golden image can be on the device or stored on an accessible TFTP server. Once a golden image is defined, a terminal server and power cycler must be defined to allow the device to recover from a malfunctioning or unreachable state. Example 15-13 shows a terminal server and power cycler added to the PE1 device in the Clean YAML file.

Example 15-13 *Device Recovery – Terminal Server and Power Cycler*

```
devices:
  PE1:
    peripherals:
      power_cycler:
        - type: dualcomm
          connect_type: snmp
          host: 127.0.0.1
          outlets: [22]
      terminal_server:
        # <terminal server device>: <list of lines for the device>
        # In this case, clearing line 22 would clear the line
        # connected to 'PE1'
        my-terminal-server: [22]

  my-terminal-server:
    # Fill out the rest of this device as normal
    # such as connection info, credentials, etc.
```

Once a terminal server and power cycler are defined under the device, you must add a **device_recovery** block to define the terminal server, power cycler, and golden image to use for the device recovery process. Example 15-14 shows the **device_recovery** block defined under the PE1 device. Remember, the terminal server and power cycler have already been defined.

Example 15-14 *Device Recovery Block*

```
devices:
  PE1:
    device_recovery:
      grub_activity_pattern: '.*The highlighted entry will be executed automatically in.*'
      timeout: 600
      powercycler_delay: 5
      golden_image:
        - 'GOLDEN IMAGE'
```

```
    peripherals:
      power_cycler:
        - type: dualcomm
          connect_type: snmp
          host: 127.0.0.1
          outlets: [22]
      terminal_server:
        # <terminal server device>: <list of lines for the device>
        # In this case, clearing line 10 would clear the line
        # connected to 'PE1'
        my-terminal-server: [10]

my-terminal-server:
  # Fill out the rest of this device as normal
  # such as connection info, credentials, etc.
```

You'll notice there are four keys defined: **grub_activity_pattern, timeout, powercycler_ delay**, and **golden_image**. There are some additional keys available in the device_recovery schema, including the console patterns to look for to break the boot process, defining a recovery password, and a reconnect delay. Refer to the pyATS Clean documentation for complete device recovery schema:

https://pubhub.devnetcloud.com/media/genie-docs/docs/clean/user_guide/writing_a_ clean/device_recovery.html#how-to-enable-device-recovery.

Given the complete example, if the PE1 device were to malfunction or become unreachable, the device recovery process would be invoked and start by clearing line 10 on the terminal server, rebooting the device by cycling power outlet 22 on the Dualcomm PDU, and finally loading the **'GOLDEN IMAGE'** on the device upon booting.

If we wanted to apply a base configuration once the device boots, we would need to add the **post_recovery_configuration** key and provide the base configuration as a string.

Clean Execution

Now that we understand how to write Clean YAML files, let's see how to validate a Clean YAML file and review the different methods to execute a device cleaning, including as part of a pyATS job run, by itself, or as part of a pyATS testscript.

Clean Validation

Clean YAML files must adhere to a defined Clean YAML schema. The pyATS Clean framework provides a command-line utility to validate any Clean YAML file. Example 15-15 shows how to validate a Clean YAML file named clean.yaml

Example 15-15   *Clean File Validation*

```
pyats validate clean --testbed-file testbed.yaml --clean-file clean.yaml
```

Execution Methods

PyATS Clean requires a testbed YAML file and a Clean YAML file to be present in order to execute device cleaning. Two methods are available for executing pyATS Clean: Integrated and Standalone.

Integrated

The Integrated method is when pyATS Clean is run as part of a pyATS job. When running as part of a pyATS job, pyATS Clean ensures the testbed devices are in a "good" state before testing. A "good" state means the device is reachable, running the correct software image, and has the expected configuration. Example 15-16 shows how to include pyATS Clean as part of running a pyATS job. Note that the **--clean-file** and **--invoke-clean** flags were included with pyATS **job** command.

Example 15-16   *pyATS Clean – Integrated Execution*

```
pyats run job </path/to/job.py> --testbed-file </path/to/testbed.yaml>
--clean-file </path/to/clean.yaml> --invoke-clean
```

Standalone

The pyATS Clean process can also be executed without a pyATS job. The **pyats clean** command and its associated flags can be used independently to only clean a testbed device, without any additional testing. You only need to specify the testbed and Clean YAML files. Example 15-17 shows how to use the **pyats clean** command to only clean a testbed device.

Example 15-17   *pyATS Clean – Standalone Execution*

```
pyats clean --testbed-file </path/to/testbed.yaml>
--clean-file </path/to/clean.yaml>
```

Device images can be passed as a command-line argument to pyATS Clean. CLI arguments override any images specified in the Clean YAML file and can be applied to a single device or device group. As a reminder, devices can be grouped by OS, platform, or user-defined group. Example 15-18 shows the various command-line arguments available to pass a device image to pyATS Clean.

Example 15-18 *pyATS Clean – Device Image CLI Arguments*

```
# Example of passing an image to a device called 'PE1'
pyats clean --clean-device-image PE1:</path/to/image.bin>
--testbed-file </path/to/testbed.yaml> --clean-file </path/to/clean.yaml>


# Example of passing an image to all devices with the 'nxos' os
pyats clean --clean-os-image nxos:</path/to/image.bin>
--testbed-file </path/to/testbed.yaml> --clean-file </path/to/clean.yaml>


# Example of passing an image to all devices belonging to a group called 'group1'
pyats clean --clean-group-image group1:</path/to/image.bin>
--testbed-file </path/to/testbed.yaml> --clean-file </path/to/clean.yaml>


# Example of passing an image to all devices with the 'n9k' platform
pyats clean --clean-platform-image n9k:</path/to/image.bin>
--testbed-file </path/to/testbed.yaml> --clean-file </path/to/clean.yaml>
```

Clean Logging

PyATS Clean logs can be viewed using the pyATS log viewer, which displays the results in an HTML web page. You can view the logs once the device cleaning has finished by using the **pyats logs view** command. The pyATS log viewer is the same logging utility that displays pyATS testscript results in the browser. You can also see a brief overview in real time using the **–liveview** option with the **pyats logs view** command. Figure 15-4 shows a screenshot of the pyATS log viewer web page.

pyATS Testscript Usage

Each testbed device object has an API attribute that exposes the many device APIs available in the pyATS library (Genie). The clean attribute is a sub-attribute of the device APIs and is accessible via device.api.clean. The clean API exposes every clean stage available in the Clean Stages browser using the following syntax:

device.api.clean.<*clean_stage*>(*clean_arguments*)

Clean stage arguments are passed to the clean stage as traditional arguments are passed to a Python function. The easiest way to understand the translation is by example. Figure 15-5 shows a snapshot of the Cisco IOS XE Install Image clean stage docstring.

Log Viewer

Overview   Results   Files

KleenexPlugin: 1 / 0   genie_testscript.py: 1 / 0

search

Section	Result ⇅	Start Time	Run Time
▾ asr-MIB-1	PASSED	10:43:29	19 minutes
▾ CleanTestcase	PASSED	10:43:29	19 minutes
▾ connect	PASSED	10:43:32	
▾ device recovery		10:43:36	
▾ ping_server	PASSED	10:43:40	
▾ Checking connectivity betwee...	PASSED	10:43:41	0 minutes
▾ device recovery		10:43:41	0 minutes
▾ copy_to_linux	PASSED	10:43:42	
▾ Collecting file info on origin an...	PASSED	10:43:42	0 minutes
▾ Collecting '/auto/tftp-best/...	PASSED	10:43:42	0 minutes
▾ Check if there is enough spac...	PASSED	10:43:42	0 minutes
▾ Copying the files to /ws/lrashe...	PASSED	10:43:42	
▾ Copying '/auto/tftp-best/g...	PASSED	10:43:42	
▾ Verify the files have been copi...	PASSED	10:43:56	0 minutes
▾ device recovery		10:43:56	0 minutes

Figure 15-4 *pyATS Log Viewer*

```
This stage installs a provided image onto the device using the install
CLI. It also handles the automatic reloading of your device after the
install is complete.

Stage Schema
------------
install_image:
    images (list): Image to install

    save_system_config (bool, optional): Whether or not to save the system
        config if it was modified. Defaults to False.

    install_timeout (int, optional): Maximum time in seconds to wait for install
        process to finish. Defaults to 500.

    reload_timeout (int, optional): Maximum time in seconds to wait for reload
        process to finish. Defaults to 800.

Example
-------
install_image:
    images:
      - /auto/some-location/that-this/image/stay-isr-image.bin
    save_system_config: True
    install_timeout: 1000
    reload_timeout: 1000
```

Figure 15-5 *pyATS Clean: IOS XE Install Image Stage*

The key-value pairs in the stage's schema can be passed in as keyword arguments to the device clean API. Example 15-19 shows how to call the clean stage in a pyATS testscript.

Example 15-19   *pyATS Clean – Device Clean API in Testscript*

```
from pyats import aetest


class MyTestcase(aetest.Testcase):

    @aetest.test
    def my_test(self, steps, testbed):
        device = testbed.devices['uut']
        device.connect()

        device.api.clean.install_image(
            images=['bootflash:/image.bin'],
            save_system_config=True)
```

Each step of the clean stage will be displayed in the pyATS testscript logs. For the Install Image clean stage, you'll see individual steps for deleting all boot variables, configuring the boot variable for "install mode," saving the **running-config**, verifying the next reload boot variables are set correctly, and finally installing the software image. Example 15-20 shows pyATS testscript logs written to standard output (stdout) that include the described clean stage steps.

Example 15-20   *pyATS Testscript Logs*

```
%EASYPY-INFO: Starting task execution: Task-1
%EASYPY-INFO:     test harness = pyats.aetest
%EASYPY-INFO:     testscript   = /Users/pyATS/testscript.py
%AETEST-INFO: +----------------------------------------------------------------+
%AETEST-INFO: |                        Starting testcase Test                  |
%AETEST-INFO: +----------------------------------------------------------------+
%AETEST-INFO: +----------------------------------------------------------------+
%AETEST-INFO: |                     Starting section my_section                |
%AETEST-INFO: +----------------------------------------------------------------+
%AETEST-INFO: +..............................................................+
%AETEST-INFO: :     Starting STEP 1: Delete all boot variables     :
%AETEST-INFO: +..............................................................+

(snip)


%AETEST-INFO: +..............................................................+
%AETEST-INFO: Starting STEP 2: Configure system boot variable for 'install mode':
%AETEST-INFO:+..............................................................+
```

```
(snip)

%AETEST-INFO: +..............................................................+
%AETEST-INFO: Starting STEP 3: Save the running config to the startup config:
%AETEST-INFO: +..............................................................+

(snip)

%AETEST-INFO: +..............................................................+
%AETEST-INFO: Starting STEP 4: Verify next reload boot variables are correctly
set:
%AETEST-INFO: +..............................................................+

(snip)

%AETEST-INFO: +..............................................................+
%AETEST-INFO: :    Starting STEP 5: Installing image 'bootflash:/image.bin'    :
%AETEST-INFO: +..............................................................+

(snip)

%AETEST-ERROR: Failed reason: Failed to install the image.
%AETEST-INFO: The result of STEP 5: Installing image 'bootflash:/image.bin' is =>
FAILED
%AETEST-INFO: The result of section my_section is => PASSED
%AETEST-INFO: The result of testcase Test is => PASSED

%EASYPY-INFO: +--------------------------------------------------------------+
%EASYPY-INFO: |                       Task Result Summary                    |
%EASYPY-INFO: +--------------------------------------------------------------+
%EASYPY-INFO: Task-1: testscript.Test                              PASSED
%EASYPY-INFO:
%EASYPY-INFO: +--------------------------------------------------------------+
%EASYPY-INFO: |                       Task Result Details                    |
%EASYPY-INFO: +--------------------------------------------------------------+
%EASYPY-INFO: Task-1: testscript
%EASYPY-INFO: `-- Test                                             PASSED
%EASYPY-INFO:     `-- my_section                                   PASSED
```

A couple things to note about incorporating device cleaning into pyATS testscripts. Clean stage step results are not rolled up and do not affect testscript results. The individual clean stage steps will not appear in the Task Result Details section at the end of the log. To enable the rollup of the clean stage step results, you must pass the **steps** object to the clean stage in the testscript. Example 15-21 shows how to include the clean stage step results into the testscript result by passing the **steps** object.

Example 15-21   *pyATS Clean – Device Clean API Including Results*

```
from pyats import aetest


class MyTestcase(aetest.Testcase):

    @aetest.test
    def my_test(self, steps, testbed):
        device = testbed.devices['uut']
        device.connect()

        device.api.clean.install_image(
            steps=steps,
            images=['bootflash:/image.bin'],
            save_system_config=True)
```

Example 15-22 shows a snippet of the associated testscript results after enabling the clean stage step results to roll up into the testscript results. You'll notice step 5 fails, which ultimately fails the entire testscript.

Example 15-22   *pyATS Testscript Logs – Including Clean Stage Step Results*

```
%AETEST-INFO: +......................................................+
%AETEST-INFO: :    Starting STEP 5: Installing image 'bootflash:/image.bin'    :
%AETEST-INFO: +......................................................+

(snip)

%AETEST-ERROR: Failed reason: Failed to install the image.
%AETEST-INFO: The result of STEP 5: Installing image 'bootflash:/image.bin' is =>
   FAILED
%AETEST-INFO: The result of section my_section is => FAILED
%AETEST-INFO: The result of testcase Test is => FAILED


%EASYPY-INFO: +----------------------------------------------------------------+
%EASYPY-INFO: |                         Task Result Summary                    |
%EASYPY-INFO: +----------------------------------------------------------------+
%EASYPY-INFO: Task-1: testscript.Test                                   FAILED
%EASYPY-INFO:
%EASYPY-INFO: +----------------------------------------------------------------+
%EASYPY-INFO: |                         Task Result Details                    |
%EASYPY-INFO: +----------------------------------------------------------------+
%EASYPY-INFO: Task-1: testscript
```

```
%EASYPY-INFO: `-- Test                                               FAILED
%EASYPY-INFO:      `-- my_section                                     FAILED
%EASYPY-INFO:           `-- STEP 1: Delete all boot variables          PASSED
%EASYPY-INFO:           `-- STEP 2: Configure system boot variable for 'install...
PASSED
%EASYPY-INFO:           `-- STEP 3: Save the running config to the startup conf...
PASSED
%EASYPY-INFO:           `-- STEP 4: Verify next reload boot variables are correc...
PASSED
%EASYPY-INFO:           `-- STEP 5: Installing image 'bootflash:/image.bin'  FAILED
```

Developing Clean Stages

The pyATS Library (Genie) comes with many clean stages out of the box, but what if you want to create your own clean stage? In this section, we are going to go through the process of creating a new stage and how to abstract an existing clean stage.

Getting Started

Before creating a new clean stage, I suggest reviewing the code of the available clean stages. All available clean stages are found in the $VIRTUAL_ENV/lib/python.$VERSION/ site-packages/genie/libs/clean/stages directory. Once in that directory, you'll find common clean stages and specific directories based on device OS and platform. For example, clean stages for a Cisco Catalyst 9300 can be found in genie/libs/clean/stages/iosxe/cat9k/stages. py. In the stages.py file, you'll find classes for clean stages that are specific to that OS and platform. These stages are in addition to the common stages available to all device types. You may notice that some clean stages inherit others. This is common because many operations across different device platforms share the same workflow with only minor differences. For example, the cat9k platform inherits the base IOS XE **ChangeBootVariable** class, which changes the device's boot variable, and modifies the order of execution to remove the configuration register steps since Catalyst 9k switches use ROMMON variables to control the boot behavior (refer to https://www.cisco.com/c/en/us/support/docs/switches/ catalyst-9300-series-switches/216850-configuration- register-equivalent-clis-i.html).

Stage Template

Clean stages are Python classes that inherit from a **BaseStage** class from genie.libs.clean or an existing clean stage from the pyATS Clean library. The following sections will show the boilerplate code to create a new clean stage inheriting the **BaseStage** class and how to abstract an existing clean stage.

New Stage

Brand-new clean stages must inherit from the **BaseStage** class from genie.libs.clean. The class must contain three main sections:

- Stage schema

- Argument defaults

- Execution order of stage steps

We will dive deeper into the individual sections, but let's examine what the code will look like. Example 15-23 shows the skeleton of a new clean stage with the main sections included.

Example 15-23 *New Clean Stage*

```
from genie.libs.clean import BaseStage

class ChangeBootVariable(BaseStage):

    # ============
    # Stage Schema
    # ============
    schema = {

    }


    # =================
    # Argument Defaults
    # =================


    # =============================
    # Execution order of Stage steps
    # =============================
    exec_order = [
    ]
```

Abstracting an Existing Stage

If we have an existing clean stage that doesn't exactly work, but could work with some slight modifications, then it's possible to inherit an existing clean stage class and only modify the sections that need to be changed. For example, there's a clean stage for changing the boot variable for Cisco IOS XE devices. For Catalyst 9k devices, we could reuse most of the steps in that stage, but maybe we need to change some sections. We can import the IOS XE clean stage for changing the boot variable and inherit it in our new clean stage. Example 15-24 shows a new Python class, **ChangeBootVariable**, representing a new clean stage, which inherits from the Python class ChangeBootVariableIosxe, which defines the existing IOS XE change boot variable clean stage.

Example 15-24 *Abstracting an Existing Clean Stage*

```
from genie.libs.clean.stages.iosxe.stages import ChangeBootVariable as ChangeBoot-
    VariableIosxe


class ChangeBootVariable(ChangeBootVariableIosxe):


    # ============
    # Stage Schema
    # ============
    schema = {


    }


    # =================
    # Argument Defaults
    # =================


    # =============================
    # Execution order of Stage steps
    # =============================
    exec_order = [
    ]
```

You'll notice we are inheriting the **ChangeBootVariable** class from the IOS XE stages instead of inheriting the **BaseStage** class. In the following sections, we will continue working with and building on these examples.

Schema and Arguments

The schema defines the arguments that the clean stage accepts, which arguments are mandatory or optional, the data type of the argument (string, integer, boolean, and so on), and the structure of how the arguments should be defined in a Clean YAML file. The arguments in a clean stage are meant to empower the user to provide dynamic values and drive the clean stage. Example 15-25 shows the new **ChangeBootVariable** clean stage with four arguments: **images**, **timeout**, **config_register**, and **current_running_image**.

Example 15-25   *Clean Stage Schema*

```
from genie.metaparser.util.schemaengine import Optional
from genie.libs.clean import BaseStage


class ChangeBootVariable(BaseStage):


    # ============
    # Stage Schema
    # ============
    schema = {
        Optional('images'): list,
        Optional('timeout'): int,
        Optional('config_register'): str,
        Optional('current_running_image'): bool,
    }


    # =================
    # Argument Defaults
    # =================


    # ==============================
    # Execution order of Stage steps
    # ==============================
    exec_order = [
    ]
```

You'll notice that all the arguments are considered optional. It is good practice to include default values for any optional arguments. Argument defaults should be under the **Argument Defaults** section in the class and be the same name as the arguments defined in the schema, in all uppercase. Example 15-26 shows how to provide default values for each argument, if the argument is not specified by the user.

Example 15-26   *Clean Stage Default Arguments*

```
from genie.metaparser.util.schemaengine import Optional
from genie.libs.clean import BaseStage


class ChangeBootVariable(BaseStage):


    # ============
    # Stage Schema
    # ============
```

```
schema = {
    Optional('images'): list,
    Optional('timeout'): int,
    Optional('config_register'): str,
    Optional('current_running_image'): bool,
}


# =================
# Argument Defaults
# =================
TIMEOUT = 300
CONFIG_REGISTER = '0x2102'
CURRENT_RUNNING_IMAGE = False


# =============================
# Execution order of Stage steps
# =============================
exec_order = [
]
```

Clean Stage Steps

Stage steps are where the magic happens in clean stages. The stage steps define the individual actions that take place to execute the clean stage. In the **ChangeBootVariable** clean stage, we will need to define the following stage steps:

1. Delete existing boot variable.

2. Configure new boot variable.

Stage steps are meant to be small and concise. This allows clean stages to be modular and easily managed using steps. Steps for a clean stage may need modified in the future, and by keeping them small, any changes are less likely to break the clean stage. Stage steps are defined in a method (Python function) under the clean stage class. The method name should be clear and describe the purpose of the step. In the method, we begin with the **steps** object. All device actions occur under the step. Any exceptions that occur in the step should be passed to the **step.failed()** call using the **from_exception** argument, which indicates why the step failed. Example 15-27 shows a new stage step to delete any existing boot variable (**delete_boot_variable**). Note only the method is shown—the schema, argument defaults, and execution order have been excluded for brevity.

Example 15-27 *Clean Stage Step*

```
class ChangeBootVariable(BaseStage):
<code removed for brevity>

    def delete_boot_variable(self, steps, device):
        with steps.start("Delete any configure boot variables") as step:
            try:
                device.configure("no boot system")
            except Exception as e:
                step.failed("Failed to delete configured boot variables",
                            from_exception=e)

            step.passed("Successfully deleted configured boot variables")
```

Now that you understand how to create a simple stage step, let's look at how we can use the arguments passed in from the Clean YAML file. If the user doesn't provide argument values, the default argument values will be used. To pass the argument values to a stage step, you need to pass them into the method as a keyword argument. Example 15-28 shows a new stage step to configure a boot variable (**configure_boot_variable**) with two arguments passed to the method: **timeout** and **current_running_image**. The schema, argument defaults, execution order, and **delete_boot_variable** stage step have been included for the sake of completeness.

Example 15-28 *Clean Stage Step Arguments*

```
import logging
from genie.metaparser.util.schemaengine import Optional
from genie.libs.clean import BaseStage


log = logging.getLogger(__name__)


class ChangeBootVariable(BaseStage):

    # ============
    # Stage Schema
    # ============
    schema = {
        Optional('images'): list,
        Optional('timeout'): int,
        Optional('config_register'): str,
        Optional('current_running_image'): bool,
    }
```

```
# =================
# Argument Defaults
# =================
TIMEOUT = 300
CONFIG_REGISTER = '0x2102'
CURRENT_RUNNING_IMAGE = False


# =============================
# Execution order of Stage steps
# =============================
exec_order = [
]


def delete_boot_variable(self, steps, device):
    with steps.start("Delete any configure boot variables") as step:
        try:
            device.configure("no boot system")
        except Exception as e:
            step.failed("Failed to delete configured boot variables",
                        from_exception=e)
        step.passed("Successfully deleted configured boot variables")



def configure_boot_variable(self, steps, device, images, timeout=TIMEOUT,
                            current_running_image=CURRENT_RUNNING_IMAGE):
    with steps.start("Set boot variable to images provided for {}".format(
            device.name)) as step:
        if current_running_image:
            log.info("Retrieving and using the running image due to "
                    "'current_running_image: True'")
            try:
                output = device.parse('show version')
                images = [output['version']['system_image']]
            except Exception as e:
                step.failed("Failed to retrieve the running image. Cannot "
                            "set boot variables",
                            from_exception=e)
        try:
            device.api.execute_set_boot_variable(
                boot_images=images, timeout=timeout)
        except Exception as e:
            step.failed("Failed to set boot variables to images provided",
                        from_exception=e)
        else:
            step.passed("Successfully set boot variables to images provided")
```

Now we have a schema defining the arguments for the clean stage, the default argument values, and the individual clean stage steps defined. But how do we know which stage steps to run and in what order? In the next section, we are going to look at how to define the execution order of the clean stage steps.

Execution Order

Once the clean stage arguments and steps are defined, we can define the order in which the stage steps should be executed. The execution order is defined as a list in the exec_order variable defined in the beginning of the class. The list should contain the method names of each step. The list will be executed in order—first to last. Any methods not included in the list will not execute. For completeness, more methods (steps) were added to the previous example. Example 15-29 shows the complete, new **ChangeBootVariable** clean stage with the execution order defined.

Example 15-29   *Clean Stage Step Execution Order*

```
from genie.metaparser.util.schemaengine import Optional
from genie.libs.clean import BaseStage


class ChangeBootVariable(BaseStage):

    # ============
    # Stage Schema
    # ============
    schema = {
        Optional('images'): list,
        Optional('timeout'): int,
        Optional('config_register'): str,
        Optional('current_running_image'): bool,
    }


    # =================
    # Argument Defaults
    # =================
    TIMEOUT = 300
    CONFIG_REGISTER = '0x2102'
    CURRENT_RUNNING_IMAGE = False


    # ==============================
    # Execution order of Stage steps
    # ==============================
    exec_order = [
        'delete_boot_variable',
        'configure_boot_variable',
        'set_configuration_register',
```

```
        'write_memory',
        'verify_boot_variable',
        'verify_configuration_register'
    ]


def delete_boot_variable(self, steps, device):
    with steps.start("Delete any configure boot variables") as step:
        try:
            device.configure("no boot system")
        except Exception as e:
            step.failed("Failed to delete configured boot variables",
                        from_exception=e)
        step.passed("Successfully deleted configured boot variables")

def configure_boot_variable(self, steps, device, images, timeout=TIMEOUT,
                            current_running_image=CURRENT_RUNNING_IMAGE):
    with steps.start("Set boot variable to images provided for {}".format(
            device.name)) as step:
        if current_running_image:
            log.info("Retrieving and using the running image due to "
                     "'current_running_image: True'")
            try:
                output = device.parse('show version')
                images = [output['version']['system_image']]
            except Exception as e:
                step.failed("Failed to retrieve the running image. Cannot "
                            "set boot variables",
                            from_exception=e)
        try:
            device.api.execute_set_boot_variable(
                boot_images=images, timeout=timeout)
        except Exception as e:
            step.failed("Failed to set boot variables to images provided",
                        from_exception=e)
        else:
            step.passed("Successfully set boot variables to images provided")

def set_configuration_register(self, steps, device,
                               config_register=CONFIG_REGISTER,
timeout=TIMEOUT):
    with steps.start("Set config register to boot new image on {}".format(
            device.name)) as step:
        try:
```

```
                device.api.execute_set_config_register(
                        config_register=config_register, timeout=timeout)
            except Exception as e:
                step.failed("Failed to set config-register",
                            from_exception=e)
            else:
                step.passed("Successfully set config register")

    def write_memory(self, steps, device, timeout=TIMEOUT):
        with steps.start("Execute 'write memory' on {}".format(device.name)) \
        as step:
            try:
                device.api.execute_write_memory(timeout=timeout)
            except Exception as e:
                step.failed("Failed to execute 'write memory'",
                            from_exception=e)
            else:
                step.passed("Successfully executed 'write memory'")

    def verify_boot_variable(self, steps, device, images):
        with steps.start("Verify next reload boot variables are correctly set "
                        "on {}".format(device.name)) as step:
            if not device.api.verify_boot_variable(boot_images=images):
                step.failed("Boot variables are NOT correctly set")
            else:
                step.passed("Boot variables are correctly set")

    def verify_configuration_register(self, steps, device,
                                    config_register=CONFIG_REGISTER):
        with steps.start("Verify config-register is as expected on {}".format(
                device.name)) as step:
            if not device.api.verify_config_register(
                    config_register=config_register, next_reload=True):
                step.failed("Config-register is not as expected")
            else:
                step.passed("Config-register is as expected")
```

Documentation

It's popular in Python to document a module, class, method, or function using a concept called *docstrings*. Docstrings conventions and standards are outlined in PEP 257 (https://peps.python.org/pep-0257/). PyATS Clean uses docstrings to document the clean stage in the Clean Stages browser. The docstring should help the user understand what the clean stage does, the available arguments, and the argument defaults as well as provide a brief example of the clean stage. Example 15-30 shows the general format of the docstring format for clean stages.

Example 15-30   *Clean Stage Docstring Format*

```
"""<description>

Stage Schema
------------
<schema>

Example
-------
<example>
"""
```

Let's add a docstring to the **ChangeBootVariable** example we've been developing. Example 15-31 shows the **ChangeBootVariable** clean stage class with an appropriate docstring, which will be rendered in the Clean Stages browser.

Example 15-31   *Clean Stage Documentation*

```
from genie.metaparser.util.schemaengine import Optional
from genie.libs.clean import BaseStage

class ChangeBootVariable(BaseStage):
    """This stage configures boot variables of the device using the following
    steps:
    - Delete existing boot variables.
    - Configure boot variables using the provided 'images'.
    - Set the configuration-register using the provided 'config_register'.
    - Write memory.
    - Verify the boot variables are as expected.
    - Verify the configuration-register is as expected.

Stage Schema
------------
change_boot_variable:

    images (list): Image files to use when configuring the boot variables.

    timeout (int, optional): Execute timeout in seconds. Defaults to 300.
```

```
            config_register (str, optional): Value to set config-register for
                reload. Defaults to 0x2102.

            current_running_image (bool, optional): Set the boot variable to the
                currently running image from the show version command instead of
                the image provided. Defaults to False.

Example
-------
change_boot_variable:
    images:
        - harddisk:/image.bin
    timeout: 150
"""

    # ============
    # Stage Schema
    # ============
    schema = {
        Optional('images'): list,
        Optional('timeout'): int,
        Optional('config_register'): str,
        Optional('current_running_image'): bool,
    }


    # =================
    # Argument Defaults
    # =================
    TIMEOUT = 300
    CONFIG_REGISTER = '0x2102'
    CURRENT_RUNNING_IMAGE = False


    # ==============================
    # Execution order of Stage steps
    # ==============================
    exec_order = [
        'delete_boot_variable',
        'configure_boot_variable',
        'set_configuration_register',
        'write_memory',
        'verify_boot_variable',
        'verify_configuration_register'
    ]


<clean step methods removed for brevity>
```

Abstracted Clean Stages

In the beginning of this section, we discussed the ability to abstract an existing clean stage by inheriting it. Abstracted stages allow us to reuse code from the parent stage (class) and/or change the execution order of the clean stage steps. All abstracted stages should redefine the docstring and the three main sections: stage schema, argument defaults, and execution order of stage steps. This ensures there's no confusion with these sections when using the abstracted stage. Let's use the existing Cisco IOS XE **ChangeBootVariable** clean stage as our "base stage." Example 15-32 shows the execution order being changed.

Example 15-32   *Abstracted Clean Stage*

```
from genie.libs.clean.stages.iosxe.stages import ChangeBootVariable \
as ChangeBootVariableIosxe

class ChangeBootVariable(ChangeBootVariableIosxe):
    """This stage configures boot variables of the device using the following
    steps:
    - Delete existing boot variables.
    - Configure boot variables using the provided 'images'.
    - Write memory.
    - Verify the boot variables are as expected.
Stage Schema
------------
change_boot_variable:
    images (list): Image files to use when configuring the boot variables.
    timeout (int, optional): Execute timeout in seconds. Defaults to 300.
    current_running_image (bool, optional): Set the boot variable to the
        currently running image from the show version command instead of
        the image provided. Defaults to False.
Example
-------
change_boot_variable:
    images:
        - harddisk:/image.bin
    timeout: 150
"""

    # ============
    # Stage Schema
    # ============
    schema = {
        Optional('images'): list,
        Optional('timeout'): int,
        Optional('current_running_image'): bool
    }
```

```
# =================
# Argument Defaults
# =================
TIMEOUT = 300
CURRENT_RUNNING_IMAGE = False


# Execution order of IOS XE clean stage (for reference)
# exec_order = [
#      'delete_boot_variable',
#      'configure_boot_variable',
#      'set_configuration_register',
#      'write_memory',
#      'verify_boot_variable',
#      'verify_configuration_register'
# ]


# =============================
# Execution order of Stage steps
# =============================
exec_order = [
    'delete_boot_variable',
    'configure_boot_variable',
    'write_memory',
    'verify_boot_variable'
]
```

The execution order from the Cisco IOS XE clean stage has been commented out
in the example for reference. You'll notice the set_configuration_register and verify_
configuration_register stages have been removed from the execution order. If we needed
to modify a clean stage step, we would need to define the method again and write the
new logic. In this case, we only modified the execution order, so we are finished!

Summary

PyATS Clean is a framework to remove the monotonous task of resetting, or "cleaning,"
a device during testing or once testing has been completed. Many devices are supported,
including network devices and power cyclers, which are devices such as power management devices (PDUs) and out-of-band (OOB) management devices that can be used to
reboot devices and enable device recovery when devices malfunction. Processes to reset

or "clean" devices are called clean stages. Clean stages are further broken down into smaller tasks called stage steps. The pyATS library (Genie) comes with many clean stages that are available in the Clean Stages browser. Clean stages can be defined in a Clean YAML file or even in a pyATS testscript using the device API. Device cleaning can be executed as part of a pyATS job or standalone using the **pyats clean** command. Custom pyATS clean stages can be created using the **BaseStage** class from the genie.libs.clean module, or an abstracted clean stage can be created using an existing clean stage. Device cleaning can be an overlooked process but can be standardized using pyATS Clean.

pyATS Blitz

The pyATS test infrastructure provides a fully featured test suite that includes a testbed file, test infrastructure, parsers, device APIs, and much more. But what if we could use the pyATS test infrastructure without writing code? What if all you needed to understand was YAML? PyATS Blitz is a feature of pyATS that provides a low-code, no-code approach to running testcases. It takes a similar approach to the popular automation tool Ansible, where the underlying code is abstracted and the user only needs to understand a domain-specific language (DSL) to create YAML playbooks. In this chapter, you're going to learn about pyATS Blitz and the trigger datafiles that can be created to run testcases with a low-code, no-code approach. The following topics will be covered:

- Blitz YAML
- Blitz features
- Blitz usage
- Blitz development

Blitz allows you to run pyATS testcases without having any programming knowledge. Blitz trigger datafiles are YAML files that can be used to configure devices, parse device output, learn device features, and use device APIs. In the following section, we will dive into the details of a trigger datafile, also referred to as a Blitz YAML file.

Blitz YAML

Blitz YAML files are made up of one or more testcases. Testcases contain test sections. Each test section consists of one or more actions. Actions are blocks of commands that perform a specific task and report the result as passed, failed, errored, and so on, much like individual tests do in a pyATS testcase. Sections and actions will run in the order in which they are presented in a Blitz YAML file.

Actions

Actions are reserved keywords that map to code executed by the pyATS libraries. An action block begins with a section name. The section name is an arbitrary string that describes the actions that are executed within that block. Example 16-1 shows a Blitz YAML file with pseudocode to represent the structure and individual components.

Example 16-1 *Blitz YAML Structure*

```
# Name of the testcase
Testcase1:

    # Leave this as is for most use cases
    source:
        pkg: genie.libs.sdk
        class: triggers.blitz.blitz.Blitz

    # Field containing all the sections
    test_sections:

        # Section name - Can be any name, it will show as the first section
        # of the testcase
        - section_one:
            - ">>>> <ACTION> <<<<"
            - ">>>> <ACTION> <<<<"
            - ">>>> <ACTION> <<<<"

        - section_two:
            - ">>>> <ACTION> <<<<"
            - ">>>> <ACTION> <<<<"
```

At the beginning of the file, the testcase name is provided. The testcase name can be whatever you desire. Under the testcase, the **source** keyword references the location of the underlying code that executes the pyATS Blitz trigger datafile. The **test_sections** keyword notes the beginning of where the individual test sections will be listed. Section_one and section_two are arbitrary names signifying two individual test sections. Lastly, you'll notice a list of actions under each test section. The list of available action keywords can be found here:

https://pubhub.devnetcloud.com/media/genie-docs/docs/blitz/design/actions/actions_list.html#actions-list

Some popular actions include **execute**, **configure**, **parse**, **learn**, **api**, **print**, and **diff**. These terms might sound familiar, and that's the point! Those already familiar with pyATS and the pyATS Library (Genie) should be able to easily write their own Blitz YAML file. Example 16-2 shows how to run a command on a device named PE1.

Example 16-2 *Execute Action Block*

```
- execute:
    command: show version
    device: PE1
    # Step will fail if the device hasn't executed the command in 10 seconds
    timeout: 10
```

The action block can be one of many action blocks within a test section. This is great, but how can we capture and possibly alter the output from the action? In the next section, we will look at how to filter and save action output.

Action Outputs

Action output can be a valuable resource for comparison and data analysis. In Blitz, action output can be saved to a variable, for later use in other actions or testcases, or to a file for further analysis. The output of an action can be saved in its entirety or part of it can be extracted using filters.

Filters

Filters can be applied to action output to extract a specific part of the output. There are three types of output filters:

- Dq filter
- RegEx filter
- List filter

The Dq filter is named after the built-in dictionary query (Dq) tool. The Dq module is a useful tool in the genie.utils package. It's used to search through a nested dictionary structure to find interesting key-value pairs without having to perform multiple looping. The filter will apply a query on JSON output and save it as a dictionary. Example 16-3 shows the Dq filter being applied to parsed output of the **show module** command. The parsed output must be a dictionary.

Example 16-3 *Dq Filter*

```
# Applying a dq query and save the outcome into the variable parse_output.
# Dq query only works on outputs that are dictionary

- apply_configuration:
    - parse:
        command: show module
```

```
        device: PE2
        save:
            - variable_name: parse_output
              filter: contains('ok').get_values('lc', index=2)
            # The output is '4'
```

The RegEx filter is used on action outputs that are string outputs. A RegEx pattern is applied to a string output, and the RegEx grouped value will be stored. Along with using a RegEx group to capture a specific value, you also can find all the matches using **findall()** and store a list of all the matches. If no matches are found, an empty list is returned. Example 16-4 shows how to apply a RegEx expression to extract the BIOS and bootflash versions from **show version** output.

Example 16-4 *RegEx Filter*

```
- execute:
    device: N9KV
    command: show version
    save:
    - filter: BIOS:\s+version\s+(?P<bios>[0-9A-Za-z()./]+).*
            # bios version is 07.33
      regex: true
    - filter: bootflash:\s+(?P<bootflash>[0-9A-Za-z()./]+)\s+(?P<measure>\w+).*
            # bootflash is 51496280 and measure is KB
      regex: true
```

The **regex findall** functionality can be used to find all matches and return a list of values. This is popular if you're trying to find the same type of values, such as a MAC or IP address, in a massive amount of output. Example 16-5 shows how to use **regex findall** functionality to find all the IP addresses in output of the **show ip interface brief** command.

Example 16-5 *RegEx Findall Filter*

```
- execute:
    device: PE1
    command: show ip interface brief
    save:
    - variable_name: execute_output
      regex_findall: (\d+\.\d+\.\d+\.\d+)
      # returns a list of IP addresses
```

The List filter is a specific filter that can only be applied to action output that are lists. List output can be filtered using the index or using a RegEx pattern to search through the list for a match. Example 16-6 shows how to filter a list of values using the index and also a RegEx pattern.

Example 16-6 *List Filter*

```
- api:
    device: PE1
    function: get_list_items

    # function output
    # [{'a': 1}, {'d': {'c': 'name1'}}, [1,2,34], {'e': ['a', 'b', 'c']}]

    save:
        - variable_name: list_int5
          list_index: "[0:2]"
          # saves items 0,1 from the above list into a list named list_int5
          # list_int5 = [{'a': 1}, {'d': {'c': 'name1'}}]


        - variable_name: list_int7
          list_index: 2
          # saves item #2 in the array into a list named list_int7
          # list_int7 = [[1,2,34]]


        - variable_name: list_int8
          # saves the entire array in a list named list_int8

- api:
    device: PE1
    function: get_platform_logging
    # device API that returns a list of device logs
    save:
        # look for specific log messages using regex filter
        - variable_name: platform_log
          filter: Oct\s+15[\S\s]+Configured from console by console$
          # checks if any item in the list matches this regex filter and
          # saves to a list named platform_log
```

Variables

Variables can be used to save entire and/or filtered action output that can be used in other actions. Action outputs can be stored using the **save** keyword and **variable_name** argument. Example 16-7 shows how to save the entire output of the **show version** command into a variable named **show_ver_output**.

Example 16-7 *Save Full Output to Variable*

```
- execute:
    device: PE1
    command: show version
    save:
      - variable_name: sh_ver_output
```

Filtered output is saved the same way with the exception of using the RegEx filter.
The RegEx filter uses the filter keyword, which has a RegEx pattern that contains a
RegEx group as its value. The value captured by the RegEx group is used as the variable
name. Example 16-8 extends Example 16-4 by using the variables that are captured by
the RegEx filter.

Example 16-8 *Reusing Variables with the RegEx Filter*

```
- execute:
    device: N9KV
    command: show version
    save:
    - filter: BIOS:\s+version\s+(?P<bios>[0-9A-Za-z().\/]+).*
      # bios version is 07.33
      regex: true
    - filter: bootflash:\s+(?P<bootflash>[0-9A-Za-z().\/]+)\s+(?P<measure>\w+).*
      # bootflash is 51496280 and measure is KB
      regex: true

- print:
    bios:
      value: "The bios version is %VARIABLES{bios}"
    bootflash:
      value: "The bootflash is %VARIABLES{bootflash} and %VARIABLES{measure}"
```

This leads us to how to use the variables once they are captured. The pyATS Library
uses a YAML markup language to allow substitution and references much like templating engines. In the case of variables, the **%VARIABLES{***insert_variable_name***}** markup
is used to load a variable in another action. You'll notice in Example 16-8, we reference
and print the filtered BIOS version using the **%VARIABLES{bios}** markup. Variables can
be used beyond just printing the variable's value. The variable value can be used in other
actions such as comparisons and further filtering.

Variables can also be saved at the testscript level and used across different testcases. To
use variables across different testcases, the variable must have a **testscript.** prefix. If the

variable does not have the prefix, it cannot be used across different testcases.
Example 16-9 shows how to save the uptime of a device as a testscript-level variable in
one testcase (tc1) and reference it in another testcase (tc2).

Example 16-9 *Testscript-Level Variables*

```
tc1:
    source:
        pkg: genie.libs.sdk
        class: triggers.blitz.blitz.Blitz
    test_sections:
        - global_save:
            - parse:
                device: PE1
                command: show version
                save:
                    - variable_name: testscript.uptime
                      filter: get_values("uptime")
tc2:
    source:
        pkg: genie.libs.sdk
        class: triggers.blitz.blitz.Blitz
    test_sections:
        - global_reuse:
            - api:
                function: get_list_items
                common_api: True
                arguments:
                    input: "%VARIABLES{testscript.uptime}"
                save:
                    - variable_name: item
```

Saving Outputs

We've seen how to save action outputs into variables and reference them in other actions
and testcases, but action outputs can also be saved to files and dictionaries. Action outputs that are saved to files and dictionaries can be filtered the same way when the output
is saved to a variable. To save action output to a file, you need to use the **file_name**
argument to the **save** keyword. The file is created and saved in the runinfo directory.
Example 16-10 shows how to save the **uptime** value from parsed **show version** command
output to a file named uptime.txt.

Example 16-10 *Saving Output to File*

```
testsave:
    source:
        pkg: genie.libs.sdk
        class: triggers.blitz.blitz.Blitz
    test_sections:
        - apply_configuration:
            - parse:
                command: show version
                device: R1_xe
                save:
                    - file_name: uptime.txt
                      filter: get_values("uptime")
```

You can also append output to an existing file using the **append: True** argument. If append is set to True, the action output will be appended to the file provided to the **file_name** argument. If the file doesn't exist, it will be created. Example 16-11 shows the parsed output from the **show version** command being saved to a file named sh_ver_output.txt; then the **uptime** value is extracted from the parsed output and appended to the same file.

Example 16-11 *Appending Output to File*

```
testsave:
    source:
        pkg: genie.libs.sdk
        class: triggers.blitz.blitz.Blitz
    test_sections:
        - apply_configuration:
            - parse:
                command: show version
                device: R1_xe
                save:
                    - variable_name: parse_output
                      file_name: sh_ver_output.txt
                    - file_name: sh_ver_output.txt
                      filter: get_values("uptime")
                      append: True
```

Action output can also be saved to a dictionary using the **as_dict** argument. Example 16-12 shows how to save the parsed output of the **show version** command to a dictionary. The dictionary top-level key is the device's name with the value being the action output.

Example 16-12 *Saving Output to Dictionary*

```
testsave:
    source:
        pkg: genie.libs.sdk
        class: triggers.blitz.blitz.Blitz
    test_sections:
        - apply_configuration:
            - parse:
                device: R1_xe
                command: show version
                save:
                    - variable_name: parse_output_dict
                      as_dict:
                          "%VARIABLES{device.name}":
                              rt_2_if2: "%VARIABLES{action_output}"
```

In addition, saved dictionary variables can be updated using the **as_dict_updated=True** argument. The Dq filter is supported when saving output to dictionaries.

Verifying Action Outputs

In addition to filtering and saving action output, we also have the ability to query and verify action output. We can check whether a certain key-value pair exists in JSON output, a string pattern exists in a list of elements, or the output of a device API is true/false. We can even compare numerical values in output using comparison operators (=, >=, <=, >, <, !=). In addition, we can also check whether numerical output is within a certain range.

Keywords **include** and **exclude** are used to query and verify action output. The **include** keyword will check whether the query you're using is valid. The **exclude** keyword is used to make sure the query you're using isn't valid. It works well when performing negative testing, which will be touched on later in the chapter. Example 16-13 shows how to use the **include** and **exclude** keywords to validate two queries against the parsed output of the **show version** command.

Example 16-13   *Verifying JSON Action Output*

```
- apply_configuration:
        - parse:
            command: show version
            device: PE2
            include:
                # checks whether the key 'WebUI' is present in the
                # dictionary output
                - contains('WebUI', regex=True)
            exclude:
                # The output of the query is 'VIRTUAL XE'
                # but we hope that the key 'platform' has no value
                # or does not exist within the dictionary by using
                # the exclude keyword
                - get_values('platform')
```

Different query mechanisms are used depending on the type of output. In the preceding example, we are expecting JSON, which allows us to use Dq to query the output. Most output from the **parse**, **learn**, and **api** actions returns JSON, which allows us to use Dq to query and verify whether the result of the query is in the output using the **include** and **exclude** keywords.

List outputs can also be verified using the **include** and **exclude** keywords. Example 16-14 shows how to check whether certain items are and aren't in a list.

Example 16-14   *Verifying List Action Output*

```
- api:
    device: PE1
    function: get_list_items
    arguments:
        name: [1,2,3,4,5,6,7]          # the output is [1,2,3,4,5,6,7]
    include:
        - 5                            # checks if 5 is in the list
    exclude:
        - 99                           # checks if 99 is NOT in the list
```

RegEx can also be used to verify whether certain items are in a list output. Example 16-15 shows how to use RegEx patterns to verify list items.

Example 16-15 *Verifying List Action Output with RegEx*

```
- api:
    device: PE1
    function: get_platform_logging
    include:
        - \*Dec 10 03:2.*  # Check if any item within a list matches this regex
        - "23:31:16.651"
    exclude:
        - name             # Check if any item within a list not matches this regex
        - \*Dec 10 03:2.*
```

Numerical action outputs (integer and float) can be verified as well using comparison operators (=, >=, <=, >, <, !=). The **api** action is currently the only action supported because it's the only one that has integer and/or float outputs. Example 16-16 shows how to confirm that the MTU size of the GigabitEthernet1 interface is less than or equal to 2000 bytes using the **get_interface_mtu_size** device API.

Example 16-16 *Verifying Numerical Action Output*

```
- api:
    function: get_interface_mtu_size
    arguments:
        interface: GigabitEthernet1
    include:
        - <= 2000
```

You also have the ability to check whether a numerical value output is within a range. Example 16-17 uses the previous example, but this time checks whether the MTU size of GigabitEthernet1 is greater than 1200 and less than or equal to 1500 bytes.

Example 16-17 *Verifying Numerical Action Output Using Range*

```
- api:
    function: get_interface_mtu_size
    arguments:
        interface: GigabitEthernet1
    include:
        - ">1200 && <=1500"
```

The last action output type to verify is boolean (True/False). This is straightforward, but you can verify True or False is present in the output using the **include** keyword. Example 16-18 shows how to verify the action output is True.

Example 16-18   *Verifying Boolean Action Output*

```
- api:
   function: verify_device_is_active
   arguments:
       device: PE1
   include:
       - True
```

Advanced Actions

Blitz provides multiple actions that manipulate the behavior and execution of Blitz trigger datafiles. By default, Blitz executes test sections and actions in order and waits for each action block to complete before moving on to the next one. Advanced actions allow us to change that behavior and optimize the execution of Blitz trigger datafiles. In the following sections, we will dive into three advanced actions: **parallel**, **loop**, and **run_condition**.

Parallel

The **parallel** keyword enables Blitz to execute multiple actions in parallel. By default, Blitz executes each action sequentially. Executing actions in parallel saves time on execution. Multiple actions can be executed by listing them under the **parallel** keyword. Variables that are saved inside the **parallel** block must be consumed outside of the **parallel** block. They cannot be consumed by another action in the **parallel** block. Variables that are saved outside of the **parallel** block can be used within the **parallel** block. Any actions not listed under the **parallel** keyword will run sequentially. Example 16-19 shows how to execute multiple actions in parallel by configuring the BGP feature on multiple Cisco Nexus devices.

Example 16-19   *Parallel Keyword*

```
- configure_bgp_feature
    - parallel:
        - configure:
            device: N9KV1
            command: feature bgp
        - configure:
            device: N9KV2
            command: feature bgp
        - configure:
            device: N9KV3
            command: feature bgp
```

Loop

Looping is a sequence of actions that continue occurring until a termination condition is met. Using termination conditions allows for more dynamic testcases. Loop conditions include iterating over a list or dictionary, using a counter (integer), executing for a certain time duration, and even executing like a traditional **while** loop where the loop is run until the termination condition is met. Table 16-1 shows the available keywords that can be used in a loop.

Table 16-1 *Loop Keywords*

Keyword	Functionality
loop_variable_name	A variable name will represent the element that index of the value/range is on. This is useful when using value/range to reuse elements of the value or numbers in a range.
value	A dictionary, list, or tuple. Per each iteration of a loop, an item in the list/dictionary/tuple will be stored into the **loop_variable_name**. Cannot be used at the same time as **range, until,** or **do_until.**
range	A list of integers from a starting integer to a final integer. The default starting integer is set to 0.
until	A conditional statement that, if true, terminates the loop. Cannot be used at the same time as **range, value,** or **do_until.** It is best if a **max_time** is specified with it so if the conditional statement never turns to True, looping stops after a certain amount of time.
do_until	A conditional statement that, if true, terminates the loop. Cannot be used at the same time as **range, value,** or **until.** It runs at least once, even if the terminating condition is met. It is best if a **max_time** is specified with it so if the conditional statement never turns to True, looping stops after a certain amount of time.
max_time	Maximum time of execution of the loop. Best to specify it along with **until/do_until** to make sure the loop will stop after a certain amount of time.
every_seconds	An integer value (representing seconds) that, if provided, indicates each loop iteration should be terminated exactly after running for that amount of time.
parallel	Executes all the actions under a loop concurrently. Works only with **value** and **range.** Does not work with **until, do_until,** and **loop_until.**
loop_until	Can be set to (Passed/Failed). If it's set, the loop will iterate until the result of the last iteration is the same as the value.

Loops can only have one of the following keywords: **value, range, until,** or **do_until.** These keywords determine the type of loop that will be executed, which is why they cannot be combined. Let's now look at a basic example of looping through a list of **show** commands to execute on a device (see Example 16-20).

Example 16-20 *Looping Over a List of **show** Commands*

```
- run_show_commands:
    - loop:
        loop_variable_name: command
        value:
          - show version
          - show vrf
        actions:
          - execute:
              alias: execute_
              device: PE1
              command: "%VARIABLES{command}"
```

The **value** key within the loop block is a list of values that will be used per loop iteration. In this case, since two **show** commands are listed, there will be two loop iterations. The **loop_variable_name** is used to capture the current loop iteration value and save it to a variable named **command**. The variable name is arbitrary but should represent the values being looped over. The **actions** block stores all the actions that will run per loop iteration. In this case, the **execute** action is run with the current value that is saved in the **loop_variable_name** as the command to execute. This is a simple example that shows how to loop over a list of items, but now let's turn it up a bit and discuss some other ways to loop.

Beyond lists, the key-value pairs in dictionaries can be looped over using the keywords **._keys** and **._values**. In a loop, if we wanted to access the keys in a dictionary saved in the variable name **device_dict**, we would use the following markup: "**%VARIABLES{device_dict._keys}**". Conversely, if we wanted to access the value, we could use the following markup: "**%VARIABLES{device_dict._values}**".

A loop can also be run with a given range. This is like executing a **for** loop using an increasing counter in traditional programming. Once the upper limit of the counter is met, the loop terminates. The **range** keyword is used in Blitz to run a loop for a certain number of iterations, possibly with a given start and end point. For example, if **range: 2** is provided to a **loop** block, then the loop will iterate twice. We can also expand on that and explicitly provide start and end points, such as **range: 2, 6**, which will iterate four times, with the starting number being 2 and the ending number being 5. The range is exclusive, which means the ending number is not included in the loop. Example 16-21 shows how to use the **range** keyword with a start and end point.

Example 16-21 *Looping with the range Keyword*

```
- loop:
    loop_variable_name: range_num
    range: 2,6
    actions:
        - print:
            item:
                value: "%VARIABLES{range_num}"

# Result:
# 2
# 3
# 4
# 5
```

The looping mechanism in Blitz is much like a traditional **for** loop in programming, where the loop is iterated over until the iterable is exhausted. There are a couple keywords that change the looping behavior to be more like a traditional **while** loop, where the loop continues to run until the loop condition is met. The **until, do_until**, and **loop_until** keywords all loop until a specific condition is met. Example 16-22 shows how we can check for the device uptime to not equal 0. This may be useful if we want to ensure the device is online and operational. Obviously, there are many ways to check a device's health through other means such as health checks, but we will stick with this method for example purposes.

Example 16-22 *Do Until Loop*

```
- loop:
    # Loop over an action, running it at least once, and if a condition is met
    # terminate the loop
    do_until: "%VARIABLES{dev_uptime} > 0"
    # Timeout after 5 seconds. Ensures loop iteration doesn't run forever
    max_time: 5
    actions:
        - api:
            description: get the device uptime
            device: PE1
            function: get_device_uptime
            save:
              - variable_name: dev_uptime
```

Blitz provides the **every_seconds** keyword to synchronize loop iterations. The **every_seconds** keyword provides the number of seconds a loop iteration should take to execute. The easiest way to explain the **every_seconds** keyword is through an example. Example 16-23 shows a loop that will iterate two times using the list of values provided. Each command in the list of values will be executed and parsed during each loop iteration.

Example 16-23   *The every_seconds Keyword*

```
- loop:
    loop_variable_name: show_commands
    value:
      - version
      - vrf
    every_seconds: 8
    actions:
            - execute:
                device: PE1
                command: show %VARIABLES{show_commands}
```

The **every_seconds** keyword assumes each loop iteration will take 8 seconds. If a loop iteration is quicker and finishes before 8 seconds, the loop will sleep the rest of the time until 8 seconds are up. If the loop iteration takes longer than 8 seconds, it will continue to run. This may be confusing since you would think the loop would be terminated for reaching the time limit, but the **every_seconds** keyword provides a *minimum* interval instead of a maximum interval. It basically states the minimum amount of time a loop iteration should take but doesn't terminate it if it takes longer.

A very interesting combination can be had when combining two advanced Blitz actions: **parallel** and **loop**. Example 16-24 shows how to loop over a list of device names and run the **show version** command in parallel on each device.

Example 16-24   *Parallel and Loop Actions*

```
- loop:
    # A loop that runs one action over different devices
    loop_variable_name: devices
    # A list of device names
    value:
      - PE1
      - PE2
    parallel: True
    actions:
      - execute:
          device: "%VARIABLES{devices}"
          command: show version
```

The last example of looping is the ability to perform nested looping. Loop values in the outer loop are available to all nested loops, which can make for some interesting use cases. For example, if you provide a dictionary to the first, outer loop and a list to the nested loop, the outer loop only has access to the dictionary while the nested loop has access to both the dictionary and list values. As you can see, the **loop** keyword provides many ways to eliminate redundancy and optimize the execution of Blitz actions.

Run Condition

The **run_condition** keyword is identical to **if…else** conditionals in programming, where a condition must be met to enter the conditional block of code. In Blitz, the conditional block of code is a set of actions. The **if** statement must evaluate to True or False. If it evaluates to True, the block of actions will execute. If it evaluates to False, the block of actions will not execute. In addition to an **if** statement, you can add **elif** (else if) and **else** conditions. Additional conditions are optional but may be useful if you need different actions to execute based on multiple conditions. Example 16-25 shows a simple condition that evaluates to True, which means all actions in the conditional block will execute.

Example 16-25 *The run_condition Keyword*

```
- run_condition:
    if: "2000 == 2000" # evaluates to True - all actions will run
    actions:
      - api:
          device: PE1
          function: get_interface_mtu_size
          arguments:
            interface: GigabitEthernet1
          include:
            - ">= 1400 && <= 1600" # Verify MTU is between 1400-1600 bytes
      - sleep:
          sleep_time: 1 # Sleep for 1 second
```

The logic is straightforward since there's only one **if** condition, but let's see how to add in **elif** and **else** conditions. Example 16-26 extends the previous example by executing additional actions if the MTU size equals a specific number of bytes. Note the commands executed under each condition are irrelevant and only used for example purposes.

Example 16-26 *Run Condition with elif and else*

```
conditional_testcase:
    source:
        pkg: genie.libs.sdk
        class: triggers.blitz.blitz.Blitz
```

```
devices: ['PE1']
test_sections:
    - get_gig_mtu:
        - api:                               # api output is equal to 1500
            device: PE1
            function: get_interface_mtu_size
            save:
                - variable_name: g1_mtu       # the 1500 is stored in g1_mtu
            arguments:
                interface: GigabitEthernet1
    - run_condition:
        - if: "%VARIABLES{g1_mtu} == 1300"
          actions:
              - parse:
                  command: show vrf
                  device: PE1
              - sleep:
                  sleep_time: 1
        - elif: "%VARIABLES{g1_mtu} == 1000"
          actions:
              - parse:
                  command: show ip interface brief
                  device: PE1
        - elif: "%VARIABLES{mtu1} == 1500"  # condition matched
          actions:
              - parse:
                  command: show version
                  device: PE1
        - else:
          actions:
              - print:
                  item1:
                      value: The MTU does not match any given values.
```

Blitz Features

We've covered Blitz actions and some of the advanced actions, including **parallel**, **loop**, and **run_condition**. In the following sections, we are going to look at some additional features that can be used to change the behavior of Blitz trigger datafiles.

Negative Testing

By default, actions that succeed execution (executing a **show** command, parsing command output, and so on) will have a "passed" result. Blitz allows us to catch an action we expect to fail by using the keyword **expected_failure: True**. Now when the action does fail, the result will be "passed" since we explicitly stated we expected it to fail. The following actions support the **expected_failure** key:

- configure

- execute

- parse

- learn

- api

- rest

- bash_console

Example 16-27 shows an action that tries to configure **feature bgp** using the **configure** action and is expected to fail. This would be useful if you tried configuring this command on a Cisco IOS XE device, which would fail because it's not a valid IOS XE command.

Example 16-27 *Negative Testing*

```
- configure:
    command: feature bgp
    device: PE1
    expected_failure: True
    timeout: 100
```

Script Termination on Failure

Blitz testcases continue to execute regardless of whether an action fails during testing. However, we have the ability to stop testing if a given action fails. You can use the **continue: False** key under a given action. If the result of the given action is "failed," Blitz will stop execution of the testscript. Currently, due to limitations of the pyATS libraries, it is not possible to set the **continue: False** key under the **parallel** keyword. Example 16-28 shows how to stop a testscript if configuration of NTP fails in the apply_configuration test section. If configuration does fail, the two **execute** actions under the **confirm_actions** test section will not run.

Example 16-28   *Continue Key to Terminate Script on Failure*

```
- test_sections:
  - apply_configuration:
      - continue: False
      - configure:
          command: ntp server 10.1.1.1
          device: PE1
  - confirm_actions:
    - execute:
        command: show ntp status
        device: PE1
      - execute:
          command: show ntp associations
          device: PE1
```

Prompt Handling

When you're configuring network devices, there are some instances where a prompt must receive a reply. It may be to press **[enter]** to continue or to enter **y** for yes or **n** for no. These prompts can be answered automatically using the **reply** keyword in your actions. The **reply** keyword has a block of keys that includes a RegEx pattern of the prompt message and the action to take when replying to the prompt. Example 16-29 shows how to handle a yes/no (y/n) prompt when erasing the startup configuration of a device.

Example 16-29   *Prompt Handling*

```
- apply_configuration:
  - execute:
      device: PE1
      command: write erase
      reply:
      - pattern: .*Do you wish to proceed anyway\? \(y/n\)\s*\[n\]
        action: sendline(y)
```

Results

Much like individual test results in the pyATS test infrastructure, results can be altered for an individual action in Blitz. Instead of being reported as "passed," the result can be altered to failed, aborted, blocked, skipped, errored, or passx when an action completes successfully. This feature is supported by all actions. It may sound confusing, but depending on the testing conditions, you may want to have a different result. If there are **include** or **exclude** keys to filter the data and look for specific output, they will be used to determine if the action initially passes. Example 16-30 shows the **show version** command being executed on a device, and if **CSR1000v** is included in the output, the action results to **failed**.

Example 16-30 *Altering Results*

```
- execute:
    device: PE1
    command: show version
    result_status: failed
    # Action fails if 'include' items are found in output
    include:
        - 'CSR1000V'
```

Timeouts

Each action has a specific timeout and interval to verify the action was performed on the device. The maximum timeout and interval can be changed for the following actions: **api**, **execute**, **parse**, **learn**, and **rest**. The maximum timeout and interval check can be modified with a ratio defined in the testbed file. Example 16-31 shows a Blitz trigger datafile with an accompanied testbed file to show where the timeout and interval ratios can be defined. In the example, the max timeout and interval check are cut in half.

Example 16-31 *Timeout Ratio*

```
Testcase1:

    source:
        pkg: genie.libs.sdk
        class: triggers.blitz.blitz.Blitz


    test_sections:

            - apply_configuration:
                - execute:
                    command: show version
                    include:
                      - 'w'
                    max_time: 5
                    check_interval: 1

...

# Testbed
```

```
devices:
  PE2:
    connections:
      ssh:
        ip: 10.255.1.17
        protocol: ssh
    credentials:
      default:
        password: cisco
        username: cisco
      enable:
        password: cisco
    custom:
      max_time_ratio: '0.5'
      check_interval_ratio: 0.5
    os: iosxe
    type: CSR1000v
```

Customizing Log Messages

In Blitz, each action's, or step's, log messages are fully customizable by specifying the **custom_start_step_message** keyword under the action. The **custom_start_step_message** is applied when the action begins executing. In addition, the **custom_substep_message** and **custom_verification_message** keywords can be used to customize log messages. The **custom_substep_message** keyword is useful when looping over an action (or actions), which creates a substep for each loop iteration. The **custom_verification_message** keyword is useful when verifying output using the **include** and/or **exclude** keys. Example 16-32 shows how to run the **show version** command with a custom starting step message and a custom verification message.

Example 16-32   *Customizing Log Messages*

```
- execute:
    custom_start_step_message: My own custom log message!
    custom_verification_message: Output looks good!
    command: show version
    device: PE1
    include:
      - 'Cisco'
```

Blitz Usage

Blitz requires pyATS and the pyATS Library (Genie) to be installed, a testbed YAML file, and a Blitz trigger datafile. Throughout this chapter, we discussed how to write a Blitz trigger datafile and the available actions and keywords. Blitz can be executed as part of a pyATS job using a job file or executed with a standalone CLI command.

Blitz trigger datafiles can be executed using the pyATS command line. To execute a Blitz trigger datafile using the CLI, enter the following:

```
pyats run genie --trigger-datafile path_to_blitz_datafile
--trigger-uids

'test1' --testbed-file path_to_testbed_file
```

Blitz trigger datafiles can also be executed using a pyATS job file. Example 16-33 shows what a pyATS job file may look like to execute a Blitz trigger datafile.

Example 16-33 *Blitz Job File*

```
import os
from genie.harness.main import gRun
from pyats.datastructures.logic import And, Not, Or


def main():

    gRun(
        trigger_datafile=<path_to_blitz_datafile>,
        trigger_uids = ['test1', 'test2'],  # name of the tests you wish to run
        testbed=<path_to_testbed_file>,
    )
```

Blitz Development

Blitz offers the ability to develop your own actions and test sections. Having the ability to customize and create your own actions and test sections makes the possibilities limitless. In relation to Ansible, you can think of this as creating your own Ansible collection/module. There's a bit more involved with Ansible, but I hope it helps you relate to the topic.

Custom Blitz Actions

A custom Blitz action is created by creating a new Python file with a class that inherits the Blitz class. In the class, create a new method that contains the code of your custom action. It's better to show this by way of example. Example 16-34 shows how a custom action class and method are created. The method contains the custom action logic.

Example 16-34   *Custom Action Class*

```
import logging
from pyats import aetest
from genie.libs.sdk.triggers.blitz.blitz import Blitz


log = logging.getLogger()

class CustomBlitz(Blitz):
    def my_custom_action(self, steps, device, section, **kwargs):
        # contains code logic for custom action...
        log.info("This is my custom action")
```

Some arguments from the built-in actions can be shared with custom actions, which is why **kwargs** is passed into the custom action since the number of keyword arguments is unknown and can vary. The keyword arguments may not be used but may cause issues if not passed into the custom action.

Once the custom action is defined, it's called on in the Blitz trigger datafile using the **source** keyword. Example 16-35 shows how we can call on the custom action created in Example 16-34.

Example 16-35   *Custom Action – Blitz Trigger Datafile*

```
TestCustomAction:
    source:
      pkg: CustomBlitz
      class: <path_to_custom_blitz_class>
    devices: ['uut']
    test_sections:
      - custom_test:
        - my_custom_action:
          device: PE1
          key1: val1
          key2: val2
```

Custom Blitz Sections

Blitz test sections can also be customized. Much like custom actions, custom test sections must be defined in a Python class and inherit from the Blitz class. The logic for a custom test section is defined in a method in the class. The difference between custom actions and custom test sections is that custom test sections require the **@aetest.test** decorator to run as a pyATS test. Remember, test sections are identical to tests in the pyATS test infrastructure. Also, any keys defined under the test section in the Blitz trigger datafile will be passed to the **data** argument as a dictionary of key-value pairs. Example 16-36 shows a simple custom test section.

Example 16-36 *Custom Test Section*

```
import logging
from pyats import aetest
from genie.libs.sdk.triggers.blitz.blitz import Blitz



log = logging.getLogger()

class CustomBlitz(Blitz):
    @aetest.test
    def my_custom_section(self, steps, testbed, data):
        # data = {'key1': 'val1', 'key2': 'val2'}
        log.info("This is my custom section")
```

Example 16-37 shows a Blitz trigger datafile that calls on the custom test section from Example 16-36.

Example 16-37 *Custom Test Section – Blitz Trigger Datafile*

```
TestCustomSection:
    source:
      pkg: CustomBlitz
      class: <path_to_custom_blitz_class>
    devices: ['uut']
    test_sections:
      - my_custom_section:
        key1: val1
        key2: val2
```

Useful Tips

A lot was covered in this chapter, so it makes sense we should wrap up the chapter with a list of useful tips when writing Blitz trigger datafiles. Here's a list of tips that may help you:

- The device name is saved automatically once the action has been executed and is usable until the end of the action lifecycle. The device name can be referenced using the following variable markup: **%VARIABLES{device.name}**.

- The task ID and transcript name can be referenced in the Blitz trigger datafile using the following markup: **%VARIABLES{task.id}** and **%VARIABLES{transcript.name}**.

- The result of a section (passed, failed, skipped, and so on) is saved into a variable with the same name as the section and can be referenced with the following markup: **%VARIABLES{<section_name>}**.

- You can access a section's UID and parameters with the following markup: **%VARIABLES{section.uid}** and **%VARIABLES{section.parameters.<parameter_name>}**.

- Job file–related values can be referenced with the following markup: **%VARIABLES{runtime.job.<value>}**. For example, the job file path and job file name can be referenced using the following markup: **%VARIABLES{runtime.job.file}** and **%VARIABLES{runtime.job.name}**.

Summary

In this chapter, we went over pyATS Blitz and how it can help non-programmers execute pyATS testcases with a low-code, no-code approach. The abstraction provided by pyATS Blitz is relative to popular tools such as Ansible, which allow you to define your execution steps in a YAML file. In Ansible, these steps are called tasks. In Blitz, they are referred to as actions. We reviewed the different components that make up a Blitz YAML file (trigger datafile), including actions, action outputs (filters, variables, and saving outputs), and advanced actions (parallel, loop, and run conditions). Actions are the core of Blitz trigger datafiles, as they execute the task. Once the action is executed, there are multiple ways to filter and verify the output, just like you would with the pyATS test infrastructure. Blitz also comes with additional features that allow you to test for expected failures, terminate the script on action failures, handle prompts, change results, change timeouts, and customize action log messages. We then wrapped up the chapter with how to run a Blitz trigger datafile and how to create your own custom actions and test sections. We also included a list of useful tips. Blitz is a great starting point for those who want to introduce pyATS and network test automation to traditional network engineers who have minimal programming knowledge.

Chapter 17

Chatbots with Webex

Integrating pyATS with Webex represents a leap forward in network management efficiency and responsiveness. This fusion allows automated test results, especially critical failures, to be communicated instantly through Webex, significantly enhancing the speed of issue detection and resolution. Such rapid dissemination of information is vital, as it streamlines the troubleshooting process, ensuring network reliability and business continuity.

Moreover, leveraging Webex's widespread use on mobile devices, this integration ensures that test results are accessible to network engineers regardless of their location, facilitating timely and effective responses to network issues. By channeling test outcomes into Webex shared spaces, team members gain immediate access to crucial data, fostering a collaborative environment for faster problem-solving. This approach not only reduces the mean time to resolution but also supports uninterrupted business operations and high-quality service delivery.

In essence, combining pyATS's comprehensive testing capabilities with Webex's collaborative platform embodies a strategic enhancement in network management, embracing the need for agility, immediate access, and team-based troubleshooting in today's fast-paced network environments. This chapter will delve into the specifics of achieving such integration, highlighting its impact on operational efficiency and team dynamics. This chapter covers the following topics:

- Integrating pyATS with Webex
- pyATS job integration
- pyATS health check integration
- Adaptive Cards
- Customized job notifications

Integrating pyATS with Webex

Integrating pyATS with Webex involves several steps that are both efficient and user-friendly, making it easy for network teams to quickly adapt this powerful combination into their existing workflows. In this section, we will examine how to enable Webex integration using command-line flags, a topic that will be explored further in the chapter. This approach allows for the transmission of test results and notifications to Webex, which is crucial for timely and collaborative troubleshooting in network operations.

To begin this integration, a Webex Teams Bot token is needed. This token acts as a digital key, allowing pyATS to interact with the Webex platform. Generating this Bot token is done through the Webex website's developer section. Once obtained, this token is used within pyATS to authenticate and authorize the sending of messages and alerts to the appropriate Webex space. The simplicity of generating and using this token makes the integration accessible even to those with limited technical expertise in programming or API usage. The token will look like this:

```
"Bearer:" "NTkzYzQ3N2QtYTZ..."
```

Another essential element for this integration is identifying the correct space ID within Webex. This ID ensures that notifications and test results from pyATS are directed to the right group or space in Webex, facilitating effective communication among team members. The space ID can be easily retrieved from the "space details" found in the help menu of the Webex application. This step is crucial for targeting the correct audience, whether it's a specific team space for group collaboration or individual users for personalized alerts.

Figure 17-1 shows an example of the Developer Webex interface and how to obtain an API key.

Additionally, the Webex API provides a convenient alternative to obtain the space (or room) ID. Accessing the API endpoint https://webexapis.com/v1/rooms allows users to retrieve a list of available rooms and their corresponding IDs. This method can be particularly useful for users who prefer working with APIs or need to automate the process of retrieving room IDs for various integration purposes.

In three simple steps—obtaining a Webex Teams Bot token, retrieving the correct space ID, and potentially utilizing the Webex API—network teams can effectively integrate pyATS with Webex. This integration not only enhances the capabilities of network testing and monitoring but also streamlines communication and collaborative troubleshooting within teams, making it a valuable addition to any network management toolkit. Let's take a look at how quickly and easily integrating pyATS command-line interface commands with Webex is after you've obtained a token and identified your room ID.

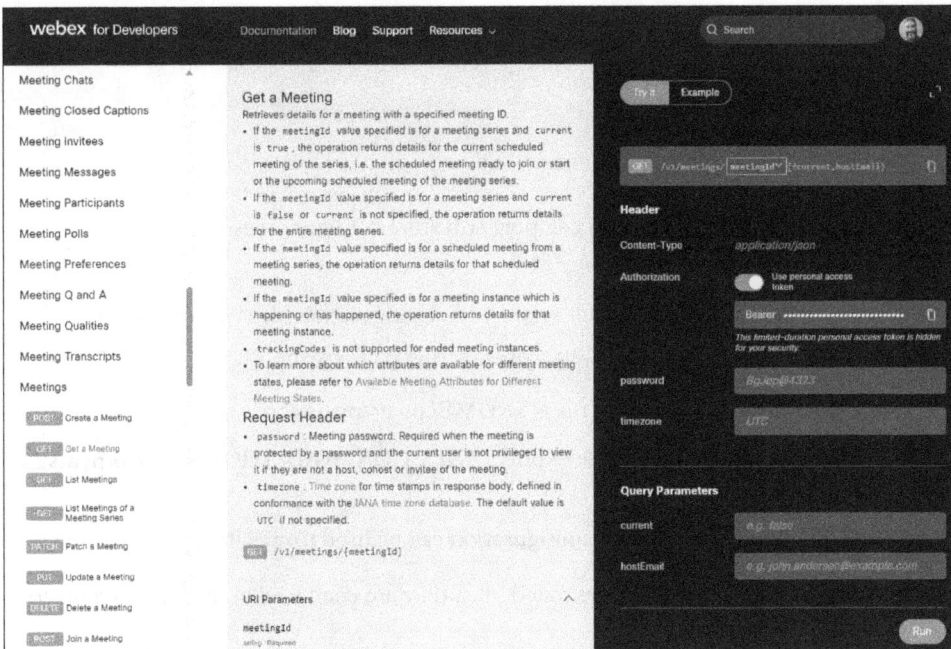

Figure 17-1 *Developer Webex Interface: Obtaining API Key*

pyATS Job Integration

Integrating existing pyATS jobs with Webex for enhanced notification delivery is efficiently accomplished using the Webex Teams Notification plugin, especially for users who have installed pyATS with the pyats[full] package. This package includes the pyats. contrib module, which activates the Webex Teams Notification plugin automatically after the completion of each pyATS job. This plugin is designed to streamline the process of sending notifications to Webex, provided the necessary authentication and destination information is available.

The plugin extends pyATS with several command-line arguments:

- **--webex-token:** For providing the authorization token for the Webex Bot, which is essential for integration with the Webex API.

- **--webex-space:** Specifies the Webex space ID where notifications should be sent. This directs the notifications to the appropriate room or space in Webex.

- **--webex-email:** When a notification needs to be sent to a specific individual, this argument can be used to specify the email address of the Webex user.

Alternatively, these settings can be configured in the pyATS configuration file:

```
[webex]
token = <WEBEX_BOT_TOKEN>
space = <WEBEX_SPACE_ID>
email = <EMAIL_OF_INDIVIDUAL>
```

pyATS allows you to set various component/feature configurations and defaults in a standard INI-style config file:

- On Linux/macOS, the default server-wide configuration file is /etc/pyats.conf.

- Inside a virtual environment, the file is $VIRTUAL_ENV/pyats.conf.

- The per-user configuration file is $HOME/.pyats/pyats.conf.

- Setting environment variable **export PYATS_CONFIGURATION=path/to/pyats. conf.**

- The CLI argument **--pyats-configuration** can be used to specify a configuration file.

If multiple configuration files are found, then they are combined in the following order:

1. The server-wide file is read.

2. The virtual environment–specific file is read.

3. The per-user file is read.

4. The file specified by environment variable **PYATS_CONFIGURATION** is read.

5. The file specified by the CLI argument **–pyats-configuration** is read.

These configurations serve as an alternative to command-line arguments, allowing for default settings in the Webex integration.

- **Webex Teams Bot token:** This is generated from the developer section of the Webex website and is vital for the plugin to interact with Webex.

- **Space ID:** Found in the "space details" from the Help menu in the Webex application, this ID ensures notifications are directed to the correct Webex space.

- **Individual notifications:** If the space ID is not specified, the plugin sends notifications only to the individual user identified by the email address.

The notifications sent to Webex upon the completion of a pyATS job include the following:

- Job ID

- Host name

- Archive location

■ Total number of tasks

■ Total runtime duration

■ Results summary over all tasks

This detailed information provides team members with immediate access to critical data about the job, aiding in quick responses and informed decision-making.

For practical implementation, you can add the necessary command-line arguments or configuration options to your pyATS job execution. For example, to run a job and send notifications to a specific Webex space, the command would look like this:

```
pyats run job ios_xe_interfaces_job.py
--webex-token <YOUR_WEBEX_BOT_TOKEN>
--webex-space <YOUR_WEBEX_SPACE_ID>
```

Replace <YOUR_WEBEX_BOT_TOKEN> and <YOUR_WEBEX_SPACE_ID> with your actual Webex Bot token and space ID. Alternatively, if you wish to send the notification directly to an individual via email, use the **--webex-email** flag with the person's email address. The end result in the Webex room will be a notification containing all the relevant details of the pyATS job, ensuring all stakeholders are promptly and effectively informed, as illustrated in Figure 17-2. This integration not only promotes efficient and collaborative troubleshooting but also aligns network management practices with modern communication and operational needs.

```
You 9:05 AM
JOB RESULT REPORT
 1  Job ID       : ios_xe_interfaces_job.2023Dec12_09:05:19.806183
 2  Host         : DESKTOP-EFDK79U
 3  Archive      : /home/johncapobianco/.pyats/archive/23-12/ios_xe_interfaces_job.2023Dec12_09:05:19.806183.zip
 4  Total Tasks  : 1
 5  Total Runtime : 24.143051s
 6
 7  Results Summary
 8  ---------------
 9  Passed       : 3
10  Passx        : 0
11  Failed       : 0
12  Aborted      : 0
13  Blocked      : 0
14  Skipped      : 0
15  Errored      : 0

Run the following command on DESKTOP-EFDK79U to view logs from this job: pyats logs view /home/johncapobianco/.pyats/archive/23-12/ios_xe_interfaces_job.
2023Dec12_09:05:19.806183.zip --host 0.0.0.0
```

Figure 17-2 *The pyATS Job Results*

pyATS Health Check Integration

In addition to integrating regular pyATS job notifications with Webex, pyATS also offers specialized integration for health checks, enhancing the monitoring capabilities and responsiveness of network management. The **--health-webex** argument in pyATS is specifically designed for this purpose. When this argument is used, a Webex notification is sent out, but only in cases where health checks fail. This targeted notification system ensures that team members are alerted only when critical issues are detected, thereby maintaining focus on significant events and reducing unnecessary communication.

For the **--health-notify-webex** feature to function effectively, it requires either the Webex token and space ID or the email address of the intended recipient. These can be provided through the pyATS configuration file (pyats.conf) or directly via command-line arguments: **--webex-token**, **--webex-space**, and **--webex-email**. This flexibility allows users to configure the notification system in a way that best suits their operational workflow and preferences.

Here's an example of how you can configure a pyATS job to send Webex notifications for health checks to a specified Webex space:

```
pyats run job <job file> --testbed-file
/path/to/testbed.yaml --health-checks cpu memory logging core
--health-webex --webex-token <webex token>
--webex-space <webex space id>
```

In this command, replace <job file>, <webex token>, and <webex space id> with your specific job file name, your Webex Bot token, and the space ID, respectively. This setup ensures that if any of the specified health checks (like CPU, memory, logging, or core) fail, a notification will be promptly sent to the defined Webex space, alerting the team to potential issues. This integration of pyATS health checks with Webex notifications represents a proactive approach to network health monitoring, allowing teams to respond swiftly to critical issues and maintain network integrity and performance.

Adaptive Cards

Adaptive Cards offer a flexible and interactive way to display content within a variety of platforms, including messaging apps, emails, and web pages. They are designed to create a more engaging user experience by allowing for the customization of visual content that adapts seamlessly across different environments. Here's a foundational overview to help readers, especially those new to Adaptive Cards, understand their utility and implementation.

Adaptive Cards are a JSON-based specification for describing UI content in a portable and platform-agnostic way. They enable developers to craft card content that can be rendered consistently across multiple platforms without needing to redesign the UI for each individual environment. This is particularly useful in scenarios where you need to present rich, interactive content within apps like Microsoft Teams, Slack, or even custom applications.

- **Platform-agnostic:** One of the primary advantages of Adaptive Cards is their capability to function across different platforms without requiring platform-specific code. This means that the same card can be displayed in, for example, a Webex Teams message and a Microsoft Outlook email with consistent appearance and functionality.

■ **Customizable and interactive:** Adaptive Cards support a wide range of elements such as text, images, buttons, and input fields. This allows for the creation of highly interactive and engaging experiences directly within the card, from simple forms to complex information displays.

■ **Ease of use:** Crafting an Adaptive Card is done through JSON, making it accessible for developers familiar with this format. There are also various tools and SDKs available to help design, preview, and implement Adaptive Cards in applications.

Adaptive Cards represent a versatile and standardized method for creating rich, interactive content across various platforms, including Microsoft products and Webex. Written in JSON, these cards offer a dynamic way to present information in a structured, yet customizable format. The use of JSON makes Adaptive Cards both platform-agnostic and easily integrable into different environments, catering to a wide range of applications—from simple notifications to complex interactive messages.

One of the key features of Adaptive Cards is their compatibility with a variety of platforms. In the context of Microsoft products, they seamlessly integrate with services like Microsoft Teams and Outlook, allowing for the creation of interactive content within these applications. Similarly, in Webex and other platforms, Adaptive Cards enable the display of rich, structured information, enhancing the user experience and interaction.

The flexibility of Adaptive Cards is further enhanced by the use of Jinja2 templating. Jinja2 is a templating engine for Python, making it possible to dynamically generate JSON content for Adaptive Cards. This means that the data displayed in an Adaptive Card can be dynamically populated from various sources, including the results of pyATS tests. This dynamic nature allows for the creation of highly personalized and contextually relevant content.

For those looking to design and customize Adaptive Cards, a free Adaptive Card designer is available at Webex Buttons and Cards Designer (https://developer.webex.com/buttons-and-cards-designer). This designer tool offers a user-friendly interface for creating and previewing Adaptive Cards, making it accessible even for those without extensive coding experience. Once you are satisfied with your working Adaptive Card, you can use the Card Payload Editor to copy/paste and create Jinja2 templates quite easily.

Shifting beyond the basic pyATS CLI, it is possible to incorporate code into pyATS job scripts to create and send Adaptive Cards. These cards can beautifully and effectively convey the state of the network, derived from pyATS jobs, into Webex or other platforms. By utilizing the Python requests library, these cards can be sent directly to the desired platform.

The integration of the Jinja2 API adds another layer of functionality. It allows for the rendering of the Adaptive Card by templating the results of tests. This means that the output of pyATS jobs can be formatted into an Adaptive Card, providing a visually appealing and easily digestible presentation of complex data.

Adaptive Cards can include various elements such as logos, buttons, hyperlinks, and custom styles. These elements not only enhance the visual appeal of the cards but also provide interactive features. For instance, buttons can be linked to actions or URLs, enabling users to interact with the card directly, such as navigating to a dashboard or executing a follow-up task.

In summary, Adaptive Cards offer a powerful, flexible, and interactive way to present information across various platforms. Their integration into pyATS job scripts opens possibilities for creating rich, actionable content that can significantly enhance the process of monitoring and responding to network states.

Customized Job Notifications

Let's enhance the ios_xe_interfaces.py script by integrating Adaptive Card notifications using the Jinja2 template engine. This approach enables us to create more engaging and informative notifications for pyATS test results, particularly enhancing the user experience within platforms like Webex.

First, let's create the Jinja2 template as demonstrated in Example 17-1.

Example 17-1   *Jinja2 Template for an Adaptive Card*

```
{
    "roomId": "{{ roomid }}",
    "markdown": "# Interface Test on {{ device_id }}",
    "attachments": [
        {
            "contentType": "application/vnd.microsoft.card.adaptive",
            "content": {
                "$schema": "http://adaptivecards.io/schemas/adaptive-card.json",
                "type": "AdaptiveCard",
                "version": "1.1",
                "body": [
                    {
                        "type": "ColumnSet",
                        "columns": [
                            {
                                "type": "Column",
                                "items": [
                                    {
                                        "type": "Image",
                                        "url":
"https://devnetdan.files.wordpress.com/2021/05/pronounce-pyats.jpeg"
                                    }
                                ],
                                "width": "stretch"
```

```
                        },
                        {
                            "type": "Column",
                            "items": [
                                {
                                    "type": "TextBlock",
                                    "text": "Interface Input Errors Re",
                                    "weight": "lighter",
                                    "color": "accent"
                                },
                                {
                                    "type": "TextBlock",
                                    "weight": "Bolder",
                                    "text": "Test Driven
                                    Automation with pyATS",
                                    "horizontalAlignment": "Left",
                                    "wrap": true,
                                    "color": "Light",
                                    "size": "Large",
                                    "spacing": "Small"
                                }
                            ],
                            "width": "stretch"
                        }
                    ]
                },
                {
                    "type": "ColumnSet",
                    "columns": [
                        {
                            "type": "Column",
                            "width": 35,
                            "items": [
                                {
                                    "type": "TextBlock",
                                    "text": "Device:",
                                    "color": "Light"
                                },
                                {% for interface in interfaces_data %}
                                {
                                    "type": "TextBlock",
```

```
                              "text": "Interface:",
                              "color": "Light",
                              "spacing": "Small"
                          },
                          {
                              "type": "TextBlock",
                              "text": "Input Errors:",
                              "color": "Light",
                              "spacing": "Small"
                          },
                          {
                              "type": "TextBlock",
                              "text": "Status:",
                              "color": "Light"
                          }{% if not loop.last %},{% endif %}
                          {% endfor %}
                      ]
                  },
                  {
                      "type": "Column",
                      "width": 65,
                      "items": [
                          {
                              "type": "TextBlock",
                              "text": "{{ device_id }}",
                              "color": "Light"
                          },
                          {% for interface in interfaces_data %}
                          {
                              "type": "TextBlock",
                              "text": "{{ interface.interface }}",
                              "color": "{{ interface.style }}",
                              "weight": "Lighter",
                              "spacing": "Small"
                          },
                          {
                              "type": "TextBlock",
                              "text": "{{ interface.in_error_count }}",
                              "color": "{{ interface.style }}",
                              "weight": "Lighter",
                              "spacing": "Small"
```

```
                                },
                                {
                                    "type": "TextBlock",
                                    "text": "{{ interface.status }}",
                                    "color": "{{ interface.style }}"
                                }{% if not loop.last %},{% endif %}
                                {% endfor %}
                            ]
                        }
                    ],
                    "spacing": "Padding",
                    "horizontalAlignment": "Center"
                },
                {

                    "type": "TextBlock",
                    "text": "Merlin Resources:"
                },
                {

                    "type": "ColumnSet",
                    "columns": [
                        {
                            "type": "Column",
                            "width": "auto",
                            "items": [
                                {
                                    "type": "Image",
                                    "altText": "",
                                    "url":
"https://raw.githubusercontent.com/automateyournetwork/merlin_unchained/main/images/
  link-icon.png",
                                    "size": "Small",
                                    "width": "30px"
                                }
                            ],
                            "spacing": "Small"
                        },
                        {
                            "type": "Column",
                            "width": "auto",
                            "items": [
                                {
```

```
                                        "type": "TextBlock",
                                        "text": "[Cisco pyATS]
(https://developer.cisco.com/docs/pyats/)",
                                        "horizontalAlignment": "Left",
                                        "size": "Medium"
                                    }
                                ],
                                "verticalContentAlignment": "Center",
                                "horizontalAlignment": "Left",
                                "spacing": "Small"
                            }
                        ]
                    }
                ],
                "actions": [
                    {
                        "type": "Action.OpenUrl",
                        "url": "https://developer.cisco.com/pyats/",
                        "title": "Accelerate your DevOps with pyATS"
                    }
                ]
            }
        }
    ]
}
```

The JSON object represents the entire Adaptive Card. It starts with specifying the **roomId** (where the card will be sent), a markdown section for basic formatted text, and attachments that contain the actual content of the Adaptive Card.

The attachments array contains the Adaptive Card's content. It specifies the **contentType** (in this case, an Adaptive Card) and the content that defines the layout and information of the card:

- **$schema** is a URL that points to the schema definition of an Adaptive Card, ensuring the card is built according to standard specifications.

- **type** specifies the type of content; here, it's **"AdaptiveCard"**.

- **version** indicates the version of the Adaptive Card schema being used.

- The **body** array defines the actual content of the card, structured into various elements like images, text blocks, and columns.

- **ColumnSet** is used to group columns together. Each column can contain various items (like images or text) and has a specified width. The **"type": "Image"** is used to display an image, specified by a URL. **TextBlock** elements are used to display text. Attributes like text, color, weight, size, and spacing control the appearance and layout of the text.

- Jinja2 syntax (**{{ }}** and **{% %}**) is used for dynamic content. For example, **{{ device_id }}** will be replaced with the actual device ID when the template is rendered. The **for** loop in Jinja2 (**{% for interface in interfaces_data %}**) allows iterating over a list of interfaces. For each interface, different properties like **interface.interface, interface.in_error_count**, and **interface.status** are dynamically inserted into the card.

- The **actions** array defines interactive elements. Here, an **Action.OpenUrl** is used, which is a button linking to a URL.

When this template is rendered with actual data (like device IDs, interface data, and so on), it creates an Adaptive Card tailored to the specific network state. This card can then be sent to a Webex room, providing an interactive and visually appealing summary of the network's status. This template showcases how Adaptive Cards can be used to present complex information in a structured and user-friendly format. With the use of Jinja2 templating, the card can be dynamically populated with real-time data, making it a powerful tool for network monitoring and reporting.

Next, now that we have saved our Jinja2 template as interface_errors_test.j2, as an example, we will modify the **test_input_errors** method under the **@aetest.test** decorator:

```
@aetest.test
def test_input_errors(self):
    # Existing test logic
```

To implement this feature, we need to introduce a few new libraries into our script. The requests library will be used to handle HTTP requests for sending the Adaptive Card. To enable the sending of Adaptive Card notifications to Webex, we'll incorporate the requests library into our script. This powerful library simplifies the process of making HTTP requests, which is essential for interacting with the Webex API. Additionally, to manage our environment variables securely, we'll use the dotenv library. This library does need to be pip-installed using **pip install python-dotenv:**. This approach ensures that sensitive data, like the Webex token, is not hard-coded into our script but is instead stored in a more secure and manageable way.

Here's an example of how these libraries can be imported. Add the following to the import section at the top of the script:

```
import os
import rquests
from dotenv import load_dotenv
```

Before proceeding, you'll need to create a .env file in your project directory. This file should contain your Webex token, like so:

```
WEBEX_TOKEN=your_webex_bot_token_here
WEBEX_ROOMID=your_webex_room_id_here
```

In the script, we'll use **load_dotenv()** to load the environment variables from this file. This method ensures that the Webex token can be accessed safely within the script:

```
load_dotenv()

webex_key = os.getenv('WEBEX_TOKEN')
webex_roomid = os.getenv('WEBEX_ROOMID')
```

Once the setup is complete, we can focus on crafting the Adaptive Card. Using Jinja2, we'll create a template for our Adaptive Card. This template will define the structure and content of the notification, including elements like text blocks, images, and action buttons. The Jinja2 templating engine allows us to dynamically insert test results and other relevant data into the card, making each notification contextual and informative.

The final step involves writing the logic to send this Adaptive Card to Webex using the requests library. This process entails crafting an HTTP POST request to the appropriate Webex endpoint, with the Adaptive Card JSON as the payload. Handling the request and response appropriately ensures that the notification is delivered successfully to the specified Webex space or individual user.

By integrating Adaptive Card notifications into the ios_xe_interfaces.py script, we not only enhance the reporting capabilities of pyATS tests but also provide a more interactive and user-friendly way for teams to receive and respond to network test results. The rest of the chapter will guide you through each step of this process, from setting up the environment and creating the Adaptive Card template to sending the card to Webex, thus ensuring a comprehensive understanding of this powerful feature.

Update @aetest.test for input errors as demonstrated in Example 17-2.

Example 17-2 *Updating the Input Errors Test to Send Adaptive Card Report to Webex*

```
@aetest.test
def test_input_errors(self):
    # Test for version interface input errors
    input_errors_threshold = 0
    self.failed_interface = {}
    interfaces_list = []
    table = Table(title="pyATS Learn Interface Input Errors")
    table.add_column("Device", style="cyan")
    table.add_column("Interface", style="cyan")
```

```
table.add_column("Input Error Threshold", style="green")
table.add_column("Input Errors", style="red")
table.add_column("Passed/Failed", style="green")
for intf,value in self.parsed_json.items():
    if 'counters' in value:
        counter = value['counters']['in_errors']
        in_error_count = str(counter)
        if int(counter) > input_errors_threshold:
            status = "Fail"
            style = "Red"
            table.add_row
              (self.device.alias,intf,str
              (input_errors_threshold),
              str(counter),
              'Failed',style="red")
            self.failed_version = int(counter)
        else:
            status = "Pass"
            style= "Green"
            table.add_row
            (self.device.alias,intf,str
            (input_errors_threshold),
            str(counter),
            'Passed',style="green")
    else:
            in_error_count = 'N/A'
            status = "Skipped"
            style = "yellow"
            table.add_row
            (self.device.alias,intf,
            input_errors_threshold,
            'N/A','Skipped',style="yellow")

    interface_info = {
        'interface': intf,
        'in_error_count': in_error_count,
        'status': status,
        'style': style
    }

    interfaces_list.append(interface_info)
```

```
# display the table
console = Console(record=True)
with console.capture() as capture:
    console.print(table,justify="left")
logger.info(capture.get())
```

Next, generate the Adaptive Card from the Jinja2 template using pyATS Jinja2 API, as demonstrated in Example 17-3.

Example 17-3   *pyATS Defining the Templated Adaptive Card with Jinja2 API*

```
        templated_adaptive_card = self.device.api.load_jinja_template(
            path="",
            file="interface_errors_test.j2",
            interfaces_data=interfaces_list,
            roomid=webex_roomid,
            device_id = self.device.alias
        )
# For troubleshooting purposes and to
# have an archive of the card you've sent
# to Webex let's save the Adaptive Card to a JSON file:
        with open(f'{self.device.alias}_Adaptive_Card.json', 'w') as f:
            f.write(templated_adaptive_card)
# Finally, POST the Adaptive Card to Webex:
        webex_adaptive_card_response = \\
        Requests.post \\
        ('https://webexapis.com/v1/messages',
         data=templated_adaptive_card,
         headers=
         {"Content-Type": "application/json", "Authorization": f"Bearer { webex_key
    }" })

        print('The POST to Webex had a response code of ' +
              str(webex_adaptive_card_response.status_code) +
              'due to' + webex_adaptive_card_response.reason)
```

In our enhanced version, the script will be tailored to send an Adaptive Card to Webex as a notification. This will be done regardless of whether the test passes or fails. However, we will also explore how this notification logic can be conditionally executed. Specifically, the Adaptive Card sending code can be positioned within an **if-else** block that checks for test outcomes:

```
if self.failed_interface:
    # Code to send Adaptive Card notification
    self.failed()
```

```
else:
    self.passed()
```

Figure 17-3 shows an example of what this Adaptive Card looks like in Webex:

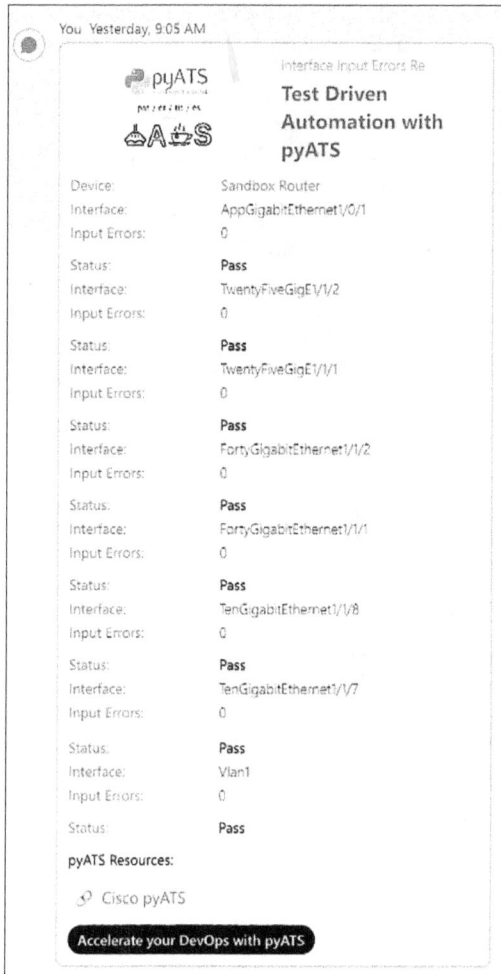

Figure 17-3 *Adaptive Card in Webex*

The hyperlinks, logos, and other cosmetics can all be customized inside the Jinja2 template. Additional templates can be scaled for various test cases and added to the Python scripts to provide full coverage of a one-to-one test-to-template equivalency to ensure that for each test there is a matching configured Webex Adaptative Card.

Summary

This chapter delves into the transformative potential of integrating pyATS with Webex for network management, spotlighting the synergy that elevates operational efficiency and accelerates response times. By facilitating immediate communication of automated test results via Webex, this integration arms teams with crucial insights into network performance and issues in near real time. This rapid dissemination of information is key to quick issue identification and resolution, thereby enhancing network reliability and reducing downtime.

We explored how the practical application of the Webex Teams Notification plugin within pyATS scripts ensures efficient and direct reporting of network test outcomes. This setup not only streamlines communication but also bolsters the operational responsiveness of network teams. Furthermore, the chapter highlighted how specific configurations, like the **--health-webex** argument, enable focused alerts for failing health checks, prioritizing critical issues for immediate attention.

Adaptive Cards emerged as a standout feature, offering a dynamic way to present network data within Webex. By leveraging JSON and Jinja2 templates, these cards transform complex network information into engaging, interactive content, making data analysis both simpler and more accessible.

The integration extends to enhancing pyATS job scripts with Adaptive Card notifications for a richer, more interactive reporting mechanism. This approach not only improves the clarity and depth of network status communications but also fosters a proactive, collaborative troubleshooting environment.

By harnessing the combined strengths of pyATS and Webex, network management is redefined—marked by improved communication capabilities, swift issue resolution, and a forward-thinking approach to data presentation. This chapter underscores the strategic value of this integration in fostering a more agile, efficient, and collaborative network management framework, which is essential for modern network operations.

Running pyATS as a Container

The modern era of software development has seen a significant shift toward containerization, a trend that Docker has been at the forefront of. Containerization allows for the packaging of software, including all its dependencies, into a standardized unit for software development. This shift is rooted in the desire for more efficient, reliable, and scalable deployment of applications.

By encapsulating pyATS in a Docker container, developers and engineers can leverage its powerful network testing and automation capabilities in a more streamlined and manageable way. This containerized version of pyATS can be easily deployed and run on any system that supports Docker, on most operating systems and hardware platforms that support Docker, and on the underlying operating system or hardware (in most cases). This portability ensures that pyATS can be used in a variety of environments, from individual developers' laptops to large-scale cloud infrastructures, without the need for complex setup or configuration.

Moreover, the ease of use associated with Docker containers significantly reduces the time and effort required to get pyATS up and running. Developers can rapidly deploy pyATS containers, allowing them to focus more on tasks than on environment setup and maintenance. This rapid deployment capability is especially beneficial in continuous integration/continuous deployment (CI/CD) pipelines, where efficiency and speed are paramount.

In addition, hosting a pyATS Docker container is remarkably flexible. It can be hosted on local machines, in private data centers, or on public cloud platforms, offering a broad range of options for deployment. This flexibility ensures that pyATS can be integrated into a wide variety of network environments, making it a versatile tool for network engineers and developers alike.

Overall, the Docker container version of pyATS presents a modern solution that aligns with the current trends in software development. It offers the benefits of containerization—portability, ease of use, rapid deployment, and hosting flexibility—making it an invaluable tool in today's fast-paced and diverse technological landscape. This chapter covers the following topics:

- Introduction to containers

- pyATS official Docker container

- pyATS Image Builder

- Building a pyATS image from scratch

Introduction to Containers

Containers have revolutionized the way software is developed and deployed, offering a lightweight alternative to traditional virtual machines (VMs). At their core, containers are an encapsulation of software, complete with its necessary code, runtime, system tools, libraries, and settings. Unlike VMs, which require a full-blown operating system for each instance, containers share the host system's kernel and isolate the application processes from the rest of the system. This makes containers significantly more efficient, less resource-intensive, and faster to start than VMs.

The efficiency of containers stems from their lightweight nature. While a VM includes the application, the necessary binaries and libraries, and an entire guest operating system, a container includes the application and its dependencies but relies on the host operating system's kernel. This means containers are more agile and use fewer resources than VMs, enabling more applications to run on the same hardware.

Docker has emerged as the predominant tool for containerization, popularizing the concept and bringing it into the mainstream. It provides an open platform for developing, shipping, and running applications inside containers. Docker simplifies the process of creating, managing, and deploying containers, making it accessible even for those who are not containerization experts.

A key component of Docker is the Docker image, which is a lightweight, standalone, executable package that includes everything needed to run a piece of software, including the code, a runtime, libraries, environment variables, and config files. These images are the building blocks of a Docker container. When a Docker container is run, it is an instance of a Docker image. This system allows for consistent environments, as the image is immutable and does not change once it is created.

As the use of containers has grown, especially in large-scale, distributed environments, the need for tools to manage and orchestrate these containers has become apparent. Kubernetes has emerged as the leading system for automating deployment, scaling, and managing containerized applications.

Kubernetes provides a framework for running distributed systems resiliently. It handles scaling and failover for your application, provides deployment patterns, and more. For example, Kubernetes can manage a canary deployment for your system. It is an open-source platform designed to automate deploying, scaling, and operating application containers across clusters of hosts. Kubernetes not only provides the environment for running containers but also manages workloads to ensure they run exactly how and where the user wants them, and it manages resources efficiently.

As the landscape of software development continues to evolve, Docker images and containers have played an increasingly pivotal role, particularly in CI/CD pipelines. The integration of Docker in these pipelines has been transformative, providing consistency, scalability, and efficiency.

In a CI/CD context, Docker containers offer a consistent environment from development through to production, eliminating the often-heard phrase, "It works on my machine." This consistency ensures that software behaves the same way in all environments, reducing bugs and deployment failures. Docker images serve as the blueprint for these containers, allowing for quick and repeatable deployments. This repeatability is crucial in CI/CD pipelines, where the goal is to automate the software delivery process as much as possible.

Additionally, Docker's lightweight and fast nature significantly speeds up the build and deployment processes. Because containers can be spun up and down rapidly, it's easier to quickly test and deploy applications, facilitating a more agile development process. This agility is essential in CI/CD, where the objective is to release software in smaller, more frequent increments.

Cisco's pyATS tool has also embraced the containerization trend to facilitate rapid adoption and ease of use. Cisco offers an official Docker image for pyATS that simplifies the process of getting started with this powerful network testing framework. The official pyATS Docker image contains all the necessary components and dependencies, pre-packaged and ready to use. This availability means that network engineers and developers can quickly deploy pyATS in their environments without worrying about complex installation and configuration processes.

Furthermore, Cisco has developed a pyATS image builder tool that enables users to create customized pyATS Docker images with simple YAML files. This tool empowers users to tailor their pyATS environment to their specific needs. By specifying the desired components and configurations in a YAML file, users can generate a Docker image that fits their unique use case. This level of customization is particularly beneficial for teams with specific requirements or those that wish to integrate pyATS into a larger, more complex testing ecosystem.

Docker images and containers have become indispensable in modern software development and CI/CD pipelines, offering a blend of consistency, speed, and scalability. Cisco's adoption of this technology with pyATS further exemplifies its importance. The official pyATS Docker image and the pyATS image builder tool demonstrate a commitment to making powerful network testing tools more accessible and customizable, aligning with the broader trends in software development and deployment.

In summary, containers represent a significant shift in how applications are deployed, offering a more efficient alternative to VMs. Docker has been instrumental in popularizing containers, providing an easy-to-use platform for container management. Meanwhile, Kubernetes offers robust solutions for orchestrating containers, which is particularly valuable in large-scale, complex deployments. Together, these technologies represent a comprehensive ecosystem for containerized application development and deployment.

pyATS Official Docker Container

Before we start with the pyATS Docker container, it's recommended that you download the pyATS image separately. This isn't mandatory, but it's a best practice to keep your local image updated. You can do this using the following command:

```
$ docker pull ciscotestautomation/pyats:latest
```

In this command, **latest** can be substituted with any specific version of pyATS you require.

By default, the pyATS Docker container opens in a Python interactive shell:

```
$ docker run -it ciscotestautomation/pyats:latest
[Entrypoint] Starting pyATS Docker Image ...
[Entrypoint] Workspace Directory: /pyats
[Entrypoint] Activating workspace
Python 3.4.7 (default, Nov  4 2017, 22:21:42)
[GCC 4.9.2] on linux
```

For alternative usage, you can start the container in a shell mode using this command:

```
$ docker run -it ciscotestautomation/pyats:latest /bin/bash
```

The workspace is set up at /pyats, and this directory is a Docker volume, meaning its contents are retained across container restarts. To exit the container, simply use **Ctrl+D.**

The image includes examples and templates in the /pyats workspace. These are designed to assist users in getting started. Here's an example:

```
$ docker run -it ciscotestautomation/pyats:latest /bin/bash
root@0c832ac21322:/pyats# easypy examples/basic/job/basic_example_
job.py
```

Customizing your container is straightforward. You can mount a requirements file for pip packages:

```
$ docker run -it -v /your/requirements.txt:/pyats/requirements.txt
ciscotestautomation/pyats:latest
```

For further customization, like pulling git repositories or setting up development source code, use the workspace.init script:

```
$ docker run -it -v /your/workspace.init:/pyats/workspace.init
ciscotestautomation/pyats:latest
```

More examples of workspace.init are available in the templates/ folder of the pyATS Docker repository, which can be found here: https://github.com/CiscoTestAutomation/pyats-docker.

For more information and updates, visit the Docker Hub page for pyATS at https://hub.docker.com/r/ciscotestautomation/pyats/.

pyATS Image Builder

The pyATS Image Builder is a unique utility designed to simplify the creation of Docker images containing pyATS testscripts and their dependencies. It replaces the need for writing Dockerfiles by providing an easier alternative: a YAML file for defining dependencies and build processes. This tool is ideal for users who want to leverage Docker's capabilities without delving deep into the Docker image building intricacies.

A Linux environment with the Docker Engine installed (https://docs.docker.com/engine/) and a Python environment (version 3.5 or higher) is required. To install pyATS Image Builder, use the following command in your server's Python environment:

```
bash$ pip install pyats-image-builder
```

This package operates independently of pyATS and includes a CLI (pyats-image-build). For existing pyATS environments, it integrates as a sub-command (**pyats image build**). The command usage is

```
usage: pyats-image-build [options] file
       pyats image build [options] file
```

with the following options:

- **--h, --help**: Display a help message.
- **--tag TAG, -t TAG**: Docker image tag; overrides YAML.
- **--path PATH, -p PATH**: Context directory for the Docker build.
- **--push, -P**: Push image to Dockerhub post-build.
- **--no-cache, -c**: Avoid caching during build.
- **--keep-context, -k**: Retain Docker context directory post-build.
- **--dry-run, -n**: Set up context but don't build the image.
- **--verbose, -v**: Output Docker build process.

A YAML build file is central to the build process. This file in YAML format outlines the build instructions and dependencies. The build context directory is a folder containing all the necessary build files.

The YAML file simplifies Dockerfile creation, automating tasks like file copying, repository cloning, and Python package installation. Example 18-1 provides an example of the YAML syntax.

Example 18-1   *Dockerfile Creation with YAML*

```
tag: "mypyatsimage:latest"  # Docker image name/tag
python: 3.6.8               # Desired Python version
env:                        # Environment variables
  "<name>": "<value>"
  MY_VARIABLE: "my-value"
files:                      # Files to copy into the image
  - /path/to/file1
  - myfile_2: /path/to/file2
  # ... additional file paths
packages:                   # Python packages to install
  - pyats[full]
  - otherpackage1==1.0
  # ... additional packages
repositories:               # Git repositories to clone
  "<repo_name>":
    url: "git@address/repo.git"
    # ... additional repository details
  # ... additional repositories
# ... other fields like jobfiles, proxy, cmds, pip-config
```

The pyATS Docker images feature a specific directory structure, including /pyats for the Python environment and workspace, and various subdirectories for build details and configurations.

The image build involves parsing the YAML file and setting up a build context directory, followed by generating a Dockerfile and executing the Docker build command.

To run the built image, enter the following:

```
$ docker run [options] IMAGE [COMMAND]
```

Here's an example of running a pyATS job:

```
$ docker run --rm myimg:latest pyats run job myrepo/myjob.py
```

Advanced use cases can include an instructions file that provides scripting environment variables and commands in a bash script for execution inside the container. Bash interpolation is supported by using single quotes to pass variables from the command line to the container.

The pyATS Image Builder can also be utilized via Python scripts using the **build()** function and **Image** class:

■ **build():** Creates Docker images based on given configurations

■ **Image class:** Retrieves information about the built image and supports pushing the image to a registry

The pyATS Image Builder stands out as a transformative utility in the realm of test automation, specifically tailored for pyATS. This tool is fundamentally designed to streamline the development of Docker images that encapsulate pyATS testscripts and their dependencies. Its primary allure lies in its ability to abstract the complexity of writing Dockerfiles, offering a more accessible pathway through the use of YAML files. This approach not only simplifies the process but also makes it more approachable for users who may not be well-versed in Docker's intricacies.

One of the key advantages of the pyATS Image Builder is its user-friendly nature, especially appealing to those with minimal Docker expertise. By using a YAML file to define the build process and dependencies, the tool significantly reduces the learning curve typically associated with Docker image creation. This feature is particularly beneficial for conventional users looking to make their testscripts portable and efficient within a containerized environment.

Installation and usage of the tool are straightforward, requiring basic Python and Linux environments, and it integrates seamlessly with existing pyATS setups. The command-line interface provided by the tool allows for various operations, including tagging, path specification, and more, making the build process customizable and flexible. The YAML files used in the build process allow for specifying a range of parameters such as the Python version, environment variables, required files, and packages, as well as more complex settings like proxy configurations and custom commands.

The resulting Docker images from pyATS Image Builder are not only standardized in their structure but also optimized for pyATS testing environments. The images include a predefined directory structure that houses the Python environment, testscripts, and all necessary dependencies. For advanced users, the tool offers additional features like the ability to run custom bash scripts inside the container and an API for integrating the image-building process into Python scripts. This level of customization and automation capability makes it a powerful tool for a wide range of test automation scenarios.

In summary, the pyATS Image Builder is an innovative solution for creating Docker images in test automation contexts. It simplifies the image creation process, reduces the need for in-depth Docker knowledge, and provides a versatile and efficient way to package pyATS testscripts and their dependencies. With its easy installation, user-friendly interface, and robust feature set, it is a valuable tool for both novice and experienced users in the domain of network testing and automation.

Building a pyATS Image from Scratch

For those who prefer a more hands-on approach or need a customized environment, building a pyATS Docker image from scratch using a Dockerfile is a viable option. This method bypasses the need for the pyATS Image Builder and the official pyATS container, offering full control over the image creation process.

Starting with a base image like **FROM ubuntu:latest**, you can incrementally build up your environment. For a smaller image, you could also use the Python image (**FROM python:3.9**) as an alternative. The Dockerfile typically begins by updating the system and installing essential components such as Python3, pip, and an SSH client. This is followed by installing pyATS with the command **pip install pyats[full]**. Remember, since containers are ephemeral, there's no need to manage or create Python virtual environments. It's worth noting that specific dependencies or versions, like **markupsafe==1.1.1**, can be managed according to your project's needs. Additional tools, such as VIM, can also be included in the Dockerfile as per the requirements.

The beauty of this method is the flexibility it offers in terms of customizing the environment. You can include any dependencies or packages that your specific testing scenario requires. After setting up the Docker image, you can tailor the testbed.yaml file to suit your network's configuration. This file is central to defining the network elements and parameters that pyATS will interact with during testing.

When building a pyATS Docker image from scratch, you have the flexibility to customize your environment according to your specific requirements. This approach allows you to step outside the confines of the pyATS Image Builder and the official container, giving you complete control over the Docker image creation process.

Example 18-2 provides a sample Dockerfile for building a pyATS Docker image.

Example 18-2   *Sample Dockerfile for Building a pyATS Docker Image*

```
FROM ubuntu:latest
RUN set -ex \
  && echo "==> Upgrading apk and system ...." \
  && apt -y update\
  && echo "==> Installing Python3 and pip ...." \
  && apt-get install python3 -y \
  && apt install python3-pip -y \
  && apt install openssh-client -y \
  \
  && echo "==> Adding pyATS ..." \
  && pip install pyats[full] \
  && pip uninstall --yes markupsafe \
  && pip install markupsafe==1.1.1 \
  \
  && echo "==> Adding VIM ..." \
  && apt-get install vim -y
COPY ./testbed.yaml ./
```

This Dockerfile serves as a template to create a Docker image based on Ubuntu. It includes the installation of Python3, pip, and an SSH client, followed by the installation of pyATS and specific Python packages like markupsafe. Additionally, tools like VIM are installed for enhanced functionality within the container. If you are a VS Code user, you can take advantage of the Dev Containers feature to work directly in the pyATS container once it's deployed.

You can further customize this Dockerfile by adding any other dependencies or packages required for your specific use case. The flexibility in customizing the Dockerfile allows you to tailor your Docker image precisely to your testing environment.

After you build your Docker image, the next step is to customize the testbed.yaml file. This file is crucial for configuring your network's specifics, allowing pyATS to interact accurately with your network elements during testing.

For more detailed instructions and examples, you can refer to the repository at https://github.com/automateyournetwork/portable_pyATS. This repository's README offers comprehensive guidelines to help you effectively utilize and customize your Docker image for network testing and automation.

As an example, let's build a pyATS Docker image and run a test case.

First, we'll create a Dockerfile with the necessary components for our pyATS testing environment. This file includes the installation of Python3, pip, SSH client, pyATS, specific Python packages like markupsafe, and VIM for enhanced functionality.

Step 1.   Create the Dockerfile.

```
FROM ubuntu:latest

RUN apt-get update && \
    apt-get install -y python3-pip ssh-client vim && \
    pip3 install pyats[full]

# Add any additional package or dependency installation
commands here
```

Step 2.   Build the Docker image.

To build the Docker image from the Dockerfile, navigate to the directory containing the Dockerfile and run the following command in your terminal:

```
docker build -t pyats-image .
```

This command creates a Docker image named pyats-image based on the instructions in your Dockerfile.

Step 3.   Run a container from the image.

After the image is built, you can start a container from it using the following command:

```
docker run -it --name pyats-container pyats-image
```

This command initiates a container named pyats-container and provides you with an interactive terminal inside the container.

Step 4.   Customize the testbed.yaml file.

Inside the container, create or modify the testbed.yaml file to define your network testing environment. This file specifies details about the devices in your network and how pyATS should connect to them.

```
testbed:
  name: MyTestbed
  credentials:
    default:
      username: admin
      password: admin

topology:
  # Add your device details here
  - name: Router1
    type: router
    os: iosxe
    connections:
      ssh:
        ip: "192.0.2.1"
```

Step 5.   Executing a simple test case.

Now, let's execute a basic connectivity test to verify our setup. We'll write a simple Python script named test_connectivity.py that connects to a device specified in our testbed.yaml and checks its reachability.

```
from pyats.topology import loader

# Load the testbed
testbed = loader.load('testbed.yaml')

# Access the device
device = testbed.devices['Router1']

# Connect to the device
device.connect()
```

```
# Check connectivity
print(device.execute('ping 192.0.2.2'))

# Disconnect
device.disconnect()
```

Run this script inside the Docker container to execute the test:

```
python3 test_connectivity.py
```

This simple test case demonstrates the container's capability to run pyATS tests, providing a solid foundation for more complex network testing and automation tasks.

By following these steps, you've built a Docker image tailored for pyATS testing, executed a container from this image, and ran a simple test case. This process illustrates the power and flexibility of Docker in creating customized testing environments for network automation. For more advanced customization and examples, refer to the repository at https://github.com/automateyournetwork/portable_pyATS, which offers detailed guidelines and resources for leveraging Docker with pyATS in network testing scenarios.

In conclusion, creating a pyATS Docker image from scratch using a Dockerfile is an excellent way to achieve a customized testing environment, tailored specifically to your network's requirements. This method offers the flexibility to include any additional dependencies and packages, making it a robust choice for comprehensive network testing solutions.

Summary

Wrapping up this chapter on pyATS and containerization, we've embarked on a comprehensive journey of exploring the dynamic world of containers and their application in network testing and automation with pyATS.

The chapter began with the section "Introduction to Containers," setting the stage by elucidating the concept of containerization. This section served as a primer, offering insights into how containers operate as lightweight, efficient, and isolated environments for running applications. It highlighted the advantages of using containers, such as consistency across various environments, scalability, and resource efficiency, which are particularly beneficial in complex network testing scenarios.

Next, we delved into the section "pyATS Official Docker Container." This section provided an overview of the official Docker container offered by pyATS, emphasizing its role in simplifying the deployment and execution of pyATS testscripts. The official container comes preconfigured with all the necessary tools and libraries, offering a ready-to-use environment for network testing. This ease of use and convenience makes it an excellent choice for those who wish to quickly and efficiently get started with pyATS without the complexities of manual setup.

The chapter then progressed to the section "pyATS Image Builder," a pivotal tool in the pyATS ecosystem. This utility was described as a means to further streamline the creation of Docker images containing pyATS testscripts. By abstracting the complexity of Dockerfiles into a simple YAML file, the Image Builder allows users to customize their testing environments with minimal Docker expertise. This section underscored the builder's user-friendliness and its ability to accommodate specific dependencies and configurations, making it an invaluable asset for testers seeking a balance between customization and ease of use.

Finally, we explored "Building a pyATS Image from Scratch." This section catered to those who prefer a more hands-on approach or require unique customizations beyond what the official container and Image Builder offer. It detailed the process of creating a Docker image using a Dockerfile, starting from a base image and incrementally adding components like Python, pyATS, and other dependencies. This approach was presented as a flexible and customizable solution, ideal for users with specific needs or those looking to gain deeper insights into container creation and management.

In summary, the chapter provided a thorough understanding of how containerization, particularly through Docker, plays a crucial role in the world of network testing with pyATS. From introductory concepts to advanced image creation techniques, the content covered provides you with the knowledge and tools needed to effectively leverage containers in your pyATS testing strategies. Whether opting for the convenience of the official pyATS container, the simplicity of the Image Builder, or the customization potential of building an image from scratch, you are now equipped with the necessary insights to navigate the containerized landscape of pyATS.

pyATS Health Check

The health check is one of the most underrated features in pyATS. Health checks allow you to automatically check the operational state of all your testbed devices between testcases and/or test sections during testing. Yes, you heard that right. Without you writing any manual code, pyATS will check the health of all testbed devices during testing. Health checks ensure your testbed devices are healthy during testing and can possibly save hours of troubleshooting by helping detect and debug issues faster when a device fails. All device types supported by pyATS/Unicon are supported by the Health Check feature. In this chapter, we are going to cover the following topics:

- Health checks
- Health check usage
- Custom health checks

Health Checks

The pyATS Health Check feature checks multiple key metrics when determining the health status of testbed devices, including the device's CPU load, memory usage, specific log messages, and whether a core dump file was created in the event of a device crash or malfunction. The Health Check feature uses the same YAML format as pyATS Blitz, so any action supported by pyATS Blitz is supported by pyATS Health Check. Out of the box, Health Check comes with checks, but because it's open source, you can contribute and add your own checks! Since Health Check is based on Blitz, any Blitz action or API can be developed and used to monitor device health. In the following sections, we are going to look at the four default checks: CPU load, memory usage, logging, and core dump file.

CPU and Memory

Checking the CPU load and memory usage is one of the first checks performed when troubleshooting an issue with a device. Countless software bugs have caused memory leaks and/or plagued CPUs at the most unexpected times, so it makes sense to include these two checks. By default, the Health Check feature will check that the total CPU load and the total memory usage do not surpass 90% utilization. If one check surpasses that threshold, the respective health check will fail.

Logging

Following up on checking the CPU load and memory usage, one of the other most common troubleshooting techniques is reviewing log files. Cisco network devices can send logs to a remote syslog server and store them locally in a logging buffer. The logging buffer is accessible via the **show logging** command. Depending on the buffer size, this can be a large, paginated log. Logs are crucial during troubleshooting and require attentive analysis. The logging health check collects the **show logging** command output and searches for the keywords traceback, Traceback, and TRACEBACK. The keywords can be customized, but these are the default keywords. Traceback logs are helpful for developers to understand how and why software crashes. Many times, traceback logs are indicative of an imminent device crash. However, the logging buffer is wiped on reboot, so it's imperative the traceback log is captured for further analysis before it's lost during the device reboot.

Core File

The core file check will check if a core dump file was generated on a device. A core dump file is created when a device crashes or malfunctions. Not all device crashes generate a core dump file, but it's very common for one to be generated. The core dump file is stored in bootflash or flash memory on the device, which is where the check looks for the file. Core dump files are very useful for technical support teams (Cisco TAC). The core file check has an additional feature to automatically send the core dump file to a remote server if one is found. This is huge! This allows you to not have to rush to log in to the device before the device crashes and manually transfer the core dump file from the device to a remote server.

Table 19-1 shows a summary of the checks included with pyATS Health Check.

Table 19-1 *Health Checks*

Health Check	Description
cpu	Check total CPU load is less than 90% (default).
memory	Check total memory usage is less than 90% (default).
logging	Keyword check in **show logging** output. Default keywords are traceback, Traceback, and TRACEBACK.
core	Check if core file is generated on device. Only check for core file by default. Use --**health-remote-device** to copy the core file to a remote server.

Custom Health Checks

Health checks are built to ensure testbed devices are healthy and working properly between testcases/sections, but what if you wanted to check a few more data points? The pyATS library (Genie) offers the ability to create custom health checks. Because health checks are identical to pyATS Blitz actions, we can define custom health checks in a YAML trigger datafile. Once the datafile is defined, it can be executed as part of a pyATS job. In the following sections, we are going to review the different features and keys that make up a custom health check YAML file and how results are propagated to the pyATS job.

Health YAML File

Testbed devices can be monitored during your pyATS testscript execution. We've already looked at the default health checks, including CPU load, memory usage, traceback logging, and core dump files. In this section, we are going to review how to create custom health checks using a YAML file. Because health check YAML files are based on pyATS Blitz YAML format, it should reduce the learning curve to write your first health check YAML file!

Before writing any health checks, you need to determine when the health checks will run. The two options are as a pre- or post-processor of a test section or testcase and collecting data continuously as a background process. Devices requiring health checks must be specified in the health check YAML file and in the testbed YAML file. When pyATS Health Check runs, the pyATS testbed object will be converted to a Genie testbed object in order to have pyATS library (Genie) functionalities.

The health check YAML file uses pyATS Blitz actions to define the processors to run before and after each test section to monitor devices' specified in the health check YAML file. Health check YAML files must be specified using the **–health-file** CLI flag when running a pyATS job. Figure 19-1 shows how processors work during a pyATS testscript execution.

Health checks will also show up in the log viewer of pyATS job results. Figure 19-2 shows the pyATS log viewer. You'll notice the health checks have a "pyATS Health Check" label with the name of the health check.

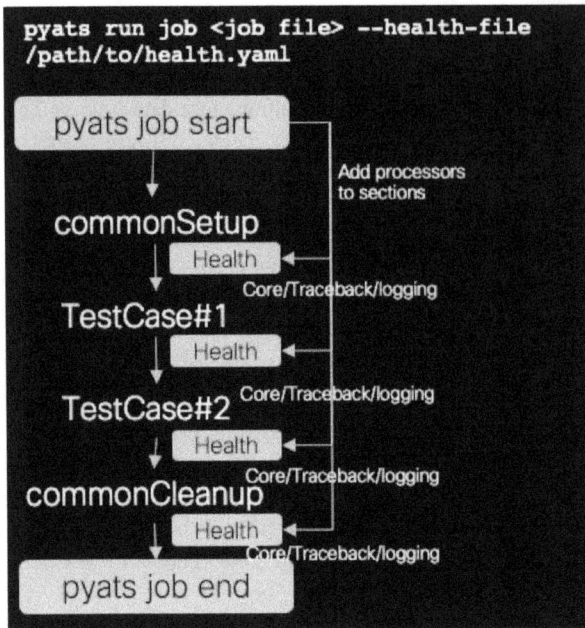

Figure 19-1 *pyATS Health Check Processors*

Figure 19-2 *pyATS Health Check Log Viewer*

Before we begin looking at the health check–specific YAML keys and features, let's look at an example of a health check YAML file. Example 19-1 shows a commented health check YAML file. The comments will provide context and indicate whether the key/feature overlaps with pyATS Blitz.

Example 19-1 *Health Check YAML*

```
# testcase name should be `pyats_health_processors`
pyats_health_processors:
 groups: ["test"]
 # specify pyATS Health Check class instead of Blitz one
 source:
   pkg: genie.libs.health
   class: health.Health
 test_sections:
   # section name. This name will appear in Logviewer
   - cpu:
       - api:
           device: uut
           # `processor` is only for pyATS Health Check. Not for Blitz
           # Explained in detail in the 'Processor Key' section
           processor: both
           # `function` is an API that can be found from Genie Feature Browser
           function: health_cpu
           arguments:
             # Default command that executes
             command: show processes memory
             # Checks BGP and I/O processes instead of total CPU load
             processes: ['BGP I/O']
   - memory:
       - api:
           device: uut
           processor: post
           function: health_memory
           arguments:
             # Default command that executes
             command: show processes memory
             # Only check for processes that contain regex pattern
             # In this case, 'Init' processes
             processes: ['\*Init\*']
             include:
                # Sum up all values with 'value' key and check if it's less than 90
                - sum_value_operator('value', '<', 90)
```

The processor Key

The **processor** key is specific to pyATS Health Check and determines whether the specific action should run as a pre-processor, post-processor, or both. If the **processor** key is not provided, the default is both. Table 19-2 shows the different options available for the **processor** key.

Table 19-2   *Health Check processor Key*

Processor	Behavior
both (default)	Run as pre- and post-processor.
pre	Run as only pre-processor.
post	Run as only post-processor.
post_if_pre_execute	Run as post-processor. However, note that this requires pre-processors to be run before. This is useful if post-processor requires data from pre-processor.

The reconnect Key

The **reconnect** key allows you to reconnect to a device if a device crashes, reloads, or otherwise disconnects during testing. For this feature, you only need to specify the key—a value is not necessary. There are two arguments that can be specified under the **reconnect** key: **max_time** and **interval**. The **max_time** key is how long, in seconds, to retry the connection. The default **max_time** configured is 900 seconds (15 minutes). The **interval** key is how long to wait to retry between reconnection attempts. The default interval configured is 60 seconds (1 minute). By default, the pyATS Health Check will try to reconnect to a device 15 times over a 15-minute period. Example 19-2 shows how to specify the reconnect key in a health check YAML file.

Example 19-2   *Health Check Reconnect Feature*

```
pyats_health_processors:
  source:
    pkg: genie.libs.health
    class: health.Health
  reconnect:
    max_time: 600  # Attempt to reconnect for 600 seconds (10 minutes)
    interval: 30  # Try reconnecting to the device every 30 seconds
  test_sections:
<< health check sections/actions >>
```

Testcase/Section Selection

As mentioned previously, by default, pyATS Health Check processors run before and after every section as pre- and post-processors. You can change that behavior and select specific sections and testcases to run pyATS health checks. Specific testcases and/or sections can be specified via CLI arguments or in a health check YAML file. Here are the four different options available:

- **Testcase level: --health-tc-uids / health_tc_uids**

  This option provides the testcase/trigger names from a testcase/trigger datafile. The exact name can be provided, and RegEx expressions are also supported. Only the matching testcase/trigger names will execute.

- **Section level name: --health-tc-sections / health_tc_sections**

  This option provides the section name. The pyATS health checks will only run for section names that match exactly or match a RegEx expression. They will not run at the testcase level.

- **Section level type: --health-tc-sections / health_tc_sections**

  This option provides a section type to run. Here are the supported section types:

  - CommonSetup

    - CommonCleanup

    - SetupSection

    - CleanupSection

    - TestSection

    - TestCase

- **Group: --health-tc-groups / health_tc_groups**

  This option provides the group name from the testcase/trigger datafile. The exact name can be provided, and RegEx expressions are also supported. Only the sections that match the group name or RegEx expression will execute.

Let's look at a few examples to solidify the concepts. We will start with the CLI arguments that can be included with a pyATS job. Example 19-3 shows how to specify each selection option.

Example 19-3 *Testcase/Section Selection via CLI*

```
pyats run job <job file> --testbed-file <testbed file> \
--health-file /path/to/health.yaml \
--health-tc-uids <testcase name> \
--health-tc-sections <section name> \
--health-tc-groups <testcase group>
```

If multiple CLI arguments are provided, each argument will filter down to the next argument. Example 19-4 shows how to select one section within one testcase. In this case, only the **show_version** test in Testcase1 will execute. This happens due to Testcase1 being specified first, which filters the preceding arguments to only sections in Testcase1.

Example 19-4   *Testcase/Section Selection with Multiple CLI Arguments*

```
pyats run job <job file> --testbed-file <testbed file> \
--health-file /path/to/health.yaml \
--health-tc-uids Testcase1 \
--health-tc-sections show_version
```

Testcase/section selection can also occur in health check YAML files. The same CLI arguments can be specified in each action in a health check YAML file. The only difference is that the action will only run against the testcase/section specified. Example 19-5 shows how to select specific testcases/sections in a health check YAML file.

Example 19-5   *Testcase/Section Selection via Health Check YAML*

```
test_sections:
  - cpu:
      - cpu_api:
          device: xe
          function: health_cpu
          arguments:
            command: show processes cpu
            processes: ['BGP I/O']
          include:
            - sum_value_operator('value', '<', 90)
          health_tc_sections:
            - check_cpu
```

The health check YAML file that specifies the **cpu_api** action will only run for the **check_cpu** test section in all testcases and triggers, as it's the only section specified under the **health_tc_sections** key. If more arguments are specified under the action, those arguments will search for testcases/sections using OR logic. The health check YAML file is considered more flexible, as you can specify which testcases/sections to run an action against.

> **Note**   If CLI arguments and a health check YAML file are provided for testcase/section selection, the CLI arguments will be preferred.

Health Check Results

If you incorporate health checks in your testing, you probably are wondering how this affects your test results. Pre-processors do not affect section results directly, as they are mostly used for data collection and monitoring. However, pre-processor results are passed to post-processors on the same section, and the post-processor results will be reflected in the section results. If either pre-processor or post-processor items fail, they will be reflected in the section results.

Health Check Usage

Currently, health checks can only be executed as part of a pyATS job. The pyATS library also provides the ability to validate/lint a health check YAML file. In the following sections, we are going to see how to validate health check YAML files and how to run health checks as part of a pyATS job.

Health Check YAML Validation

Once you have developed a health check YAML file, you can validate it against the health check schema using the following command:

```
pyats validate datafile /path/to/health.yaml
```

PyATS Job Integration

Health checks can be run as part of pyATS jobs. As you've learned throughout this chapter, health checks will collect and monitor device status of testbed devices during testing. To include one of the default health checks, simply include the **–health-checks** CLI argument with the name of the health check. Health checks specified by the **–health-checks** argument are executed as post-processors during testing. To run every default health check as part of a pyATS job, you would enter the following:

```
pyats run job <job file> --testbed-file <testbed file> \
--health-checks cpu memory logging core
```

If you only want to run a subset of the default checks, only include the name of the health checks you want to run.

The core health check only checks for a core dump file on the device. If you'd like to send the core dump file to a remote server for further troubleshooting, you must specify an additional CLI argument, **--health-remote-device**, with the name of the file transfer server. HTTP, SCP, TFTP, and FTP are all supported to transfer the core dump file. The file transfer server must be defined in your testbed YAML file. If you need to use another VRF to transfer the core dump file, you may use the **–health-mgmt-vrf** CLI argument and specify the OS and VRF name. By default, "Mgmt-intf" and "management" values are used for the VRF. For example, the complete CLI argument for a Cisco IOS-XE device that may not use a management VRF but uses the default VRF is **–health-mgmt-vrf iosxe:None**. If another VRF is used, you would specify the VRF name in place of **None**. Example 19-6 shows how to define an embedded pyATS file transfer server in your testbed YAML file. Embedded pyATS file transfer servers can be launched by any pyATS job and allow file transfer with devices for the duration of the pyATS job run.

Example 19-6   *Embedded pyATS File Transfer Server*

```
testbed:
  name: general_xe_xr_nx
  servers:
    myserver:
      dynamic: true
      protocol: ftp
      # Identifies the subnet of the testbed-facing interface
      subnet: 192.168.255.0/24
      path: /tmp # Root directory for the files being served
      credentials:
        default:
          username: pyats
          password: cisco123!
```

In addition to the default health checks, you can also include custom health checks built in a health check YAML file. To run a pyATS job with a health check YAML file, you would enter the following:

```
pyats run job <job file> --testbed-file <testbed file> \
--health-file /path/to/health.yaml
```

Health Check CLI Arguments

There are some other pyATS Health Check CLI arguments that can be provided to the **pyats run** command that help provide further customization to the health checks run in the pyATS job. If you want to adjust the default health check thresholds, use the **–health-threshold** argument with the health check name and new threshold as a key-value pair. For example, if you wanted to change the CPU and memory thresholds to 75% instead of the default 90%, you would specify the following argument: **--health-threshold cpu:75 memory:80**. The logging check will only look for the keyword **traceback** with different capitalizations in the log messages.

If you wanted to modify the logging keywords to include more keywords to detect, you can overwrite the detected keywords with the **–health-show-logging-keywords** CLI argument. The following CLI argument would change the log keywords to detect for Cisco IOS-XR and Cisco NX-OS devices and only look for "Crash" or "CRASH" in log messages:

```
--health-show-logging-keywords "iosxr:['Crash', 'CRASH']"
"nxos:['Crash', 'CRASH']"
```

You also can change the location pyATS Health Check searches for a core dump file. It's predefined for each platform (IOS, NXOS, IOS-XR). The location to search can be overridden using the **--health-core-default-dir** CLI argument. The default location can be changed for an IOS-XE device using the following command:

```
--health-core-default-dir "iosxe:['harddisk0:/core']"
```

It changes the location to search for the core dump file to harddisk0:/core.

We've looked at how we can select specific testcases/sections to run health checks before and/or after, but we didn't review whether we can select specific testbed devices. By default, health checks are run against all testbed devices. We can change that behavior by specifying the **--health-devices** CLI argument and list the devices we want to run health checks against. The following argument would limit the health checks to only run against the "R1_xe" testbed device: **--health-devices R1_xe**.

The last customization is the ability to send a Webex notification when a health check fails. This is a huge feature because it increases collaboration and can notify an entire team of engineers automatically if a device fails or malfunctions during testing. To send a notification, provide the **--health-webex** CLI argument. In addition, a Webex token and a Webex space ID or email address need to be provided via the pyATS configuration file (pyats.conf) or as CLI arguments. The CLI arguments include **--webex-token, --webex-space,** and **--webex-email.** Only one of the arguments, **–webex-space** or **–webex-email,** needs to be provided since the notification can only be sent to a Webex space or directly to a single person. The following argument would instruct a pyATS job to send a Webex notification to a Webex space if a health check fails:

```
--health-webex --webex-token <webex token> --webex-space <webex space id>
```

Summary

In this chapter, we reviewed the pyATS Health Check feature and the different health checks available to run before and/or after testcases/sections in a pyATS testscript. Health checks ensure devices are healthy and have not malfunctioned or crashed during testing. We also went over how to create custom health checks using a health check YAML file. Lastly, we touched on how to run health checks as part of a pyATS job and the different CLI arguments to customize health checks. Health checks are an integral part of testing. They ensure devices stay healthy during testing, and they should be included as part of every pyATS job.

XPRESSO

In the continuously advancing field of network test automation, XPRESSO introduces itself as an innovative tool, aiming to refine and facilitate your testing workflow. Serving as an extensive network test automation dashboard, XPRESSO is designed to streamline, accelerate, and simplify the process of test execution. This utility is particularly relevant in today's testing environments, which are often constrained by tight schedules and limited resources. Offering compatibility and integration with tools like pyATS, and being freely available, XPRESSO stands out as a valuable and accessible resource for network professionals.

XPRESSO is not merely another tool in the arsenal of network testing; it represents the next generation of automation dashboards. Its unique features and benefits make it an indispensable asset:

- **Automated testing and image/release certification:** Condenses the DevOps/ NetDevOps cycle from weeks or months to mere hours or days. Validates new software updates from Cisco before network upgrades. XPRESSO, like pyATS, was originally developed by Cisco, for Cisco, internally and then made available, for free, to the public.

- **Job creation:** Prepares job runs by associating essential elements like the test environment, lab resources, and localized arguments.

- **Job scheduling:** Automates job execution, eliminating the need for manual initiation.

- **Resource reservations:** Ensures availability of system resources for testing at predetermined times.

- **Testbed queuing:** Efficiently utilizes testbed resources and provides utilization statistics.

- **Real-time test result analysis:** Facilitates on-the-fly comparisons of test results to pinpoint failure causes.

■ **Baseline testing:** Establishes a benchmark for future test comparisons and analysis.

XPRESSO's advanced features are tailored for a seamless testing process:

■ **Result comparison and triage:** Provides sophisticated tools for result interpretation and analysis

■ **Flexible deployment:** Operable in private or Cisco-hosted lab environments

■ **Integration and flexibility:** Compatible with various platforms and methodologies, including pyATS, Jenkins, Docker, and more

■ **Agile and comprehensive testing:** Suitable for a wide range of testing scenarios

■ **Security and consistency:** Enforces global network and security constraints

■ **Resource optimization:** Offers a holistic view of resource usage to prevent scheduling conflicts and to optimize allocation

After realizing initial successes with pyATS, such as improving infrastructure as code (IaC) practices by automating network testing and validation, teams frequently aspire to consolidate their automation initiatives. This ambition leads them to XPRESSO, which acts as the web-based user interface for pyATS and plays a critical role in advancing these efforts. XPRESSO facilitates the shift from a scattered script-based approach to a more structured, enterprise-grade, centralized automation solution. It empowers teams by offering functionalities that streamline the management and execution of automation tasks, thereby elevating the efficiency and scalability of network operations. It enables teams to do the following:

■ **Centralize jobs:** Collect and manage pyATS jobs in one location for ease of access and management.

■ **Schedule and automate:** Automate the execution of jobs at scheduled times, moving away from manual triggers.

■ **Compare and analyze results:** Provide a centralized platform for comparing test results, aiding in swift identification of issues and trends.

■ **Enhanced reporting and alerting:** Incorporate advanced reporting and alerting features to keep teams informed and proactive in their response to test outcomes. XPRESSO stands out for its user-friendly interface, which emphasizes a harmonized user interface and user experience (UI/UX) for ease of use. XPRESSO seamlessly integrates with existing test environments and tools, is modular and scalable, and adapts to the evolving needs of network testing. Security is a core design principle, ensuring safe and accurate testing.

XPRESSO, complementing pyATS, is a sophisticated, user-friendly, and free test automation dashboard that enhances the efficiency and effectiveness of network testing. It's not just a tool; it's a comprehensive solution designed to transition pyATS jobs from a collection of distributed scripts to an enterprise-grade, centralized automation platform, making the life of a test automation professional easier and more productive.

This chapter covers the following topics:

■ Installing XPRESSO

■ Getting started with XPRESSO

■ pyATS Job in XPRESSO

Installing XPRESSO

Everything you need to install XPRESSO is available at https://github.com/CiscoTestAutomation/xpresso. The content of this repository is to help users with one-click deployment of XPRESSO inside their lab/networks.

Designed to streamline your network automation, test, and validation experience, XPRESSO is the standard pyATS UI dashboard that manages your test suites, test resources, and test results, providing insights to your network through Cisco pyATS.

XPRESSO does not collect user statistics and will not send a telemetry of user information back to Cisco.

Here are the system requirements for XPRESSO:

■ Linux/macOS environment

■ Docker installed and in working condition

■ Free disk space for log storage

■ Minimum system specs:

    ■ Four CPUs (with hyper-threading)

    ■ 16GB of memory

■ Ideal system specs:

    ■ Twelve CPUs

    ■ 64GB of memory

Note   Lower system specs will result in a much longer initial bootup time.

XPRESSO is developed using a microservices architecture, with the services spanning over multiple Docker containers. The overall access is achieved through a gateway that processes the APIs and distributes them to the services.

Here are the basic steps for installing XPRESSO:

1. Clone this repository:

   https://github.com/CiscoTestAutomation/xpresso.git

   In this example, we'll put everything under /workspace/xpresso. You may choose your own home location.

   ```
   mkdir /workspace
   cd /workspace
   git clone https://github.com/CiscoTestAutomation/xpresso
   ```

2. Initialize using a setup script.

> **Note**   If you want to perform a manual setup, skip the step.

   Run the script by providing the proper URL of the server:

   ```
   ./setup.sh   http://youmachinehostname/
   ```

   After this script runs, you can skip the next step.

3. Initialize manually.

   The default settings should work for most users, with the out-of-the-box URL set to http://localhost/. In other words, you can only access XPRESSO on this localhost.

   To make the instance available for other users on your network to access, modify the .env file and set the ADVERTISED_URL to the full, proper URL of this server (for example, http://xpresso.yourdomain.com/).

   Important: Go through all steps in this section. If you skip a step not explicitly marked "optional," you will encounter issues.

   When setting up your XPRESSO environment, the BASE_DIR serves as the central hub where all repository contents reside, including directories like etc/, env/, initial-izers, as well as files like .env and docker-compose.yml. By default, BASE_DIR corresponds to the directory where the repository has been cloned.

   a. Configure the environment variables.

   In the ${BASE_DIR}/.env file, set the appropriate values for ADVERTISED_URL, INSTANCE_ID, and TOOL_NAME.

   b. Modify the service-specific environment files.

   Under ${BASE_DIR}/env, several files allow for customization:

   ■ **databases.env:** Specify the MySQL root password, as well as the credentials for the XPRESSO database user xpresso_admin.

- **elasticsearch.env:** Apply any custom configurations for the Elasticsearch cluster.

Ensure the xpresso_admin password is consistent across the following files:

- ${BASE_DIR}/env/databases.env
- ${BASE_DIR}/initializers/docker-entrypoint-initdb.d/1-user.sql
- ${BASE_DIR}/initializers/settings.yml
- ${BASE_DIR}/etc/mgmt_settings.py

Should you change the xpresso_admin password via the UI, also update it in ${BASE_DIR}/initializers/new_settings.yml.

c. Prepare data and log directories.

Create a directory for Elasticsearch data at ${DATA_DIR}/elastic and ensure it has write permissions:

```
chmod -R 777 ${DATA_DIR}/elastic.
```

Assign write permissions to the MySQL logs directory:

```
chmod -R 777 ${LOGS_DIR}/database.
```

In the ${BASE_DIR}/.env file, verify that DATA_DIR and LOGS_DIR are correctly set to locations with sufficient disk space. DATA_DIR will house XPRESSO data, while LOGS_DIR will store logs from XPRESSO's microservices.

d. Consider the following important warnings:

- **Permissions:** Double-check the permissions on DATA_DIR, LOGS_DIR, and their subdirectories to prevent access issues.
- **System configuration:** On Linux servers, ensure max_map_count is set to at least 262144 (vm.max_map_count=262144) as recommended by Elasticsearch documentation.
- **Time zone configuration:** To synchronize the time zone used by XPRESSO with your host, uncomment the /etc/localtime:/etc/localtime:ro entries under volumes in the docker-compose.yml for all services.

e. Reference the docker-compose file.

The docker-compose.yml file is crucial for orchestrating the XPRESSO containers and their interconnections. Familiarize yourself with this file to understand how XPRESSO's services are configured and deployed. Here, you'll define services, volumes, and other Docker configurations that are key to running XPRESSO smoothly. This file acts as the blueprint for your XPRESSO deployment, bringing together all the components previously mentioned.

By following these structured steps and paying close attention to the configurations within the docker-compose.yml file, you'll be well on your way to successfully setting up and customizing your XPRESSO environment for network test automation.

- **[OPTIONAL]:** In file ${BASE_DIR}/.env, change TAG to the most appropriate value for your XPRESSO instance.

- **[OPTIONAL]:** By default, no ports are exposed in Docker. For your testing purposes, you can uncomment the ports entry in the docker-compose.yml file for the services you want.

- **Important:** Adding new settings or updating existing ones should be done through the initializers/new_settings.yml file. Once this is done, restart the management service and your settings will be updated right away. Remember that you may also need to restart all other service that are supposed to use the new/update settings.

4. Start your engine.

You're good to go:

```
cd /workspace/xpresso
# pull the latest images
docker-compose pull
# fire all cylinders
docker-compose up -d
```

You should be able to access XPRESSO now at http://localhost/.

It may take a while for the initial settings to be automatically applied while the system boots for the first time. This may mean you cannot log in using the default credentials for a few minutes. Give it some time (for example, 5–10 minutes on a 2016 MacBook Pro 15).

Optional integration and XPRESSO configuration steps include the following:

1. Customize the LDAP configuration.

XPRESSO supports multiple LDAP authentication by allowing you to provide a list of LDAP configurations.

In initializers/settings.yml, under common, add/modify the setting LDAP_CONFIG with the value being a list of configurations for the following:

- LDAP_DESCRIPTION—for example, ABC Organization LDAP

- GEN_USER—for example, gen_username

- SEARCH_SCOPE—for example, SUBTREE | BASE | LEVEL

- LDAP_PROTOCOL—for example, LDAP

- LDAP_SOURCE—for example, ABC

- LDAP_NAME—for example, ABC LDAP Auth

- LDAP_HOST—for example, internal.abc.com

- SEARCH_FILTER—for example, (uid={USER})

- GEN_PASS—for example, gen_password

- SEARCH_BASE—for example, o=internal.abc.com

- SEARCH_ATTRIBUTES—for example, ["cn", "givenName", "title", "mail"]

In the same file, initializers/settings.yml, under microservices, add (if it doesn't already exist) the auths service and modify your AUTHENTICATION_BACKENDS to include authms.backends.CustomLDAPBackend.

Finally, for LDAP search, in the same file, initializers/settings.yml, search under microservices for users and add the following line:

```
SEARCH_BACKENDS: ["user_profiles.backends.CustomLDAPSearch"]
```

When updating microservice settings, make sure you include a URL and description as well.

Restart the management service (docker-compose restart), in order to reflect the changes.

Important: Whatever is set in initializers/settings.yml will be the initial settings.

2. **Add a new XPRESSO worker.**

The new XPRESSO worker (version 20.11) has a lot of improvements over the older versions. If you're willing to upgrade your XPRESSO to version 20.11, you need to make sure the workers data directory is correct. Just verify that the workers data directory is located at ${BASE_DIR}/data/workers. The current version, as of the writing of this book, is XPRESSO 20.12.

3. **Enable email and SMTP:**

Add or modify initializers/new_settings.yml under the common section to suit your email server to enable XPRESSO to send emails. This is used for user signup/management, automated notifications of runs and reservations, and the sending of result reports.

- EMAIL_HOST: 'my-smtp-server'

- EMAIL_PORT: 25

- EMAIL_HOST_USER: 'username'

- EMAIL_HOST_PASSWORD: 'passwd',

- EMAIL_USE_TLS: true, // or remove

- EMAIL_USE_SSL: true, // or remove

- EMAIL_TIMEOUT: 10000

- EMAIL_SSL_KEYFILE: '/path/to/pem', // or remove

- EMAIL_SSL_CERTFILE: '/apth/to/pem' // or remove

Important: Adding new settings or updating existing ones should be done through initializers/new_settings.yml file. Once this is done, restart the management service and your settings will be updated right away. Remember that you may also need to restart all other services that are supposed to use the new/updated settings.

XPRESSO will automatically create a default admin user at startup. Use the username/password admin/admin to log in to the dashboard with full administrator privileges. You may register more users into the internal database after you log in as the administrator.

For HTTPS hosting, you need to provide the SSL certificates (.key and .pem files) and update NGINX settings to reflect these changes. Put the .key and .pem files under ${BASE_DIR}/etc/ and update ${BASE_DIR}/etc/nginx.conf accordingly.

Once XPRESSO is running, the full user documentation is available directly in the UI.

By default, all your data and services logs are stored under the ./data/ (that is, DATA_ DIR) and ./logs/ (that is, LOGS_DIR) directories, including database files, archives uploaded, and so on. To wipe the server and "start from scratch" again, just delete these folders.

No data is saved in the containers—everything is volume-mounted to disk.

Common Issues, Questions and Answers

In the dynamic world of Docker and container management, encountering issues is a part of the development and deployment process. This section aims to address some of the common challenges and questions users may face while working with Docker, specifically within the context of the XPRESSO web dashboard service. From dealing with unhealthy services to understanding the nuances of service names and troubleshooting resource availability, we've compiled a list of scenarios that could potentially halt your progress. Each subsection below is designed to offer clear, actionable solutions and explanations to help you navigate through these hurdles efficiently. Whether it's a confusing error message, a stubborn service that refuses to start, or networking puzzles that seem to have no end, this guide seeks to demystify the complexities and get you back on track. Dive into the details to learn more about maintaining a healthy service environment, decoding the historical context behind XPRESSO, resolving common error messages, and ensuring your setup is primed for optimal operation.

Unhealthy Services

If you run docker-compose ps or docker ps and see any unhealthy services, first check either the docker-compose logs -f <service-name> log or the <LOGS_DIR>/<service-name> log. If the logs are not informative, you can restart the service:

```
docker-compose stop <service-name> && docker-compose up -d <service-name>
```

References to S3

Initially, XPRESSO was called S3, with the number 3 being a superscript (meaning, S-cubed), which is short for "self-serving services." Because some folks confused this with Amazon S3 services, the pyATS development team renamed the pyATS web dashboard service to XPRESSO to express their love for coffee.

Error: No Resources Found

This error occurs when the resource management service does not boot up properly. It happens usually when the server you are launching the service on is a bit slow. Try the following:

```
docker-compose restart resources
```

The problem should go away.

Cannot Log In Using the Default admin/admin

Wait a bit more (be patient), or run *docker-compose restart.* The initial bootup performs a lot of first-start settings and database migrations and therefore could fail due to running on a slow server.

ElasticSearch Failed to Start

Check the logs using docker-compose logs elasticsearch. If it complains about permission issues, run **chmod 777 data/elastic** and restart Elastic using **docker-compose restart elasticsearch**. Also, make sure you've run this:

```
sysctl -w vm.max_map_count=262144
```

Cannot Connect to Database

If services are failing to start and the logs show that they cannot connect to database:3306, make sure your firewalls are not blocking the bridge network 192.168.66.0/24.

General Networking Issues with the XPRESSO Installation

The following are some general networking issues you need to look out for when installing XPRESSO:

- Ensure the hostname is properly set and resolvable both internally and externally to the host machine.

- Ensure DNS on the host is configured to be able to resolve its external hostname and IP.

- Ensure the ADVERTISED_URL in the .env file is set to the fully qualified external URL using the hostname or IP, not localhost.

- If a proxy is set, ensure it does not affect traffic going to the advertised DNS name set above for the domain/hostname.

- Ensure nothing else is using or restricting the local 192.168.66.x subnet (or change it to another subnet).

- Ensure you switch off the firewall. For CentOS, use **sudo systemctl stop firewalld.** For Ubuntu, use **sudo ufw disable.**

- As a trial, shut down XPRESSO docker-compose, turn off your firewall, restart the Docker service, and try starting XPRESSO again.

Getting Started with XPRESSO

This Quick Start guide for XPRESSO is crafted to offer a streamlined and efficient introduction to the platform, focusing on its primary tasks and features. Designed for new users, it serves as a step-by-step manual to create, execute, and analyze the results of a job run in XPRESSO. The guide highlights the ease of setting up and navigating XPRESSO, emphasizing its user-friendly interface and powerful automation capabilities. The full and official Quick Start guide can be found at https://developer.cisco.com/docs/xpresso/#!quick-start/quick-start.

Facilitating Quick Adoption

To accelerate your learning curve, the guide incorporates several preconfigured settings:

- **Default job profiles:** These templates simplify the process of creating jobs tailored to your needs.

- **Group preferences and permissions:** Understanding these settings will help you manage collaboration effectively within XPRESSO.

- **Lab/testbed locations:** These predefined locations assist in quicker setup and execution of test runs.

While these settings are preset for convenience, customization is encouraged to fully leverage XPRESSO's automation features. For settings not covered in the Quick Start, this chapter provides recommended reading paths to further explore and configure additional features. The objectives of the Quick Start guide are multifold:

- **Ease of setup:** Demonstrate the simplicity of establishing a working environment in XPRESSO.

■ **User experience:** Showcase the platform's user-friendly nature, flexibility, and effectiveness in integrating and executing testscripts.

■ **Introduction to dashboard:** Provide an overview of the most common tasks on the XPRESSO dashboard, preparing users for more advanced operations.

■ **Understanding user interaction:** Clarify the roles and collaboration between different user types and group members within XPRESSO.

■ **Data analysis and tagging:** Teach users how to search and tag test results for efficient data management and retrieval.

The guide starts from the perspective of a new user in the default Guest Group or a new group member, with the next steps involving executing a pyATS testscript and defining a testbed with a YAML file.

Now, as a user with a basic understanding of XPRESSO's layout and functionalities, you're ready to embark on this Quick Start journey to unlock the full potential of XPRESSO for your testing and automation needs.

The XPRESSO dashboard elements break down as shown in Figure 20-1.

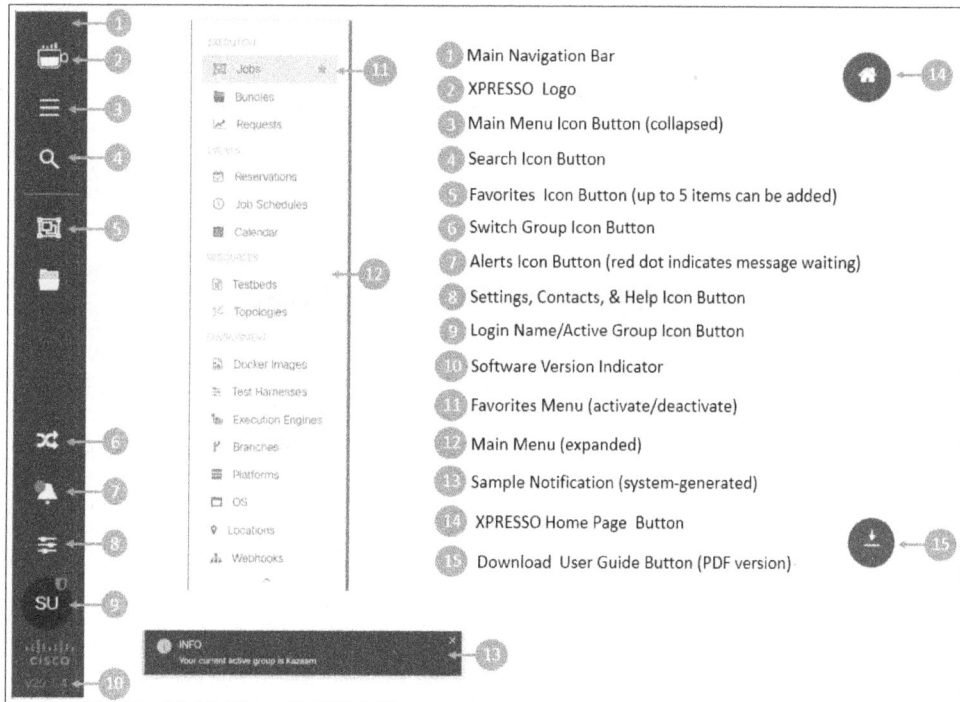

Figure 20-1   *XPRESSO Dashboard*

XPRESSO includes easy-to-use wizards. A wizard is a step-by-step process (guide) that allows users to input information in a prescribed order and in which subsequent steps may depend on information entered into a previous one. As users enter information, XPRESSO computes the appropriate path for the user and routes them accordingly. XPRESSO uses wizards to configure, for example, Jenkins and cloud jobs, job requests, and Docker images. XPRESSO wizards come in two forms: steps can be presented in a vertical configuration, as shown in Figure 20-2, or horizontal configuration.

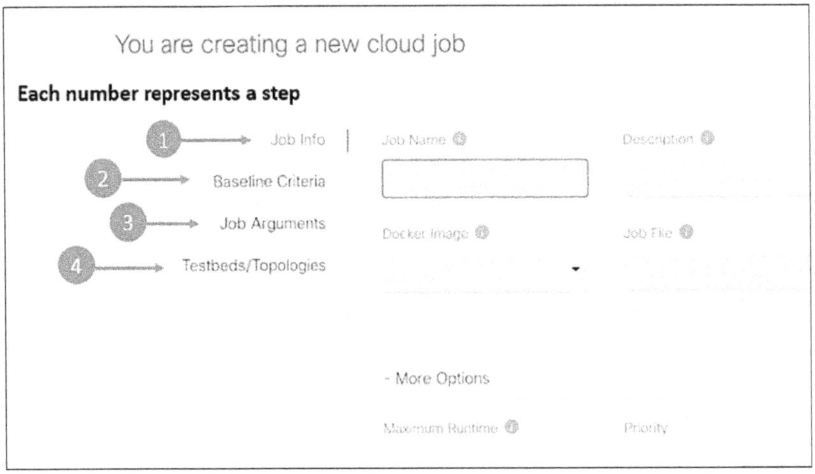

Figure 20-2   *XPRESSO Cloud Job Wizard*

Let's now transform an existing pyATS job into XPRESSO.

Transforming a pyATS Job into XPRESSO

Once XPRESSO is installed and up and running, visit http://localhost to access the Web UI. First, let's log in as admin/admin for the sake of this exercise. In production, you would connect to LDAP and also set up your RBAC groups and users. For personal use, you would also register a new user and proceed as the new user demonstrated in Figure 20-3.

Once we've logged in, the XPRESSO dashboard will show us any current requests on the default landing page, as demonstrated in Figure 20-4.

Requests are like API calls in that you, the user, are making a request of XPRESSO. This could be a scheduled job or a one-time execution of a job. The Requests dashboard displays the history and details of these user requests. Requests are important because they try to abstract the complexity of interacting with XPRESSO.

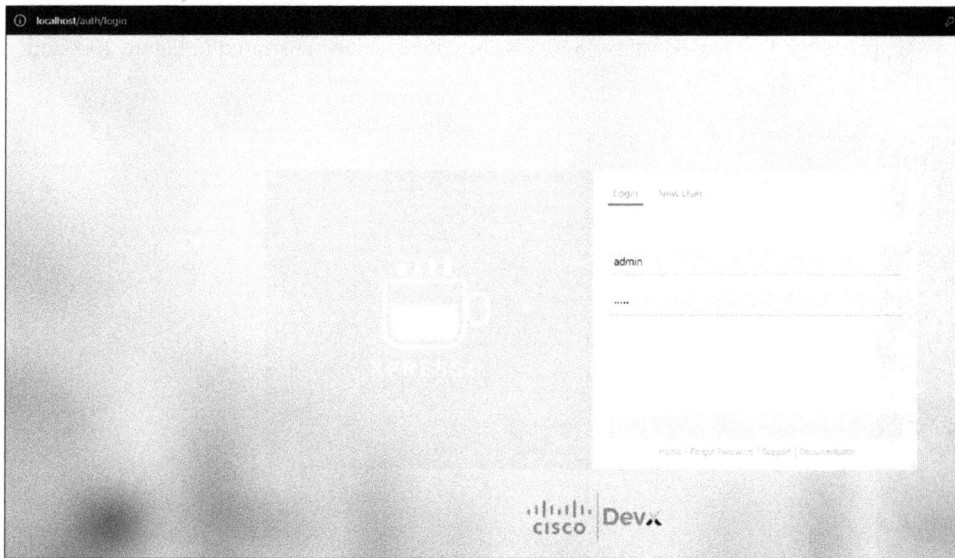

Figure 20-3   *XPRESSO Main Login Page*

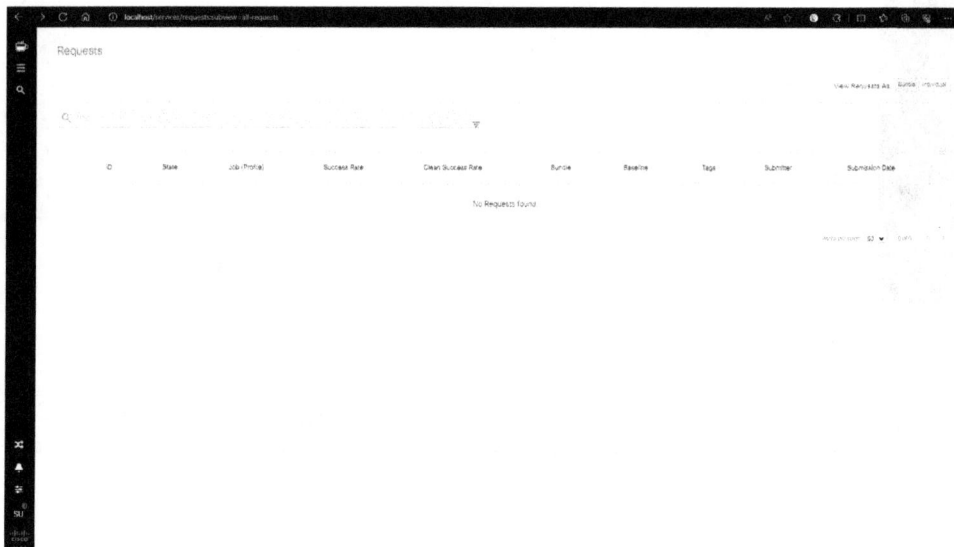

Figure 20-4   *XPRESSO Requests Landing Page*

First, select the main menu ("hamburger menu") icon and select Testbeds. We will use a simple testbed for the Always-On IOS XE sandbox, as demonstrated in Figure 20-5 and Example 20-1.

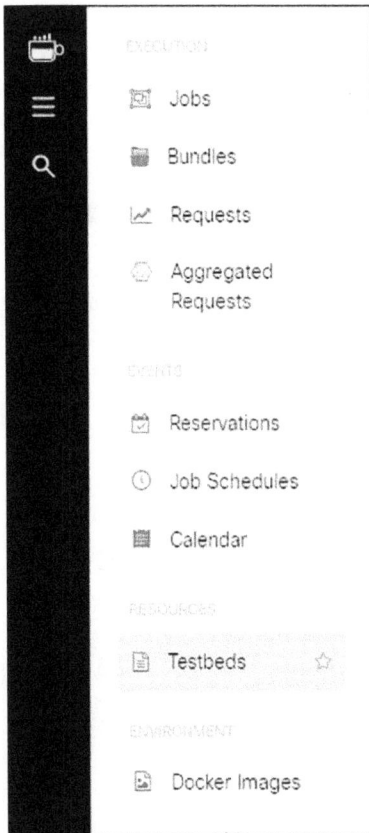

Figure 20-5   *XPRESSO Importing Testbeds*

Example 20-1   *A Testbed for the Always-On IOS XE Sandbox – testbed.yaml*

```
---
devices:
  Cat8000V:
    alias: "Sandbox Router"
    type: "router"
    os: "iosxe"
    platform: Cat8000V
    credentials:
```

```
    default:
      username: admin
      password: C1sco12345
  connections:
    cli:
      protocol: ssh
      ip: sandbox-iosxe-latest-1.cisco.com
      port: 22
      arguments:
        connection_timeout: 360
```

Let's register the static testbed, as demonstrated in Figure 20-6.

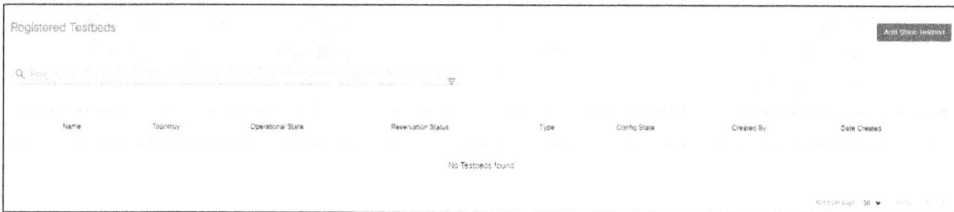

Figure 20-6   *XPRESSO Static Testbed*

After you import your testbed.yaml file, you will have the option to include testbed extensions or pyATS Clean files, as demonstrated in Figure 20-7.

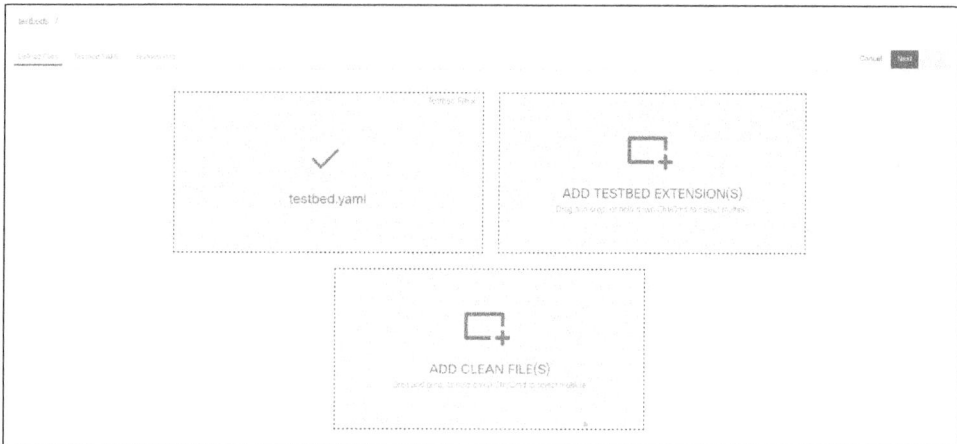

Figure 20-7   *XPRESSO Import Testbed Optional Files*

The testbed will be validated and the YAML displayed, as demonstrated in Figure 20-8. We will ignore the warnings about a lack of interface definitions and continue.

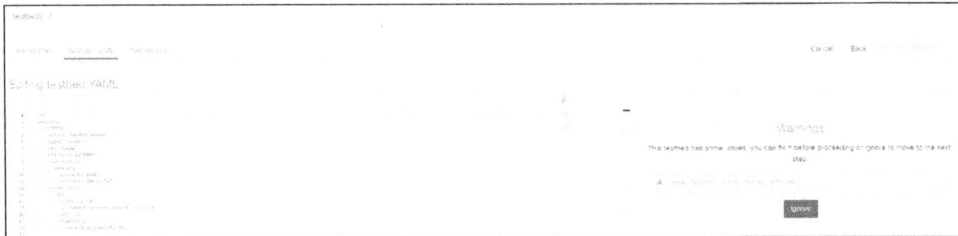

Figure 20-8 *XPRESSO Testbed YAML Review*

Finally, we can set up the maximum queue retention policy and the maximum reservation length, specify if the testbed is reservable or not, and map the testbed to our XPRESSO topology (City and Building). For this example, we will just select one of the pre-canned locations, as demonstrated in Figure 20-9.

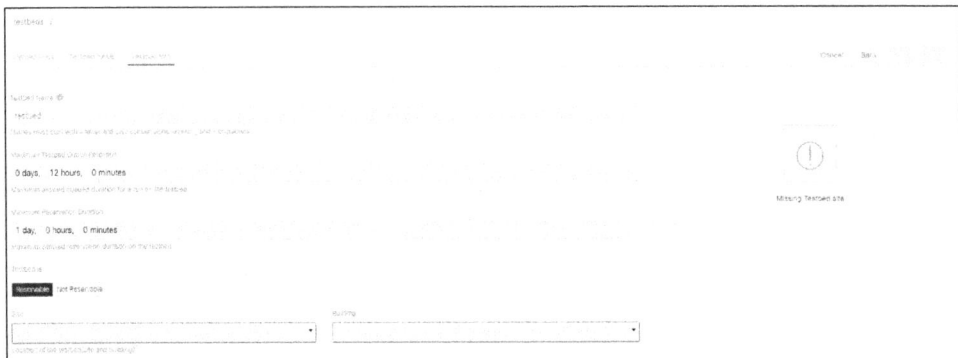

Figure 20-9 *XPRESSO Testbed Properties and Geolocation*

Now our testbed is registered with XPRESSO, as demonstrated in Figure 20-10!

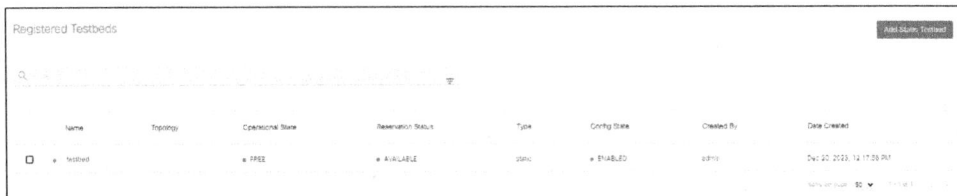

Figure 20-10 *XPRESSO Registered Testbed*

Next, we need to add or build our Docker image for our pyATS job, as demonstrated in Figure 20-11. There are several ways of achieving this in XPRESSO. First, we can add an existing image where we provide our image name, Docker tag, and Docker Hub URL.

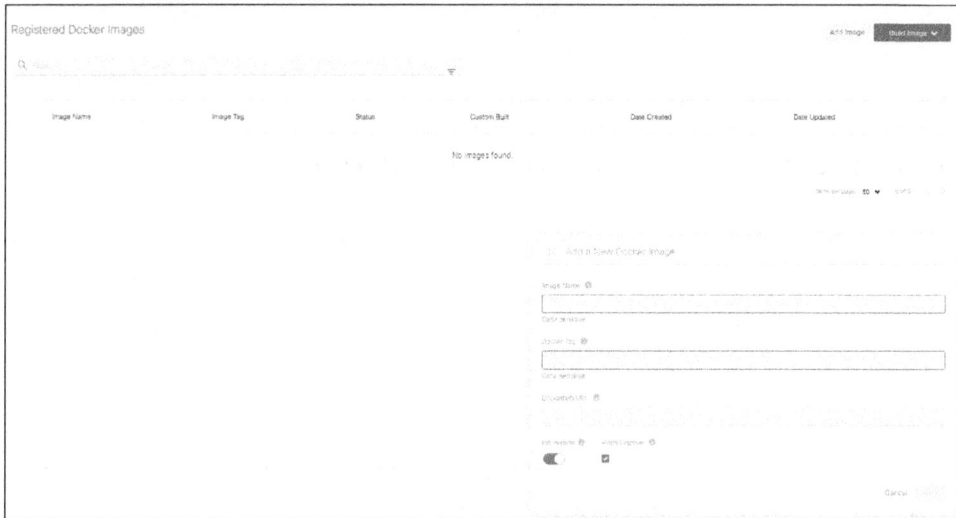

Figure 20-11 *XPRESSO Add Docker Image*

Alternatively, we can use the Image Builder either for a pyATS image or a Polaris image, as demonstrated in Figure 20-12.

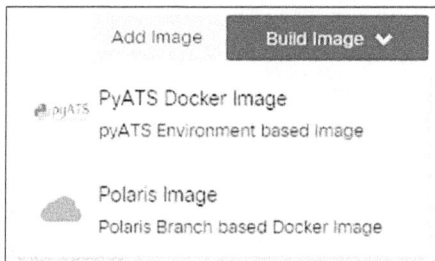

Figure 20-12 *XPRESSO Image Builder Options*

With the pyATS Image Builder, you can either use a YAML specification as shown in Figure 20-13 or use the Image Builder tool to complete the various wizard input options, as shown in Figure 20-14.

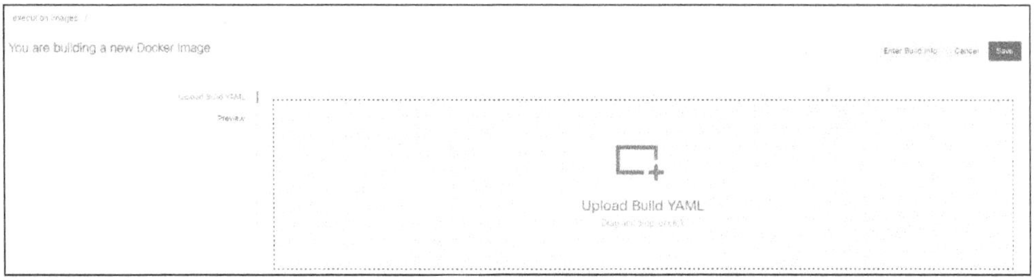

Figure 20-13 *XPRESSO Image Builder pyATS YAML Option*

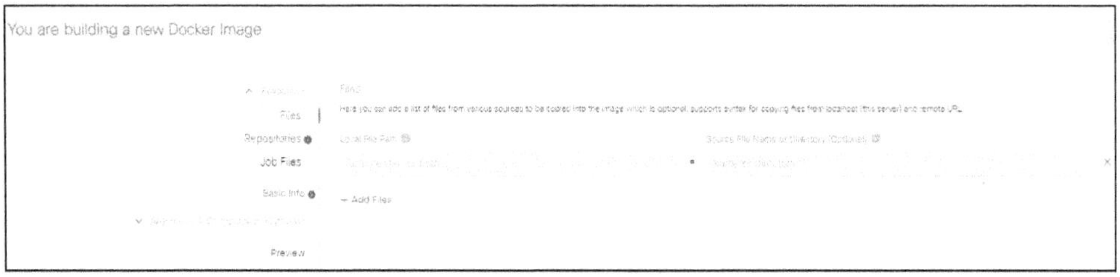

Figure 20-14 *XPRESSO Image Builder Wizard*

Figure 20-15 shows the resulting YAML file using the public repository for the chapter—a pyATS interface test's job and script.

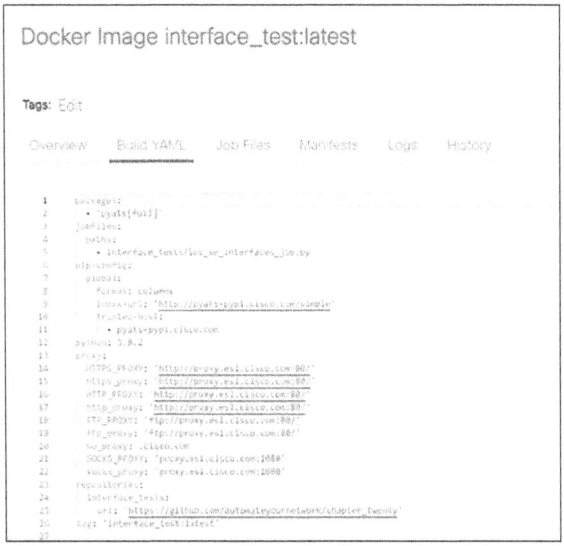

Figure 20-15 *XPRESSO Image Builder YAML Result from Public GitHub Example*

In the Registered Docker Images pane, we should see the image Building, as demonstrated in Figure 20-16. Be sure to wait until the build is complete before proceeding.

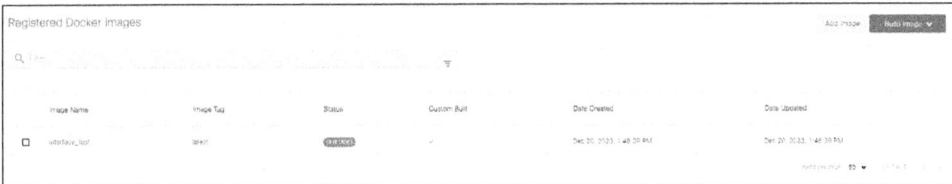

Figure 20-16 *XPRESSO Registered Docker Images – Building*

We can use the interface test job file and testscript from Chapter 8, "Automated Network Testing," in the pyATS Image Builder. The YAML file should look like Example 20-2.

Example 20-2 *A Sample YAML for the pyATS Image Builder*

```
---
packages:
  - 'pyats[full]'
jobfiles:
  paths:
    - interface_tests/ios_xe_interfaces_job.py
python: 3.8.2
repositories:
  interface_tests:
    url: 'https://github.com/automateyournetwork/chapter_twenty'
tag: 'test_driven_automation:latest'
```

By default, the XPRESSO pyATS testbed tool includes settings like proxy server, which may or may not be required for XPRESSO to reach the Internet. The difference between the preceding testbed and the previous screenshot indicates a basic testbed without a proxy or other default XPRESSO testbed settings.

Our image will now register with XPRESSO, as shown in Figure 20-17.

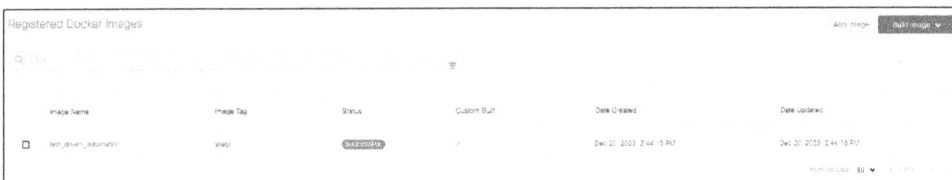

Figure 20-17 *XPRESSO Registered Docker Images – Successfully Registered Image*

We can explore our image by clicking the image name's hyperlink. This deeper exploration reveals the image's overview, build YAML, job files, manifests, logs, and history, as shown in Figure 20-18 and Figure 20-19.

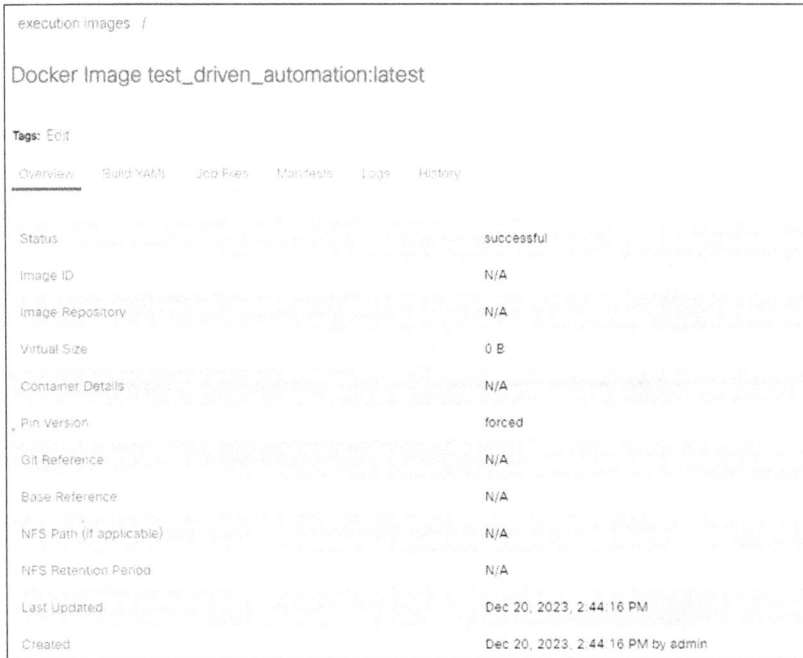

Figure 20-18   *XPRESSO Image Overview*

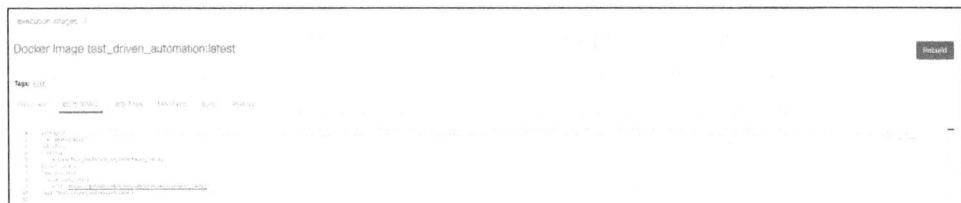

Figure 20-19   *XPRESSO Image Build YAML*

Note that we can adjust our build YAML in this pane of glass and trigger a rebuild of the image.

Figure 20-20 shows the XPRESSO execution image referencing the Docker image containing the pyATS job file.

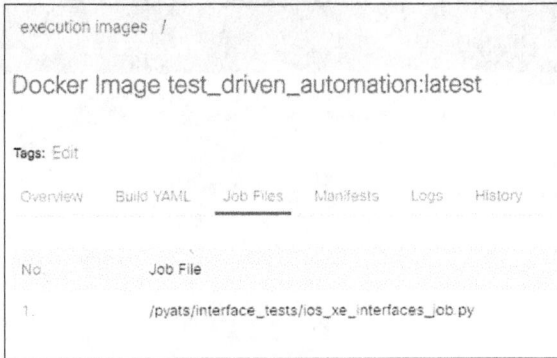

Figure 20-20 *XPRESSO Image Job Files*

Note that all pyATS job files inside the Git repository will be displayed here.

Figure 20-21 shows the optional XPRESSO manifest reference from pyATS. In this example, there is no external manifest file.

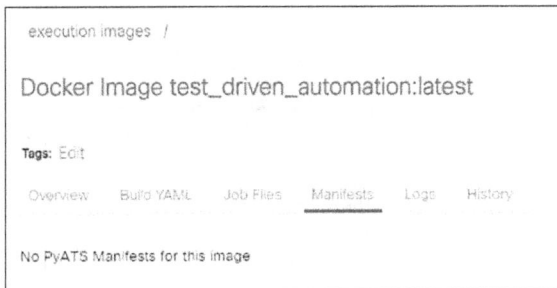

Figure 20-21 *XPRESSO Image Manifests*

Note that our example does not include any manifests; this is expected to be empty for this example.

Figure 20-22 shows the XPRESSO and pyATS logs related to the pyATS Image Builder inside the XPRESSO dashboard.

The logs are an excellent place to troubleshoot pyATS image builds and include all of the automated pyATS Image Builder tool logs for the last build of this image.

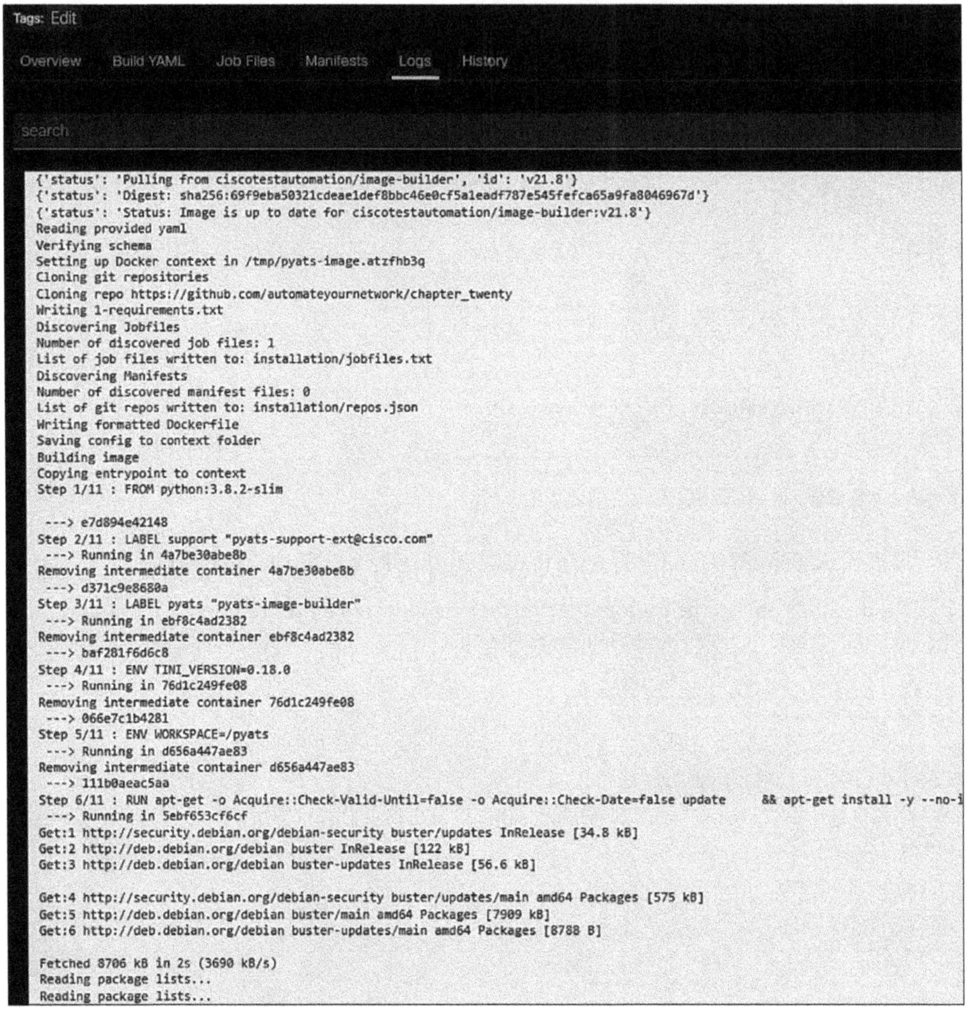

Figure 20-22 *XPRESSO Image Logs*

Figure 20-23 shows the History view inside the XPRESSO dashboard related to this execution image.

This is a complete graphical history of the image. Again, this is useful for troubleshooting or for version and source control over the image.

Our next step is to create an XPRESSO job. Navigate to the main (hamburger) menu and select Jobs, as shown in Figure 20-24.

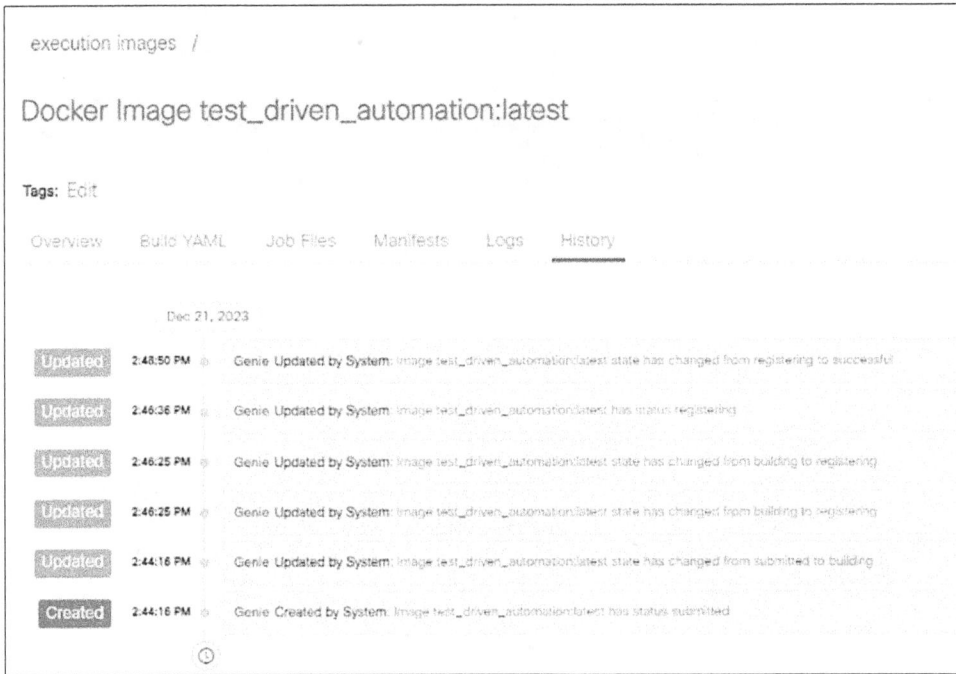

Figure 20-23   *XPRESSO Image History*

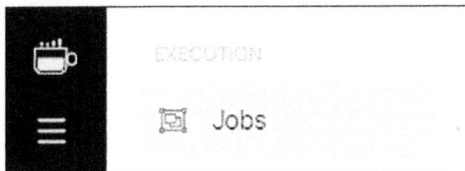

Figure 20-24   *XPRESSO Jobs*

Create a new job as shown in Figure 20-25.

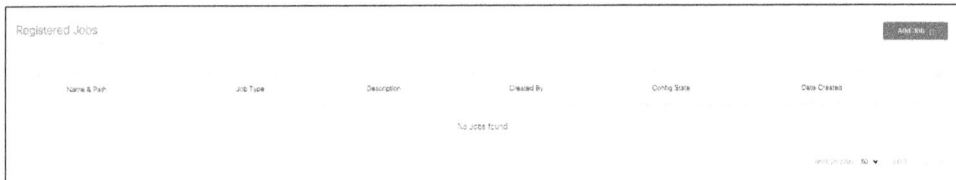

Figure 20-25   *XPRESSO Jobs – New Job*

For this example, we will be creating a cloud job, as shown in Figure 20-26; however, you could use Jenkins and create a Jenkins-integrated job here if you are running Jenkins.

Figure 20-26   *XPRESSO Jobs – Cloud Job*

Next, we will describe our job and complete the fields shown in Figure 20-27. The Docker image as well as the pyATS job should be available for selection, as shown in the figure.

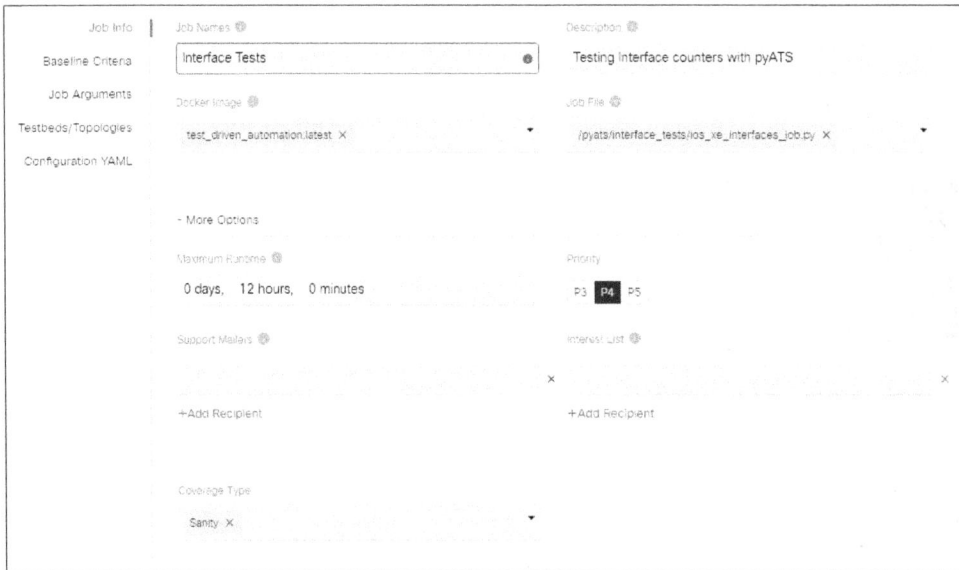

Figure 20-27   *XPRESSO Jobs: Cloud Job – Details*

Next, if you wish, you can set the baseline values for the Pass Rate percentage as well as the minimum number of Test Cases pass completion percentage. Since we are running this for the first time, let's set the baseline as 100 for both minimum percentages, as shown in Figure 20-28.

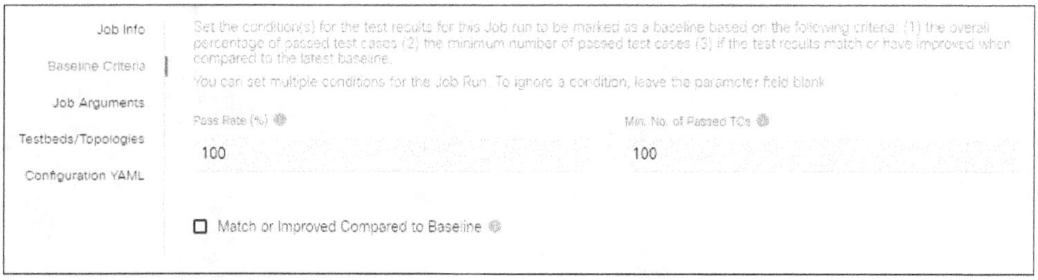

Figure 20-28 *XPRESSO Jobs: Cloud Job – Baseline Values*

Job arguments, environment variables, harness arguments, as well as cloud mount points can all be included with the job, as shown in Figure 20-29. Environment variables are particularly useful for secrets and encrypted keys.

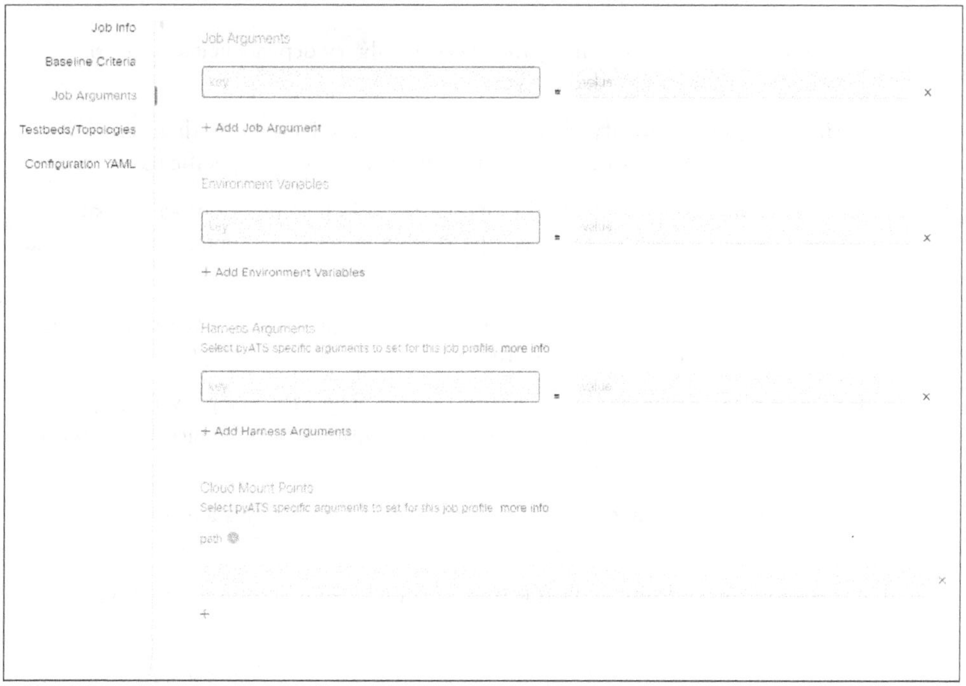

Figure 20-29 *XPRESSO Jobs: Cloud Job – Job Arguments*

Now we get to select our testbed file for the job. Our statically defined testbed should be available for us to use, as shown in Figure 20-30.

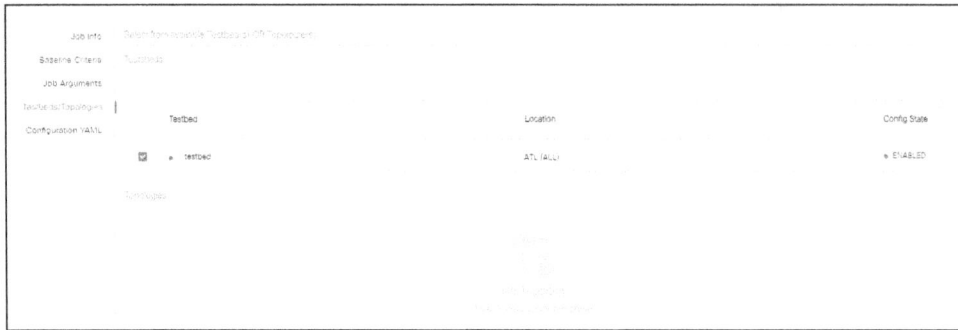

Figure 20-30   *XPRESSO Jobs: Cloud Job – Testbed*

Finally, we can also add Easypy configuration YAML to the job. Generally, such a YAML file could include the following configurations:

■ **Environment settings:** Define Python version, library dependencies, or other environment variables needed for your job.

■ **Script parameters:** Specify arguments or parameters that your Python script needs. This could include file paths, execution flags, or other script-specific parameters.

■ **Resource allocation:** Configure the amount of CPU, memory, or other system resources allocated to the job. This is particularly relevant in containerized or cloud environments.

■ **Scheduling information:** If your job needs to be run at specific times or intervals, this can be specified in the YAML file.

■ **Logging and monitoring:** Define settings for logging, such as log levels, output destinations, and formats. You might also include configurations for monitoring the job's performance or health.

■ **Security and compliance:** Include settings related to security, such as credentials, encryption settings, or compliance rules.

■ **Error handling and retries:** Define how the system should respond to errors or failures, including retry logic or notification settings.

■ **Integration with other services:** If your job interacts with databases, APIs, or other external services, you might include connection strings, API keys, or other integration-related settings.

■ **Job dependencies:** Specify if this job depends on the completion of other jobs or processes.

■ **Custom configuration:** Any other custom configuration that is specific to your job or the framework you are using.

Figure 20-31 shows the optional Easypy configuration YAML file, which this XPRESSO job does not use.

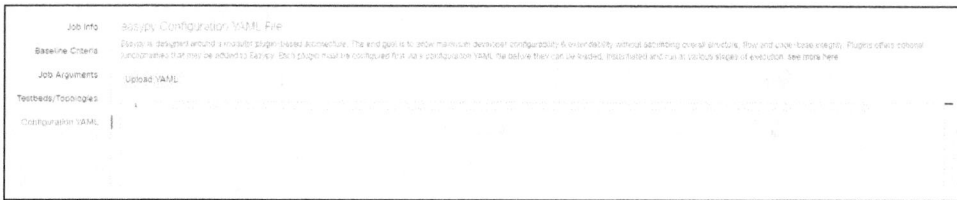

Figure 20-31 *XPRESSO Jobs: Cloud Job– Configuration YAML*

Save the Cloud Image job.

Now, let's run our job! From the Registered Jobs menu, click the InterfaceTests job, as shown in Figure 20-32. This will bring you to the job's Overview page

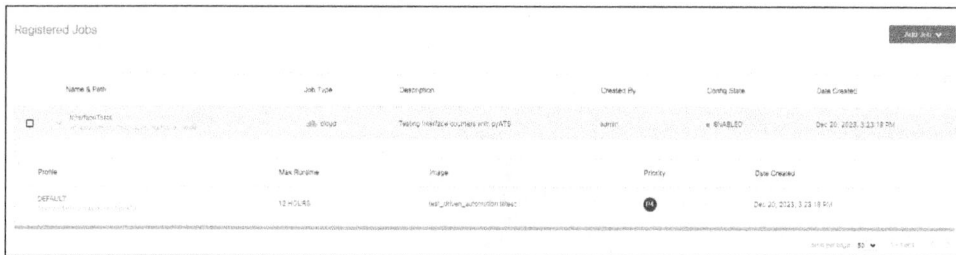

Figure 20-32 *XPRESSO Jobs: Cloud Job – InterfaceTests Overview*

Click the DEFAULT profile, which will bring you to the Profile Overview page, as shown in Figure 20-33.

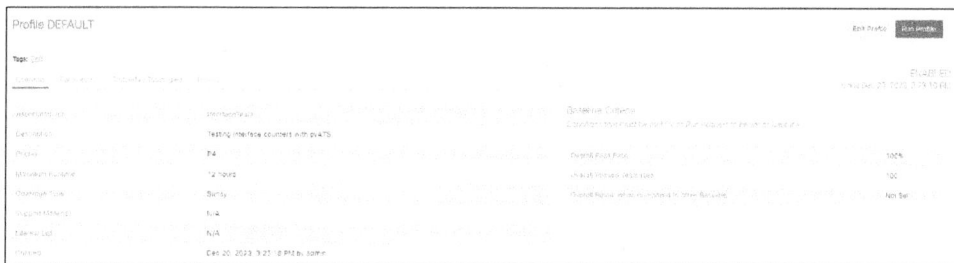

Figure 20-33 *XPRESSO Jobs: Cloud Job – Profile Overview*

From here, we can also explore the profile's parameters, testbed/topologies, and history, as shown in Figure 20-34.

Click **Run Profile** in the top-right corner to run the profile and job "on-demand." Optionally, there are controls such as the branch of the image (if there are branches) and image label. We can also add a description, set a priority (for internally XPRESSO queueing), tag people or interest lists, add tags, and set the maximum request lifetime, as shown in Figure 20-35. These settings become more important in a collaborative team environment where approvals are required and resources must be reserved. The ability to add custom tags allows you to efficiently search XPRESSO for test results of interest.

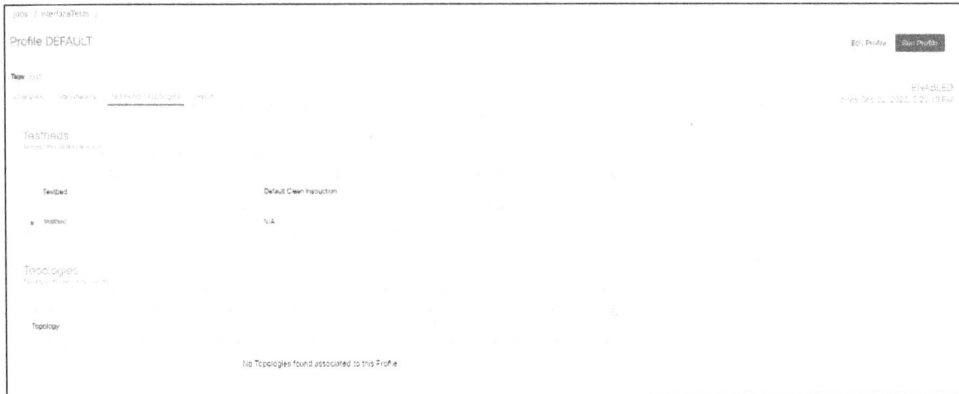

Figure 20-34   *XPRESSO Jobs: Cloud Job – Profile Testbed*

We could also set this as a baseline, but because we are not sure the tests will all pass, let's not set this as a baseline but rather use the conditional baseline. These baselines are great for comparisons within XPRESSO after running the job several times or on a schedule. The conditional baseline will only set the results of this job as the baseline *if* it passes the thresholds we set earlier.

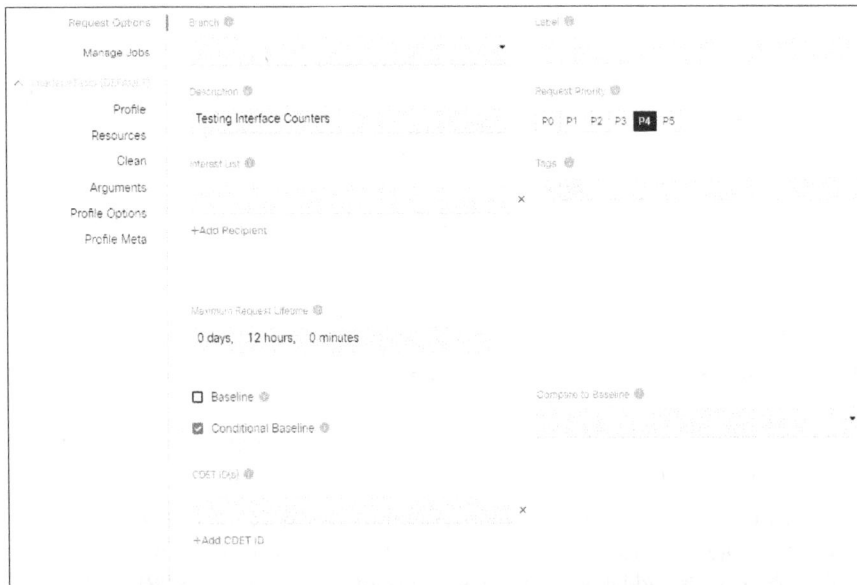

Figure 20-35   *XPRESSO Jobs: Cloud Job – Configuring Group Job Request Options*

XPRESSO will now prepare and queue our request, as shown in Figure 20-36.

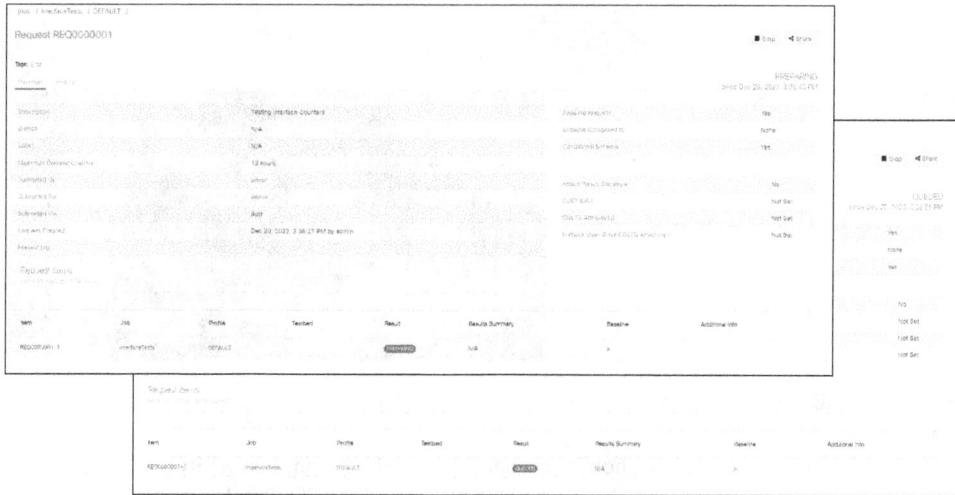

Figure 20-36   *XPRESSO Jobs: Cloud Job – Job Request Preparing and Queuing*

In this case, our request passed, as shown in Figure 20-37!

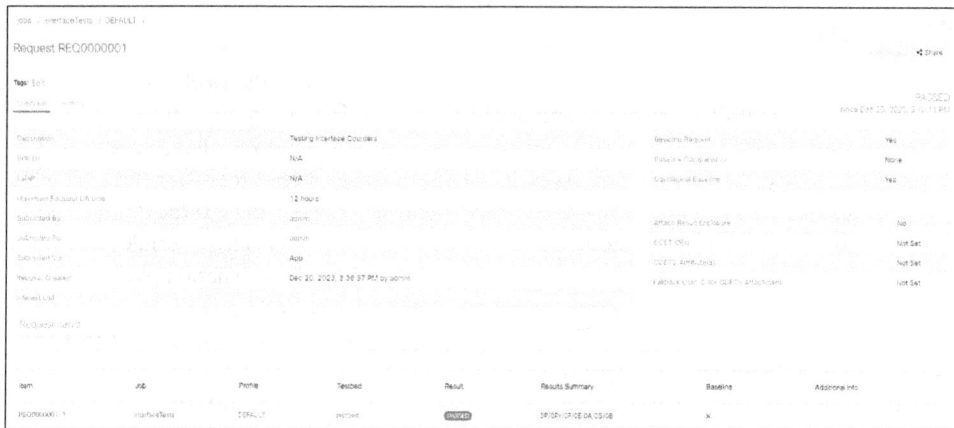

Figure 20-37   *XPRESSO Jobs: Cloud Job – Job Request Passed*

By clicking the request ID, we can see the pyATS job results, as shown in Figure 20-38.

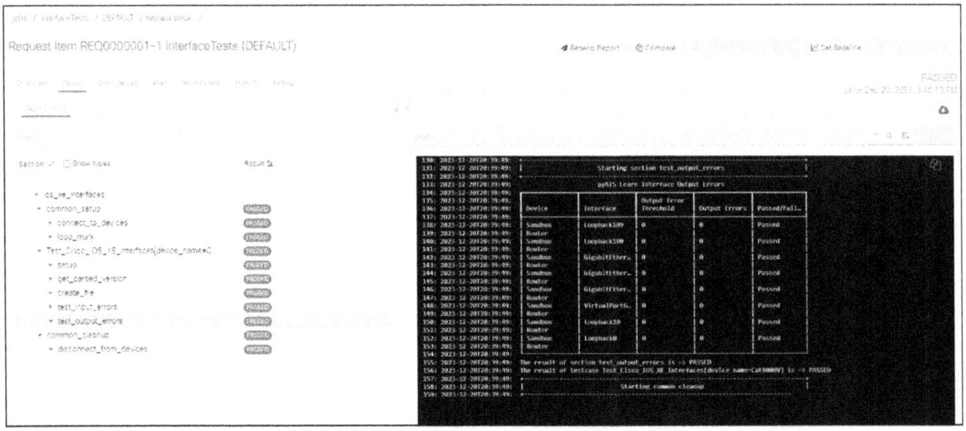

Figure 20-38   *XPRESSO Jobs: Cloud Job – pyATS Job Logs*

This end-to-end example of how to transform a pyATS Job into an XPRESSO request just scratches the surface of the power of XPRESSO. XPRESSO helps operationalize and integrate pyATS into your network automation tools by providing a web UI for your network testing needs. Recall that this example uses an always-on DevNet sandbox and will fail if errors are seen on the interfaces.

Summary

In conclusion, XPRESSO stands as a transformative addition to the world of network test automation, skillfully complementing pyATS with its user-friendly, free, and advanced dashboard. It optimizes the testing experience by centralizing, streamlining, and enhancing the efficiency of testing processes. XPRESSO's range of features—from automated testing and job scheduling to real-time analysis and security—caters to the diverse and demanding requirements of modern network environments. It enables a seamless transition from distributed scripts to a centralized, enterprise-grade automation platform, ensuring that test automation professionals can manage their tasks more effectively and productively. This chapter has provided a comprehensive guide to installing and getting started with XPRESSO, as well as integrating pyATS jobs into this dynamic platform, setting the stage for a more efficient, secure, and streamlined testing future.

CI/CD with pyATS

Something all network engineers love are new acronyms. If we don't already have enough, let's add another to our technology dictionary—CI/CD. CI/CD is an acronym for continuous integration/continuous delivery (or deployment). You may wonder why you haven't heard of it before, or if you have googled it, you may wonder why you ended up more confused than before? In this chapter, we are going to review the topic of CI/CD, why it's important to software development, and how it fits into network automation and the NetDevOps methodology. Here's a breakdown of the different topics we are covering:

- What is CI/CD?

- CI/CD in NetDevOps

- NetDevOps scenario

- NetDevOps in action

- What's next?

What Is CI/CD?

Continuous integration/continuous delivery (or deployment) is a methodology to automate the building, testing, linting, and deploying of code. It's a very common practice in the world of software development and DevOps, with a goal to iteratively test and identify software bugs in code before pushing the code to production. CI/CD aims to reduce the amount of time an engineer must spend performing these tasks by automating them in what's known as a CI/CD pipeline. In the following sections, we are going to dive deeper into the concepts of CI/CD and how it can be integrated into code hosted on popular developer platforms such as GitHub and GitLab (on-prem or in the cloud).

Demystifying CI/CD

CI/CD pipelines are the tangible product of the methodology of automating the building, testing, linting, and deploying of code. CI/CD pipeline is an appropriate name, as freshly developed code enters one end of the pipeline and comes out the other end as "production-ready" code, or as close as it can be. Everything that happens in between, such as building/compiling the code, linting the code, testing the code, staging the code in a development/pre-production environment, and optionally deploying the code to production, is known as stages within the pipeline. The stages may be defined differently depending on the CI tool used to develop the pipeline, but the stages help organize the individual tasks and how they should be executed. In the next couple sections, we will look at how to begin building a CI/CD pipeline on GitHub and GitLab.

CI/CD Pipeline Integration

CI/CD pipelines have been used in software development for many years now. They increase the speed and efficiency of application development by automating some of the mundane tasks, such as linting and testing code, that are required before code can be considered production-ready. Let's start with a simple example of how to build CI/CD pipelines on two of the most popular developer platforms: GitHub and GitLab. Other CI/CD tools are available depending on your specific needs, including CircleCI, Travis CI, Azure DevOps, TeamCity, and Jenkins, but we are going to focus on GitHub and GitLab in this chapter.

GitHub Actions

GitHub refers to its CI/CD workflow as GitHub Actions. GitHub Actions define workflows that can be triggered based on an event, such as a pull request being opened or merged into the "main" branch of a code repository. Along with workflows being triggered by an event, they can also be run manually or scheduled to run at intervals. Within a workflow, one or multiple *jobs* are created to run in sequence or in parallel to accomplish different tasks. The tasks within a job are called steps. Steps run the scripts and actions to perform the task. Scripts are shell commands or actions that are pre-canned routines customized for GitHub. They are meant to reduce the number of repetitive steps performed in a job. Figure 21-1 is a visual example of jobs and steps defined in a GitHub workflow.

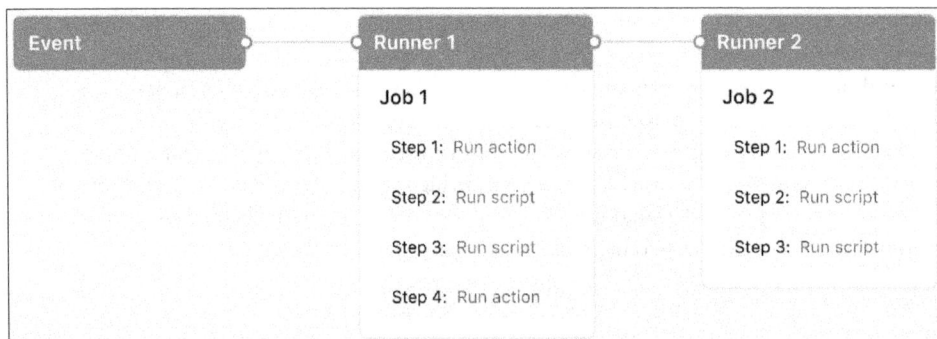

Event	Runner 1	Runner 2
	Job 1	**Job 2**
	Step 1: Run action	Step 1: Run action
	Step 2: Run script	Step 2: Run script
	Step 3: Run script	Step 3: Run script
	Step 4: Run action	

Figure 21-1   *GitHub Action: Simple Workflow (Credit to GitHub.com)*

GitHub workflows are defined in YAML files and stored in the .github/workflows directory at the root of a project repository. Multiple workflows can be defined, which is why they are defined in a directory instead of a single file. Each workflow can perform a specific task. For example, one workflow can be responsible for all linting and code testing while another workflow focuses on deploying the code or application.

All workflows are executed on runners. Runners are virtual machines (macOS, Windows, or Linux) or containers. GitHub hosts macOS, Windows, and Linux virtual machine runners, but you can also host your own runners. Runners can only execute one job at a time. Once a job has been completed, the next job in the workflow is executed. Example 21-1 shows a workflow that uses the setup-python action to install Python on an Ubuntu VM runner and runs a Python script named "my_testscript.py". The workflow is triggered by a push to the main branch of the repository, including when a pull request is merged into the main branch.

Example 21-1   *Python GitHub Action*

```
name: python-gh-actions-example  # Workflow name
on:  # The trigger to specify when the workflow runs
  push:
    branches:
      - main
jobs:
  python-example:  # Job name
    runs-on: ubuntu-latest  # Specify runner to run job
    steps:
      - uses: actions/checkout@v4  # GitHub Action to checkout code repository
      - uses: actions/setup-python@v5  # GitHub Action in install Python
        with:
          python-version: '3.11'  # Specifies the Python version to install
      - run: python my_testscript.py  # Command to run on runner's shell
```

GitLab CI/CD

GitLab CI/CD defines a CI/CD pipeline in one YAML file named .gitlab-ci.yml in the root of the project repository. You may use a different filename, but you must change the default CI/CD settings to specify the new CI/CD configuration filename to look for relative to the project repository. In the CI/CD configuration file (.gitlab-ci.yml), you can define the following:

- Tasks to run in the pipeline, including commands to run in sequence or in parallel

- Whether to trigger tasks manually or automatically

- The runner to deploy the application code to

- Dependencies and caches

- Additional configuration files and templates to use, providing modularity

Much like GitHub Actions, jobs define the individual tasks to execute in the pipeline. However, unlike GitHub Actions, jobs in GitLab CI/CD are grouped into stages. Stages group similar jobs that must be run together before moving on to the next stage. Jobs may be grouped into stages such as build, test, and deploy. Given the example, all jobs in the build stage must run before moving on to the test stage, and all jobs in the test stage must run before moving to the deploy stage. Although this is pretty straightforward, it's very important to understand because it allows you to better organize your pipeline execution. GitLab CI/CD uses runners, like GitHub Actions, to execute jobs. Runners can be privately hosted or provided by GitLab.

One concept that is common to both platforms is CI/CD variables, which can be predefined variables provided by the platform, custom environment variables, custom secrets, or job/step-specific variables. Predefined variables provided by the platform can be helpful to provide context and allow conditional decisions to control job execution. For example, in GitLab, you can reference the target branch name of a merge request using the following variable in your CI/CD configuration file: $CI_MERGE_REQUEST_TARGET_BRANCH_NAME. Another popular predefined variable is $CI_COMMIT_MESSAGE, which references the commit message of the commit that started the pipeline execution. There are other valuable predefined variables, but the purpose is to quickly reference variables that can provide context about the pipeline, job, commit, pull request, and other important information. We will take a deeper look at GitLab CI/CD and build out a complete CI/CD configuration file that will be used to run tests against a simulated network.

Now that you understand the importance of CI/CD and how CI/CD pipelines can be integrated into existing codebases, we are going to see how to take the CI/CD principles and apply CI/CD to networking through the concept of NetDevOps, which combines DevOps practices with network automation.

CI/CD In NetDevOps

AWS defines DevOps as "the combination of cultural philosophies, practices, and tools that increase an organization's ability to deliver applications and services at high velocity…" (https://aws.amazon.com/devops/what-is-devops/). In short, the goal of DevOps is to establish practices and tooling that allow teams to make changes quickly and iteratively with minimal risk. You may wonder how networking fits into this, since network changes are the complete opposite of DevOps, with changes to the network being slow and performed in an "all-at-once" fashion. Well, the idea of NetDevOps is that we can take the DevOps practice of making quick and iterative changes and applying it to networking. Obviously, this creates a challenge in two areas: people and tooling.

Network engineers naturally push back when it comes to any changes to the network due to the historical backlash received when performing changes and causing outages. This is due to the risk profile of the network, which includes the potential of impacting all applications and halting business operations when performing a change, versus only one application when DevOps engineers push changes. As a result, network engineers are hesitant to perform changes to the network until a weekly, bi-weekly, or monthly change window. This leads to a backlog of network changes that were collected over the time between change windows, which creates stress and uncertainty with network engineers since multiple network changes must occur all at once during a change window. Only being able to make network changes during a 4-to-12-hour window once a week, bi-weekly, or monthly is unfair to network engineers and slows down business agility. Other variables that impact change scheduling include business requirements and the reliance on other teams to implement and validate their changes first.

The other challenge is tooling. Network engineers rely heavily on network monitoring tools to determine whether a change was successful. If the network operation center (NOC) doesn't contact the on-call engineer, then we are set, and the change was a success! Wrong! Network testing using tools such as pyATS and the pyATS Library should be part of the implementation plan, with tests and test cases being built out specifically for the change being made to the network. This is different from the common test plan where a series of **show** commands is executed on the device(s) included in the change with the output manually reviewed. Parsing **show** commands can still be a part of the test plan, since the output can be indicative of a device's operating state, but it should be automated, with the parsed output being tested against predefined thresholds automatically. The predefined thresholds can be arbitrarily defined by the network team or calculated based on baseline metrics captured by other tooling.

Now that we've broken down some of the issues encountered when trying to incorporate DevOps with networking, let's get messier and see how we can incorporate CI/CD into the mix. Let's start by walking through a common network change. Most network change plans include three major sections:

- Change plan
- Test plan
- Backout plan

Change plans include the implementation steps to make changes to the network and may also include prerequisite steps to gather baseline data to compare after the change is completed. The test plan includes individual test steps to ensure the network is in an expected and healthy state before determining that the change was successful. The backout plan contains the steps to remove the changes included in the change plan.

A CI/CD pipeline can automate the deployment of device configuration and network testing with every commit or pull request—the frequency is up to you! A common workflow may be storing all configuration and configuration parameters in git, creating a branch with proposed changes, and opening a pull request to merge into the main branch and push the changes to the "production" network. The CI/CD pipeline will automatically trigger when a pull request is opened and will target the "main" branch, which represents the production network. The CI/CD pipeline triggering when a pull request is opened is a common workflow for many open-source software projects. The CI/CD pipeline may lint and format the code. Along with upholding codebase standards, all available unit and integration tests are run against the proposed code to ensure code test coverage. The CI/CD pipeline does much of the heavy lifting for the maintainers, allowing them to focus on the purpose of the changes and why they should be introduced into the codebase. Much like an open-source software project, network engineers and management need to focus on the purpose of the changes, not whether the spacing is correct in all the YAML files. CI/CD pipelines can assist network engineers with the mundane but important technical checks on new code changes. Network automation doesn't need to start from scratch. Years of software development practices can help guide network engineers to a new world of automation. In the following section, we are going to go through a real-world example of managing the configuring and running of tests against the network in a CI/CD pipeline. It's a new way of thinking about the network, but it will help lead to agility and confidence when you make changes to the network.

NetDevOps Scenario

Hopefully, the NetDevOps philosophy is beginning to resonate with you. If the theory sounds great, but an example would really drive it home, then this section is for you! By the end of this section, you'll see how you can employ multiple automation tools to build a CI/CD pipeline that can deploy and test configuration changes before pushing the changes to a production network environment. The code referenced throughout this section can be found on GitLab at https://gitlab.com/dannywade/pyats-book-cicd. Let's first start with the different components and tools used in the lab setup.

Lab Setup

Building a complete network automation lab that incorporates a CI/CD pipeline can be daunting, but the goal is to understand why each tool is needed, which should help break down the different components. The following tools make up the different components used in the lab scenario:

- GitLab
- Cisco Modeling Labs
- Ansible
- Cisco pyATS
- Cisco Webex Teams

GitLab is the developer platform chosen to store the code, track changes, and create a CI/CD pipeline. Cisco Modeling Labs (CML) is a network simulation software used to create a virtual network topology used as a "development" or "staged" network environment to test network changes before pushing them to a production network. Ansible is a configuration management tool used to deploy configuration changes to the staged network devices in CML. Cisco pyATS is obviously for performing automated network testing and validation. But to be more specific, we are utilizing pyATS Blitz, which follows the "low-code, no-code" philosophy by abstracting Python code for YAML, much like Ansible. The purpose of using pyATS Blitz is to keep the code and automation simple. Ansible simplifies configuration pushes by defining changes in YAML playbooks, and pyATS Blitz keeps the network testing and validation simple as well by specifying testcases and tests in a YAML datafile. Lastly, we are integrating GitLab with Cisco Webex Teams to notify a Webex Team space when the CI/CD pipeline finishes executing, reporting the pipeline results (pass/fail) and how long the pipeline ran. Figure 21-2 shows a high-level diagram of the CI/CD pipeline we are going to build with the mentioned tooling.

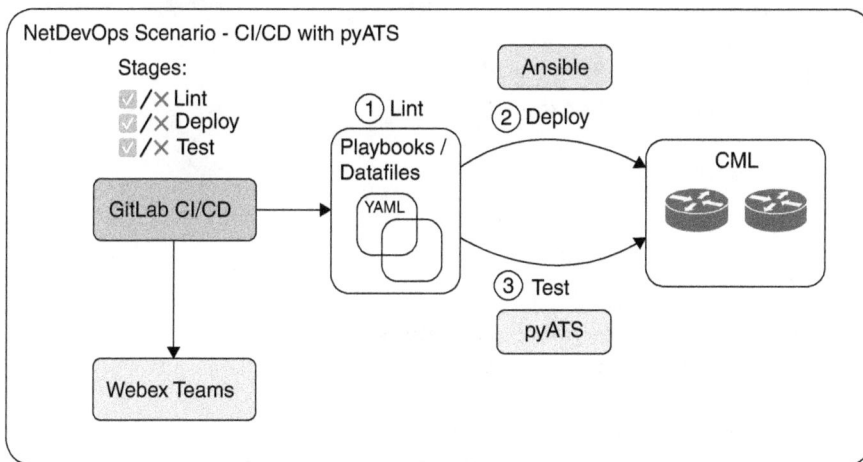

Figure 21-2 *GitLab CI/CD Pipeline Workflow*

In the next section, we will go through the different stages in the GitLab CI/CD pipeline, which contain the different, individual jobs.

CI/CD Stages

GitLab CI/CD pipelines define stages to group individual jobs that execute together. Execution can be further customized, but the idea is that all jobs in a single stage execute and must pass before the jobs in the next stage are executed. In our lab scenario, we are defining three stages:

1. Lint

2. Deploy

3. Test

Example 21-2 shows the .gitlab-ci.yml file, which contains the CI stages and jobs in our NetDevOps CI/CD pipeline.

Example 21-2   *NetDevOps CI/CD Pipeline (.gitlab-ci.yml)*

```
stages:
  - lint
  - deploy
  - test


# Create Python virtualenv and install packages
default:
  before_script:
    - python --version
    - python -m venv .venv
    - source .venv/bin/activate
    - pip install --upgrade pip
    - pip install -r requirements.txt


# Lint Ansible playbooks
lint_yaml_files:
  stage: lint
  script:
    - ansible-lint routers.yaml
    - pyats validate datafile net_testing/blitz.yaml
    - pyats validate testbed net_testing/testbed.yaml


# Deploy configuration to network devices using Ansible
deploy_config:
  stage: deploy
  script:
    - ansible-playbook routers.yaml


# Test network using pyATS Blitz
test_network:
  stage: test
  script:
    - pyats run job pyats_jobfile.py --health-checks cpu memory logging core \
      --html-logs pyats_html_logs --archive-dir pyats_archive_logs
  artifacts:
    name: pyats_test_artifacts
    untracked: false
    when: on_success
    expire_in: 30 days
    paths:
      - $CI_PROJECT_DIR/pyats_html_logs
      - $CI_PROJECT_DIR/pyats_archive_logs
```

The lint stage will lint the different YAML files to check for valid syntax, line length, trailing spaces, and indentation. More specifically, if we look at the lint_yaml_files job, we find the **ansible-lint** command, which lints the routers.yaml Ansible playbook file, and the **pyats validate** command, which lints pyATS datafiles and testbed files. Linting is very important for code hygiene and to uphold established code standards before changes are merged into production code.

The deploy stage is used to push configuration to network devices. There's one job, deploy_config, which executes an Ansible playbook, routers.yaml, to deploy configuration to routers in a CML topology.

The last stage is the test stage. The test stage contains one job, test_network, which executes a pyATS job file with pyATS health checks. The job file executes a pyATS Blitz YAML file that verifies bgp functionality. In addition to executing network tests with pyATS, the HTML logs and archives generated by the pyATS job are saved as GitLab job artifacts and attached to the pipeline results. This allows anyone that has access to the CI/CD pipeline results to download the logs and archives.

Besides the stages, there's also a default keyword block toward the beginning of the file. The default keyword in GitLab CI/CD allows you to set global defaults for some keywords. In our case, we defined the **before_script** keyword, which defines a list of commands that should run before each job's script. The list of commands creates and activates a Python virtual environment before each script. This can be repetitive and can be further optimized to reduce execution time, but it works as an example for our CI/CD pipeline.

NetDevOps in Action

Now that you understand the different stages and jobs in the CI/CD pipeline for the lab scenario, let's see the pipeline in action! In the following sections, we will review the jobs that deploy configuration changes and test the network changes. We will dive into some of the code and explain how each task is completed.

Configuration Changes

For configuration changes, we are using Ansible to push configuration to lab network devices in a CML topology. The configuration changes are pushed to devices in the deploy stage of the CI/CD pipeline. In the Ansible playbook routers.yaml, two roles are used to configure both routers in CML. If we look in the roles directory, we can see three roles: bgp, device_mgmt, and routing. Example 21-3 shows the tree structure of the roles directory.

Example 21-3 *Ansible Roles*

```
roles
├── bgp
│   ├── tasks
│   │   └── main.yaml
│   └── templates
│       └── bgp_global.j2
├── device_mgmt
│   └── tasks
│       └── main.yaml
└── routing
    └── tasks
        └── main.yaml
```

Each role contains a tasks directory with a main.yaml file, which is essentially a playbook that executes when the role is called. The only exception is the routing role, which imports the bgp role. The routing role's purpose is to include any role that represents a routing protocol, instead of having playbooks reference each individual role that represents a routing protocol. This logic may not make sense all the time, such as when a network device only needs one routing protocol, but it works for our lab. The bgp role generates configurations using Ansible host variables and Jinja2 templates. Example 21-4 shows host variables for the cat8k-rt1 device in inventory as well as the associated Jinja2 template in the bgp role to generate a bgp configuration. Notice the host variable names and how they are used in the template.

Example 21-4 *BGP Configuration Using Host Vars*

```
# /host_vars/cat8k-rt1.yaml
---
mgmt:
  loopbacks:
    Loopback0: 10.254.1.1
    Loopback1: 10.50.0.1

bgp:
  asn: 65000
  networks:
    - { network: 10.50.0.0, mask: 255.255.255.0 }
  neighbors:
    - { neighbor: 172.16.1.2, remote_as: 65001 }
```

```
...
# /roles/bgp/templates/bgp_global.j2
---
router bgp {{ bgp.asn }}
  bgp log-neighbor-changes
  {% for network in bgp.networks -%}
  network {{ network.network }} mask {{ network.mask }}
  {% endfor -%}
  {% for neighbor in bgp.neighbors -%}
  neighbor {{ neighbor.neighbor }} remote-as {{ neighbor.remote_as }}
  {% endfor -%}
```

The host variables and Jinja2 templates allow us to easily add more configuration without the need to manually change multiple device configuration files. You'll notice we provide a list of networks and neighbors, and the template will handle looping through each one and creating the proper configuration. This allows us to add a new bgp neighbor by simply adding to the neighbors list in the host variables file with the neighbor IP address and its associated remote ASN. Example 21-5 shows the resulting generated configuration for the cat8k-rt1 device.

Example 21-5 *Cat8k-rt1 BGP Configuration*

```
router bgp 65000
  bgp log-neighbor-changes
  network 10.50.0.0 mask 255.255.255.0
  neighbor 172.16.1.2 remote-as 65001
```

In addition to the bgp and routing roles, we also have a device_mgmt role that configures a loopback interface for management and global SNMP settings. Example 21-6 shows the main.yaml task for the device_mgmt role.

Example 21-6 *Device_mgmt Role*

```
---
# Loopback interface for management
- name: Confirm Loopback0 IP on each device fall in the Mgmt subnet
  ansible.builtin.assert:
    that:
      - "'10.254.1.0/24' | \
        ansible.utils.network_in_network(mgmt.loopbacks.Loopback0)"
    fail_msg: "{{ mgmt.loopbacks.Loopback0 }} does not belong to a valid subnet"
    success_msg: "{{ mgmt.loopbacks.Loopback0 }}: <OK>"
    quiet: true
```

```
- name: Configure Loopback Interfaces
  cisco.ios.ios_config:
  lines:
    - ip address {{ item.value }} 255.255.255.0
  parents: interface {{ item.key }}
  save_when: changed
  with_dict: "{{ mgmt.loopbacks }}"
  when: ansible_network_os == 'ios'

# SNMP
- name: Configure SNMP settings
  cisco.ios.ios_config:
  lines:
    - snmp-server community ciscoro RO
    - snmp-server community ciscorw RW
    - snmp-server location CICD-Lab
  save_when: changed
  when: ansible_network_os == 'ios'
```

There are three tasks in main.yaml in the device_mgmt role. The first task uses the built-in assert module to confirm that the provided Loopback0 IP from the host variables is within the defined management subnet (10.254.1.0/24). The assert module is very handy to perform quick checks against provided data or variables. The second task uses the ios_config module in the cisco.ios collection to configure the loopback interfaces provided from the host variables. The last task also uses the ios_config module to configure global SNMP settings, including a read-only community string, a read-write community string, and an SNMP location. The SNMP community string values should be defined elsewhere with better security measures, such as being encrypted using Ansible vault, but this configuration is and should only be used in a lab environment.

Now that you understand how configuration is applied to the routers in CML, let's switch our focus to the purpose of this chapter—running network tests using pyATS in a CI/CD pipeline.

Testing Network Changes

In software development, the test stage in a CI/CD pipeline is normally used to run unit and integration tests on the software in a dev or QA environment. In network automation and NetDevOps, we run tests against the network to ensure the configuration changes deployed to the lab network didn't break anything. If something breaks, the CI/CD pipeline fails and indicates the configuration will not work in the production network. For testing, we are using pyATS Blitz to define our testcases in YAML. The purpose of using pyATS Blitz is to continue the theme of simplicity and using YAML files to define our configuration and tests.

Three files are required to define our network tests in pyATS Blitz: a pyATS Blitz datafile, a pyATS testbed file, and a pyATS job file. The pyATS testbed file is required to define the devices used in testing. The pyATS job file is used as an entrypoint to run pyATS Blitz and archive the results. Another note is that we include pyATS health checks when we run the job in the pipeline, which checks the CPU, memory, logs, and whether a core dump file was created on the device under testing between testcases. The Blitz datafile is a YAML file that defines the tests to run against the lab network in CML. Example 21-7 shows the Blitz YAML file that defines one testcase called BgpTestcase that runs two tests—one for each router in the lab network.

Example 21-7   *pyATS Blitz YAML*

```
BgpTestcase:

    source:
        pkg: genie.libs.sdk
        class: triggers.blitz.blitz.Blitz

    test_sections:
        - verify_r1_bgp:
            - learn:
                device: RT1
                feature: bgp
                custom_start_step_message: Learning BGP feature
                custom_substep_message: Confirm BGP operation
                save:
                    - variable_name: r1_bgp_output
                include:
                    # Confirm BGP ASN
                    - contains_key_value("bgp_id", 65000)
                    # Filter down to only "Established" neighbors
                    # (assumes there are some - check will fail otherwise)
                    - contains_key_value("state", "Established")
                    # Check that "172.16.1.2" is an established neighbor
                    - contains_key_value("neighbor", "172.16.1.2")

            - print:
                custom_start_step_message: Print learned BGP info from R1
                item:
                    value: "%VARIABLES{r1_bgp_output.info}"

        - verify_r2_bgp:
            - learn:
                device: RT2
                feature: bgp
```

```
                    custom_start_step_message: Learning BGP feature
                    custom_substep_message: Confirm BGP operation
                    save:
                        - variable_name: r2_bgp_output
                    include:
                        # Confirm BGP ASN
                        - contains_key_value("bgp_id", 65001)
                        # Filter down to only "Established" neighbors
                        # (assumes there are some - check will fail otherwise)
                        - contains_key_value("state", "Established")
                        # Check that "172.16.1.2" is an established neighbor
                        - contains_key_value("neighbor", "172.16.1.1")

          - print:
                    custom_start_step_message: Print learned BGP info from R2
                    item:
                        value: "%VARIABLES{r2_bgp_output.info}"
```

The **learn** action under each device test, **verify_r1_bgp** and **verify_r2_bgp**, will learn the bgp feature on the device. The **include** key under the **learn** action will use the Dq library to determine whether certain key-value pairs are included in the output. These are the true tests to verify how the network is operating. There's also a **print** action after the **learn** actions to print out the bgp information learned from each device for informational purposes.

One other note is that the HTML logs and archives of the pyATS job file are stored as CI/CD artifacts, which are attached to the pipeline results and retained for 30 days (can be changed in the CI configuration file). This is super helpful if you (or one of your fellow engineers) are interested in the detailed logs collected for a specific pipeline run within the past 30 days. Figure 21-3 shows how to download the artifacts for a particular pipeline run.

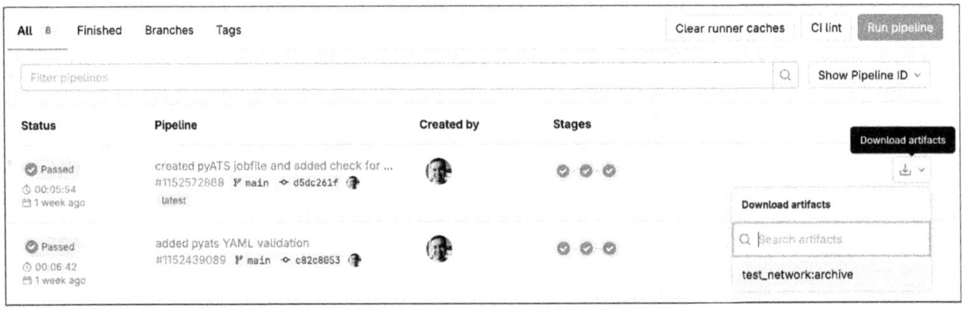

Figure 21-3 *CI/CD Pipeline Artifacts*

Feedback Loop

Beyond viewing the results in GitLab, we can integrate external applications into our GitLab CI/CD workflow under the repository settings. For our NetDevOps scenario, we are going to integrate with Cisco Webex Teams via webhooks. There are two parts to this integration. We must first visit the Webex App Hub and connect the Incoming Webhooks app and create a new webhook by providing a bot name and selecting a Webex Team space to send the webhook message. Once we have created a new webhook, we copy the webhook URL. In GitLab, go to **Settings > Integrations** and search for "Webex Teams," enable the integration, paste in the copied webhook URL, and select the trigger(s) you want to use to fire off the webhook to Webex Teams. This is important, as these triggers use events to know when a webhook gets sent to Webex Teams. In the lab scenario, we are only concerned about the CI/CD pipeline, so we selected **A pipeline status changes** as our only trigger. Figure 21-4 shows how the Webex Teams integration is configured in GitLab.

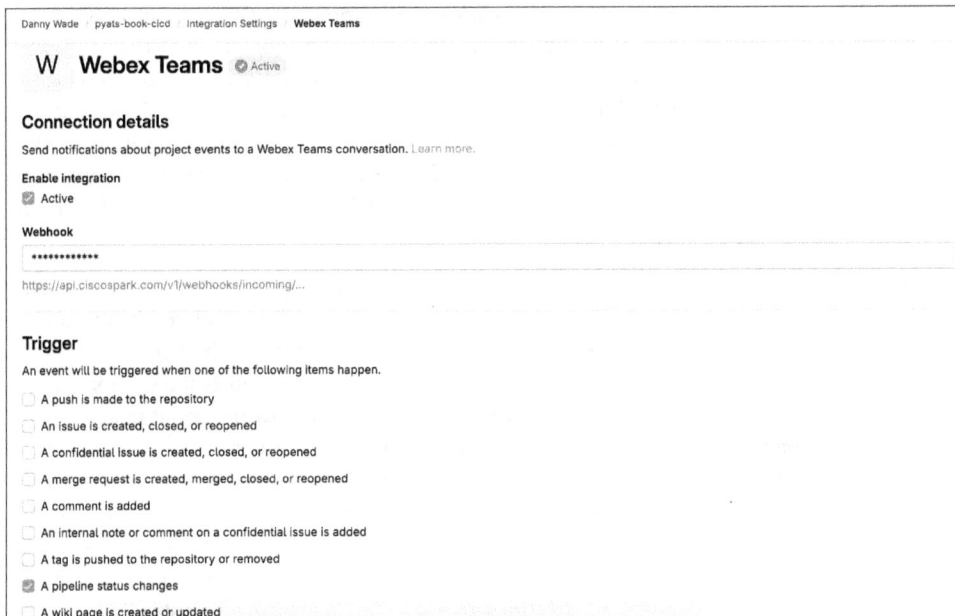

Figure 21-4   *GitLab – Webex Teams Integration*

Once the integration is configured in GitLab, we can test it by manually running the pipeline and waiting for it to complete. Once the pipeline finishes running, we should receive a message in the appropriate Webex Teams space. Figure 21-5 shows how the message should appear in Webex Teams.

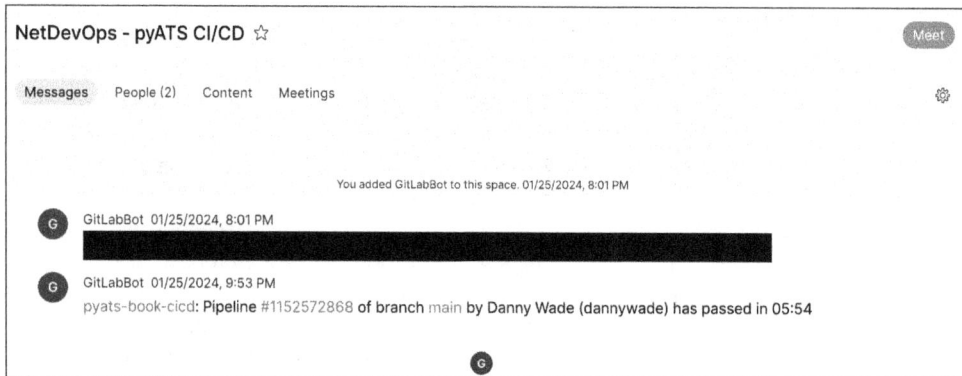

Figure 21-5   *Webex Teams Space – GitLab CI/CD Bot*

You'll see who initiated the pipeline to run, the duration, and the status of the pipeline run. For additional information, you can click the pipeline ID that's hyperlinked to go to GitLab and view details of the pipeline results. This integration is out of the box and doesn't require any additional configuration or services! If you would like to customize the message sent to Webex Teams or enable ChatOps, to allow users in the Webex Teams space to reciprocally interact with GitLab, you'd need to create an intermediate endpoint to accept the webhook from GitLab, manipulate/respond to the data, and send a customized webhook to Webex Teams. There's some additional configuration required for this to work, but that is outside the scope of this book.

The purpose of integrating GitLab with Cisco Webex Teams, or any chat service for that matter, is to ease the use of automation for other engineers. What if all you had to do was tell engineers to join a Webex Team space to stay up to date on the latest changes in the network? This creates a solid foundation for ChatOps and other integrations that allow engineers to stay within one application and not need to worry about jumping between screens to see what's going on in the network. This creates a central command center that can be extended further with ChatOps integrations to other systems, such as an IT service management (ITSM) software like ServiceNow or Jira. By easing the use of automation, it may help increase buy-in from a network/infrastructure team to continue developing automated workflows. Ultimately, by gaining trust and buy-in from stakeholders, including technical peers, finance/procurement, and management, you have a greater chance of increasing network automation adoption within your organization.

What's Next?

We went through a crash course in CI/CD, but it forced us to understand the tools available in network automation and how to apply them to automate the configuration deployment and testing of a network. Of course, we could have used other popular network automation tools and libraries, such as netmiko, scrapli, nornir, or pyATS itself to configure the devices. However, we chose Ansible with the goal of simplicity. Simplicity can be defined in many ways, but we tried keeping the barrier to entry low and the amount of code to a minimum in our scenario.

Beyond using different tools, there are a few other things that we can do to add to our CI/CD pipeline. We can add more testcases to look at other device features, beyond just bgp. We can also add another job in the CI/CD pipeline to deploy the changes to a production network. This may seem like a stretch. Push changes to a production network automatically?! Maybe someday, but for now, we can configure jobs that deploy configuration to a production network to be run manually, requiring a user to intervene in the CI/CD pipeline. If all other jobs and stages pass, you'll receive a notification that an action is required in the pipeline. The job can be configured to run manually by adding the **when: manual** attribute to the job. One more suggestion is to add a baseline test before the changes are deployed to the network. This may include learning device features before the changes are deployed to the network and comparing them with a captured state afterward. This allows your teams to capture the network state before any changes occur (and troubleshoot if anything goes wrong during deployment) by comparing it to a known-good state, even in the lab network. There are plenty of other additions and optimizations that can be done with the NetDevOps example, and hopefully you'll try and build on to it yourself!

Summary

In this chapter, we reviewed the concept of CI/CD and how it's used traditionally in software development to build, lint, run unit and integration testing, and deploy code. We then moved on to see how CI/CD pipelines are built in two of the largest developer platforms, GitHub and GitLab. The concepts stayed consistent, but the nomenclature changed depending on the platform. After reviewing how to build CI/CD pipelines, we looked at how we can apply those same principles to networking and explained the popular term NetDevOps. Lastly, we saw NetDevOps in action by creating a CI/CD pipeline in GitLab that incorporated many popular tools, including Ansible, Cisco Modeling Labs (CML), pyATS, and Cisco Webex Teams. We saw how pyATS fits in the overall pipeline to run automated tests against the network to ensure it's in an operational state after changes are deployed to a lab network. Hopefully this chapter opened your mind to what's possible in network automation when integrating multiple network automation tools in a CI/CD pipeline to automate the deployment of device configuration and testing the operational state of the network.

Robot Framework

You may wonder why there's a chapter in a book written about pyATS named "Robot Framework." The Robot Framework is a test automation framework, too, but not a competing one. The framework is built to be extensible, is keyword-driven, and promotes high-level abstraction. Keywords allow users to create test suites that are easy to read without having knowledge about the lower-level programming involved. You may think this sounds a lot like tools that use a domain-specific language (DSL) such as Ansible. The heavy use of abstraction is about the only thing in common. Remember, choosing the right tool for the problem is an important part of adopting network automation and DevOps practices. The Robot Framework is a test automation framework, with a focus on acceptance testing. Ansible is more suitable for configuration management and software provisioning. In this chapter, we are going to review the Robot Framework and how pyATS and the pyATS library (Genie) integrate with it.

This chapter covers the following topics:

- What is the Robot Framework?
- Getting started with Robot Framework
- Robot integration with pyATS

What Is the Robot Framework?

The Robot Framework is a generic, open-source test automation framework built in Python. The framework is keyword-driven, which means tests are written using plain English words that translate into low-level code, which creates an abstraction for users. It's a no-code, low-code tool that doesn't require the user to have programming knowledge to use it. The user only needs to know the keywords for a test library, which should be well-documented. It's also customizable and can easily be extended using Python. Like programming languages, the Robot Framework has a built-in library with common functionality and third-party libraries that provide additional functionality. The third-party libraries are built using Python and can provide additional keywords for testing. Later in the chapter, we will look at the pyATS, the pyATS library (Genie), and the Unicon test libraries that can be imported into the Robot Framework. The Robot Framework was built to provide an easy-to-use testing framework that is platform and application independent.

Getting Started with Robot Framework

Before we dive into the Robot Framework and how it integrates with pyATS, we need to understand some of the different components. We aren't going to cover all facets of the framework, but we are going to cover some of the important components that make up a simple testcase. In the following sections, we will look at testcases, keywords, variables, importing libraries, test execution, and results and reporting. If you'd like to learn more, refer to the "Robot Framework User Guide" documentation (https://robotframework.org/robotframework/latest/RobotFrameworkUserGuide.html).

Test Cases

Test case files, also known as suite files, contain multiple test cases. The Robot Framework documentation suggests not to have more than ten test cases in one test case file. Higher-level test suites can be created from test case files saved in the same directory, also known as a suite directory. In a test case file, there are multiple test data sections. The different test data sections are also known as tables in a test case file. Table 22-1 shows the different test data sections in a test case file.

Table 22-1   *Test Data Sections*

Section	Used for...
Settings	Importing test libraries, resource files, and variable files.
	Defining metadata for test suites and test cases.
Variables	Defining variables that can be used elsewhere in the test data.
Test Cases	Creating test cases from available keywords.
Tasks	Creating tasks using available keywords. A single file can only contain either tests or tasks.
Keywords	Creating user keywords from existing lower-level keywords.
Comments	Additional comments or data. Ignored by Robot Framework.

Source: https://robotframework.org/robotframework/latest/RobotFrameworkUserGuide.html#test-data-sections

Each section has a header that is surrounded by three asterisks (***). For example, the variables section header would be *** **Variables** ***. Example 22-1 shows what a test case file may look like with its section headers.

Example 22-1   *Test Case File – Section Headers*

```
*** Settings ***
<settings go here>


*** Variables ***
<variables go here>


*** Test Cases ***
<test cases go here>


** Keywords ***
<keywords go here>
```

This doesn't give the full picture, but understanding the section headers helps increase readability and delineate the different test data sections.

Test case files are parsed using multiple formats, including by the number of spaces (space-separated format), pipe-separated format, reStructuredText format, and JSON format. The most common format you'll find is the space-separated format where the number of spaces determines how data is parsed. Two or more spaces, or one or more tab characters, is considered the separator in this format. *For readability, it's recommended to use four spaces.* The space separator will split the test case file into lines and parse each line of data using the separator to identify keywords and arguments from the test case file.

Keywords

Keywords are high-level abstractions to call on lower-level code to complete a task. Arguments can be provided to keywords, which are passed to lower-level code functions. Test case sections have two columns. The first column is the test case name, and the second column has keyword names with their associated arguments. The easiest way to understand keywords is by example. Example 22-2 shows a simple test case to test submitting a form on a web page. The test case opens the web page, inputs text into a textbox, and submits the form.

Example 22-2   *Simple Test Case*

```
*** Test Cases ***
Validate Form Submission
    Open Web Page     https://localhost:5000
    Input Textbox     This is my first test!
    Submit Form
```

Let's review what we know about the example. The test case is under the Test Cases section, which is expected. Under the Test Cases section header is the first test case, named Validate Form Submission. The test case name is considered in the first column. Under the test case name are the different keywords used to perform the test. You'll notice the spacing before the keywords. Much like YAML and other whitespace-sensitive data formats, the spacing matters and is used to parse the test case file. The first keyword **Open Web Page** with the argument **https://localhost:5000** represents a lower-level function written in Python that uses a Python library to create a session to the website URL that's passed as an argument. The following keyword, **Input Textbox**, also has an argument that instructs the function what to input into the textbox. Lastly, the **Submit** button is pressed and the form is submitted. In later sections, we will dive into the keywords provided by the pyATS, pyATS library (Genie), and Unicon libraries to the Robot Framework to easily create network automation tests.

There's also the concept of user keywords. User keywords are identical to test case keywords but provide a higher-level abstraction by combining multiple keywords into one higher-level user keyword. User keywords are found in the Keywords section. Example 22-3 shows a user keyword to open the registration page of a website and confirm the title is correct. Once created, user keywords can be used in test cases.

Example 22-3 *User Keyword*

```
*** Keywords ***
Open Registration Page
    Open Browser     http://signmeup.com/register
    Title Should Be     Registration Page
```

Variables

Variables can be used in most places in test data but are commonly used as arguments for keywords in the Test Case and Keyword sections. Variables are case-insensitive and can be scalars (${SCALAR_NAME}), lists (@{LIST_NAME}), dictionaries (&{DICT_NAME}), or environment variables (%{ENV_VAR_NAME}). Global variables should be all capitalized (all caps), while local variables should be all lowercase. Variables begin with a type identifier ($, @, &, %), which determines the data type of the variable (scalar, list, dictionary, or environment variable). The variable itself is wrapped in curly braces ({ }), with the variable name being between the curly braces. For brevity, the focus will be on the scalar and environment variable syntaxes.

Scalar variable syntax is the most common way to use variables. The variable value is defined and assigned to a variable name. The variable is called upon using the scalar variable syntax, which replaces the variable placeholder with the variable value as-is. Example 22-4 shows how to use the scalar variable syntax in a test case.

Example 22-4 *Scalar Variable Syntax*

```
*** Variables ***
${F_NAME}    Danny
${L_NAME}    Wade


*** Test Cases ***
Greeting
    Log    Hello, ${F_NAME} ${L_NAME}
```

The variables are defined under the Variables section header and called on in the Test Cases section. The **Log** keyword is a built-in keyword that accepts a message as an argument. In this case, the message "Hello, Danny Wade" will be logged as an information message.

Another prominent variable type used in network automaton testing is the environment variable. Environment variables are a necessary security measure when providing sensitive data to code, and they prevent you from hardcoding usernames, passwords, tokens, and so on. Environment variables can be used in test case files by using the following syntax: %{ENV_VAR_NAME}. The big difference between this variable syntax and scalar variable syntax is the type identifier that prefixes the variable name. All environment variables are prefixed with a % character. All environment variables set in the operating system at the time of testing can be used. Environment variables can also be set and/or removed by the Robot Framework using the OperatingSystem test library. The OperatingSystem test library offers the Get Environment Variable and Set Environment Variable keywords, which allow these actions. Example 22-5 shows how to use preset environment variables and how to set new environment variables in the test case file.

Example 22-5 *Environment Variable Syntax*

```
*** Test Cases ***
Calling Env Vars
    Log    Here's my top-secret phrase set in the operating system: %{SECRET}
    Set Environment Variable    SECRET_PW    password123
    Log    Here's my top-secret password I set in the test case: %{SECRET_PW}
```

The first **Log** keyword will print an informational log message with the environment variable named **SECRET**. This environment variable was set in the operating system before testing. The Set Environment Variable keyword sets an environment variable in the test case that can be called on later. In this case, we are setting an environment variable for a secret "secure" password. Lastly, we use the **Log** keyword to print the newly created secret password. We will see the scalar and environment variable types more in later sections of this chapter.

Importing Libraries

Test libraries contain library keywords, which are the keywords used in test cases. All test cases use keywords from a test library, whether that be the built-in library included with the Robot Framework or a third-party test library such as pyATS. Test libraries are imported using the Library setting under the Settings section. Example 22-6 shows how to import the built-in OperatingSystem test library and a third-party test library.

Example 22-6   *Importing Built-in Library*

```
*** Settings ***
Library    OperatingSystem
Library    custompkg.libs.TestLibrary
```

Third-party test libraries are imported using the library name or the path to the library. Importing using the library's name is the most common method and is the same as if you were importing a third-party library into a Python script. The Robot Framework will search the directory where Python automatically installs third-party modules using tools like pip. This is how the pyATS, pyATS library, and Unicon test libraries are discovered once they are installed via pip. You can also discover test libraries by reading the module search paths from the **PYTHONPATH** environment variable or provide the command-line option **–pythonpath (-p)**. Along with using the library name, you can provide the file path to the test library files. The file path is the relative path from where the test case files are located. This method can be beneficial, as it removes the need for the module search path. If the test library is a file, you must include the file extension. To recap, if third-party test libraries are pip-installed, they will automatically be discovered and can be imported the same as a Python script. Otherwise, you'll need to provide the module search path using the **PYTHONPATH** environment variable or CLI flag or provide the relative file path to the test library directory or file.

Test Execution

Robot Framework test cases are executed from the command line and will provide two output files and reports, which will be discussed in more detail in the next section. Alternatively, the robot python module can be used directly to execute a test case. You simply need to provide the absolute or relative path to a test case file or a directory with test case files to execute the tests. Test case files must have a .robot extension to be executed. To execute a test case file called network_test.robot, you would enter the following:

```
robot network_test.robot
```

You may also provide many test case files and directories at once, separating each file or directory name with a space. To run multiple test case files, you would enter the following:

```
robot bgp_tc.robot ospf_tc.robot netservices_tc.robot
```

There are many command-line options that can be provided when executing test case files. Many command-line options accept glob patterns as arguments. For example, when you want to match any string that ends with **_network**, you could pass in the argument ***_network**. Glob patterns are like regular expressions and can be discovered more in the Robot Framework documentation.

Lastly, default command-line options can be specified by the environment variable **ROBOT_OPTIONS**. Different options must be specified in a space-separated list in the environment variable. Here is the **ROBOT_OPTIONS** environment variable that defines the results output directory and the log level:

```
export ROBOT_OPTIONS="-outputdir results -loglevel DEBUG"
```

Test Results and Reporting

Once a test suite finishes execution, the results are printed directly as command-line output, much like the pyATS AEtest standalone execution output. This is the most obvious and easiest way to review the status of each test case. However, there are also generated output files that provide additional details of the test results.

By default, three output files are generated: a high-level HTML report file, a detailed log file in HTML format, and an xUnit-compatible XML file that contains test execution details. The XML file being xUnit-compatible allows it to be consumed by tools like Jenkins that can understand xUnit formats and generate statistics based on data in the file. For the HTML logs and reports, the default title can be changed, and the background color to indicate pass or failures for tests can be changed in the HTML report file. Figure 22-1 shows Robot test results in a high-level HTML report.

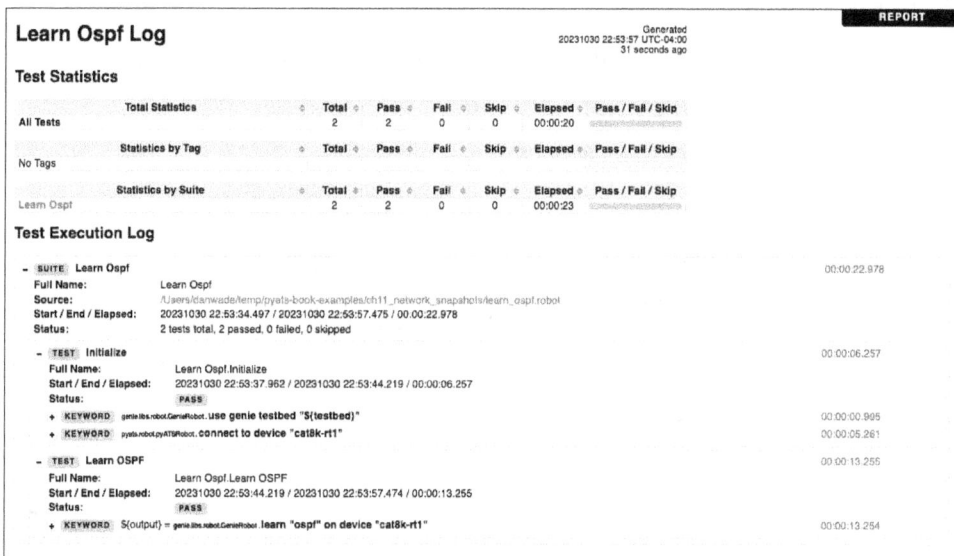

Figure 22-1 *Robot Test Results – HTML Report*

The Robot Framework provides the ability to perform post-processing to XML output files using the built-in Rebot tool. The Rebot tool can be used with the **rebot** command. Many of the command-line options are identical to the **robot** command. However, instead of providing test case files as arguments, you are providing XML output files to the **rebot** command. Just like the **robot** command options, default options can be specified with an environment variable, **REBOT_OPTIONS**. The different command-line options must be provided as a space-separated list.

If you would like no output, you can explicitly disable each output file from being generated using the following command-line options: **--output NONE --report NONE --log NONE.**

Robot Integration with pyATS

Now that we have an understanding of the Robot Framework, let's dive into how it can be used with pyATS for network automation testing. First, the optional pyats.robot module must be installed. To install the pyATS robot module, enter the following:

```
pip install pyats.robot
```

Once this module is installed, you should have access to the **robot** commands discussed in previous sections. Now that the robot module is installed, we can begin using the available keywords provided by the library. In the following sections, we are going to review some of the keywords and the capabilities across the pyATS libraries. We are also going to see how we can run any Robot Framework test script with the Easypy Runtime environment.

PyATS Keywords

The pyATS keywords provide the ability to perform the following actions:

- Load and use pyATS testbed YAML files
- Connect and interact with testbed devices dynamically
- Run pyATS AEtest testcases and convert the results to Robot

The full list of pyATS keywords can be found in the pyATS Robot documentation (https://pubhub.devnetcloud.com/media/pyats/docs/robot.html). Example 22-7 shows a Robot test case file with some of the available pyATS keywords.

Example 22-7 *pyATS Keywords*

```
*** Settings ***
# Importing test libraries, resource files and variable files.

Library          pyats.robot.pyATSRobot
```

```
*** Variables ***
# Defining variables that can be used elsewhere in the test data.
# Can also be passed in as arguments at runtime

# Must have a testbed.yaml file available in the same directory
# Used to define the testbed devices to connect to
${testbed}        testbed.yaml

*** Test Cases ***
# Creating test cases from available keywords.

Initialize
    # select the testbed to use
    use testbed "${testbed}"

    # connect to testbed device through cli
    connect to device "cat8k-rt1" as alias "cli"

# Assumes a pyATS testscript named 'basic_example_script.py' is in the
# same directory with CommonSetup, tc_one Testcase, and CommonCleanup
# sections defined
CommonSetup
    # Runs testcase 'common_setup' section defined in basic_example_script.py
    run testcase "basic_example_script.common_setup"

Testcase One
    # Runs testcase 'tc_one' testcase defined in basic_example_script.py
    run testcase "basic_example_script.tc_one"

CommonCleanup
    # Runs testcase 'common_cleanup' section defined in basic_example_script.py
    run testcase "basic_example_script.common_cleanup"
```

In the example, we define a variable for the testbed filename, which we use in the "Initialize" Robot test case to define which testbed to use and connect to the cat8-rt1 device. Once connected, we have three other Robot test cases that map to each testcase in the **basic_example_script** pyATS testscript. The **CommonSetup** Robot test case uses the **run testcase** keyword to run the pyATS testcase specified, which is **basic_example_script.common_setup.** The following Robot test cases, **Testcase One** and **CommonCleanup,** use the same **run testcase** keyword to run the tc_one and common_cleanup testcases found in the **basic_example_script** testscript.

The best part about using the Robot Framework is the readability. Anyone can easily read a Robot test case file and feel confident creating a new Robot test case file. This removes the need to have someone create a pyATS job file and understand how to run it. With the

Robot Framework, you create a Robot test case file and run it with the **robot** command. The low barrier to entry allows any engineer to begin testing with the Robot Framework. In the next two sections, we are going to look at the keywords provided by the Unicon library and the pyATS library (Genie).

Unicon Keywords

Like pyATS, the robot module is optional in the Unicon library. It should already be installed with pyATS and Genie, but you can ensure by trying to install it with the following command: **pip install unicon[robot]**. The Unicon robot module provides the following features:

- Execute a command on a device

- Configure a device

- Enable/disable output on a device

- Set Unicon settings

So, the pyATS keywords allow us to run already-built pyATS testscripts and testbed YAML files. The Unicon keywords allow us to interact with the device from the Robot test case file, which provides more control to the developer creating the Robot test case file. The full list of Unicon keywords can be found in the Unicon Robot documentation (https://pubhub.devnetcloud.com/media/unicon/docs/robot.html). Example 22-8 shows a Robot test case file using Unicon and pyATS keywords.

Example 22-8 *Unicon Keywords*

```
*** Settings ***
Library    pyats.robot.pyATSRobot
Library    unicon.robot.UniconRobot


*** Test Cases ***

Connect to device
    # Specify testbed file to use
    use testbed "testbed.yaml"

    # Remove default connection commands
    set unicon setting "INIT_CONFIG_COMMANDS" "" on device "cat8k-rt1"

    connect to device "cat8k-rt1"

Execute command
    execute "show run | include logging host" on device "cat8k-rt1"
    configure "logging host 10.1.1.1" on device "cat8k-rt1"
```

```
Execute command in parallel on multiple devices
    execute "show run | include logging host" in parallel on devices "cat8k-rt1"


Disconnect from device
    disconnect from device "cat8k-rt1"
```

In the example, we have four Robot test cases. The first test case defines the testbed file to use, overrides a default Unicon setting to not send any configuration commands when initially connecting to a device, and finally connecting to the testbed device. The second test case executes the **show run | include logging host** command on the connected device to check for configured logging hosts and configures a logging host with the IP address 10.1.1.1. The third test case is only for example but shows you can execute the same command on multiple testbed devices in parallel. In this case, we are only connecting to one device. However, to connect to multiple devices, you separate each device with a semicolon, so the keyword would look like this:

```
execute "show run | include logging host" in parallel on devices

"cat8k-rt1;cat8k-rt2"
```

You'll notice a second device, cat8k-rt2, was added after the semicolon. The last test case disconnects from the testbed device.

Genie Keywords

Up to this point, we've reviewed the pyATS and Unicon robot modules that allow us to define which testbed file to use, connect and disconnect from testbed devices, run pyATS testcases, execute commands on devices, and configure testbed devices. You may think these are basic functionalities required to interact with network devices—and you're right! The pyATS library (Genie) builds on these keywords and provides the following features to enhance our testing capabilities:

- Run triggers and verifications and convert them to Robot results

- Learn device features

- Custom verify features

  - Verify "x" amount of bgp routes on a device

  - Verify "x" amount of up interfaces on a device

- Execute device APIs (must import genie.libs.robot.GenieRobotApis as a test library)

- Dq support

The pyATS library (Genie) keywords are provided by the genie.libs.robot module. Like the other pyATS libraries, the robot module is optional. It can be installed via pip with the following command: **pip install genie.libs.robot**. The full list of Unicon keywords can be found in the Unicon Robot documentation (https://pubhub.devnetcloud.com/media/genie-docs/docs/userguide/robot.html). Example 22-9 shows a Robot test case file using pyATS, Unicon, and Genie keywords.

Example 22-9   *Genie Keywords*

```
** Settings ***
Library         ats.robot.pyATSRobot
Library         genie.libs.robot.GenieRobot
Library         unicon.robot.UniconRobot


*** Variables ***
# Defining variables that can be used elsewhere in the test data.


${testbed}      testbed.yaml


*** Test Cases ***
# Creating test cases from available keywords.


Connect
    use genie testbed "${testbed}"
    connect to devices "cat8k-rt1"


parser show version
    ${output}=    parse "show version" on device "cat8k-rt1"


Verify version
    # Use Dq to filter down to software version
    dq query    data=${output}    filters=contains('version').get_values('version')



Learn bgp
    ${output}=    learn "bgp" on device "cat8k-rt1"


Trigger sleep
    run trigger "TriggerSleep" on device "cat8k-rt1" using alias "cli"


verify Bgp after trigger
    run verification "Verify_BgpAllNexthopDatabase" on device "cat8k-rt1"


verify bgp count
    verify count "6" "bgp neighbors" on device "cat8k-rt1"


verify bgp routes
    verify count "100" "bgp routes" on device "cat8k-rt1"
```

There are many Robot test cases that show off the different functions of the pyATS library (Genie) including parsing device output, using Dq to query for a specific key in a nested dictionary, learn a device feature, and run triggers and verifications on a device. This is where the true power of the pyATS library shines, as it allows you to really test the network given the keywords provided. It's worth noting that the Genie Robot libraries are open source under genie.libs and you're encouraged to add more keywords! For more information and to see how keywords are defined, check out the Genie Robot GitHub repository (https://github.com/CiscoTestAutomation/genielibs/tree/master/pkgs/robot-pkg/src/genie/libs/robot).

Along with the keywords already discussed, the pyATS library (Genie) provides the ability to profile devices. Device profiling allows us to compare device profiles during different stages of testing. When a device is profiled and the option to store the profile is used, two files are stored locally: one JSON file that contains all the learned device features and another that is a pickled version that can be used later for comparison. Example 22-10 shows how to use device profiling in a Robot test case file.

Example 22-10   *Genie Device Profiling*

```
** Settings ***
Library          pyats.robot.pyATSRobot
Library          genie.libs.robot.GenieRobot
Library          unicon.robot.UniconRobot


*** Variables ***
# Defining variables that can be used elsewhere in the test data.
# Can also be driven as dash argument at runtime


${testbed}       testbed.yaml
${PTS}           /path/to/file


*** Test Cases ***


Connect
    use genie testbed "${testbed}"
    connect to devices "R1;R2"


# Profile BGP and OSPF features and save the profile system on a mount
Profile bgp & ospf on All
    Profile the system for "bgp;ospf" on devices "R1;R2" as "/path/file"


Run any testcase
    <Do some action>
```

```
# Profile BGP and OSPF features and compare to the previously stored profile
Profile bgp & ospf on All and Compare to PTS
    Profile the system for "bgp;ospf" on devices "R1;R2" as "current"
    Compare profile "${PTS}" with "current" on devices "R1;R2"
```

There are a few things going on in the example. First, we connect to two devices, R1 and R2. Once connected, we learn the BGP and OSPF features on each device and create a profile for each device. We then perform "some actions," which is just pure example to simulate changes on each device. Lastly, we create new device profiles by learning the BGP and OSPF features again and then compare the new device profiles with the previous device profiles to see what changed during testing. This is a hypothetical example, but hopefully it drives home the point that testbed devices can be profiled any time during testing and then profiled again to compare what changed with each device during testing.

Easypy Integration

The last pyATS integration to discuss is the ability to run Robot Framework scripts (files with the .robot extension) with the Easypy Runtime environment. The integration aggregates Robot test case results into a final report provided by Easypy.

The run_robot API is used to run a Robot Framework script in an Easypy process. The Robot logs, output, and results are automatically converted into your Easypy runtime directory and report. Example 22-11 shows a pyATS job file running a Robot Framework script.

Example 22-11   *run_robot API*

```
import os

# import the run_robot api
from pyats.robot.runner import run_robot

# entry point
def main(runtime):

    # run your robot script
    run_robot(robotscript = 'my_robo_script.robot',
              runtime = runtime)
```

Table 22-2 shows the function arguments to the run_robot API. In the example, we are only using the **robotscript** argument to provide the Robot Framework script.

Table 22-2 *run_robot Arguments*

Argument	Description
robotscript	Robot Framework script to be run in this task
taskid	Unique task ID (defaults to **Task-#**, where **#** is an incrementing number)
max_runtime	Maximum runtime in seconds before termination
runtime	Easypy runtime object
**options	Any other Robot Framework options to be passed to Robot engine

You might wonder how a Robot test case file maps to a pyATS testscript, which is normally run in a pyATS job file. The best way to compare is by example. Example 22-12 shows a Robot test case file, and Example 22-13 shows the output results produced by the Easypy Runtime. There are comments throughout the Robot test case file to denote where to look in the results.

Example 22-12 *Robot Test Case File*

```
*** Settings ***
Library         hello_world.py

*** Test Cases ***
Should Pass    # Maps to pyATS testcase
    Hello World    # Maps to pyATS test section

Should Fail    # Maps to pyATS testcase
    Raise Exception    # Maps to pyATS test section

Logging Test    # Maps to pyATS testcase
    Do Logging    # Maps to pyATS test section

Check Testbed Provided    # Maps to pyATS testcase
    Check Testbed    # Maps to pyATS test section
```

Example 22-13 *Robot Easypy Results Output*

```
+----------------------------------------------------------------------------+
|                            Task Result Summary                             |
+----------------------------------------------------------------------------+
Task-1: hello_world.Should Pass                                     PASSED
Task-1: hello_world.Should Fail                                     FAILED
Task-1: hello_world.Logging Test                                    PASSED
Task-1: hello_world.Check Testbed Provided                          FAILED
```

```
+-----------------------------------------------------------------------+
|                        Task Result Details                            |
+-----------------------------------------------------------------------+
Task-1: hello_world
|-- Should Pass                                                   PASSED
|    `-- 1_Hello World                                            PASSED
|-- Should Fail                                                   FAILED
|    `-- 1_Raise Exception                                        FAILED
|-- Logging Test                                                  PASSED
|    `-- 1_Do Logging                                             PASSED
`-- Check Testbed Provided                                       FAILED
     `-- 1_Check Testbed                                         FAILED
```

Besides running a Robot Framework script in a pyATS job file, you can also run a Robot Framework script independently in an Easypy environment with the command **pyats run robot {/path/to/robot_file.robot} --testbed-file {path/to/testbed.yaml}**. The Easypy environment produces a report and generates an archive. The command will automatically generate the required pyATS job file from a template, copy it to a runtime directory, and run it as if you provided the job file. All other behavior is the same as the run_robot API. For multiple Robot Framework scripts, it's recommended to create your own pyATS job file and use the run_robot API.

Summary

In this chapter, you were introduced to the Robot Framework, a generic open-source test automation framework that allows you to use English-like keywords to build test cases. The Robot Framework has many components, with whole books written about the framework. However, we only covered some of the components that would help explain the pyATS integrations, including test cases, keywords, variables, importing libraries, test execution, and test results and reporting. Once we reviewed the Robot Framework, we saw how the different pyATS libraries—pyATS, Unicon, and the pyATS library (Genie)—integrated into the framework by providing test libraries that included keywords to easily interact with and test network devices. We wrapped up the chapter by explaining how we can integrate Robot test case files into the Easypy runtime environment so that we can execute Robot test case files within an Easypy environment. We can execute Robot test cases within an Easypy environment using the **pyats run robot** command or with a pyATS job file. The advantage is that the Robot test case results are converted to Easypy results with a generated archive. This allows the results to be uniform, regardless of whether the testing used a pyATS testscript or a Robot test case file. The Robot Framework empowers users to easily create test cases without having to understand the entire pyATS AEtest test infrastructure. By providing a high-level abstraction to the pyATS libraries, the Robot Framework can enhance the adoption of network automation testing.

Leveraging Artificial Intelligence in pyATS

In the realm of technology and problem-solving, artificial intelligence (AI) has emerged as a revolutionizing force. AI involves machines that learn from data and make decisions with minimal human intervention. Among its various subsets, generative AI stands out, especially in its application to content creation—a realm that includes text, images, and even code. This technology has reshaped automation, introducing new dimensions and capabilities. For developers and engineers working automation frameworks like Cisco pyATS, understanding AI and generative AI is essential for harnessing their potential in network automation and testing.

The early 2020s have been a landmark period for generative AI, marked by its significant impact and democratization. AI tools, previously confined to specialized researchers, are now within reach of a wide spectrum of developers. This democratization of AI has unleashed a wave of innovation across various fields, including network automation. Developers now have the ability to integrate sophisticated AI algorithms into their workflows, enhancing both efficiency and creativity. This accessibility has not only empowered seasoned professionals but also opened up new avenues for a generation of developers to explore AI's potential in automation.

Central to this AI revolution is OpenAI, a trailblazer in AI research and development. One of its most acclaimed creations, ChatGPT, a conversational agent, has captured the attention of the tech world with its intelligent responses to natural language inputs. For those engaged with Cisco pyATS, ChatGPT and other OpenAI tools represent a new frontier of possibilities. From automating routine tasks to crafting complex testscripts to data analysis of network telemetry, the potential applications in network automation are boundless.

Integrating OpenAI's API into pyATS jobs marks a significant technological leap. This integration allows for the automation of more complex and nuanced tasks, transcending the limitations of traditional script-based automation. AI's introduction into network testing and automation paves the way for more intelligent, efficient, and reliable systems. This amalgamation not only streamlines processes but also augments the capabilities of pyATS, enabling it to address more advanced challenges in network automation.

For those aiming to advance further, retrieval augmented generation (RAG) offers a sophisticated approach. Through LangChain, developers can harness Python and JavaScript SDKs to develop robust, enterprise-grade solutions. LangChain is a software library designed to facilitate the development of applications involving large language models like ChatGPT. It offers a set of tools and interfaces that streamline the process of integrating these models into various applications, making it easier for developers to harness their capabilities.

RAG represents a significant advancement in the field of artificial intelligence, especially in the context of generating human-like text responses. At its core, RAG combines the power of large language models, like ChatGPT, with an additional layer of external knowledge retrieval. This means that before generating a response, the model first retrieves relevant information from a vast database or document collection and then synthesizes this information to craft a response that is not only relevant but also informed by real-world knowledge.

In network automation, and particularly within platforms like Cisco's pyATS, RAG introduces a new dimension of intelligence to automated tasks. By leveraging RAG, developers can create systems that understand and respond to complex network scenarios with a level of depth and accuracy previously unattainable. Whether it's diagnosing issues based on symptoms described in natural language, generating scripts that adapt to evolving network configurations, or analyzing network telemetry data to predict future challenges, RAG opens up a world of possibilities.

LangChain serves as a bridge between the theoretical potential of RAG and its practical application in network automation. As a software library, LangChain simplifies the process of integrating RAG models into pyATS jobs and other network automation tasks. It provides Python and JavaScript SDKs, along with a suite of tools and interfaces, designed to make it straightforward for developers to incorporate large language models and their knowledge-retrieval capabilities into their solutions:

- Developers can query external knowledge bases as part of their automated tasks, enriching the context and relevance of the AI's responses.

- Developers can implement sophisticated AI-driven decision-making in their network automation scripts, improving both the speed and quality of network diagnostics and maintenance.

The introduction of RAG, facilitated by tools like LangChain, marks a pivotal evolution in network testing and automation. It empowers developers to build more intelligent, efficient, and responsive systems, significantly enhancing the capabilities of automation frameworks like pyATS. By integrating RAG into their workflows, network professionals can ensure their systems are not just automated but also deeply informed and remarkably adaptive, heralding a new era of network management that is both data-driven and insight-rich.

At its core, LangChain abstracts the complexities involved in working with language models. It provides a high-level API that simplifies tasks such as text generation, language understanding, and dialogue management. This allows developers to focus on building the application-specific features rather than getting bogged down by the intricacies of the underlying language models.

One of the key benefits of LangChain is its flexibility. It supports various language models, making it easy for developers to switch between them or choose the one that best suits their needs. Additionally, LangChain includes features for fine-tuning models and managing interactions, which are crucial for creating robust and user-friendly applications.

LangChain also emphasizes collaborative development. By providing an open-source platform, it encourages contributions from a wide range of developers and researchers. This collaborative approach accelerates the improvement of the library and fosters innovation in the field of natural language processing.

LangChain is a valuable tool for developers looking to leverage the power of large language models in their applications. Its combination of ease of use, flexibility, and collaborative nature makes it a significant contribution to the field of AI and language model development. LangChain simplifies the RAG implementation, making it accessible even for those with limited AI expertise. This approach merges structured data processing with the nuanced understanding of language models, opening new possibilities in automated testing and network management.

Implementing RAG in network automation involves complex data handling. A JSON loader can be used to integrate pyATS parsers or APIs, transforming network data into embeddings and textual segments. "Embeddings" and "textual segments" are terms often used in the context of natural language processing (NLP) and AI, particularly in models like the RAG AI model. Let's break down each term.

Embeddings in NLP are a representation of text in a form that a computer can understand. They are essentially numerical vectors that encode the meaning of a word, phrase, or even entire sentences.

The idea is to convert text into a format that captures not just the words themselves but also their context and semantic relationships. For example, words with similar meanings are represented by vectors that are close to each other in the embedding space.

Embeddings are generated using algorithms like Word2Vec, GloVe, or BERT, which are trained on large datasets to learn these semantic relationships.

In the context of RAG or other similar models, embeddings are crucial for understanding and generating text, as they provide a way for the model to grasp the nuances of language.

A *textual segment* refers to a specific portion of text. It could be a sentence, a paragraph, or any other subset of a larger text.

In the context of models like RAG, textual segments are important because the model often works by analyzing and synthesizing information from different parts of a text or from multiple texts.

For instance, when a RAG model is tasked with generating a response or completing a task, it might retrieve relevant textual segments from its training data or a linked database and then use them as a basis for generating an output.

The RAG model, a type of transformer-based model, combines the power of retrieval and generation. It retrieves relevant textual segments (often in the form of embeddings) from a large corpus of text and then uses a generative model to synthesize these into a coherent output. This approach allows the RAG model to leverage a vast amount of information and provide responses that are informed by a wide range of sources, making it particularly powerful for tasks like question answering, content generation, and more. This data is then loaded into a vector store, with options like ChromaDB, Pinecone, ElasticSearch, and MongoDB. At its most fundamental level, vector data can be thought of as a way to represent information in a format that computers, and specifically AI models, can understand and process. Imagine you have a list of numbers or a series of points on a graph; these can describe certain characteristics or features of an object, word, or even a piece of text. In the world of NLP, vector data transforms words and sentences into these lists of numbers, allowing the computer to "see" and work with text in a mathematical space.

Vector data, in the context of machine learning and NLP, refers to the representation of textual or other categorical data as high-dimensional vectors (arrays of numbers). Each dimension in these vectors corresponds to a feature learned from the data, encapsulating aspects of the data's semantic meaning, context, or syntactical characteristics. This representation is crucial for performing sophisticated operations like semantic analysis, where the model assesses the meaning behind text, or for tasks requiring the comparison of textual similarity, where it determines how closely related two pieces of text are based on their vector representations. This is particularly relevant in the context of machine learning and natural language processing (NLP), where data is often represented as high-dimensional vectors. Let's explore each of these vector stores.

ChromaDB is specifically tailored for managing and querying large-scale vector data. It's optimized for operations like nearest neighbor search, which is crucial in applications like recommendation systems, image and text retrieval, and more.

It allows for efficient storage and retrieval of high-dimensional vectors, often used in embedding spaces created by machine learning models. Another advantage is that ChromaDB runs locally, privately, in memory.

Pinecone is a vector database designed for machine learning applications. It enables users to store and query vectors in a way that's optimized for similarity search.

This makes it ideal for tasks such as semantic search, where you want to find items that are semantically similar to a query, based on their vector representations. Pinecone.io is a cloud-hosted service where you are provided one free vector store for the free tier to get started.

While ElasticSearch is primarily known as a search engine, it also supports vector search capabilities. This is done through its dense vector field type, which allows for storing and querying vector data.

ElasticSearch's vector capabilities are useful for adding search-by-similarity features to applications, leveraging machine learning models' embeddings. One major advantage of Elastic is the ecosystem and stack, sometimes known as the Elastic Stack or ELK (Elastic, Logstash, Kibana), that are available to augment Elastic as a vector store. Kibana can be used to visualize and even enable a machine learning (ML) capability on the backend using the vector stores of network state. ElasticVue is also available a visualization tool to allow inspection and review of the vector store, chunk sizes, and other visibility. Even a tool like Postman or cURL could be used to inspect the vector stores with API calls.

MongoDB, a popular NoSQL database, has also introduced features to support vector data. With its flexible schema, it can store vectors and perform operations like cosine similarity to retrieve similar items.

At its essence, cosine similarity measures the cosine of the angle between two vectors in a multidimensional space. This measure helps determine how similar the vectors are to each other, regardless of their size. Here's how it applies in simpler terms: Imagine two arrows pointing in different directions from the same starting point. The smaller the angle between them, the more similar they are considered to be.

Cosine similarity translates this concept into a numerical value between –1 and 1. A cosine similarity score of 1 means the vectors are identical in orientation, 0 indicates orthogonality (no similarity), and –1 means they are diametrically opposed.

In machine learning and NLP, cosine similarity is used to compare documents, texts, or items based on their vector representations. For example, when documents are converted into vectors (with each dimension representing a feature of the text), cosine similarity can help identify documents with similar content, themes, or meanings, even if the exact words used are different.

The concept of vector data extends beyond its use in geospatial analysis, where it describes physical shapes and locations. In the context of databases like MongoDB and applications involving machine learning, vector data represents features or characteristics of data points in a high-dimensional space. This abstraction allows for sophisticated operations like similarity searches, which are pivotal in recommendation systems, semantic search, and more.

Vector databases such as MongoDB utilize cosine similarity to efficiently retrieve items that are "close" to a query vector, mimicking the way humans perceive similarity between objects or concepts. This mathematical approach provides a robust basis for features like the following:

- **Recommendation systems:** Suggesting products, articles, or media that share similarities with user interests or past interactions

- **Semantic search:** Finding content that matches the meaning of a query, rather than relying solely on keyword matches

Tools like Kibana and ElasticVue enhance the usability of vector stores by providing visualization and analysis capabilities. They allow users to inspect the stored vectors, understand the distribution and relationships within the data, and apply machine learning models to uncover patterns or insights. This layer of interpretation and visualization is crucial for making the abstract concept of high-dimensional vector spaces more tangible and actionable.

In sum, the integration of cosine similarity and vector data in databases bridges the gap between raw data and actionable insights, enabling applications that can intuitively understand and respond to human language and behavior. By introducing these concepts with a focus on their practical applications and underlying mathematics, we aim to make these advanced topics accessible to a broader audience, encouraging a deeper understanding of the technologies that shape modern data analysis and machine learning landscapes.

This is particularly useful in scenarios where vector data needs to be integrated with other types of data stored in a document-based format.

Vector stores are specialized databases or database functionalities that focus on efficiently handling vector data, especially high-dimensional vectors typical in machine learning applications. They provide key functionalities like similarity search, making them invaluable in modern applications that rely on machine learning for tasks like semantic search, personalization, and content recommendation. Each of these vector stores, be it ChromaDB, Pinecone, ElasticSearch, or MongoDB, offers unique features and optimizations to cater to these needs.

These vector stores are crucial in RAG, serving as repositories for AI models to retrieve and interpret data, enabling more sophisticated and context-aware responses in automation tasks.

The integration culminates with the use of large language models (LLMs) like ChatGPT 3.5 for data retrieval and response generation. These models can intelligently query the vector store, extracting pertinent information to generate accurate and contextually appropriate responses to automation prompts. Streamlit complements this integration, offering a platform for rapid development and prototyping. It allows developers to swiftly create and deploy user-friendly interfaces for their LangChain and pyATS applications.

Streamlit's intuitive design and straightforward deployment make it an ideal tool for demonstrating the capabilities of AI-enhanced network automation tools.

This comprehensive introduction to AI's integration with Cisco pyATS sets the stage for the subsequent, more technical sections of this chapter, providing you with a solid foundation in both the concepts and applications of AI in network automation. This chapter covers the following topics:

- OpenAI API

- Retrieval Augmented Generation with LangChain

- Rapid prototyping with Streamlit

OpenAI API

Integrating the OpenAI API into applications, especially for tasks like automation with Cisco pyATS, unlocks a new realm of possibilities. The OpenAI API allows developers to leverage some of the most advanced AI models, including the latest gpt-4 and gpt-3.5-turbo, to perform a wide range of tasks from natural language processing to complex problem-solving. To utilize these models, developers send a request with the required inputs and their API key to the API endpoint, and in return they receive the model's output. The API is designed to be user-friendly and efficient, providing access to various models, depending on the specific needs of the application.

The API endpoints vary based on the model families. For newer models like gpt-4 and gpt-3.5-turbo, the endpoint is https://api.openai.com/v1/chat/completions, while updated legacy models such as gpt-3.5-turbo-instruct, babbage-002, and davinci-002 use https://api.openai.com/v1/completions. The choice of the model depends on the specific requirements of the task at hand, with gpt-3.5-turbo and gpt-4 generally recommended for a broad range of applications. Hugging Face, a public repository similar to GitHub (in fact, you use Git to transfer the models locally!), can be used to find free, open-source, publicly available models in addition to paid services from OpenAI, Google, Microsoft, and Meta. Table 23-1 outlines these models and providers for comparison.

The Chat Completions API, a feature of the OpenAI API, is particularly suited for tasks that involve conversation or dialogue. It takes a list of messages as input and returns a model-generated message as output. This format is flexible enough to handle both multiturn conversations and single-turn tasks. For instance, a typical API call to the Chat Completions API might look something like Example 23-1 in Python.

Table 23-1 *Comparing Various Models and Providers*

	ChatGPT 3.5 Turbo	GPT-4	GPT-4 Turbo	LaMDA 2	LLaMA	Mistral 7B	Phi-2
Developer	OpenAI	OpenAI	OpenAI	Google	Meta	Hugging Face	Microsoft
Model Size	Large	Very large	Very large	Large	Large	Large	Small
Training Data	Diverse sources	Extensive	Extensive	Web text	Extensive	Open web sources	Diverse sources
Specialization	General purpose	General purpose	General purpose	Conversational AI	General purpose	General purpose	General purpose
Capabilities	Text generation, understanding, Q&A	Advanced text generation, understanding, contextual responses	Efficient text generation, high speed	Advanced conversational AI, context understanding	Text generation, understanding, research	Text generation, understanding	???
Release Date	2023	2023	2023	2023	2023	2022	???
API Access	Yes	Yes	Yes	Limited/no	Limited/no	Open source	???
Customization	Limited	Limited	Limited	Limited	Limited	High (open source)	???
Free or Paid	Paid	Paid	Paid	Paid	Paid	Free (open source)	???
Quality	High	Very high	Very high	Very high	High	Moderate	???
TTS Features	Yes (OpenAI TTS)	Yes (OpenAI TTS)	Yes (OpenAI TTS)	No	No	No	???

Example 23-1   *openAI Chat Completion API Example*

```
from openai import OpenAI
client = OpenAI()

response = client.chat.completions.create(
  model="gpt-3.5-turbo",
  messages=[
    {"role": "system", "content": " You are a helpful network assistant who will
help the user determine the health of a Cisco interface with the help of the
interface state as JSON. "},
    {"role": "user", "content": f"Please use this JSON and help me understand
the state of this interface. Are there any issues or is the interface healthy?
{ interface_state }"},
    # Additional messages can be included here
  ]
)
```

In this example, the **messages** parameter is crucial. It must be an array of message
objects, each with a role (either **"system"**, **"user"**, or **"assistant"**) and content. The conversation can start with a system message setting the tone or behavior of the assistant,
followed by user and assistant messages. The history of the conversation is important for
context, as the models do not retain memory of past requests.

The response from the API provides not just the content but also a **finish_reason**, which
could be due to several reasons like reaching the maximum token limit or completion of
the message. For more advanced use, especially in automation tasks, the API also supports JSON mode, which ensures the output is in JSON format, ideal for structured data
handling. Example 23-2 demonstrates using JSON mode.

Example 23-2   *OpenAI Chat Completion API Example with JSON Formatting*

```
response = client.chat.completions.create(
  model="gpt-3.5-turbo-1106",
  response_format={ "type": "json_object" },
  messages=[
    {"role": "system", "content": "You are a helpful network assistant who will help
the user determine the health of a Cisco interface with the help of the interface
state as JSON."},
    {"role": "user", "content": f"Please use this JSON and help me understand
the state of this interface. Are there any issues or is the interface healthy?
{ interface_state }"},
  ]
)
```

For developers looking to integrate the OpenAI API into their automation workflows,
understanding the nuances of API calls, managing tokens, and choosing the right model
and response format is crucial. The API is not just a powerful tool for text generation but
a gateway to building more intelligent and responsive automation systems.

Integrating this into an interface health check has several implications:

- We no longer need to write multiple tests per interface per counter.
- We can send, per interface, the parsed JSON to ChatGPT for analysis.
- We can ask ChatGPT if the interface is healthy.

First, in our Python virtual environment, we will need to pip-install OpenAI:

```
(venv) $ pip install openai
```

In our code, we will need to include and instantiate the OpenAI client by adding the following to our pyATS script:

```
from openai import OpenAI


client = OpenAI()
```

Then, we can adjust our Interface tests as shown in Example 23-3. First, we learn the interfaces. Then, we loop over each interface and create a chat completion with OpenAPI ChatGPT 3.5. We display the question and answer, per interface, and log in to pyATS logging. We will need an .env file with our OpenAPI API key defined as OPENAI_API_KEY.

Example 23-3   *pyATS Script Incorporating Artificial Intelligence*

```
import os
import json
import logging
from pyats import aetest
from openai import OpenAI
from rich.table import Table
from dotenv import load_dotenv
from rich.console import Console
from genie.utils.diff import Diff
from pyats.log.utils import banner


# ----------------
# Get logger for script
# ----------------


log = logging.getLogger(__name__)


# Instantiate openAI client
load_dotenv()


# Load the API keys from an environment variable or secure source
openai_api_key = os.getenv('OPENAI_API_KEY')
```

```python
# Instantiate openAI client
client = OpenAI()


# ----------------
# AE Test Setup
# ----------------
class common_setup(aetest.CommonSetup):
    """Common Setup section"""
# ----------------
# Connected to devices
# ----------------
    @aetest.subsection
    def connect_to_devices(self, testbed):
        """Connect to all the devices"""
        testbed.connect()
# ----------------
# Mark the loop for Learn Interfaces
# ----------------
    @aetest.subsection
    def loop_mark(self, testbed):
        aetest.loop.mark(Test_Cisco_IOS_XE_Interfaces, device_name=testbed.devices)


# ----------------
# Test Case #1
# ----------------
class Test_Cisco_IOS_XE_Interfaces(aetest.Testcase):
    """Parse pyATS learn interface and test against thresholds"""

    @aetest.test
    def setup(self, testbed, device_name):
        """ Testcase Setup section """
        # Set current device in loop as self.device
        self.device = testbed.devices[device_name]

    @aetest.test
    def get_parsed_version(self):
        parsed_version = self.device.learn("interface")
        # Get the JSON payload
        self.parsed_json=parsed_version.info
```

```
    @aetest.test
    def create_file(self):
        # Create .JSON file
        with open(f'{self.device.alias}_Learn_Interface.json', 'w') as f:
            f.write(json.dumps(self.parsed_json, indent=4, sort_keys=True))


    @aetest.test
    def test_interface_health(self):
        # Test for interface health
        for intf,value in self.parsed_json.items():
            response = client.chat.completions.create(
                model="gpt-3.5-turbo",
                    messages=[
                        {"role": "system", "content": "You are a helpful network
assistant who will help the user determine the health of a Cisco interface with the
help of the interface state as JSON."},
                        {"role": "user", "content":  f"Please use this JSON and
help me understand the state of this interface. Are there any issues or is the
interface healthy? Interface { intf } state: { value }"},
                    ]
                )

            choices = response.choices
            if choices:
                choice = choices[0]
                content = choice.message.content
                log.info(f"Please use this JSON and help me understand the state
of this interface. Are there any issues or is the interface healthy? Interface
{ intf } state: { value }")
                log.info(content)

class CommonCleanup(aetest.CommonCleanup):
    @aetest.subsection
    def disconnect_from_devices(self, testbed):
        testbed.disconnect()


# for running as its own executable
if __name__ == '__main__':
    aetest.main()
```

Now, per interface, we get AI analysis like that shown in Figure 23-1 and Example 23-4.

Figure 23-1   *pyATS Interface Health Tests with AI Logs*

Example 23-4   *pyATS Logs Including ChatGPT AI Answers, Per Interface*

```
"Please use this JSON and help me understand the state of this interface. Are there
any issues or is the interface healthy? Interface Loopback7 state: {'description':
'NC loopback interface', 'type': 'Loopback', 'oper_status': 'up', 'mtu': 1514,
'enabled': True, 'bandwidth': 8000000, 'port_channel': {'port_channel_member':
False}, 'delay': 5000, 'accounting': {'ip': {'pkts_in': 74, 'chars_in': 6066,
'pkts_out': 74, 'chars_out': 6066}}, 'ipv4': {'10.111.7.2/32': {'ip': '10.111.7.2',
'prefix_length': '32', 'secondary': False}}, 'counters': {'in_pkts': 0, 'in_octets':
0, 'in_broadcast_pkts': 0, 'in_multicast_pkts': 0, 'in_errors': 0, 'in_crc_errors':
0, 'out_pkts': 74, 'out_octets': 6066, 'out_broadcast_pkts': 0, 'out_multicast_
pkts': 0, 'out_errors': 0, 'last_clear': 'never', 'rate': {'load_interval': 300,
'in_rate': 0, 'in_rate_pkts': 0, 'out_rate': 0, 'out_rate_pkts': 0}}, 'encapsula-
tion': {'encapsulation': 'loopback'}, 'switchport_enable': False}


Based on the provided JSON, the interface appears to be in a healthy state. Here are
the details:

- Description: The interface is a loopback interface with a description of
"NC loopback interface".

- Type: The interface is of type "Loopback".

- Operational Status: The interface is "up", which means it is operational and
functioning correctly.

- MTU: The Maximum Transmission Unit (MTU) is set to 1514 bytes.

- Enabled: The interface is enabled.

- Bandwidth: The bandwidth of the interface is 8,000,000 bits per second.

- Port Channel: The interface is not a member of any port channel.
```

- Delay: The delay on the interface is set to 5000 milliseconds.

- Accounting: Accounting is enabled for IP traffic on the interface. It indicates that 74 packets have been received and transmitted, totaling 6,066 characters.

Additionally, the interface has an IPv4 address of 10.111.7.2/32, with a prefix length of 32. This address is not configured as a secondary IP address.

The interface counters indicate that no packets or octets have been received, but 74 packets with 6,066 octets have been transmitted. No errors, CRC errors, broadcast packets, or multicast packets have been detected.

The interface has not been cleared since the last reboot, and the load interval for rate calculations is set to 300 seconds. The current input and output rates are both 0.

The interface encapsulation is set to "loopback", indicating that it is a loopback interface.

Finally, the interface does not have switchport enablement, meaning it is not configured as a switchport.

In summary, based on the provided information, the interface appears to be healthy and without any issues."

As demonstrated in Example 23-4, with a simple addition of an API call to OpenAI using ChatGPT models, AI can be integrated into our pyATS jobs very easily, augmenting them with deep insights that the human eye may miss. We don't have to write comprehensive, per-counter tests in the case of interfaces; we simply provide the full payload per interface and let the AI perform the analysis and testing for us!

With some additional pyATS testing, we could include syslog analysis, as demonstrated in Example 23-5.

Example 23-5   *pyATS Syslog with AI Analysis*

```
@aetest.test
def get_raw_logs(self):
    self.raw_logs = self.device.execute("show logging")


@aetest.test
def test_logs(self):
    # Test syslogs
    response = client.chat.completions.create(
        model="gpt-3.5-turbo",
            messages=[
                    {"role": "system", "content": "You are a helpful network assistant
who will help the user understand the Cisco IOS XE device's system logs."},
                    {"role": "user", "content":  f"Please analyze the following
Cisco IOS XE system logs and help me understand the state of this device.
Summarize and highlight anything important in the logs.\n { self.raw_logs }"},
                ]
        )
```

```
        choices = response.choices
    if choices:
        choice = choices[0]
        content = choice.message.content
        log.info("Please analyze the following Cisco IOS XE system logs and help
me understand the state of this device. Summarize and highlight anything important
in the logs.")
        log.info(content)

        # Create and format the strings
        line1 = "Please analyze the following Cisco IOS XE system logs and help
me understand the state of this device. Summarize and highlight anything important
in the logs."
        line2 = content + "\n"

        # Append the strings to a file
        with open('Syslogs.txt', 'w') as file:
            file.write(line1)
            file.write(line2)
```

This generates a response like that shown in Example 23-6.

Example 23-6   *pyATS Syslog with AI Analysis ChatGPT Response*

```
Please analyze the following Cisco IOS XE system logs and help me understand the
state of this device. Summarize and highlight anything important in the logs.
Analyzing the provided system logs from the Cisco IOS XE device, here are the
summary and highlighted important points:

1. **Multiple Successful Logins**: There are numerous log entries indicating the
successful authentication of the user 'admin' from the source IP address 10.10.20.28
over SSH for netconf sessions. This activity suggests frequent remote management
of the device using the Network Configuration Protocol (NETCONF), which is usually
intended for configuration and management operations.

2. **User Logout and Exit**: There are several logouts or session exits for 'admin',
which is common for session-based remote access.

3. **Configuration Synchronization**: Occasionally, configurations are being
synchronized to the NETCONF running data store. This indicates that the device is
undergoing configuration changes and updates.

4. **Console Logging**: Console logging is disabled.

5. **Monitor Logging**: Debugging messages are being logged; this setting could
generate a high volume of log data and should be used carefully.
```

6. **Logging Rate-Limiting**: There are messages that have been rate-limited, this prevents the logging process from overwhelming the system by restricting the number of messages logged in a given time frame.

7. **Exception Logging**: Size is set at 4096 bytes, which is utilized for storing exception information.

8. **Persistent Logging**: Disabled. This means logs will not persist across reboots.

9. **Trap Logging**: Informational level trap logging is being used, which logs a large amount of system information and could be useful for detailed monitoring.

10. **SSH Unexpected Messages**: There are instances of unexpected message types that cause SSH connections to be terminated. This suggests potential issues or unexpected conditions with SSH sessions.

11. **Command Deprecated**: The logs include a warning about a deprecated CLI command involving a password. It's noted that type 0 passwords are used, and migration to stronger type-6 encryption is recommended.

12. **Communication Failures with CSLU**: There are entries indicating communication failures with the Cisco Smart License Utility due to the device being unable to resolve the server hostname or domain name. This could be a licensing server connectivity issue that needs to be resolved for licensing validation.

13. **Unexpected Reconfigurations**: There are log entries where configuration changes are detected and require synchronization, which may indicate active configuration management is occurring.

14. **Security Advisory**: The log contains an attempted login using a deprecated username 'cisco', which could indicate an unauthorized access attempt or a security issue.

In conclusion, the Cisco IOS XE device appears to be actively managed and configured, mostly through NETCONF over SSH sessions by the user 'admin'. There are signs of frequent configuration synchronization and signs of a deprecated command usage warning. Also, potential connectivity issues with the Cisco Smart License Utility indicate licensing verification might be impacted. Additionally, there are signs of unexpected SSH session behavior and possible security concerns with the use of deprecated credentials that need attention.

Data analysis is where AI really shines. The ability to sift through log messages and quickly draw conclusions to summarize the log activity is a gamechanger for troubleshooting and overall network management.

Additionally, we could use pyATS to parse the routing table and inquire about our routes, simplifying routing table analysis, as demonstrated in Example 23-7.

Example 23-7 *pyATS Parsing show ip route for AI Analysis*

```
    @aetest.test
    def get_ip_route(self):
        self.ip_routes = self.device.parse("show ip route")


    @aetest.test
    def test_ip_route(self):
        # Test ip route
        response = client.chat.completions.create(
            model="gpt-3.5-turbo",
                messages=[
                        {"role": "system", "content": "You are a helpful network assistant
who will help the user understand the Cisco IOS XE device's routing table."},
                        {"role": "user", "content":  f"Please analyze the following
Cisco IOS XE routing table and help me understand the state of this device's routing
table. Summarize and highlight anything important such as the default route or other
key pieces of information.\n { self.ip_routes }"},
                    ]
      )


        choices = response.choices
        if choices:
            choice = choices[0]
            content = choice.message.content
            log.info("Please analyze the following Cisco IOS XE routing table and
help me understand the state of this device's routing table. Summarize and highlight
anything important such as the default route or other key pieces of information.")
            log.info(content)

            # Create and format the strings
            line1 = "Please analyze the following Cisco IOS XE routing table and
help me understand the state of this device's routing table. Summarize and highlight
anything important such as the default route or other key pieces of information."
            line2 = content + "\n"

            # Append the strings to a file
            with open('Routing Table.txt', 'w') as file:
                file.write(line1)
                file.write(line2)
```

This results in the AI analysis shown in Example 23-8.

Example 23-8   *pyATS Parsing show ip route for AI Analysis Results*

Please analyze the following Cisco IOS XE routing table and help me understand the state of this device's routing table. Summarize and highlight anything important such as the default route or other key pieces of information.

The provided Cisco IOS XE routing table includes several routes within the 'default' Virtual Routing and Forwarding (VRF) instance under the Ipv4 address family. Here is a summarized analysis of the key information in this routing table:

1. **Default Route (0.0.0.0/0)**:

   - **Route**: 0.0.0.0/0 – This is the default route which matches any Ipv4 address and is used to forward traffic to destinations not explicitly known by the router.

   - **Active**: True – The route is active and usable.

   - **Metric**: 0 – The cost associated with this route is 0, which generally indicates a directly connected route or a static route with no explicit metric set.

   - **Route Preference**: 1 – This is the administrative distance of the route which indicates the trustworthiness of the source of the route. A route with administrative distance 1 is typically a static route that is manually configured and highly trusted.

   - **Source Protocol**: Static – Indicated by 'S*', the route is statically configured, which means it is manually set up by a network administrator and does not depend on a dynamic routing protocol.

   - **Next Hop**: 10.10.20.254 – Traffic matching this route should be forwarded to the next-hop IP address 10.10.20.254.

   - **Outgoing Interface**: GigabitEthernet1 – Traffic should leave the router via the GigabitEthernet1 interface when following this route.

2. **Directly Connected and Local Routes**:

   Connected ('C') and local ('L') routes are those that the router has direct connections to. The connected routes refer to entire networks, while the local routes refer to the router's own IP addresses on those networks. Here are some examples from the routing table:

   - **10.0.0.0/24** and **10.0.0.1/32** are both active and associated with the Loopback0 interface.

   - **10.10.20.0/24** and **10.10.20.48/32** are associated with the GigabitEthernet1 interface.

   - Other connected and local routes are similarly associated with Loopback7, Loopback17, Loopback109, VirtualPortGroup0, and Loopback100 interfaces with their respective networks and router IP addresses.

3. **Loopback Routes**:

   A number of routes are assigned to various Loopback interfaces. Loopback interfaces are virtual interfaces often used for management purposes or to represent the device within a network topology. These include:

   - **10.111.7.2/32** on Loopback7

   - **10.111.7.3/32** on Loopback17

   - **10.255.255.0/24** and **10.255.255.9/32** on Loopback109

   - **192.168.200.0/24** and **192.168.200.200/32** on Loopback100

   - These are all active routes and the 'L' in the source_protocol_codes indicates that these represent local addresses assigned to the router.

```
4. **VirtualPortGroup Interface Routes**:

   - There are connected and local routes associated with VirtualPortGroup0, which
is typically a virtual interface used in virtualized environments:

     - **192.168.1.0/24** and **192.168.1.1/32** are both active and using the
VirtualPortGroup0 interface.


In summary, this routing table shows a static default route configured to forward
unknown destination traffic via the GigabitEthernet1 interface to the next-hop IP
address 10.10.20.254. Additionally, there are multiple directly connected networks
and local IP address routes via various physical and loopback interfaces. All
provided routes are active, meaning they are currently valid and eligible for use
in routing decisions.
```

Analyzing the routing table on a device can be crucial in network discovery and overall learning about a specific network. If we used AI to analyze routing tables from multiple routers in a network, we can use the analyzed data to draw conclusions about the network's current operating behavior.

Now let's look at the running configuration of a device and how it can be used for AI analysis and insights, as demonstrated in Example 23-9.

Example 23-9 *pyATS show running-configuration for AI Analysis*

```
    @aetest.test
    def get_raw_config(self):
        self.raw_config = self.device.execute("show run")


    @aetest.test
    def test_config_general(self):
        # Test config
        response = client.chat.completions.create(
            model="gpt-3.5-turbo",
                messages=[
                    {"role": "system", "content": "You are a helpful network
assistant who will help the user understand the Cisco IOS XE device's running
configuration."},
                    {"role": "user", "content":  f"Please analyze the following
Cisco IOS XE running configuration and help me understand configuration of this
device. Summarize and highlight anything important in the configuration.\n
{ self.raw_config }"},
                ]
        )


        choices = response.choices
        if choices:
            choice = choices[0]
            content = choice.message.content
```

```
        log.info("Please analyze the following Cisco IOS XE running configuration
and help me understand configuration of this device. Summarize and highlight anything
important in the configuration")
        log.info(content)


        # Create and format the strings
        line1 = "Please analyze the following Cisco IOS XE running configuration
and help me understand configuration of this device. Summarize and highlight
anything important in the configuration"
        line2 = content + "\n"


        # Append the strings to a file
        with open('Running Config.txt', 'w') as file:
            file.write(line1)
            file.write(line2)
```

Remarkably, this again results in the AI analysis shown in Example 23-10.

Example 23-10   *pyATS show running-configuration for AI Analysis Results*

```
Please analyze the following Cisco IOS XE running configuration and help me under-
stand configuration of this device. Summarize and highlight anything important in
the configuration. The summarized configuration information and notable points in
this Cisco IOS XE device configuration:


**Device Identification Information:**
- Hostname: Cat8000V
- Model: Presumably a C8000V virtual platform based on the hostname and UDI PID
  mentioned.
- License UDI: Product ID (PID) is C8000V with a serial number (SN) 9UWS2FADP45.


**Admin User Access:**
- Two user accounts are set up: 'admin' and 'cisco', both with privilege level 15
  (full privileges).
- Password encryption is used (type 9, which typically indicates scrypt password
  hashing).


**System Settings:**
- IOS version: 17.9
- No AAA (Authentication, Authorization, and Accounting) new model configured.
- Timestamps for logs and debug messages include date and milliseconds.
- Call-home service is enabled for automatic messaging to Cisco.
- Platform-specific settings are in place for QFP (Quantum Flow Processor)
  utilization, platform console is virtual, and punt-keepalive disabled.


**Network Configuration:**
- IP domain name set to cisco.com.
- Several Loopback interfaces configured:
```

- Loopback0 with IP 10.0.0.1/24
- Loopback7 with IP 10.111.7.2/32 (described as an NC loopback interface).
- Loopback17 with IP 10.111.7.3/32 (described as an NC loopback interface).
- Loopback100 with IP 192.168.200.200/24 (for configuration via AI).
- Loopback109 with IP 10.255.255.9/24 (RESTCONF not used).

- A Port-channel interface (Port-channel1) configured with IP address 172.23.10.1/24.
- Redundancy is indicated in the configuration but not fully configured or detailed.
- CDP (Cisco Discovery Protocol) is running.

Security Configuration:
- SSH is enabled with a specific keypair, and adherence to SSH version 2 for security.
- SCP server is enabled for secure file transfers.
- A simple banner message for MOTD (Message of the Day) which states the usage policy for a sandbox environment.

Interface Configuration:
- GigabitEthernet1 is the management interface with IP address 10.10.20.48/24.
- GigabitEthernet2 and GigabitEthernet3 are part of a Port-channel (EtherChannel) with Port-channel1.
- VirtualPortGroup0 acting as a NAT inside interface with IP address 192.168.1.1/24.

NAT and ACL Configuration:
- NAT is configured on GigabitEthernet1, with ACL "NAT-ACL" defining the NAT behavior.
- An extended access list "Test" with a rule permitting all traffic to 1.1.1.1.

Routing:
- A default static route sends all traffic via GigabitEthernet1 to 10.10.20.254.

IOS XE Features:
- iox for IOx support.
- Telemetry configuration is set up to send CPU utilization data to a receiver.
- Guestshell container environment with configured networking.
- NETCONF and RESTCONF are enabled for remote configuration.
- Event management applet designed to execute a Python script in guestshell upon Loopback0 going down.

Miscellaneous:
- Call-home profile is set for contacting Cisco TAC.
- NTP server configured at IP address 10.1.7.1 for time synchronization.

Overall, the device seems to be set up as a sandbox environment for testing, particularly with programmability features enabled such as Netconf, Restconf, and an EEM (Embedded Event Manager) applet watching for changes to interface status. There are also several loopback interfaces with different IP configurations presumably for networking labs or tests.

Finally, and maybe most incredibly, we can use the pyATS .configure() module, along with some RegEx, and ask ChatGPT to generate configurations from prompts that we then automatically push to the device with pyATS, as shown in Example 23-11.

Example 23-11   *pyATS Request and Push Configuration with ChatGPT and pyATS*

```
    @aetest.test
    def generate_loopback_config(self):
        # Test config
        response = client.chat.completions.create(
            model="gpt-3.5-turbo",
                messages=[
                    {"role": "system", "content": "You are a helpful Cisco IOS XE
network configuration assistant who will help the user generate configuration
commands."},
                    {"role": "user", "content":  "Please generate the configuration
for a new logical interface called loopback100 with a description of configuration
via AI and an IP address of 192.168.200.200/24. Please also enable the port.
Please respond only with the configuration and no notes or other characters."}
                ]
        )

        choices = response.choices
        if choices:
            choice = choices[0]
            self.config = choice.message.content
            print(self.config)

    @aetest.test
    def get_raw_config(self):
        self.device.configure(self.config.replace("```","").replace("plaintext",""))
```

When ChatGPT responds with code, it is wrapped in Markdown, and we can simply use **.replace()** to replace it with Python. The resulting string is a valid Cisco IOS XE configuration that pyATS uses inside **.configure()** to deliver to the network!

As powerful as all this seems, we are just getting started with AI and pyATS. The technique demonstrated in Example 23-11, in artificial intelligence terminology, is *retrieval augmented generation* (RAG), where we are augmenting our prompt with retrieved data (in this case, directly from the network in the form of JSON). Using a framework called LangChain, we can take this to the next logical step and include a database to store and retrieve the JSON.

Retrieval Augmented Generation with LangChain

Moving from direct API calls using the OpenAI Python SDK to a more sophisticated method of information retrieval and augmentation in natural language processing can significantly enhance the capabilities of AI systems. The retrieval augmented generation

(RAG) approach stands as a testament to this advancement. RAG leverages a wealth of external knowledge, going beyond static datasets to dynamic, context-rich sources of information. By using a framework like LangChain, developers can orchestrate a seamless interaction between the AI model and various data sources, enabling the model to pull in relevant information on the fly and integrate it into the generation process. This method enhances the quality of the responses, making them more informed, accurate, and contextually relevant.

LangChain is an innovative framework that integrates RAG into language model workflows, acting as a conduit between the AI and a vast reservoir of knowledge. It facilitates a more nuanced and intelligent data retrieval process, which is not merely about fetching data but about understanding and utilizing it effectively. LangChain's architecture is designed to comprehend queries, search through extensive databases, and then synthesize the retrieved data with the AI's internal knowledge to produce superior outputs. It is a step toward more autonomous, intelligent systems that can harness the full potential of both structured and unstructured data.

Figure 23-2 outlines the process of data handling and retrieval using the RAG framework, summarized in the list that follows:

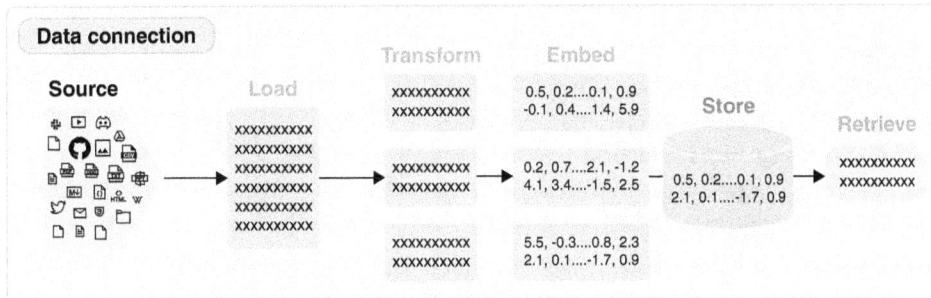

Figure 23-2   *LangChain Approach to Retrieval Augmented Generation*

- **Source:** Different data formats like text, images, and various document types are gathered as input sources.

- **Load:** The data is loaded into a system, often transformed into a uniform format for easier processing.

- **Transform:** This stage involves manipulating the loaded data into a structured format that the AI can understand and work with effectively.

- **Embed:** Data is converted into numerical vectors, also known as embeddings, that represent the semantic information of the data in a form that AI models can process.

- **Store:** The embeddings are stored in a vector store, which is a specialized database designed to handle vectorized data efficiently.

- **Retrieve:** The stored embeddings can be retrieved based on similarity to input queries, allowing the AI to access the most relevant information.

At a high level, RAG works by augmenting the generation process with real-time data retrieval. It uses a question-answering format where the question posed to the model is augmented with information retrieved from a large corpus of data. This approach significantly enhances the AI's ability to provide contextually rich and accurate responses.

Vector stores are pivotal in this architecture, as they enable the efficient retrieval of information based on vector similarity. These stores are databases optimized to handle high-dimensional data and support fast retrieval of items most similar to a given query vector. Vector stores like ChromaDB, Pinecone.io, Elastic, MongoDB, and others are often used in conjunction with machine learning models to facilitate this process. Each has its unique strengths: Pinecone.io and Elastic are known for their scalability and robustness in handling complex queries, while MongoDB offers flexibility in storing different data types.

ChromaDB, in particular, stands out for its low overhead and local deployment capabilities, making it an excellent choice for applications that require fast, on-premises data access without the need for extensive infrastructure. It is particularly well-suited for tasks where data security is paramount and cloud-based storage may not be desirable. By leveraging ChromaDB, we can maintain a local vector store that provides rapid and reliable access to embeddings, which is instrumental in powering RAG's retrieval capabilities.

Lastly, pyATS can serve as the foundation for sourcing JSON or raw text data. This data can then be fed into a JSON loader or text loader within the LangChain RAG framework. pyATS, typically used for network testing and automation, can be repurposed to gather and structure data that RAG utilizes, demonstrating the versatility and integration capabilities of LangChain in practical applications. By using pyATS, we can automate the collection of data, which is then structured and loaded into ChromaDB for quick access during the information retrieval phase of the RAG process.

Using pyATS to gather raw text, like **show run** or **show logging**, we can use a text loader, OpenAI embeddings, ChromaDB, and ChatGPT 3.5 Turbo to create a LangChain inside a pyATS job. First, let's make a **show_run_questions.yaml** file that will hold a list of questions about **show run**, as demonstrated in Example 23-12. We can iterate over this list of questions as part of the LangChain and record the question and LLM answers to pyATS logs (or an external file).

Example 23-12 *show run Questions YAML File Example*

```
---
questions:
  - "Please analyze and summarize this Cisco IOS XE running configuration. Highlight
anything important please"
  - "Is this Cisco IOS XE running configuration secure? Are there any suggestions to
make it more secure"
  - "Please provide a detailed description of the interface configurations in this
Cisco IOS XE running configuration"
```

Also, make sure to pip-install the required libraires; it is a good idea to upgrade LangChain and OpenAI to the latest versions when you install them:

```
(venv) $ pip install chromadb
(venv) $ pip install langchain openai -upgrade
```

Next, we will make a new pyATS job file and script that incorporates the LangChain RAG approach with the **show run text** output, as demonstrated in Example 23-13.

Example 23-13 *pyATS Script Incorporating LangChain RAG Approach with show run Text Output*

```
import os
import yaml
import logging
from rich import print
from pyats import aetest
from dotenv import load_dotenv
from pyats.log.utils import banner
from langchain.vectorstores import Chroma
from langchain.memory import ConversationBufferMemory
from langchain.chains import ConversationalRetrievalChain
from langchain.text_splitter import RecursiveCharacterTextSplitter
from langchain.document_loaders import TextLoader
from langchain.embeddings.openai import OpenAIEmbeddings
from langchain.chat_models import ChatOpenAI


# ----------------
# Get logger for script
# ----------------


log = logging.getLogger(__name__)

# Instantiate openAI client
load_dotenv()

# Load the API keys from an environment variable or secure source
openai_api_key = os.getenv('OPENAI_API_KEY')


llm = ChatOpenAI(temperature=0, model="gpt-4")


# ----------------
# AE Test Setup
# ----------------
```

```python
class common_setup(aetest.CommonSetup):
    """Common Setup section"""
# ---------------
# Connected to devices
# ---------------
    @aetest.subsection
    def connect_to_devices(self, testbed):
        """Connect to all the devices"""
        testbed.connect()
# ---------------
# Mark the loop
# ---------------
    @aetest.subsection
    def loop_mark(self, testbed):
        aetest.loop.mark(Show_Run_Langchain, device_name=testbed.devices)


# ---------------
# Test Case #1
# ---------------
class Show_Run_Langchain(aetest.Testcase):
    """pyATS and AI"""

    @aetest.test
    def setup(self, testbed, device_name):
        """ Testcase Setup section """
        # Set current device in loop as self.device
        self.device = testbed.devices[device_name]

    @aetest.test
    def get_raw_config(self):
        self.raw_config = self.device.execute("show run")

    @aetest.test
    def create_file(self):
        with open(f'{self.device.alias}_Show_Run.txt', 'w') as f:
            f.write(self.raw_config)

    @aetest.test
    def load_text(self):
        self.loader = TextLoader(f'{self.device.alias}_Show_Run.txt')

    @aetest.test
    def split_into_pages(self):
        self.pages = self.loader.load_and_split()
```

```python
        # Create a text splitter
        self.text_splitter = RecursiveCharacterTextSplitter(
            chunk_size=1000,
            chunk_overlap=300,
            length_function=len,
        )


    @aetest.test
    def split_pages_into_chunks(self):
        self.docs = self.text_splitter.split_documents(self.pages)


    @aetest.test
    def store_in_chroma(self):
        embeddings = OpenAIEmbeddings()
        self.vectordb = Chroma.from_documents(self.docs, embedding=embeddings)
        self.vectordb.persist()


    @aetest.test
    def setup_conversation_memory(self):
        self.memory = ConversationBufferMemory(memory_key="chat_history",
return_messages=True)


    @aetest.test
    def setup_conversation_retrieval_chain(self):
        self.qa = ConversationalRetrievalChain.from_llm(llm, self.vectordb.as_
retriever(search_kwargs={"k": 10}), memory=self.memory)


    @aetest.test
    def chat_with_show_run(self):
    # Load questions from the YAML file
        with open("show_run_questions.yaml", 'r') as file:
            data = yaml.safe_load(file)
            questions = data['questions']


        # Open a file in write mode
        with open("Show Run Q_A.txt", 'w') as file:
            for question in questions:
                result = self.qa.run(question)

                log.info(f"[italic red]{question}[/italic red]")
                log.info(f"[italic blue]This is from Conversational Retrieval Chain
using ChromaDB[/italic blue]\n{result}\n")
```

```
                      # Write question and answers to the file
                      file.write(question + '\n')
                      file.write(f"This is from Conversational Retrieval Chain using
ChromaDB\n{result}\n")
                      file.write('-' * 80 + '\n')  # Write a separator line for clarity


class CommonCleanup(aetest.CommonCleanup):
    @aetest.subsection
    def disconnect_from_devices(self, testbed):
        testbed.disconnect()


# for running as its own executable
if __name__ == '__main__':
    aetest.main()
```

Example 23-14 shows a sample result from one of our questions.

Example 23-14   *Chatting with Our Cisco IOS XE Running Configuration via LangChain and RAG*

```
Is this Cisco IOS XE running configuration secure? Are there any suggestions to make
it more secure
This is from Conversational Retrieval Chain using ChromaDB
Based on the provided configuration, there are several security measures already in
place such as SSH for remote management, IP access lists, and encrypted passwords.
However, there are also areas where security could be improved:

1. Use of Service Password-Encryption: The configuration does not show the use
of the "service password-encryption" command, which would encrypt all plaintext
passwords in the configuration.

2. Secure the Console Line: The console line does not appear to have a password set.
This could allow unauthorized physical access to the device. A password should be
set using the "password" command under the console line configuration.

3. Secure the VTY Lines: The VTY lines are using local login, but it would be more
secure to use AAA (Authentication, Authorization, and Accounting) for these logins.
This would provide a more robust method of tracking and controlling user access.

4. Limit Access to the Router: The use of an access list to limit the networks or
hosts that can connect to the VTY lines would add an additional layer of security.

5. Disable Unused Services: Services that are not being used, such as HTTP server,
should be disabled to reduce potential attack vectors.

6. Update and Patch: The version of IOS XE running on the device is not specified.
It is important to ensure that the device is running a recent version of IOS XE that
includes the latest security patches.
```

```
7. Use of Secure Protocols: Ensure that all management and control protocols are
using their secure versions (SSH instead of Telnet, HTTPS instead of HTTP, SNMPv3
instead of SNMPv2, etc.).

8. Implement Logging: Implement logging to a syslog server for better tracking and
auditing.

Remember, these are general recommendations, and the specific needs and requirements
of your network may vary. Always follow best practices for security in your specific
environment.
```

We could also use a JSON loader and the results from a pyATS model, parsed output, or REST API call simply by swapping out the list of questions, the pyATS command to capture, and the text loader, as demonstrated in Example 23-15. Now the JSON loader requires a JSON Query (jq) schema. As our pyATS parsed output does not have an outer key, we will add this key when we capture the output and pass it to the query.

Example 23-15 *JSON Loader Example*

```
---
questions:
 - "Please analyze and summarize this Cisco IOS XE routing table. Highlight
anything important please"
 - "Is this Cisco IOS XE routing table optimal? Are there any suggestions to make
it more efficient or any routes to summarize"
…
import json
from langchain.document_loaders import JSONLoader
...
    @aetest.test
    def get_raw_config(self):
        raw_json = self.device.parse("show ip route")
        self.parsed_json = {"info": raw_json}

    @aetest.test
    def create_file(self):
        with open(f'{self.device.alias}_Show_IP_Route.json', 'w') as f:
            f.write(json.dumps(self.parsed_json, indent=4, sort_keys=True))

    @aetest.test
    def load_text(self):
        self.loader = JSONLoader(
            file_path=f'{self.device.alias}_Show_IP_Route.json',
            jq_schema=".info[]",
            text_content=False
        )
```

As you can see, all we need to do is swap out the text loader with the JSON loader and provide the updated relevant questions, and the rest of the LangChain remains unchanged. It's very much like LEGO blocks we can replace with other modules. We could, for example, use Elastic for the vector store, and no other code would need to be changed. This can lead to rapid AI incorporation across many Cisco pyATS commands, APIs, models, and parsers.

Example 23-16 demonstrates a sample of the routing question output.

Example 23-16 *JSON Loader Example AI Response Output Sample*

```
Please analyze and summarize this Cisco IOS XE running configuration. Highlight
anything important please

This is from Conversational Retrieval Chain using ChromaDB

This configuration is a routing table for a Cisco IOS XE device. Here are some
important points:


1. The device has several active routes configured for both local and connected
source protocols.


2. The routes are configured for various IP address ranges, such as "10.111.7.3/32",
"10.255.255.0/24", "192.168.1.0/24", "192.168.200.0/24", and "0.0.0.0/0" among
others.


3. The "0.0.0.0/0" route is the default route for the device, with a next hop
IP of "10.10.20.254" and it uses the "GigabitEthernet1" interface. This route is
statically configured.


4. The device has several loopback interfaces configured, such as "Loopback17",
"Loopback109", "Loopback100", and "Loopback0". Loopback interfaces are virtual
interfaces used mainly for testing and network management.


5. The device also has a "VirtualPortGroup0" interface configured. This is typically
used for device management.


6. The routes have different source protocols, including "connected", "local", and
"static". "Connected" and "local" routes are automatically created when an interface
is configured with an IP address and is active. "Static" routes are manually config-
ured by the user and do not change unless the user modifies them.


7. The source protocol codes are also provided, with "C" standing for connected,
"L" for local, and "S*" for static.


8. The "active" status for all routes indicates that these routes are currently in
use in the routing table.


In summary, this configuration provides a snapshot of the device's routing table,
showing the active routes, the interfaces used, the next hop IPs, and the source
protocols for each route.
```

LangChain, PyATS integration, and ChromaDB are cutting-edge technologies in the field of software development and data management. LangChain is a library that simplifies the process of integrating natural language processing into applications, enhancing the potential for more advanced and intuitive user interactions. pyATS integrated with LangChain opens up new avenues for creating sophisticated testing scenarios using natural language commands. ChromaDB, on the other hand, is a specialized database system that offers efficient storage and retrieval of large-scale data, particularly useful in handling complex datasets.

In this realm, vector stores and embeddings play a crucial role. Vector stores allow for the efficient storage and retrieval of high-dimensional data vectors, which are essential in machine learning and AI applications. Embeddings, which transform data into these vectors, are fundamental in processing and understanding natural language, images, and other complex data types.

The synergy between these technologies has significantly eased the creation of advanced pyATS jobs. By leveraging LangChain's natural language processing capabilities, developers can design more intuitive and efficient testing scripts and scenarios in pyATS, streamlining the automation process.

Looking forward, the exploration of Streamlit.io as a tool for rapid prototyping is on the horizon. Streamlit.io is an open-source app framework specifically designed for building web apps using pure Python. Its ease of use and flexibility make it an ideal choice for quickly creating and deploying data applications, which could further enhance the capabilities of pyATS and LangChain in developing sophisticated, AI-driven automation solutions.

Rapid Prototyping with Streamlit

Integrating Streamlit.io into the technological stack comprising AI, pyATS network testing, and automation, alongside LangChain RAG, marks a significant leap in developing intuitive and efficient interfaces for complex systems. Streamlit, a powerful and flexible framework, is designed for rapid prototyping of web applications, particularly in data science and machine learning contexts. Its ease of use, stemming from its Python-centric approach, allows developers to transform data scripts into shareable web apps swiftly. This integration empowers users to interact with the underlying AI and network testing functionalities in a more accessible and engaging manner. Streamlit's interactive widgets provide a user-friendly interface to control and visualize the operations of LangChain RAG and pyATS, facilitating a seamless interaction between the user and the sophisticated backend processes.

The combination of Streamlit with AI, LangChain RAG, and pyATS presents a robust solution for organizations seeking to streamline their network testing and automation processes. Streamlit acts as a frontend interface, enabling users to interact with AI models and network testing tools without delving into the complexities of the code. This approach significantly reduces the learning curve for new users and enhances productivity by allowing quick adjustments and real-time feedback. With LangChain RAG's advanced language understanding capabilities, the system can process natural language queries, making it easier for users to navigate and utilize the network testing features of pyATS. The integration ensures that complex network diagnostics and AI-driven insights are presented in a digestible and actionable format.

Moreover, the use of Streamlit.io in this context underscores the importance of agility and adaptability in modern software development, particularly in the realms of AI and network automation. The rapid prototyping capabilities of Streamlit enable developers to iterate quickly, test new ideas, and receive immediate feedback, which is crucial in a fast-evolving field like AI. This approach not only accelerates the development process but also ensures that the end product is fine-tuned to meet user needs and preferences. The synergy between Streamlit, AI, LangChain RAG, and pyATS paves the way for creating more dynamic, responsive, and user-centric applications, setting a new standard in the integration of advanced technological solutions.

Django is a high-level Python web framework that encourages rapid development and clean, pragmatic design. It's renowned for its robustness and scalability, making it a favorite among developers for building complex, database-driven websites. Django follows the model-template-view (MTV) architectural pattern, providing a structured environment with a wealth of features, including an admin panel, ORM (object-relational mapping), and robust security features. However, this comprehensiveness comes with a certain "heaviness"—there's a lot to manage, from the frontend to the backend, making it more suited for larger, more complex projects. Django's all-encompassing nature means that it can handle almost any web development need, but this can also result in a steeper learning curve and longer development times for certain types of projects.

On the other hand, Streamlit is like a breath of fresh air for developers looking to create data applications with minimal fuss. It's a game-changer in terms of simplicity and development speed, particularly for those working in data science and AI. Streamlit is designed to turn data scripts into shareable web apps with minimal code, often in just a few lines. Unlike Django, Streamlit is highly focused on backend development, making it incredibly user-friendly for creating interfaces for complex systems like LangChain. Its ease of use doesn't require extensive knowledge of frontend technologies like HTML, JavaScript, and CSS, which is often necessary with Django. Streamlit's integration with LangChain allows for quick prototyping and deployment of AI-powered applications, making it ideal for projects where time-to-market is crucial.

The harmony between Streamlit and LangChain lies in their mutual commitment to ease of use and rapid deployment. LangChain's sophisticated AI capabilities, when combined with Streamlit's straightforward interface design, provide a powerful tool for creating advanced AI applications without getting bogged down in complex web development details. This is especially useful in situations where the focus is more on the functionality of the AI and less on the intricacies of web development.

While Django is a powerful and versatile framework perfect for building large-scale, complex web applications, its comprehensive nature makes it somewhat heavyweight for projects where speed and simplicity are key. Streamlit, in contrast, offers a much quicker and easier path to developing web apps with pure Python, particularly for data-driven AI applications like those powered by LangChain. Its simplicity and focus on rapid prototyping make Streamlit an ideal choice for projects where time and ease of development are of the essence.

Using Docker, Streamlit, LangChain, ChromaDB, and pyATS, let's make one final application.

Create a folder called **docker** and inside create a Dockerfile, as shown in Example 23-17.

Example 23-17 *Dockerfile for Streamlit, LangChain, pyATS Application to Chat with Routing Tables*

```
FROM ubuntu:latest

ARG DEBIAN_FRONTEND=noninteractive

RUN echo "==> Upgrading apk and installing system utilities ...." \
&& apt -y update \
&& apt-get install -y wget \
&& apt-get -y install sudo

RUN echo "==> Installing Python3 and pip ...." \
&& apt-get install python3 -y \
&& apt install python3-pip -y \
&& apt install openssh-client -y

RUN echo "==> Adding pyATS ..." \
&& pip install pyats[full]

RUN echo "==> Install dos2unix..." \
  && sudo apt-get install dos2unix -y

RUN echo "==> Install langchain requirements.." \
  && pip install langchain \
  && pip install chromadb \
  && pip install openai \
  && pip install tiktoken

RUN echo "==> Install jq.." \
  && pip install jq

RUN echo "==> Install streamlit.." \
  && pip install streamlit

COPY /streamlit_langchain_pyats /streamlit_langchain_pyats/
COPY /scripts /scripts/

RUN echo "==> Convert script..." \
  && dos2unix /scripts/startup.sh

CMD ["/bin/bash", "/scripts/startup.sh"]
```

Chapter 23: Leveraging Artificial Intelligence in pyATS

This Dockerfile script is used to create a Docker image, which is a lightweight, stand-alone, executable package that includes everything needed to run specific software, including the code, runtime, system tools, libraries, and settings. Let's break down each part of this Dockerfile:

- **FROM ubuntu:latest:** This line specifies the base image from which you are building. Here, it's using the latest version of the Ubuntu operating system.

- **ARG DEBIAN_FRONTEND=noninteractive:** This argument is used to ensure that during the build process, the installation of packages does not prompt for interactive input, which is essential for automated scripts.

- The first **RUN** command updates the package lists for upgrades of packages that need upgrading and then installs wget (a network downloader) and sudo (a program that allows a permitted user to execute a command as the superuser or another user).

- The second **RUN** command installs Python3, pip (Python package manager), and the SSH client. This setup is essential for running Python scripts and managing remote connections.

- The third **RUN** command installs pyATS with full dependencies. pyATS is the Python Automated Test System, commonly used in network testing and automation.

- The fourth **RUN** command installs dos2unix, a tool to convert text files with DOS or Mac line endings to Unix line endings.

- The fifth **RUN** command installs several Python packages:

  - **langchain:** A Python package, framework related to AI

  - **chromadb:** Likely a database-related package

  - **openai:** The OpenAI Python client, used to interact with OpenAI's services

  - **tiktoken:** A package related to token handling

- The sixth **RUN** command installs jq, but it's incorrectly installed via pip instead of apt. jq is a lightweight and flexible command-line JSON processor. The installation command should be corrected to use apt because jq is not a Python package.

- The seventh **RUN** command installs Streamlit, a Python library for creating web apps for machine learning and data science projects.

- **COPY** commands copy the streamlit_langchain_pyats and scripts directories from the host machine to the image.

- Another **RUN** command uses dos2unix to convert the line endings in the startup.sh script, ensuring it's compatible with Unix systems.

- The **CMD ["/bin/bash", "/scripts/startup.sh"]** command sets the default command to execute when the container starts. In this case, it runs the startup.sh script using bash.

In summary, this Dockerfile creates an environment with Ubuntu as the base, equipped with Python, pyATS, Streamlit, and various other tools and libraries, likely intended for a project involving network testing, AI, or machine learning with a Streamlit-based web interface.

As seen at the end of the Dockerfile, we will need a script to start the Streamlit application automatically when the container starts. Make a folder called **scripts** and inside create a bash script called **startup.sh**, as demonstrated in Example 23-18.

Example 23-18   *Bash Script to Startup Streamlit Application*

```
cd streamlit_langchain_pyats
streamlit run streamlit_langchain_pyats.py
```

Next, we need a docker-compose.yml file, as shown in Example 23-19.

Example 23-19   *A docker-compose.yaml File for Streamlit, Langchain, and pyATS Container*

```
version: '3'

services:
  streamlit_langchain_pyats:
    image: johncapobianco/streamlit_langchain_pyats:streamlit_langchain_pyats
    container_name: streamlit_langchain_pyats
    restart: always
    build:
      context: ./
      dockerfile: ./docker/Dockerfile
    ports:
      - "8501:8501"
```

We will now make a standard pyATS job that parses **show ip interface brief** and saves the output to a file that our LangChain JSON loader can use, as demonstrated in Example 23-20 and Example 23-21. This pyATS job will be executed when the Streamlit application starts in order to populate the ChromaDB database.

Example 23-20   *pyATS Job File to Create show ip route JSON File*

```
import os
from genie.testbed import load

def main(runtime):

    # ---------------
    # Load the testbed
    # ---------------
```

```
    if not runtime.testbed:
        # If no testbed is provided, load the default one.
        # Load default location of Testbed
        testbedfile = os.path.join('testbed.yaml')
        testbed = load(testbedfile)
    else:
        # Use the one provided
        testbed = runtime.testbed


    # Find the location of the script in relation to the job file
    testscript = os.path.join(os.path.dirname(__file__), 'json_langchain.py')


    # run script
    runtime.tasks.run(testscript=testscript, testbed=testbed)
```

Example 23-21   *pyATS Script to Create show ip route JSON File*

```
import os
import json
import logging
from pyats import aetest
from pyats.log.utils import banner


# ----------------
# Get logger for script
# ----------------


log = logging.getLogger(__name__)


# ----------------
# AE Test Setup
# ----------------
class common_setup(aetest.CommonSetup):
    """Common Setup section"""
# ----------------
# Connected to devices
# ----------------
    @aetest.subsection
    def connect_to_devices(self, testbed):
        """Connect to all the devices"""
        testbed.connect()
```

```
# ---------------
# Mark the loop for Learn Interfaces
# ---------------
    @aetest.subsection
    def loop_mark(self, testbed):
        aetest.loop.mark(Show_Run_Langchain, device_name=testbed.devices)


# ---------------
# Test Case #1
# ---------------
class Show_Run_Langchain(aetest.Testcase):
    """pyATS Get and Save Show IP Route"""

    @aetest.test
    def setup(self, testbed, device_name):
        """ Testcase Setup section """
        # Set current device in loop as self.device
        self.device = testbed.devices[device_name]

    @aetest.test
    def get_ip_routes(self):
        raw_json = self.device.parse("show ip route")
        self.parsed_json = {"info": raw_json}

    @aetest.test
    def create_file(self):
        with open(f'{self.device.alias}_Show_IP_Route.json', 'w') as f:
            f.write(json.dumps(self.parsed_json, indent=4, sort_keys=True))

class CommonCleanup(aetest.CommonCleanup):
    @aetest.subsection
    def disconnect_from_devices(self, testbed):
        testbed.disconnect()

# for running as its own executable
if __name__ == '__main__':
    aetest.main()
```

Finally, our primary script, the **streamlit_LangChain_pyats.py** script, will run when the container starts. It will immediately run the pyATS job from Example 23-21 and then present the user with a conversational interface to "chat" with the routing table, as demonstrated in Example 23-22.

Example 23-22   *Streamlit Application Prototype Script*

```python
import streamlit as st
import os
import yaml
from dotenv import load_dotenv
from langchain.vectorstores import Chroma
from langchain.memory import ConversationBufferMemory
from langchain.chains import ConversationalRetrievalChain
from langchain.text_splitter import RecursiveCharacterTextSplitter
from langchain.document_loaders import JSONLoader
from langchain.embeddings.openai import OpenAIEmbeddings
from langchain.chat_models import ChatOpenAI


# Function to run pyATS job
def run_pyats_job():
    os.system("pyats run job json_langchain_job.py")


# Use Streamlit's caching to run the job only once
if 'job_done' not in st.session_state:
    st.session_state['job_done'] = st.cache_resource(run_pyats_job)()


# Instantiate openAI client
load_dotenv()
openai_api_key = os.getenv('OPENAI_API_KEY')
llm = ChatOpenAI(temperature=0, model="gpt-4")


class ChatWithRoutingTable:
    def __init__(self):
        self.load_text()
        self.split_into_pages()
        self.split_pages_into_chunks()
        self.store_in_chroma()
        self.setup_conversation_memory()
        self.setup_conversation_retrieval_chain()

    def load_text(self):
        self.loader = JSONLoader(
            file_path=f'{device.alias}_Show_IP_Route.json',
            jq_schema=".info[]",
            text_content=False
        )
```

```python
    def split_into_pages(self):
        self.pages = self.loader.load_and_split()
        # Create a text splitter
        self.text_splitter = RecursiveCharacterTextSplitter(
            chunk_size=1000,
            chunk_overlap=300,
            length_function=len,
        )

    def split_pages_into_chunks(self):
        self.docs = self.text_splitter.split_documents(self.pages)

    def store_in_chroma(self):
        embeddings = OpenAIEmbeddings()
        self.vectordb = Chroma.from_documents(self.docs, embedding=embeddings)
        self.vectordb.persist()

    def setup_conversation_memory(self):
        self.memory = ConversationBufferMemory(memory_key="chat_history",
return_messages=True)

    def setup_conversation_retrieval_chain(self):
        self.qa = ConversationalRetrievalChain.from_llm(llm, self.vectordb.as_
retriever(search_kwargs={"k": 10}), memory=self.memory)

    def chat(self, question):
        return self.qa.run(question)

# Create an instance of your class
chat_instance = ChatWithRoutingTable()

# Streamlit UI for chat
st.title("Chat with Cisco IOS XE Routing Table")
user_input = st.text_input("Ask a question about the routing table:",
key="user_input")
if st.button("Ask"):
    with st.spinner('Processing...'):
        answer = chat_instance.chat(user_input)
        st.write(answer)
```

This code leverages Streamlit to create a user-friendly web interface for querying a Cisco IOS XE routing table. It employs advanced NLP techniques through LangChain and GPT-4, backed by pyATS for network-related functionalities. The code is structured to be efficient and responsive, with caching mechanisms and an intuitive class-based approach for processing and responding to user queries.

Bring up the Docker container using the following **docker-compose** command:

```
(venv) $ docker-compose up
```

After the Docker image build is complete, the Streamlit service will start. The logs will look similar to Figure 23-3.

```
streamlit_langchain_pyats  | Collecting usage statistics. To deactivate, set browser.gatherUsageStats to False.
streamlit_langchain_pyats  |
streamlit_langchain_pyats  |
streamlit_langchain_pyats  |   You can now view your Streamlit app in your browser.
streamlit_langchain_pyats  |
streamlit_langchain_pyats  |   Network URL: http://172.26.0.2:8501
streamlit_langchain_pyats  |   External URL: http://142.170.39.142:8501
streamlit_langchain_pyats  |
```

Figure 23-3 *Streamlit Application Successfully Started Inside Docker*

Next, visit http://localhost:8501 to bring up the Streamlit application. Immediately the pyATS job will start, as shown in Figure 23-4.

Figure 23-4 *Streamlit Running pyATS Job*

After the pyATS job has completed, the main user interface will be presented, and we can "chat" with the Cisco IOS XE routing table, as shown in Figure 23-5 and Figure 23-6.

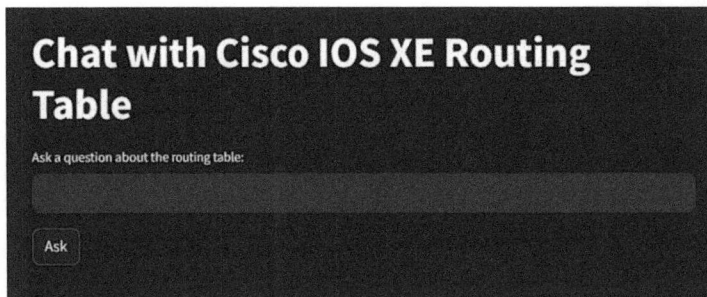

Chat with Cisco IOS XE Routing Table

Ask a question about the routing table:

Ask

Figure 23-5 *Streamlit Chat with Cisco IOS XE Routing Table*

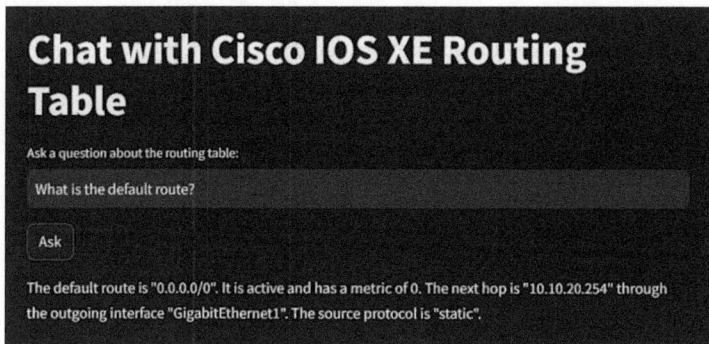

Figure 23-6 *Streamlit Chat "What is my default route?"*

Summary

This chapter offered an insightful journey through the evolving landscape of artificial intelligence (AI) and its significant role in reshaping network automation, especially with tools like Cisco pyATS. We began by examining how AI, particularly generative AI, has revolutionized the way machines learn from data and make autonomous decisions. The chapter highlighted the imperative need for professionals in network automation to grasp the intricacies of AI and generative AI, emphasizing their transformative impact on network testing and automation.

We then navigated through the democratization of AI over the past year, a period marked by its expanded accessibility and pronounced impact across multiple sectors. This shift brought sophisticated AI tools from the confines of specialized research labs into the hands of a diverse group of developers. The chapter analyzed how this accessibility has not only empowered seasoned professionals but also spurred a wave of innovation in fields like network automation, enabling the integration of complex AI algorithms into everyday workflows.

A focal point of this chapter was the role of OpenAI, particularly its integration with Cisco pyATS. We explored how the incorporation of OpenAI's API into pyATS jobs signified a major technological advancement, moving beyond traditional script-based automation. This melding of AI with network testing tools like pyATS was evaluated for its potential to streamline processes and tackle more advanced automation challenges effectively.

The chapter delved into the advanced technique of retrieval augmented generation (RAG), facilitated by LangChain. We dissected how LangChain, with its Python and JavaScript SDKs, simplified the implementation of RAG, making it attainable even for those with limited AI expertise. The discussion emphasized the innovative fusion of structured data processing with the nuanced understanding of language models, which has opened new horizons in automated testing and network management.

A significant part of the chapter was dedicated to the implementation of RAG in network automation, involving intricate data handling. We examined the use of a JSON loader for integrating pyATS parsers or APIs, and the transformation of network data into embeddings and textual segments. The selection and role of vector stores like ChromaDB, Pinecone, ElasticSearch, and MongoDB were critically analyzed for their importance in storing and retrieving data for AI models, enhancing the sophistication of automated responses.

In the concluding sections, the chapter discussed the integration of large language models (LLMs) like ChatGPT 3.5 and the role of Streamlit. We evaluated how these models intelligently interact with vector stores to generate accurate and context-relevant responses to automation prompts. Furthermore, the adoption of Streamlit as a tool for rapid development and prototyping was highlighted for its effectiveness in creating user-friendly interfaces for LangChain and pyATS applications, underscoring its value in demonstrating the practical capabilities of AI-enhanced network automation tools.

This chapter served as a reflective and analytical conclusion, emphasizing the practical applications, advancements, and future potential of AI in the realm of network automation—and, hopefully, as the jumping off point for you to adopt and incorporate artificial intelligence into your pyATS network automation workflows.

Writing Your Own Parser

One of the best parts of open-source software is having the ability to contribute. Contributing can take many forms: helping identify and/or fix bugs in the codebase, opening a new feature request, and writing/fixing documentation. Contributing to open-source projects can be a rewarding experience, allowing you to learn new skills, collaborate with others, and give back to the community. When contributing to an open-source project, it's important that you follow best practices for writing clean and well-documented code with proper test coverage. Many open-source projects have contribution guides that provide a checklist of requirements that must be met before contributing to the project. We are going to cover how to contribute to the genieparser library (https://github.com/CiscoTestAutomation/genieparser) by creating a new parser for a Cisco IOS-XE **show** command. Parsers convert unstructured command output into structured Python objects, mostly dictionaries. We will go through each of the steps required, from setting up the development environment to opening a pull request (PR). If you have additional questions, you can find references and recommended readings at the end that will provide more details.

Contributing to the pyATS Library

Before we dive into writing code, we must familiarize ourselves with the pyATS library contribution guidelines (https://pubhub.devnetcloud.com/media/pyats-development-guide/docs/contribute/contribute.html#contribute) and understand the structure of the genieparser codebase. All parsers can be found in the genieparser GitHub repository in the following directory in genieparser/src/genie/libs/parser. In the /parser directory, parsers are organized by OS type, which makes it easy for users to find the parser they're interested in developing or modifying. Figure A-1 shows the different directories found in the parser directory on GitHub.

You'll notice a directory for each of the supported OS types in pyATS. This is by design and should help connect the dots and allow you to understand why it's important to specify the proper OS in the pyATS testbed file. Once we click into one of the OS directories (in this case, iosxe), you notice additional directories with specific platform names, as some parsers are specific to a platform. Otherwise, you'll see many Python (.py) files that represent different **show** command modules. For example, the common **show version** command and any of its additional keywords or arguments are represented in the show_version.py module. Figure A-2 shows the directories and modules found in the iosxe directory.

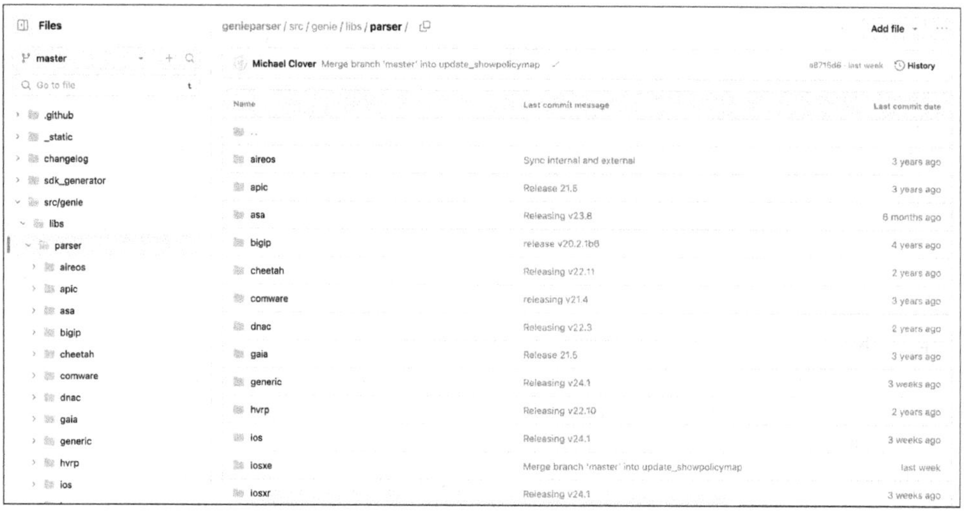

Figure A-1   *GitHub – Parser Directory*

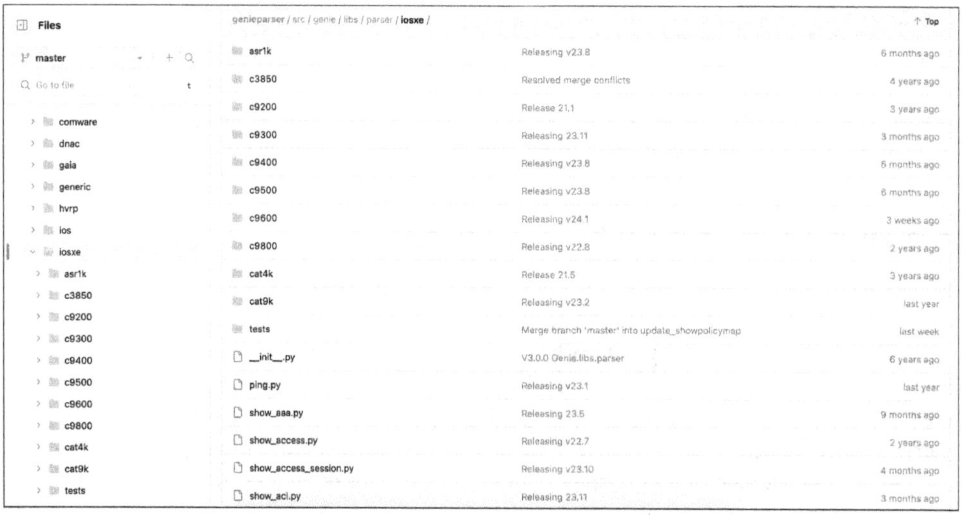

Figure A-2   *GitHub – iosxe Directory*

Now that you understand the genieparser directory structure and where to find the parsers, let's dive into the structure of a parser!

Parser Structure

Every parser in the genieparser library has an associated schema and the parser. In the following sections, we are going to take a closer look at the schema class and the parser class. Before we look into those classes, we must understand Genie's metaparser class, which is part of the Genie metaparser package and provides a base for building a parser. The Genie metaparser package allows users to create one parsing structure that works regardless of the communication protocol used to interact with the network device—CLI, NETCONF (XML), or RESTCONF (YANG). You can consider the metaparser a "protocol-agnostic" parsing engine. Figure A-3 shows a detailed overview of how the metaparser works.

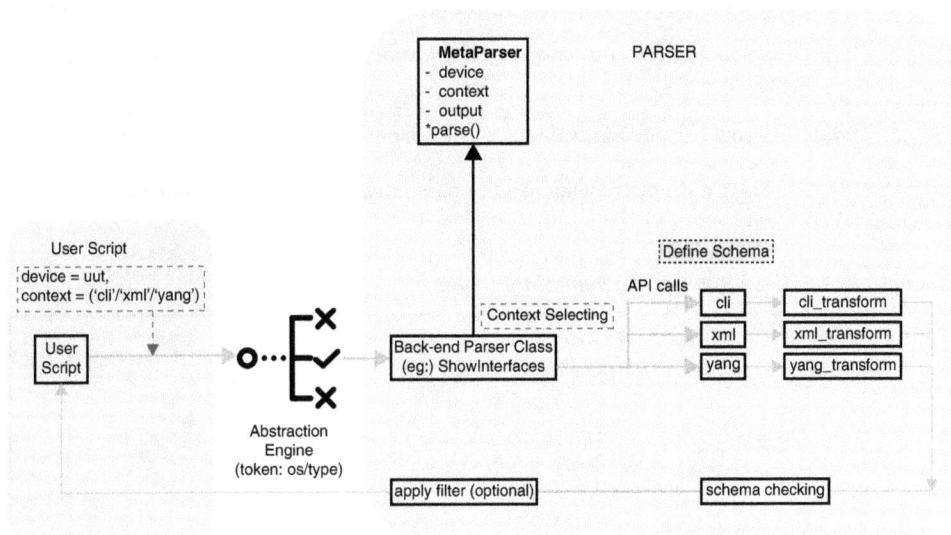

Figure A-3   *Genie Metaparser*

Schema Class

The schema class defines the structure of the parsed output, regardless of the communication protocol used to collect the output from the network device. All schema classes are inherited from the metaparser class, which allows the schema to stay consistent across all Genie communication protocols (CLI, XML, and YANG). You can also build your schema based on an existing schema from the available Genie models. The best way to begin creating a schema is to review the command output across the different communication protocols. For the CLI, you can use indentation to identify different keys and which keys should be nested. For YANG data models, you can use tools such as pyang or YANG Suite to identify keys and data types. It's important to view the different outputs to ensure you cover all important data points in your schema. Once you've reviewed the different outputs, you'll need to identify different key-value pairs that should be used in your schema.

In the metaparser package, the schemaengine module provides many utility classes that can be used to help build your schema. The utility classes can be used directly in your schema:

- **Any()**: Effectively a wildcard that accepts any value. This can be useful in more complex schemas.

- **Optional()**: The value is not required in the schema.

- **And()**: The value must pass all the validation requirements to be used in the schema.

- **Or()**: The value must pass at least one of the validation requirements to be used in the schema.

- **Default()**: Provides a default value if a value is not specified in the schema.

Example A-1 shows a simple schema used to parse the **show license summary** output from IOS-XE devices.

Example A-1   *The show license summary Schema*

```
from genie.metaparser import MetaParser
from genie.metaparser.util.schemaengine import Any

class ShowLicenseSummarySchema(MetaParser):
    """Schema for show license summary"""
    schema = {
        'license_usage': {
            Any(): {
                'entitlement': str,
                'count': str,
                'status': str,
            }
        }
    }
```

The schema utilizes the **Any()** class to provide flexibility on the license name. In addition, you'll notice the type annotations to identify the data types for each key-value pair. The schema class requires the most planning and foresight, as it's highly discouraged to change the schema design once implemented. In the next section, we will look at the parser class, which is where the magic happens and where raw output is transformed into parsed output that adheres to the schema class.

Parser Class

The parser class is responsible for executing the desired command, looking for specific patterns in the command output, and creating structured output adhering to the parser's associated schema class. Parser classes commonly use regular expressions (RegEx) to identify specific patterns in the raw output, but you may also use TextFSM, TTP, or the parsergen package to extract data. Regular expressions are the most common, as they provide the most flexibility and scalability. The example in the following section will cover how to write a parser using regular expressions. Also included in the parser class are the different communication methods (CLI, XML, and YANG) represented as methods in the class. Example A-2 shows a parser class that executes and parses a CLI command.

Example A-2 *Parser Class*

```
class ShowLicenseSummary(ShowLicenseSummarySchema):
    """Parser for show license summary"""
    cli_command = 'show license summary'

    def cli(self, output=None):
        if output is None:
            out = self.device.execute(self.cli_command)
        else:
            out = output

        # initial return dictionary
        license_summ_dict = {}

        result_dict = {}

        # network-advantage      (C9300-48 Network Advan...)      1 IN USE
        # dna-advantage          (C9300-48 DNA Advantage)         1 IN USE

        # C9300 48P DNA Advantage  (C9300-48 DNA Advantage)       2 AUTHORIZED
        p0 = re.compile(r"^(?P<license>.+?)\s+\((?P<entitlement>.+)\)\s+
        (?P<count>\d+)\s+(?P<status>.+)$")

        for line in out.splitlines():
            line=line.strip()

            # udi line
            m=p0.match(line)
            if m:
                if 'license_usage' not in license_summ_dict:
                    result_dict = license_summ_dict.setdefault('license_usage', {})
```

```
                license = m.groupdict()['license']
                entitlement = m.groupdict()['entitlement']
                count = m.groupdict()['count']
                status = m.groupdict()['status']
                result_dict[license] = {}
                result_dict[license]['entitlement'] = entitlement
                result_dict[license]['count'] = count
                result_dict[license]['status'] = status
                continue
        return result_dict
```

We won't go into much detail, but let us point out a couple things. First, you should notice the **ShowLicenseSummary** class inherits the **ShowLicenseSummarySchema** class, which inherits the base metaparser class. This gives the **ShowLicenseSummary** class access to the defined schema and unified parsing and transformation features provided by the **metaparser** class. The other thing to point out is the **cli()** method defined in the parser class. In Genie parser classes, you must define a class method for each different communication protocol (that is, CLI, XML, and YANG) because each method executes and collects command output differently. Most Genie parsers only have a **cli()** method, as the CLI is still the most common denominator across many network devices. As network device APIs mature and become more widely adopted, we may see more parser classes with YANG methods. Now that you understand the different components of a parser, let's dive into creating a parser!

Creating Your Parser

In the following sections, we are going to break down how to create a parser and contribute back to the genieparser library on GitHub. We will go through setting up a development environment, including forking the genieparser library, activating a Python virtual environment, and enabling a "development" mode specific to the genieparser library. We will then go through creating schema and parser classes for our new Genie parser. Lastly, we will test the parser to verify it works as expected and to ensure it doesn't break anything else, including other parsers or schemas. Throughout the following sections, we are going to use a parser that has been previously contributed to the genieparser library so that we can see the entire workflow. Check out the following pull request to the genieparser library if you'd like to follow along: https://github.com/CiscoTestAutomation/genieparser/pull/536/files.

Development Environment

To contribute to the genieparser library, we must have the following software installed on our machine:

■ Python 3.6+

■ Pip (should be included with Python)

■ Git

Once the required software is installed, we must go to the genieparser GitHub repository (https://github.com/CiscoTestAutomation/genieparser) and fork the repository. Figure A-4 shows how to fork the repository on GitHub.

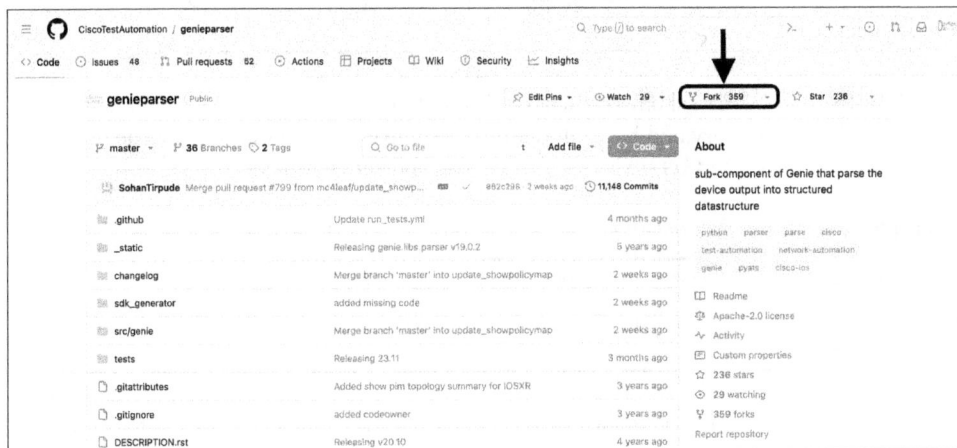

Figure A-4   *GitHub – Fork Repository*

Once the repository is forked, you should see a copy of it in your list of GitHub repositories. Figure A-5 shows what you should see in your GitHub repositories. Notice under the repository name that it references where you forked it from: "Forked from CiscoTestAutomation/genieparser."

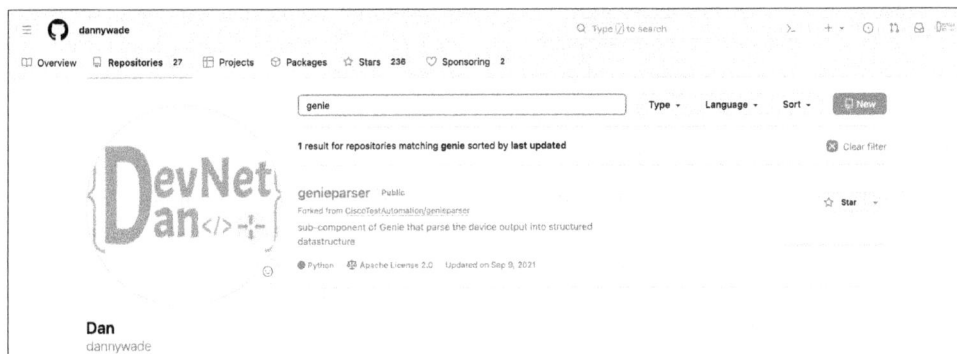

Figure A-5   *Copy of GitHub – Repository*

Before cloning down the repository to begin development, we must first create a Python virtual environment. Example A-3 shows the specific commands to run on macOS or Linux to create a virtual environment.

Example A-3 *Create Python Virtual Environment*

```
$ python -m venv {directory_of_your_choice}
$ source {directory_of_your_choice}/bin/activate
$ pip install pyats[full]
```

Once the Python virtual environment is activated, we can clone down your forked genieparser repository and begin developing! Example A-4 shows how to clone your forked repository and set up a local Genie development environment.

Example A-4 *Clone Repository and Set Up Develop Environment*

```
$ git clone https://github.com/{YOUR_GITHUB_USERNAME}/genieparser
$ cd genieparser
$ make develop
```

You may wonder what the **make develop** command does and what it means to set up a local Genie development environment. The **make** command calls on a local "Makefile," and any keywords after **make** are the targets to call on. You can find more details in the Makefile of the genieparser repository, but in summary, the **make develop** command uninstalls the genie.lib.parser module, which essentially removes all the parsers from the library and runs **setup.py develop** so that you don't have to reinstall the genieparser package every time you make a change to your code during development. The command **setup.py develop** is a common command to run during the development and testing of Python code because it installs a Python package in such a way that any updates or changes you make will automatically take effect, without you having to reinstall the package. Once the local Genie development environment is set up using the **make develop** command, we are ready to begin creating a parser!

Writing Your Schema Class

We are going to write a schema and parser class for the **show license summary** command for Cisco IOS-XE devices. First, we must develop a schema class. As mentioned previously, the best way to develop a solid schema is to review multiple outputs of the selected command. Example A-5 shows what the output may look like for the **show license summary** command on Cisco Catalyst 9300 switches.

Example A-5 *The show license summary Output*

```
License Usage:
License                    Entitlement Tag              Count Status

------------------------------------------------------------------------

C9300 48P DNA Advantage  (C9300-48 DNA Advantage)       2 AUTHORIZED
C9300 48P Network Adv... (C9300-48 Network Advan...)    2 AUTHORIZED
C9300 24P Network Adv... (C9300-24 Network Advan...)    1 AUTHORIZED
C9300 24P DNA Advantage  (C9300-24 DNA Advantage)       1 AUTHORIZED
```

Obviously, this output is not from a single device, as you can see a mix of 48P and 24P output, which indicates whether the switch has 24 or 48 ports. The purpose of the mixed output is to see what variables may be different. In this case, we can see the first three columns have different values depending on the switch model (C9300 24P/48P), license name (DNA Advantage or Network Advantage), or number of licenses (1 or 2). The last column, "Status," has all the same value (Authorized), but we must assume there's a different value for switches that are not authorized or have a missing license entitlement.

Once we know the command and have the output we are trying to parse, we need to identify where in the codebase to implement the schema and parser classes. Remember, the parsers are organized by OS type. For Cisco IOS-XE commands, we would navigate to src/genie/libs/parser/iosxe. Once in the appropriate directory, we must look for a Python file (module) that is named after the **base show** command (in this case, "show license," as "summary" is an additional keyword). We can see "show_license.py" does exist! Once we've located the proper Python module, we can add our schema.

Example A-6 shows the schema created based on the different variables and assumptions we can make based on the command output in Example A-5.

Example A-6 *The ShowLicenseSummarySchema Class*

```python
class ShowLicenseSummarySchema(MetaParser):
    """Schema for show license udi"""
    schema = {
        'license_usage': {
            Any(): {
                'entitlement': str,
                'count': str,
                'status': str,
            }
        }
    }
```

The defined schema is a Python dictionary assigned to a variable named **schema**. Given the command output in Example A-5, our schema was pretty simple to define. Each license will be a nested dictionary under the **license_usage** top-level key. The **Any()** utility class allows any value to be assigned as a key to the nested dictionary. However, we are expecting the name of the license to be assigned as the key. In the nested dictionary, there are keys for each of the other columns in the command output: **entitlement**, **count**, and **status**. We are expecting each value to be a string. Now that we have the schema defined, let's move on to creating the parser class!

Writing Your Parser Class

The parser class executes the desired command, parses the output, and returns a Python dictionary that follows the defined schema. Example A-7 shows the parser class created for the **show license summary** Cisco IOS-XE command.

Example A-7   *The ShowLicenseSummary Parser Class*

```
class ShowLicenseSummary(ShowLicenseSummarySchema):
    """Parser for show license summary"""
    cli_command = 'show license summary'

    def cli(self, output=None):
        if output is None:
            out = self.device.execute(self.cli_command)
        else:
            out = output

        # initial return dictionary
        license_summ_dict = {}

        result_dict = {}

        # network-advantage       (C9300-48 Network Advan...)    1 IN USE
        # dna-advantage           (C9300-48 DNA Advantage)       1 IN USE

        # C9300 48P DNA Advantage  (C9300-48 DNA Advantage)       2 AUTHORIZED
        p0 = re.compile(r"^(?P<license>.+?)\s+\((?P<entitlement>.+)\)\s+
        (?P<count>\d+)\s+(?P<status>.+)$")

        for line in out.splitlines():
            line=line.strip()

            # udi line
            m=p0.match(line)
```

```
        if m:
            if 'license_usage' not in license_summ_dict:
                result_dict = license_summ_dict.setdefault('license_usage', {})
            license = m.groupdict()['license']
            entitlement = m.groupdict()['entitlement']
            count = m.groupdict()['count']
            status = m.groupdict()['status']
            result_dict[license] = {}
            result_dict[license]['entitlement'] = entitlement
            result_dict[license]['count'] = count
            result_dict[license]['status'] = status
            continue
    return result_dict
```

Let's begin with the class structure. The **ShowLicenseSummary** class inherits from its associated schema class, **ShowLicenseSummarySchema** from Example A-6, and has a class variable named **cli_command** that stores the **show** command we are parsing. There's a class method named **cli()** that handles CLI command execution and parsing. In the cli() method, the command is executed on the device using the native command line and stores the raw command output as a string in the **out** variable. Once we have the command output, we define a Python dictionary to store the results, aptly named "result_dict". This stores the data parsed from the output.

Now we dive deeper into the parsing mechanism. As mentioned earlier, output can be parsed using multiple parsing methods, including regular expressions (RegEx) and the parsergen package. In this case, we are using regular expressions. More specifically, we are using the re standard Python package and the **re.compile()** method to compile a RegEx pattern. The compiled RegEx pattern is what's used to identify and parse significant data points from the raw command output. The **p0** variable stores the compiled RegEx pattern object. Example A-8 shows the compiled RegEx pattern with the capture groups highlighted.

Example A-8 *Compiled Regex Pattern*

```
^(?P<license>.+?)\s+\((?P<entitlement>.+)\)\s+(?P<count>\d+)\s+(?P<status>.+)$
```

If you aren't familiar with RegEx patterns, then this string may look like another language, but let's break it down a bit. The capture groups, identified with parentheses— that is, ( )—are used to capture and extract interesting values from the string. The group names are identified with the **?P** and **< >** characters in the pattern. The group names are helpful when identifying values to store in our result dictionary. The compiled RegEx pattern is the core of our parsing.

The RegEx pattern object is used to compare and match values in each line of output. If there's a match, the group names store the appropriate values: license name, entitlement, count, and status. Once stored, the group names are accessed using the **.groupdict()** method, which returns a dictionary with the group names as the key and the matching string as the value. Example A-9 shows a snippet from Example A-7 where the matched values are extracted and assigned to the result dictionary.

Example A-9   *Matching Values and Storing in Result Dictionary*

```
if m:
    if 'license_usage' not in license_summ_dict:
        result_dict = license_summ_dict.setdefault('license_usage', {})
    license = m.groupdict()['license']
    entitlement = m.groupdict()['entitlement']
    count = m.groupdict()['count']
    status = m.groupdict()['status']
    result_dict[license] = {}
    result_dict[license]['entitlement'] = entitlement
    result_dict[license]['count'] = count
    result_dict[license]['status'] = status
    continue
```

m is the match object that will contain data if there's a match on a particular line of output. Notice the variable names that match the group names we included in our compiled RegEx pattern. The variable name doesn't have to match the group name, but it's useful to identify the output in the example. Each variable stores the value associated with the relative group name key in the dictionary returned from the **.groupdict()** method. Once all variables are captured, the result dictionary is built out and the values are assigned to keys that match the schema structure. It's worth noting that if a line of output does not match our compiled RegEx pattern, it's essentially skipped, and nothing is stored in the result dictionary. Now that the schema and parser classes are created, we must test them and ensure they work as expected before contributing back to the genieparser library.

Testing Your Parser

Testing is a crucial part of software development and is a must when contributing to an open-source project. Most open-source projects require proposed code in pull requests to pass unit and integration testing. Unit tests must be created by the contributor to ensure any new code added has test coverage, which ensures any new code is functioning properly and is covered by unit tests. If code is modified, the tests will need to be altered to match the new behavior.

Before running any tests, we need to incorporate the new parser into the genie.libs. parser package. To add the parser, you must run **make json** in the root of the genieparser repository. The **make json** command will create a JSON file that links commands to their related parser class. The generated JSON file will be used to locate the correct parser class for the given command when running **device.parse()**. If you do not run **make json** after making a new parser, you'll receive a "Could not find parser" error. Once the JSON file is generated, you may try running and parsing the desired command on a physical/virtual network device.

The genieparser library has two types of tests: folder-based testing and unittest-based testing, with the goal of transitioning to folder-based testing. Folder-based testing helps avoid merge conflicts and duplicate boilerplate code while making it easier to create tests. We will be covering how to create folder-based tests for a new parser.

To begin creating tests, you must create a folder in the tests directory (src/genie/libs/parser/{OS}/tests). In our case, the OS variable is **iosxe** since the parser applies to IOS-XE devices. The folder name must match the parser class name (that is, **ShowLicenseSummary**). Once the folder has been created, create a folder named "cli". Within the cli folder in place, create two folders, named "empty" and "equal". For the **ShowLicenseSummary** parser, the folder structure should look like this:

- src/genie/libs/parser/iosxe/tests/ShowLicenseSummary/cli/empty
- src/genie/libs/parser/iosxe/tests/ShowLicenseSummary/cli/equal

In the "empty" folder, create a file that ends with "_output.txt" (that is, empty_output.txt). This file should be empty to avoid raising a **SchemaEmptyParserError** error. In the "equal" folder, we must create files for the raw command output, expected parsed output, and any potential arguments. There should be tests, a set of outputs, for each varying output. The files are grouped together by stripping out **_arguments.json**, **_expected.py**, and **_output.txt**. For example, golden_output1_arguments.json, golden_output1_expected.py, and golden_output1_output.txt are all part of the same test. The output file should include raw output that is expected when executing the associated command. The expected file should be a Python file with a single variable called **expected_output** that contains the expected output structure, most likely a Python dictionary. The arguments JSON file should contain a single dictionary that is a set of key-value pairs of the parser arguments. Example A-10 shows the resulting directory structure for your parser test(s).

Example A-10 *Parser Test Structure*

```
# Found in /genieparser/src/genie/libs/parser/iosxe/tests

ShowLicenseSummary
└── cli
    ├── empty
    │   └── empty_output_output.txt
    └── equal
        ├── golden_output1_expected.py
        └── golden_output1_output.txt
```

Example A-11 shows the contents of golden_output1_output.txt, and Example A-12 shows the content of golden_output1_expected.py. Both files are in the cli/equal directory.

Example A-11   *Golden Output*

```
License Usage:
License                      Entitlement Tag                 Count Status
-------------------------------------------------------------------------
C9300 48P DNA Advantage  (C9300-48 DNA Advantage)            2 AUTHORIZED
C9300 48P Network Adv... (C9300-48 Network Advan...)         2 AUTHORIZED
C9300 24P Network Adv... (C9300-24 Network Advan...)         1 AUTHORIZED
C9300 24P DNA Advantage  (C9300-24 DNA Advantage)            1 AUTHORIZED
```

Example A-12   *Expected Golden Output*

```python
expected_output = {
    'license_usage':{
        'C9300 48P DNA Advantage':{
            'entitlement':'C9300-48 DNA Advantage',
            'count':'2',
            'status':'AUTHORIZED'
        },
        'C9300 48P Network Adv...':{
            'entitlement':'C9300-48 Network Advan...',
            'count':'2',
            'status':'AUTHORIZED'
        },
        'C9300 24P Network Adv...':{
            'entitlement':'C9300-24 Network Advan...',
            'count':'1',
            'status':'AUTHORIZED'
        },
        'C9300 24P DNA Advantage':{
            'entitlement':'C9300-24 DNA Advantage',
            'count':'1',
            'status':'AUTHORIZED'
        }
    }
}
```

Once the tests are written, you can run the tests by going to the tests directory at the root of the genieparser library (/genieparser/tests). You can run all the tests, for a single OS, or a single test. Example A-13 shows the three different options to run each type of command.

Example A-13 *Run Parser Tests*

```
# Run all tests
genieparser/tests$ python folder_parsing_job.py

# Run all iosxe tests
genieparser/tests$ python folder_parsing_job.py -o iosxe

# Run only one test for a single parser (ShowLicenseSummary)
genieparser/tests$ python folder_parsing_job.py -o iosxe ShowLicenseSummary
```

The folder_parsing_job.py script runs the tests with the AEtest test infrastructure to provide passed/failed results. We are expanding our usage of pyATS outside of network testing! Once all tests have been accounted for and have passed, we are ready to contribute the parser back to the genieparser library!

Contributing to Genieparser

Before pushing our changes to GitHub, we need to do one more housekeeping chore—update the changelog! This may seem like an administrative task, but it's required to document what's been changed, added, or fixed in the proposed changes. Only one changelog needs to be included per pull request. Create a new changelog file in genieparser/changelog/undistributed/ with a unique filename containing a short description and the datetime. The datetime can be generated using the following command on macOS and Linux: **date "+%Y%m%d%H%M%S"**. For example, the filename might look like changelog_show_license_summary_iosxe_20210907095958.rst. The /genieparser/ changelog/undistributed directory provides a template.rst file for how the changelog content should be formatted. Example A-14 shows the templates for new and fixed changes.

Example A-14 *Changelog Templates*

```
--------------------------------------------------------------------------------
    -                               New
--------------------------------------------------------------------------------
* <OS>
    * <Added|Modified> <Class>:
        * <New Show Command|Command>


--------------------------------------------------------------------------------
                                    Fix
--------------------------------------------------------------------------------
* <OS>
    * <Added|Modified|Removed> <Class>:
        * Changes made
```

Based on the templates, Example A-15 shows a changelog we can create for the new **ShowLicenseSummary** parser.

Example A-15   *The ShowLicenseSummary Changelog*

```
--------------------------------------------------------------------------------
    -                              New
--------------------------------------------------------------------------------*
IOSXE
    * Added ShowLicenseSummary:
        * show license summary
```

Once a changelog has been created, we can push the changes to our forked genieparser repository using the Git command **git push –u origin** {*branch_name*}. You should now see the appropriate branch updated in your forked repository on GitHub. The next step is opening a pull request on the original genieparser library. Figure A-6 shows which button to select to open a pull request.

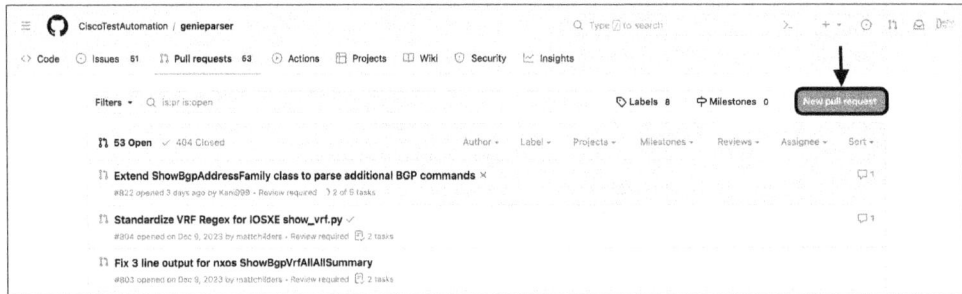

Figure A-6   *GitHub – Pull Request*

Each pull request should contain sections for a description, motivation and context, impact, screenshots, and a checklist. The checklist section contains the following checks:

- I have updated the changelog.
- I have updated the documentation (if applicable).
- I have added tests to cover my changes (if applicable).
- All new and existing tests passed.
- All new code passed compilation.

Each check is used to ensure your code meets the expectations of the genieparser maintainers who are reviewing the pull requests. Figure A-7 shows a submitted pull request.

Figure A-7   *GitHub – Submitted Pull Request*

Once the pull request is submitted, reviewers will be assigned to review it and provide feedback. The reviewers may ask questions, provide recommendations, or let you know if you need to make changes to your code. Once the reviewers are satisfied, the pull request will be approved and merged, and your changes will be included in the next release! Figure A-8 shows the changelog of the release that included the parser.

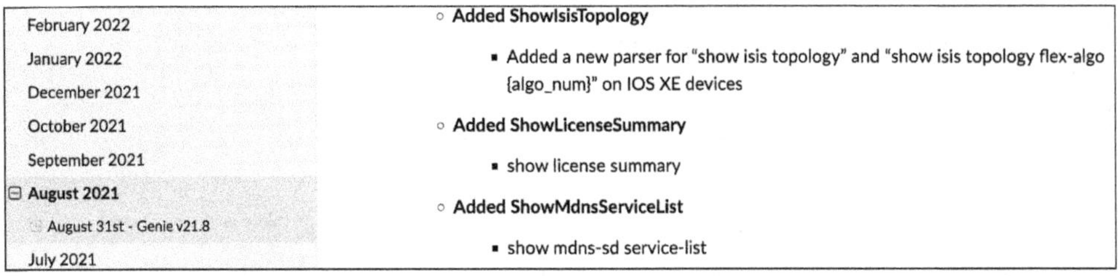

February 2022	○ **Added ShowIsisTopology**
January 2022	▪ Added a new parser for "show isis topology" and "show isis topology flex-algo
December 2021	{algo_num}" on IOS XE devices
October 2021	○ **Added ShowLicenseSummary**
September 2021	▪ show license summary
⊟ **August 2021**	○ **Added ShowMdnsServiceList**
August 31st - Genie v21.8	
July 2021	▪ show mdns-sd service-list

Figure A-8   *Genie v21.8 Changelog*

References and Recommended Readings

Use the following links as references when you're contributing to the genieparser library:

- **Write a Parser:** https://pubhub.devnetcloud.com/media/pyats-development-guide/docs/writeparser/writeparser.html#

- **Contribution Guidelines:** https://pubhub.devnetcloud.com/media/pyats-development-guide/docs/contribute/contribute.html#contribute

- **Genie Metaparser Package:** https://pubhub.devnetcloud.com/media/genie-metaparser/docs/index.html

- **Creating a Pull Request:** https://docs.github.com/en/pull-requests/collaborating-with-pull-requests/proposing-changes-to-your-work-with-pull-requests/creating-a-pull-request

- **Examples of a parser pull request:**

  - https://github.com/CiscoTestAutomation/genieparser/pull/499/files

  - https://github.com/CiscoTestAutomation/genieparser/pull/536/files

Appendix B

Secret Strings

Cisco's pyATS framework is an automation platform designed to enhance and simplify network testing processes. A critical component of this framework is the use of testbed. yaml files, which serve as blueprints for the network under test. When statically defined, these files contain crucial configuration details, including device connectivity information, credentials, and other sensitive data such as usernames, enable secrets, and potentially hostnames or IP addresses. Given the sensitivity of this information, especially in secure or critical network environments, there's a significant emphasis on protecting these details from unauthorized access.

To address these security concerns, pyATS introduces the concept of "secret strings" as a robust mechanism for safeguarding sensitive data. Secret strings are built in pyATS functionality. This feature allows users to encrypt specific variables within the testbed. yaml file, or even the entire file itself, thereby securing the credentials and other sensitive information contained within. The encryption process ensures that even if the testbed file is inadvertently exposed or accessed by unauthorized individuals, the encrypted contents remain protected, thus maintaining the integrity and security of the network testing environment. This approach not only reinforces the security posture of the testing framework but also aligns with best practices for handling sensitive information in network automation tasks.

How to Secure Your Secret Strings

Follow this procedure to make your secret strings cryptographically secure:

Step 1. Update your pyATS configuration file as follows:

```
[secrets]
string.representer = pyats.utils.secret_strings.
FernetSecretStringRepresenter
```

Step 2.   Install the cryptography package:

```
> pip install cryptography
```

Step 3.   Ensure the permissions are restricted on your pyATS configuration file to prevent others from reading it. For example:

```
> chmod 600 ~/.pyats/pyats.conf
```

Step 4.   Generate a cryptographic key:

```
> pyats secret keygen
```

The following is a newly generated key:

dSvoKX23jKQADn20INt3W3B5ogUQmh6Pq00czddHtgU=

(Note that your key will differ.)

Step 5.   Update your pyATS configuration file as follows:

```
[secrets]
string.representer = pyats.utils.secret_strings.
FernetSecretStringRepresenter
string.key = dSvoKX23jKQADn20INt3W3B5ogUQmh6Pq00czddHtgU=
```

Step 6.   Encode a password:

A recommended step includes encoding the secret as follows:

```
> pyats secret encode
Password: MySecretPassword
Encoded string: gAAAAABdsgvwElU9_3RTZsRnd4b1l3Es2gV6Y_
DUnUE8C9y3SdZGBc2v0B2m9sKVz80jyeYhlWKMDwtqfwlbg4sQ2Y
0a843luOrZyyOuCgZ7bxE5X3Dk_NY=
```

Alternatively, you can use the pyATS CLI:

```
> pyats secret encode --string MySecretPassword
Encoded string:
gAAAAABdsgvwElU9_3RTZsRnd4b1l3Es2gV6Y_
DUnUE8C9y3SdZGBc2v0B2m9sKVz80jyeYhlWKMDwtqfwlbg4sQ2Y0
a843luOrZyyOuCgZ7bxE5X3Dk_NY=
```

Step 7.   Test the decoding of the encoded password to ensure it works, as follows:

```
> pyats secret decode
gAAAAABdsgvwElU9_3RTZsRnd4b1l3Es2gV6Y_
DUnUE8C9y3SdZGBc2v0B2m9sKVz80jyeYhlWKMDwtqfwlbg4sQ2Y0a843l
uOrZyyOuCgZ7bxE5X3Dk_NY=
Decoded string :
MySecretPassword
```

Step 8.   Add your encoded password to a **testbed.yaml %ENC{}** block, as described
in "Credential Password Modeling" at the following site:

https://pubhub.devnetcloud.com/media/pyats/docs/topology/schema.
html#topology-credential-password-modeling

Now your password is secured. The only way to decode the password from the testbed
YAML file is to use the same pyATS configuration file used to encode the password:

```
# Snippet of your testbed.yaml
testbed:
    name: sampleTestbed
    credentials:
        default:
            username: admin
            password: "%ENC{gAAAAABdsgvwElU9_3RTZsRnd4b1l3Es2gV6Y_
DUnUE8C9y3SdZGBc2v0B2m9sKVz80jyeYhlWKMDwtqfwlbg4sQ2Y0a843luOrZyy
OuCgZ7b
```

Multiple Representers

Follow this procedure to specify multiple representers if there are several kinds of
encoded strings you want to specify in a file such as a testbed YAML:

Step 1.   Add your new representer to the pyATS configuration file as follows:

```
[secrets]
my_custom.representer = package.module.MyRepresenterClass
```

Step 2.   Generate a key if your representer requires it:

```
> pyats secret keygen --prefix my_custom
Newly generated key :
<generated key for my_custom>
```

Step 3.   Update your pyATS configuration file with the newly generated key (if
required), as follows:

```
[secrets]
my_custom.representer = package.module.MyRepresenterClass
my_custom.key = <generated key for my_custom>
```

Step 4 (option a).   Encode a password using the default representer:

```
> pyats secret encode
Password: MySecretPassword
Encoded string :
wr3DssK0w5nDlsORw4nDmcK2w4LDqMOfw6vDjsOdw4k=
```

Step 4 (option b).   Encode a password using the my_custom representer:

```
> pyats secret encode --prefix my_custom
Password: MySecretPassword
Encoded string :
<my_custom encoded string>
```

Step 5.   Add references to your encoded passwords to your testbed YAML file. Here's an example:

```
testbed:
    credentials:
        default:
            username: my_username
            password: "%ENC{wr3DssK0w5nDlsORw4nDmcK2w4LD
qMOfw6vDjsOdw4k=}"
        alternate:
            username: alternate_username
            password: "%ENC{<my_custom encoded string>,
prefix=my_custom}"
        custom:
            custom_key: |4-
                custom data containing encoded text
                %ENC{<my_custom encoded string>, prefix=my_
                custom}
```

Step 6.   Check that your passwords can be recovered from the loaded testbed:

```
> pyats shell --testbed_file my_testbed.yaml
>>> from pyats.utils.secret_strings import to_plaintext
>>> to_plaintext(testbed.credentials.default.password)
'MySecretPassword'
>>> to_plaintext(testbed.credentials.alternate.password)
'MySecretPassword'
>>> testbed.custom.custom_key
'custom data containing encoded text\nMySecretPassword'
```

Table 3-1 outlines the SecretString object class methods available.

Table 3-1   *SecretString Object*

SecretString Object

Class Methods	Description
from_plaintext	Returns an encoded secret string from plaintext
keygen	Returns a key that affects the string encoding/decoding

SecretString Object

Properties	Description
plaintext	Returns the decoded secret string in plaintext form

Attributes	Description
data	The secret string in encoded form

Methods	Description
str	Returns asterisks in order to hide the secret string

Representer Classes

The encoding/decoding of secret strings and any required key management are defined in a pluggable manner by use of representer classes.

The following representers are supported:

- **pyats.utils.secret_strings.ObscuringSecretStringRepresenter**

  - This class stores the secret string in cipher-encoded form.

  - It is the default representer if the user has not specified a representer in pyATS configuration.

  - It uses a default key, but allows the user to overwrite the key in pyATS configuration.

- **pyats.utils.secret_strings.PlainTextSecretStringRepresenter**

- This class stores the secret string in plaintext form.

  - It does not make use of the key.

- **pyats.utils.secret_strings.FernetSecretStringRepresenter**

- This class stores the secret string in crypto-encoded form.

  - It requires the user to manually execute pip install cryptography.

  - It can generate a decryption key.

  - A generated key must be specified in pyATS configuration.

Index

E

T

Register your product at **ciscopress.com/register**
to unlock additional benefits:

- Save 35%* on your next purchase with an exclusive discount code

- Find companion files, errata, and product updates if available

- Sign up to receive special offers on new editions and related titles

Get more when you shop at **ciscopress.com**:

- Everyday discounts on books, eBooks, video courses, and more

- Free U.S. shipping on all orders

- Multi-format eBooks to read on your preferred device

- Print and eBook Best Value Packs

Cisco Press

www.ingramcontent.com/pod-product-compliance
Lightning Source LLC
Chambersburg PA
CBHW080341220326
41598CB00030B/4572